Ore Geology and Industrial Minerals
An Introduction

GEOSCIENCE TEXTS

SERIES EDITOR

A. HALLAM
Lapworth Professor of Geology
University of Birmingham

GEOSCIENCE TEXTS

Ore Geology and Industrial Minerals
An Introduction

ANTHONY M. EVANS
BSc, PhD, MIMM, FGS
Formerly Senior Lecturer in Mining Geology
University of Leicester

THIRD EDITION

**Blackwell
Science**

To Jo, Nick, Caroline and Jason

© 1980, 1987, 1993 by
Blackwell Science Ltd
Editorial Offices:
Osney Mead, Oxford OX2 0EL
25 John Street, London WC1N 2BL
23 Ainslie Place, Edinburgh EH3 6AJ
350 Main Street, Malden
    MA 02148 5018, USA
54 University Street, Carlton
    Victoria 3053, Australia
10, rue Casimir Delavigne
    75006 Paris, France

Other Editorial Offices:
Blackwell Wissenschafts-Verlag GmbH
Kurfürstendamm 57
10707 Berlin, Germany

Blackwell Science KK
MG Kodenmacho Building
7–10 Kodenmacho Nihombashi
Chuo-ku, Tokyo 104, Japan

First published in 1980 under the title
An Introduction to Ore Geology
Chinese edition 1985
Reprinted 1982, 1983, 1984
Second edition 1987
German edition 1992
Malaysian edition 1989
Japanese edition 1989
Third edition 1993
Reprinted 1994, 1995, 1996, 1997, 1998

Set by Semantic Graphics, Singapore
Printed at Alden Press Limited,
Oxford and Northampton, Great Britain

DISTRIBUTORS

Marston Book Services Ltd
PO Box 269
Abingdon
Oxon OX14 4YN
(*Orders*: Tel: 01235 465500
              Fax: 01235 465555)

USA
Blackwell Science, Inc.
Commerce Place
350 Main Street
Malden, MA 02148 5018
(*Orders*: Tel: 800 759 6102
                  617 388 8250
              Fax: 617 388 8255)

Canada
Login Brothers Book Company
324 Saulteaux Crescent
Winnipeg, Manitoba R3J 3T2
(*Orders*: Tel: 204 224-4068)

Australia
Blackwell Science Pty Ltd
54 University Street
Carlton, Victoria 3053
(*Orders*: Tel: 3 9347 0300
              Fax: 3 9347 5001)

A catalogue record for this title
is available from the British Library

ISBN 0-632-02953-6

Library of Congress
Cataloging in Publication Data

Evans, Anthony M.
    Ore geology and industrial minerals/Anthony M.
    Evans, – 3rd ed.
        p. cm. – (Geoscience texts)
    Rev. ed. of: An Introduction to ore geology, 2nd ed. 1987.
    Includes index.
    ISBN 0-632-02953-6
    1. Ore deposits. 2. Industrial minerals.
        I. Evans, Anthony M. Introduction to ore geology.
        II. Title III. Series.
    QE390.E92   1993
    553′.1 – dc20

For further information on
Blackwell Science, visit our website:
www.blackwell-science.com

Contents

Preface to the third edition

This edition, like the second, is an enlarged and extensively revised work. I blame much of the increased size, after all this is an introductory text only, on many of my readers, reviewers and translators. These, almost without exception, have ignored the plea at the end of the preface to my first edition and have called for additions, changes but rarely for deletions! Once again I am in their debt for approving letters, good reviews and their many helpful comments. I am particularly grateful to the many lecturers in North America and the UK who returned my questionnaire concerning industrial minerals. These respondents voted overwhelmingly for the inclusion of sections on this topic and for the mode of presentation that I had tentatively suggested. Of this group I would like to give sincere thanks to Dr Bladh of Wittenberg University, Ohio and Dr Garlick of Humboldt University, California for the considerable thought and time they put into their replies.

All this encouragement has led me to develop Chapter 1 into an overview of mineral economics, to emphasize the non-metallurgical applications of metallic elements at various points in the book and to include two chapters devoted entirely to industrial minerals. The first of these (Chapter 20) illustrates in a little depth some chosen examples of industrial minerals (and bulk materials) that possess contrasting chemical and physical properties as well as having different modes of formation, uses and financial values. The second chapter (Chapter 21) contains summary details of other industrial mineral commodities to make the reader aware of the potential of many common non-metalliferous resources.

In turning my hand to writing about industrial minerals I have been ably assisted and encouraged by Professor Peter Scott of Camborne School of Mines in Cornwall, and Professor Ansel Dunham and Mr Michael Whateley of Leicester University's Geology Department. Mr David Highley of the British Geological Survey also gave me invaluable help, particularly in the sphere of mineral statistics. Without the help of these good friends this text would contain many more sins of commission and omission than are no doubt still present. Many of my other colleagues at Leicester have good naturedly allowed me to pester them with questions in my search for enlightenment on various, to me, dark problems. I would also like to thank all those in industrial circles who have encouraged me to proceed to a third edition, in particular Professor Colin Bristow who supplied me with invaluable data, which I have incorporated into Chapter 1.

Apart from the new material discussed above I have included a description of hydraulic fracturing, hypothermal and epithermal gold mineralization and introduced new material into most chapters of this book. This work is, however, an introductory text and therefore does not deal with the esoteric subjects, lists of which one or two reviewers have drawn up and then proceeded to deplore their absence. This game is better played in assessing the merits of advanced geology texts!

Finally I am happy once more to confess my overwhelming debt to my loving wife who has encouraged and helped me at every stage in the preparation of this third edition, especially through the hiatus of major surgery.

Anthony Evans
Burton on the Wolds
January 1992

Preface to the second edition

This revision appears in response to what the media are pleased to call popular demand. The publishers and I were quite astonished by the impressive sales figures for the first edition, the flattering reviews, the 'fan mail' from places as far apart as France, California, Japan, New Zealand and Spain and the offers to translate it into both French and Japanese.

I would like to express my thanks to the many readers who have been kind enough to comment on the first edition, instead of making the usual exclamation marks in the margins of their copies when they objected to my prose, or caught me out in some fact, or disagreed with my interpretation of the evidence. Many of what I hope will be seen as improvements to the text owe their presence to the kindness of readers and reviewers, and I hope that none of them will feel that any of their constructive criticism has been ignored.

I have attempted a thorough revision and many sections have been rewritten. A chapter on diamonds has been added to meet requests. Chapters on greisen and pegmatite deposits have also been added, the former in response to the changing situation in tin mining following the recent tin crisis and the latter in response to suggestions from geologists in a number of overseas countries. Some chapters have been considerably expanded and new sections added; in particular on disseminated gold deposits and unconformity-associated uranium deposits. The chapter on ore genesis has been enlarged and I am grateful to Dr A.D. Saunders for his comments on it.

To emphasize still further the importance of viewing mineral deposits from an economic standpoint, I have expanded Chapter 1 considerably and I am grateful to Mr M.K.G. Whateley for reviewing it. I have continued my policy of the first edition of peppering the text with grade and tonnage figures and other allusions to mineral economics in a further attempt to create commercial awareness in the tyro.

As in the first edition bibliographic references generally direct attention to works in English. The student should note that this, in itself, is misleading; for much significant work in the field is written in French, German, Russian and other languages. But works in English are much more widely accessible and the main aim has been to help the reader find works that will amplify the discussions this book has begun.

Much of the success of the first edition was due to Sue Aldridge's fine artwork and I am deeply grateful to her for the pains she has taken. Once again I lovingly acknowledge the encouragement, editorial and typing skills which my wife has contributed and without which this edition would still be awaiting attention.

Anthony Evans
Burton on the Wolds
July 1986

Preface to the first edition

This book is an attempt to provide a textbook in ore geology for second and third year undergraduates which, in these days of inflation, could be retailed at a reasonable price. The outline of the book follows fairly closely the undergraduate course in this subject at Leicester University which has evolved over the last 20 years. It assumes that the student will have adequate practical periods in which to handle and examine hand specimens, and thin and polished sections of the common ore types and their typical host rocks. Without such practical work students often develop erroneous ideas of what an orebody looks like, ideas often based on a study of mineralogical and museum specimens. In my opinion, it is essential that the student handles as much run-of-the-mill ore as possible during his course and makes a start on developing such skills as visual assaying, the ability to recognize wall rock alteration, using textural evidence to decide on the mode of genesis, and so on.

In an attempt to keep the reader aware of financial realities I have introduced some mineral economics into Chapter 1 and sprinkled grade and tonnage figures here and there throughout the book. It is hoped that this will go some way towards meeting that perennial complaint of industrial employers, that the new graduate has little or no commercial awareness, such as a realization that companies in the West operate on the profit motive. This little essay into mineral economics only scratches the surface of the subject, and the intending practitioner of mining geology would be well advised to accompany his study of ore geology by dipping into such journals as *World Mining*,* the *Mining Journal*, the *Engineering and Mining Journal* and the *Mining Magazine*,[†] to watch the latest trends in metal and mineral prices and to gain knowledge of mining methods and recent orebody discoveries.

In order to produce a reasonably priced book, a strict word limit had to be imposed. As a result, the contents are necessarily selective and no doubt some teachers of this subject will feel that important topics have either received rather scanty treatment or have been omitted altogether. To these folk I offer my apologies, and hope that they will send me their ideas for improving the text, always remembering that if the price is to be kept down additions must be balanced by subtractions!

I would like to thank Mr Robert Campbell of Blackwell Scientific Publications for his help and encouragement, and not least for his tact in leaving me to get on with the job. My colleagues Dr J.G. Angus and Dr J. O'Leary read some of the chapters and made helpful suggestions for their improvement, and I thank them for their kindness. To my wife I owe an inestimable debt for the care with which she checked my manuscript and then produced the typescript.

* No longer available.
[†] *Industrial Minerals* should be added to this list.

Units and abbreviations

Note on units

With few exceptions the units used are all SI (Système International), which has been in common use by engineers and scientists since 1965. The principal exceptions are: (a) for commodity prices still quoted in old units, such as Troy ounces for precious metals and the short ton (= 2000 lb); (b) when there is uncertainty about the exact unit used, e.g. tons in certain circumstances might be short or long (2240 lb); (c) degrees Celsius (centigrade)—geologists do not seem to be able to envisage temperature differences in degrees kelvin! (neither do meteorologists!); and (d) centimetres (cm), which like °C refuses to die because it is so useful!

SI prefixes commonly used in this text are k = kilo-, 10^3; M = mega-, 10^6 (million); G = giga-, 10^9 (billion is never used as it has different meanings on either side of the Atlantic).

Some abbreviations used in the text

ASTM	American Society for Testing Materials
BS	British Standard
CIF	Carriage, insurance and freight
CIPEC	Conseil Inter-governmental des Pays Exportateurs de Cuivre (Intergovernmental Council of Copper Exporting Countries)
EEC	European Economic Community; this is the correct name of what is sometimes referred to as the EC or European Community
FOB	Freight on board
MEC	Market economy countries
OECD	Organization for Economic Cooperation and Development
OPEC	Organization of Petroleum Exporting Countries
PGM	Platinum group metals
REE	Rare earth elements
REO	Rare earth oxides
t p.a.	Tonnes *per annum*
t p.d.	Tonnes *per diem*

Note on the USSR

As many references in this book are concerned with production statistics that cannot be attributed readily to the individual republics of the former union, I have kept this abbreviation as a description of the geographical area that once made up the now disbanded Soviet Republics.

Part 1
Principles

'Here is such a vast variety of phenomena and these many of them so delusive, that 'tis very hard to escape imposition and mistake'

These words, written about ore deposits by John Woodward in 1695, are every bit as true today as when he wrote them.

1 / Some elementary aspects of mineral economics

Ore, orebodies, industrial minerals, gangue and protore

'What is ore geology?' Unfortunately, it is not possible to give an unequivocal answer to this question if one wishes to go beyond saying that it is a branch of economic geology. The difficulty is that there are a number of distinctly different definitions of ore. A definition which has been current in capitalist economies for nearly a century runs as follows: 'Ore is a metalliferous mineral, or an aggregate of metalliferous minerals, more or less mixed with gangue, which from the standpoint of the miner can be won at a profit, or from the standpoint of the metallurgist can be treated at a profit. The test of yielding a metal or metals *at a profit* seems to be the only feasible one to employ.' Thus wrote J.F. Kemp in 1909. There are many similar definitions of ore which all emphasize (a) that it is material from which we extract a metal, and (b) that this operation must be a profit-making one. Economically mineable aggregates of ore minerals are termed orebodies, oreshoots, ore deposits or ore reserves.

The words ore and orebody have, however, been undergoing slow and confusing transitions in their meanings, which are still not complete, and the tyro must read the context carefully to discern the sense in which a particular writer is using these words. For example, in Craig (1989) the ore minerals are defined as those from which metals are extracted, e.g. chalcopyrite and galena from which we extract copper and lead respectively, and many authors use this term as a synonym for opaque minerals, which is actually a better description for them since they include pyrite and pyrrhotite, minerals that are discarded in the processing of most ores. Craig is by no means alone. Most economic geologists concerned with the extractive industries divide the materials they exploit into either ore minerals or industrial minerals. Nevertheless recent definitions of ore include both groups, so what are industrial minerals?

'Industrial minerals have been defined as any rock, mineral or other naturally occurring substance of economic value, exclusive of metallic ores, mineral fuels and gemstones' (Noetstaller 1988). They are therefore minerals where either the mineral itself, e.g. asbestos, baryte, or the oxide or some other compound derived from the mineral has an industrial application (end use) and they include rocks, such as granite, sand, gravel and limestone that are used for constructional purposes (these are often referred to as aggregates or bulk materials), as well as the more valuable minerals with specific chemical or physical properties, such as fluorite, phosphate, kaolinite and perlite.

Although practically all industrial minerals (e.g. halite, $NaCl$) contain metallic elements they are frequently and confusingly termed non-metallics, e.g. Harben & Bates (1984). To add to the reader's confusion it must now be noted that many 'metallic ores', such as bauxite, ilmenite, chromite and manganese minerals, are also important raw materials for industrial mineral end uses. In Fig. 1.1 some of the end uses of bauxite are displayed to illustrate this point and to give an idea of the diversity of end uses that characterize man's utilization of industrial minerals. Depending on how far the path of a mineral through industrial uses can be traced, so the number of known uses increases. It has been estimated that halite is the starting point of about 18 000 end uses! In view of all the above discussion, how do we now define ore?

Two very useful discussions of this subject are to be found in Lane (1988) and Taylor (1989). Taylor's discussion is an easier and better introduction for the beginner, Lane should be read by all industrial and mining geologists and advanced students. Taylor favours a wide and inclusive definition that will survive being 'blown about by every puff of economic wind', such as changes in market demand, commodity prices, mining costs, taxes, environmental legislation and other factors: 'ore is rock that may be, is hoped to be, will be, is or has been mined; and from which something of value may be (or has been) extracted'. This is very similar to the official UK Institution of Mining and Metallurgy (IMM) definition: 'Ore is a solid naturally-occurring mineral aggregate of economic interest from which one or

3

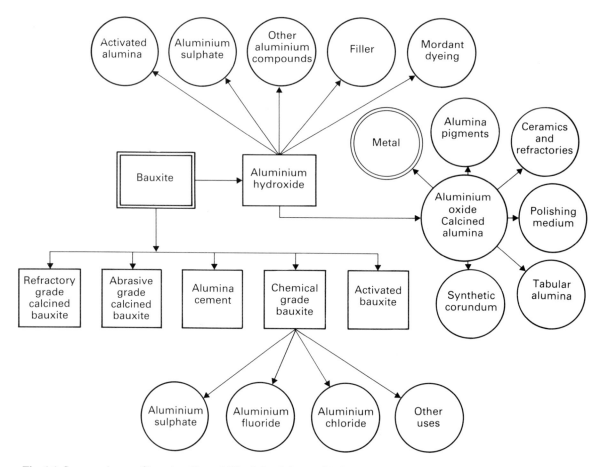

Fig. 1.1 Some end uses of bauxite. About 90% of aluminium oxide is used for the manufacture of aluminium metal but the other end uses are expanding rapidly, particularly in ceramics and refractories. (Modified from Anon. 1977.)

more valuable constituents may be recovered by treatment'. Both these definitions cover ore minerals and industrial minerals and imply extension of the term orebody to include economic deposits of industrial minerals and rocks. This is the sense in which these terms will normally be used in this book, except that they will be extended to include the instances where the whole rock, e.g. granite, limestone and salt, is utilized and not just a part of it. Lane prefers the use of the term mineralized ground for such comprehensive usage of the word ore as that given in the definitions by Taylor and the IMM, and he would restrict ore to describing material in the ground that can be extracted to the overall economic benefit of a particular mining operation, governed by the financial determinants at the time of examination.

A further complication, which we may note in passing, is that in socialist economies ore is often defined as mineral material that can be mined for the benefit of mankind. Such an altruistic definition is necessary to cover those examples in both capitalist and socialist countries where minerals are being worked at a loss. Such operations are carried on for various good or bad reasons depending on one's viewpoint! These include a government's reluctance to allow large isolated mining communities to be plunged into unemployment because a mine or mines have become unprofitable, a need to earn foreign currency and other reasons.

A definition about which there is little argument is that of gangue. This is simply the unwanted material, minerals or rock, with which ore minerals are usually intergrown. Mines commonly possess mineral processing plants in which raw ore is milled before the separation of the ore minerals from the gangue minerals by various processes, which provide ore concentrates, and tailings which are made up of the gangue.

Another word that must be introduced at this stage is protore. This is mineral material in which an initial but uneconomic concentration of metals has occurred that may by further natural processes be upgraded to the level of ore.

The relative importance of ore and industrial minerals

There has always been an aura of romance about metallic deposits, especially those of gold and silver, which has stimulated the writing of heroic narratives such as that of Jason's search for the Golden Fleece (undoubtedly an expedition to recover placer gold from the Black Sea region) right up to the recent novels of Joseph Conrad and Hammond Innes. Wars have been fought over metallic deposits and new finds quickly reach the headlines—'gold rush in Nevada', 'silver fever in Mexico', 'nickel rush in Western Australia', but never talc fever or sulphur stampede! The poor old industrial minerals tend to be overlooked by the public and cursorily treated in many geological textbooks, which commonly focus on metallic ores and fossil fuel deposits to the virtual exclusion of the industrial minerals; and yet, in the form of flints and stone axes, bricks, pottery, etc., these were the first earth resources to be exploited by man. Today industrial minerals permeate every segment of our society (McVey 1989). They occur as components in durable and non-durable consumer goods. In many industrial activities and products, from the construction of buildings to the manufacture of ceramic tables or sanitary ware, the use of industrial minerals is obvious but often unappreciated. With numerous other goods, ranging from books to pharmaceuticals, the consumer frequently is unaware that industrial minerals play an essential role.

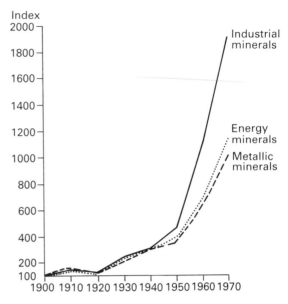

Fig 1.2 Growth comparisons for mineral products. In each case the index for 1900 is 100. (After Anon. 1977.)

Both Bristow (1987a) and Noetstaller (1988) have drawn attention to the increasingly rapid growth in the production of industrial minerals compared with metals. Table 1.1 shows the growth rate of industrial minerals compared with two other mineral products, and Fig. 1.2 shows how the world production of industrial minerals is outstripping that of metals. The relative positions in terms of tonnage and financial value appear in Table 1.2.

Bristow also has made the interesting remark that at some point in time during the development of

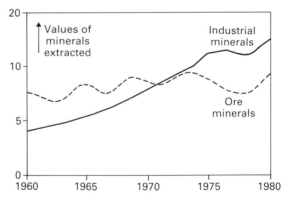

Fig. 1.3 Spanish mining production. Mineral values are in millions of constant pesetas indexed to 1970. (After Bristow 1987.)

Table 1.1 Average growth rates in world production. (After *The Economist World Business Cycles* 1982)

Product	1966–73 (%)	1973–80 (%)
Crude oil	70	7
Industrial minerals	29	16
Metals	54	7

Table 1.2 Tonnage and value of mineral products in 1983. (From Noetstaller, 1988)

Category of solid minerals	World production		Value of output	
	10^3 t	%	10^6 US$	%
Industrial minerals	11 798 630.0	72	129 147.3	40
Solid fuel minerals	4 004 287.4	24	122 285.0	38
Metals and ores[a]	543 580.6	4	39 007.3	13
Precious minerals	14.0	1	30 341.3	9
Totals	16 346 512.0	100	320 781.0	100

[a] Iron is included in this figure as iron ore.

an industrialized country, industrial minerals become more important in terms of value of production than metals. This happened in the UK in the nineteenth century, in the USA early in this century, in Spain in the early seventies (Fig. 1.3) and in younger economies, like Australia's, it is only just happening. The time of the crossover, Bristow suggested, may be a rough measure of the 'industrial maturity' of that country, and that in nearly all mature industrialized countries the value of industrial mineral production is very much greater than

that of ore minerals (Fig. 1.4). The world production of some individual mineral commodities ranked in order of tonnage produced is given in Table 1.3 and that of some other metals in Table 1.4.

Graphs of world production of the traditionally important metals (Figs 1.5–1.7) show interesting trends. The world's appetite for the major metals appeared to be almost insatiable after World War Two, and post-war production increased with great rapidity; however, in the mid 1970s an abrupt slackening in demand occurred, triggered by the

Table 1.3 World production of some mineral commodities in 1987. Metals are marked with italics. (Compiled from various sources and with considerable help from Mr D.E. Highley of the British Geological Survey)

Rank	Commodity	Tonnage (Mt)	Rank	Commodity	Tonnage (Mt)
1	Aggregates	10 250	23	Fluorite	4.8
2	Coal	4656	24	Feldspar	4.6
3	Crude oil	2838	25	Baryte	4.2
4	Portland cement	1033	26	*Titania*	4.2
5	*Pig iron*	508	27	Asbestos	4.1
6	Clay	400	28	Fuller's earth	3.6
7	Silica	200	29	*Lead*	3.4
8	Salt	177	30	Nepheline syenite	3.2
9	Phosphate	144	31	Borates	2.7
10	Gypsum	84	32	Perlite	2.4
11	Sulphur	54	33	Diatomite	1.9
12	Potash	31	34	*Zirconium minerals*	0.85
13	Sodium carbonate (trona)	30	35	*Nickel*	0.80
14	*Manganese ore*	22	36	Graphite	0.63
15	Kaolin	21	37	Vermiculite	0.54
16	*Aluminium*	16.2	38	Sillimanite minerals	0.50
17	Magnesite	12.3	39	*Magnesium*	0.33
18	*Chromium* ores and concentrates	10.8	40	Mica	0.27
19	*Copper*	8.7	41	*Tin*	0.19
20	Talc	7.4	42	Strontium minerals	0.18
21	*Zinc*	7.2	43	Wollastonite	0.12
22	Bentonite	6.4			

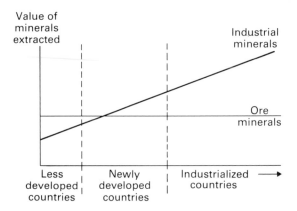

Fig. 1.4 Relative importance of industrial and ore mineral exploitation in evolving economies. (After Bristow 1987a.)

Table 1.4 World production of selected metals in 1987

Metal	Amount[a]
Molybdenum	0.089
Antimony	0.059
Tungsten	0.040
Uranium	0.038
Vanadium	0.032
Cadmium	0.019
Lithium	0.007
Mercury	0.006
Silver	14 133
Gold	1610
Platinum group	264

[a] In Mt except silver, gold and the platinum group metals (t).

coeval oil crisis but clearly continuing up to the present day. These curves suggest that consumption of major metals is following a wave pattern in which the various crests may not be far off in time. Lead, indeed, may be over the crest. Various factors are probably at work here; recycling, more economical use of metals and substitution by ceramics and plastics—industrial minerals are much used as a filler in plastics. Production of plastics rose by a staggering 1529% between 1960 and 1985 and a significant

fraction of the demand behind this is attributable to metal substitution. In Table 1.5 the increases in production of selected metals and industrial minerals provide a striking contrast and one that explains why for some years now many large metal mining companies have been moving into industrial mineral production, an example being the RTZ Corporation, probably now the world's largest min-

Table 1.5 Increases (%) in world production of some metals and industrial minerals, 1973–1988; metals are given in italics. Recycled metal production is not included

Aluminium	27.5
Feldspar	81.5
Lead	− 5.5
Phosphate	42.5
Sulphur	19.0
Trona	44.0
Cobalt	34.6
Gold	8.8
Mica	18.9
PGM	80.8
Talc	44.0
Zinc	26.7
Copper	16.2
Gypsum	37.6
Molybdenum	8.3
Potash	39.1
Tantalum	143.0
Diatomite	29.1
Iron ore	12.0
Nickel	16.8
Silver	13.9
Tin	− 9.8

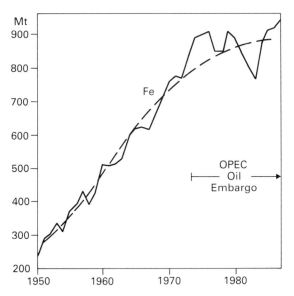

Fig. 1.5 World production of iron ore from 1950 to 1987 with general trend superimposed. (After Lofty *et al.* 1989.)

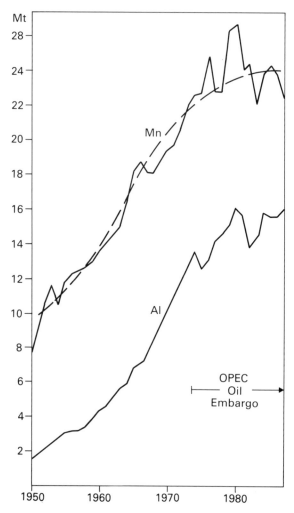

Fig. 1.6 World production of manganese and aluminium from 1950 to 1987 with general trend for manganese superimposed. (After Lofty *et al.* 1989.)

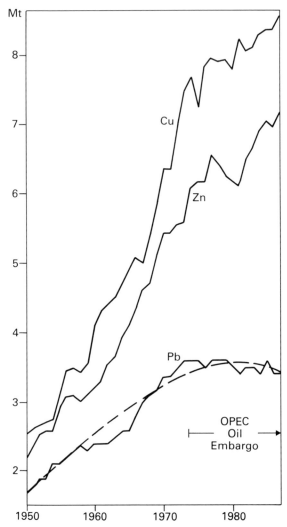

Fig. 1.7 World production of copper, zinc and lead from 1950 to 1987. General trend for lead superimposed. (After Lofty *et al.* 1989.)

ing company, which in 1989 derived 30% of its net earnings from industrial mineral operations compared with 58.4% from metal mining. Are we soon to pass onwards from the Iron Age into a ceramic–plastic age? Readers are urged to monitor this *tentative* prophecy by keeping these graphs up-to-date using data from the same or a similar source, which includes production from eastern-bloc countries as well as that from non-communist countries; be warned, some compilations ignore this latter production but still pose as world production figures. A factor of small but growing importance is the demand for non-ferrous metals in the non-

OECD countries; this has grown by over 6% per annum during the present decade, compared with less than 1% in the OECD countries and it may increase sufficiently in the coming decade to influence present trends in demand for these metals. This demand too should be monitored. Finally, although the increase in demand for the major, high-tonnage production metals is decreasing at the present time, the future is bright for certain minor, low-tonnage metals such as cobalt, platinum group metals (PGM), rare earth elements (REE), tantalum and titanium.

Commodity prices—the market mechanism

Most mineral trading takes place within the market economies of the non-communist world and the prices of minerals or mineral products are governed by the factors of supply and demand. If consumers want more of a mineral product than is being supplied at the current price, this is indicated by their 'bidding up' the price, thus increasing the profits of companies supplying that product and, as a result, resources in the form of capital investment are attracted into the industry and supply expands. On the other hand if consumers do not want a particular product its price falls, producers make a loss and resources leave the industry.

World markets

Modern transport leads to many commodities having a world market; a price change in one part of the world affects the price in the rest of the world. Such commodities include wheat, cotton, rubber, gold, silver and base metals. These commodities have a wide demand, are capable of being transported and the costs of transport are small compared with the value of the commodity. The market for diamonds is worldwide but that for bricks is local.

Formal organized markets have developed in various civilizations. In the thirteenth century England began to build up her large export trade in raw wool to the neighbouring continental countries and it was extended by the subsequent development of the chartered companies. These were based in London, and merchants gathered there to buy and sell the produce transported by the companies' ships (Harvey 1985). Later with the expansion of trade following the Industrial Revolution in the eighteenth century, the UK became the greatest exporting and importing country in the world and commodity markets developed further. In these markets buying and selling takes place in a recognized building, business is governed by agreed rules and conventions, and usually members only are allowed to engage in transactions. Base metals are traded on the London Metal Exchange, gold and silver on the London Bullion Market. Similar markets exist in many other countries, e.g. the New York Commodity Exchange—Comex. Because these markets are composed of specialist buyers and sellers and are in instant communication with each other, prices are sensitive to any change in worldwide supply and demand.

Futures dealings on these markets, although often misrepresented as sheer gambling, enable buyers and sellers to protect themselves from heavy losses through price changes. When a quantity of metal is bought for delivery that day, the deal is known as a spot transaction and the price is the spot price. When the seller contracts to deliver at a later date the price agreed upon is the future or forward price. These dealings normally help to even out price fluctuations, but speculators can trigger off violent price fluctuations.

Another example of the usefulness of future markets can be illustrated by recording the action of Echo Bay Mines. In 1979 this company's silver property was almost worked out but the company had a highly skilled work force. It therefore purchased the Lupin gold property from INCO (International Nickel Company of Canada). This left Echo Bay with a large debt to service. In order to reduce this the company sold forward a third of its 1983 production and at the start of 1984 about 20% of that year's planned production.

Sometimes a company may not be able to deliver the product it has contracted to sell, as for example when it is affected by a prolonged strike, it then declares *force majeure*—a legal term excusing it from completing its side of the bargain.

The prices of some metals on Comex and the London Metal Exchange are quoted daily by many newspapers, whilst more comprehensive guides to current metal and mineral prices can be found in the *Engineering and Mining Journal, Industrial Minerals,* the *Mining Journal, Erzmetall* and other technical journals. Short and long term contracts between seller and buyer may be based on these fluctuating prices. On the other hand, the parties concerned may agree on a contract price in advance of production, with clauses to allow for price changes because of such factors as inflation or fluctuations in currency exchange rates. Contracts of this nature are still very common in the case of iron and uranium ore and industrial minerals production. However, there is now a tendency towards the development of a global market for iron ore, but pricing mechanisms are still separate in the three principal markets: Japan, North America and Western Europe. The bases for price negotiations in these three markets are described by Barrington (1986), who also provides a clear short summary of the form of sales contracts which the tyro will find very informative. Whatever the form of sale is to be, the mineral economists of a mining company must try to forecast demand for, and hence the price of, the mine product, well in advance of mine development. A useful recent discussion of mineral markets can be found in Gocht *et al.* (1988).

Forces determining prices

Demand and supply

Demand is defined by economists as the quantity of a good, i.e. a commodity, product or service which satisfies a human need, that buyers will purchase at a given price over a certain period of time (Harvey 1985). Demand may change over a short period of time for a number of reasons. Where one good substitutes for another to a significant extent and the price of this latter falls, then the substituting good becomes relatively expensive and less of it is bought. Copper and aluminium are affected to a degree in this way. Similarly when goods are complementary a change in the price of one may affect the demand for the second. For example if car prices fall more are bought and the demand for tyres and petrol increases. A change in technology may increase the demand for a metal, e.g. the use of titanium in jet engines, or decrease it, e.g. tin—development of thinner layers of tin on tinplate and substitution (see Table 1.5). The expectation of future price changes or shortages will induce buyers to increase their orders to have more of a good in stock.

Supply refers to how much of a good will be offered for sale at a given price over a set period of time. This quantity depends on the price of the good and the conditions of supply. High prices stimulate supply and investment by suppliers to increase their output. A fall in price has the opposite effect and some mines may be closed completely or put on a care-and-maintenance basis in the hope of better times in the future. Conditions of supply may also change fairly quickly through: (a) changes owing to abnormal circumstances, such as natural disasters, war, other political events, fire and strikes at the mines of big suppliers; (b) impoved techniques in exploitation; and (c) discovery and exploitation of large new orebodies.

Government action

Governments can act to stabilize or change prices. Stabilization may be attempted by building up a stockpile, although the mere building up of a substantial stockpile increases demand and tends to push up the price! With a substantial stockpile in being, sales from the stockpile can be used to prevent prices rising significantly and purchases for the stockpile may be used to prevent or moderate price falls. As commodity markets are worldwide it is in most cases impossible for one country acting on its own to control prices. Groups of countries have attempted to exercise control over copper prices (CIPEC) in this way but with little success. The International Tin Council, operating through the London Metal Exchange, stabilized the price of tin fairly successfully for several decades but eventually succumbed when, on top of other difficulties, an increasing flood of tin came on to the market in the 1980s from a non-member of the Council—Brazil.

Brazilian production rose from 6000 t in 1981 to 26 514 t in 1985 and in August of that year the ITC Buffer Stock Manager was forced to cease trading. The price plummeted from about US$13 500 t^{-1} to under US$6000 t^{-1}. This had a devastating effect on countries such as Bolivia and Malaysia; in Malaysia over 200 mines were forced to close, and closures occurred in all the tin producing countries. Brazilian production in 1988 was 44 000 t, nearly a quarter of world production, and with no increase in consumption, despite the low price (about US$7100 in November 1989), there is little hope of a significant rise in the foreseeable future.

Stockpiles also may be built up by governments for strategic reasons, and this, as mentioned above, can push up prices markedly. A material needed for military purposes is considered strategic and a material is termed critical if future events involving its supply from abroad threaten to inflict serious damage on a nation's economy (Anon. 1987, Clark & Reddy 1989). Clearly materials classified in these categories will vary from country to country. Metals such as platinum, manganese and chromium are considered critical in the USA, but in the Republic of South Africa (RSA), a major source of all three, they are not. Metals are by no means the only critical mineral products for the USA and other industrialized nations. A very important industrial mineral group is the sillimanite minerals from which refractory bricks, ladles and tubes for steel manufacture are made. Later decisions to run down strategic stockpiles can have a crushing effect on market prices. Stockpiling policies of some leading industrialized nations are discussed by Morgan (1989).

An action that will increase consumption of platinum and rhodium is the adoption of new regulations on car exhausts by the EEC countries.

This, it is estimated, will increase consumption of platinum in Europe by 145% by 1993. Comparable actions by governments stimulated by environmental lobbies will no doubt occur in the coming years.

Governments may also alter the relative prices of products to secure greater use of one of them. For example to conserve its North Sea oil supplies the British Government could give the national coal producer, British Coal, or coal consumers such as National Power, a subsidy. In contrast a high tax could be imposed on the producers or consumers of oil, but the UK government has no conservation policy for energy minerals.

For these and other reasons nations need to formulate mineral policies to safeguard their economies. Japan is the best example of a highly industrialized nation that has to import nearly all its raw materials for energy and industrial production. Diversified sources of supply have been developed so that political risks are hedged and Japan has a very far sighted, closely integrated mineral policy. By comparison, that of the USA is full of contradictions and *non sequiturs* (Wolfe 1984) and those of some EEC countries are little better.

Cartels

The attempt by CIPEC to control copper prices was an attempt to set up a *cartel*—an agreement to restrict the production or sales of a good and set prices not related directly to costs of production and distribution. The Organization of Petroleum Exporting Countries (OPEC) has operated what at times has been a more successful cartel but the most effective has been that covering the international sale of diamonds. Only a tiny fraction of world production of natural diamonds is not marketed by the Central Selling Organization (CSO) which is controlled by De Beers, itself a subsidiary of the Anglo-American Corporation of the RSA. The CSO policy is to maintain a stable diamond price by withholding sales if the price is weak and increasing them if prices rise excessively. The CSO has been remarkably successful in this regard apart from the boom-and-bust cycle of 1979–82 (Wolfe 1984). This was a period of considerable uncertainty in financial circles. The average price of oil was increased by 9% at the end of 1980 to approximately US$35 a barrel after having doubled in 1979. The price had already risen from US$1.70 in 1970. OPEC congratulated itself on its restraint in view of the world recession! Inflation was rampant and many investors rushed

almost blindly into various markets in their search for inflation-proof investments for their money. Prices of some precious goods rocketed. The diamond market indicator '1 carat D-flawless brilliant' rose to about $65 000—completely out of line with the supply and cost of production. Then, like silver, it came crashing down, being quoted at about $19 000 in mid 1982. This wild swing might have been even more pronounced had not the CSO released more diamonds in an attempt to prevent the wild upward price rise. Cartels rarely work for long (see previous section). The CSO in the future will be handicapped by (a) the potential development of huge new mines, (b) the development of synthetic gem-quality diamonds and (c) high world interest rates which have to be paid on the money CSO has expended on buying up international production, much of which is stored in vaults in Johannesburg where it earns no revenue and provides jobs for security guards!

The cartels discussed so far are sometimes termed *artificial cartels* and most of those set up in the minerals industry have been a flop. They tend to conform to the same general pattern (Youngquist 1990). At the start the cartel pushes up the price above what the normal world price would be. This encourages more production by marginal producers and smaller suppliers, who are outside the cartel, as well as substitution and conservation by the end users. The cartel then finds it necessary to hold down supply by members agreeing on individual production quotas. For political and/or economic reasons some individual members then tend to cheat and the cartel falls apart. This is how OPEC fell into disarray in the mid 1980s, leading to a considerable drop in oil prices. This whole sorry story of broken promises was succinctly summarized by Youngquist who pointed out that eventually the marginal non-OPEC producers will deplete their resources, and the present somewhat unsuccessful artificial cartel will evolve into a *natural* one as world oil production becomes concentrated in those countries bordering the Persian Gulf. These are all, at the moment, OPEC members. A *natural cartel* is then one that arises when a particular mineral resource is concentrated in one or two countries, which may then be able to control the world price. Platinum is not far from providing an example, with the bulk of the world's reserves being in the RSA. Cobalt is another example, but the nations producing this metal are in urgent need of foreign exchange and are unlikely to cut off supplies. Should this happen for cobalt or another mineral product, then the consuming nations will have to pay exorbitant prices or develop a substitute.

Recycling

Recycling is already having a significant effect on some product prices. Economic and particularly environmental considerations will lead to increased recycling of materials in the immediate future. Recycling will prolong resource life and reduce mining wastes and smelter effluents. Partial immunity from price rises, shortages of primary materials or actions by cartels will follow. A direct economic and environmental bonus is that energy requirements for recycled materials are usually much lower than for treating ores—in the case of aluminium 80% less electricity is needed.

In the USA the use of ferrous scrap as a percentage of total iron consumption rose from 35 to 42% over the period 1977–87 and aluminium from 26 to 37%; but copper has remained mainly in the range 40–45% and zinc 24–29% (Kaplan 1989). It must be noted that the end uses and life cycles of products can place severe limitations on the annual percentage of a commodity that can be recycled. Aluminium in beer cans is soon available for recycling but that in window frames may not be available for a generation or so. Antimony used in lead acid batteries is readily reclaimable, that used in flame retardants is unlikely to be recycled. The potential for recycling platinium, chromium and cobalt (at present 10–12%) is promising (van Rensberg 1986).

Contrary to the case for metals the potential for recycling industrial minerals is much lower and is confined to a few commodities, such as bromine, fluor-compounds, industrial diamonds, iodine, and feldspar and silica in the form of glass; so prices will be affected less by this factor (Neotstaller 1988). Owing to the large volume and low value of demolition materials, the degree of recycling depends not only on where and in what condition and quantity they occur, but also on the materials themselves. Both asphalt (tarmacadam) roads and concrete from roadways and buildings can now be crushed, screened, mixed with new binders and rolled or pressed back into place, but in the USA such recycling is still no more than 10% of available wastes (Wilson 1989). In Germany much more recycling of road material is carried out (Smith 1987).

Substitution and new technology

Both substitution and new technology may lead to a diminution in demand. We have already seen great changes, such as the development of longer lasting car batteries that use less lead, substitution of copper and plastic for lead water pipes and a change to lead-free petrol; all of which have contributed to a downturn in the demand for lead (Fig. 1.7). A factor that affected all metals was the OPEC shock in 1973 (Figs 1.5–1.7), which led to huge increases in the price of oil and other fuels, pushed demand towards materials with a low sensitivity to high energy costs and favoured the use of lighter and less expensive substitutes for metals (Cook 1987). The substitution of natural diamonds by synthetic ones is steadily growing (see Chapter 8).

In the past, base metal producers have spent vast sums of money on exploration, mine development and production but have paid too little attention to the defence and development of markets for their products (Davies 1987, Anthony 1988). Producers of aluminium, plastics and ceramics on the other hand have promoted research for new uses including substitution for metals. Space will allow me to cite a few examples only. Tank armour is now frequently made of multilayer composites—metal, ceramic and fibres, ceramic-based engine components are already used widely in automobiles and it has been forecast that by AD 2030 90% of engines used in cars, aeroplanes and power stations will be made from novel ceramics. A useful article on developments in ceramic technology is that by Wheat (1987).

Metal and mineral prices

Metals

Metal prices are erratic and hard to predict (Figs 1.8–1.10). In the short run prices fluctuate in response to unforeseen news affecting supply and demand, e.g. strikes at large mines or smelters, unexpected increases in warehouse stocks. This makes it difficult to determine regular behaviour patterns for some metals. Over the intermediate term (several decades) the prices clearly respond to rises and falls in world business activity, which is some help when attempting to forecast price trends (Figs 1.9, 1.10). The OPEC shock of 1973, which has been mentioned above, besides setting off a severe recession, led to less developed countries building

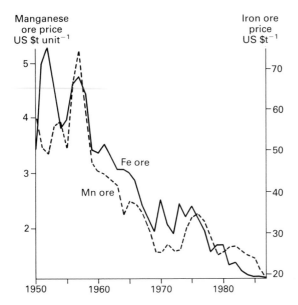

Fig. 1.8 Yearly average price of iron and manganese ores for 1950–1988. The iron ore price is for 61.5% (of iron) Brazilian ore CIF German ports expressed in constant 1980 US dollars. (Prior to 1965, Liberian ore.) The manganese ore price is for 46–48% Indian ore CIF US ports, expressed in constant 1980 US dollars per metric unit (10 kg manganese content in the ore). (Source Lofty *et al.* 1989.)

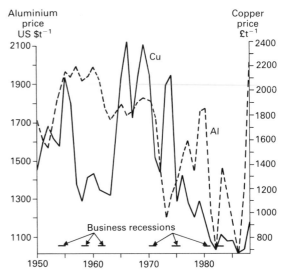

Fig. 1.9 Yearly average producer price of unalloyed aluminium ingot on the New York Market expressed in constant 1980 US dollars and the yearly average price of electrolytic wire bars of copper on the London Metal Exchange expressed in constant 1980 pounds sterling. Both graphs cover the period 1950–1988. (Source Lofty *et al.* 1989.)

up huge debts in order to pay the increased costs of energy. This involved reducing their living standards and purchasing fewer durable goods. At the same time many metal producing, developing countries, such as Chile, Peru, Zambia and Zaïre, increased production irrespective of metal prices in order to earn hard currencies for debt repayment. A further aggravation from the supply and price point of view has been the large number of significant mineral discoveries since the advent of modern exploration methods in the 1950s (Fig. 1.11). Metal explorationists have, to a considerable extent, become victims of their own success. It should be noted that the fall-off in non-gold discoveries from 1976 onwards is largely due to the difficulty explorationists now have in finding a viable deposit in an increasingly unfavourable economic climate.

Despite an upturn in price for many metals during the last few years (1986–89) the general outlook is not promising for most of the traditional metals, in particular iron, manganese (Fig. 1.8), lead (Fig. 1.10), tin and tungsten. Some of the reasons for this prognostication have been discussed above. It is the

Fig. 1.10 Yearly average domestic prices of pig lead and prime western grade zinc for 1950–1988 on the New York Market expressed in constant 1980 US dollars. (Source Lofty *et al.* 1989.)

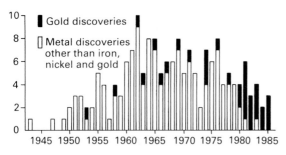

Fig. 1.11 Significant metallic orebody discoveries in non-Communist countries. Significant discoveries are relatively low-cost producers that have a potential to generate over US$1000 million in gross revenue. (After Cook 1987.)

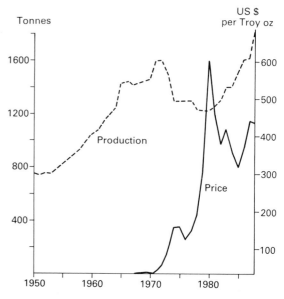

Fig. 1.12 World production of gold from 1950 to 1988 and the actual yearly average price in US dollars per Troy ounce (i.e. no correction for inflation). (Source Lofty *et al.* 1989, and the *Mining Annual Review* 1989.)

minor metals, such as titanium, tantalum and others, that are likely to have a brighter future. For a similar view and a discussion of the underlying causes see Kelly (1990), but for a more bullish view on the major metals see Green (1989) and for price trends over a longer period (1880–1980) see Slade (1989).

Gold has had a different history since World War Two. From 1934 to 1972 the price of gold remained at US$35 per Troy ounce. In 1971 President Nixon removed the fixed link between the dollar and gold and left it to market demand to determine the daily price. The following decade saw gold soar to a record price of US$850 an ounce, a figure inconceivable at the beginning of the 1970s; it then fell back to a price no higher in real terms than that of the 1930s (Fig. 1.12). Citizens of many countries were again permitted to hold gold either as bars or coinage and many have invested in the metal. Unfortunately for those attempting to predict future price changes, demand for this metal is not determined so much by industrial demand but by fashion and sentiment— two notoriously variable and unpredictable factors! The main destinations of gold at the present day are carat jewelry and bars for investment purposes. Bar hoarding, led by Japan and Taiwan, jumped by 77% in 1988 compared with the previous year and reached a record 474 t! Carat jewelry in that year absorbed 1484 t and industrial users took up only 199 t; dentistry accounted for another 59 t (Jacks 1989).

The rise in the price of gold since 1971 has led to a great increase in prospecting and the discovery of many large deposits (Fig. 1.11). This trend is continuing at an increasing rate and gold production, since reaching a low point in 1979, has been increasing rapidly (Fig. 1.12)—will fashion and

sentiment absorb such an annually sustained increase in supply? Many mining companies have adopted a cautious approach and are not opening new deposits without being sure that they could survive on a price of around US$250 per ounce.

Industrial minerals

Most industrial minerals can be traded internationally. The exceptions are the low value commodities, such as sand, gravel and crushed stone, which have a low unit value and are produced mainly for local markets. Minor deviations from this statement are beginning to appear, however, such as crushed granite being shipped from Scotland to the USA, sand from Western Australia to Japan and filtration sand and water from the UK to Saudi Arabia! Lower middle unit value minerals from cement to salt can be moved over intermediate to long distances provided that they are shipped in bulk by low cost transport. Nearly all industrial minerals of higher unit value are internationally tradeable, even when shipped in small lots.

Minerals with a low unit value will increase greatly in cost to the consumer with increasing distance to the place of use. Consequently commodities of low unit value are normally of little or no value unless available close to a market. Exceptions to this rule may arise in special circumstances, e.g. the southeastern sector of England (including London) where demand for aggregates cannot now be met from local resources. Considerable additional supplies now have to be brought in by rail and road over distances in excess of 150 km. For minerals of high unit value such as industrial diamonds, sheet mica and graphite, location is largely irrelevant.

Like metals industrial minerals respond to changes in the intensity of business activities, but as a group to nothing like the extent shown by metals and their prices are generally much more stable (see Table 1.6). A more comprehensive table is given by McCarl (1989). One reason for the greater stability of many industrial mineral prices is their use or partial use in consumer non-durables, for which consumption remains comparatively stable during recessions, e.g. potash, phosphates and sulphur for fertilizer production, and diatomite, fluorspar, iodine, kaolin, limestone, salt, sulphur, talc, etc. used in chemicals, paint, paper, rubber and so on. The value of an industrial mineral depends largely on its end use and the amount of processing it has undergone; with more precise specifications of chemical purity, crystalline perfection, physical

form, hardness, etc. the price goes up; for this reason many minerals have quite a price range, e.g. kaolin to be used as coating clay on paper is four times the price of kaolin for pottery manufacture.

Individual commodities show significant price variations related to supply and demand. Potash over the last 40 years is an example. When supplies were plentiful, such as after the completion of several large Canadian mines, prices were depressed, whereas when demand has outstripped supply prices have shot up.

According to Noetstaller (1988), already discovered world reserves of most industrial minerals are adequate to meet the expected demand up to at least AD 2000 and so no significant increases in real long-term prices are expected. Exceptions to this are likely to be sulphur, baryte, talc and pyrophyllite. Growth rates are expected to rise steadily, rates exceeding 4% per annum are forecast for nine industrial minerals and 2–4% for 29 others. These figures may well prove to be conservative estimates. Contrary to metals, the recycling potential of industrial minerals is, with some exceptions, low and competing substitutional materials are frequently less efficient (e.g. calcite for kaolinite as a cheaper paper filler) or more expensive.

Table 1.6 Long term price trends of some major industrial minerals (based on Noetstaller 1988). Prices are ex-mine or processing plant

Commodity	Average annual price in constant US$ t^{-1}		
	1965	1975	1983
Asbestos	313	273	276
Baryte	81	76	78
Diatomite	160	151	204
Feldspar	29	33	35
Fluorspar	144	148	181
Graphite flake	336	495	606
Gypsum	12	9	9
Kaolin	75	77	90
Mica (ground)	153	122	130
Perlite	25	24	33
Phosphate rock	21	43	24
Potash	132	171	146
Sand and gravel	3.16	2.94	3.25
Stone (crushed)	4.57	4.23	4.27
Stone (dimension)	122	133	139
Sulphur	64	53	91
Talc	69	33	94
Vermiculite	52	72	96

The role of the firm

In the private sector a firm can operate as a sole proprietor, partnership, private company, public company or cooperative society. A new firm commonly commences as one of the first three.

The sole proprietor

The one-man firm is the oldest form of business organization. In the minerals industry it still flourishes in the form of prospectors, consultants and small mine operators. The owner may of course have quite a number of employees but, in general, he or she will be restricted in the amount of capital it is possible to raise.

The partnership

Partnerships of two or more people make more capital available to increase the size of the firm. Each partner provides part of the capital and shares the profits on an agreed basis. Again the amount of capital available is likely to be inadequate for modern large scale business.

The joint-stock company

The oldest known joint-stock company appears to have been the Société de Bazacle formed at Toulouse, France initially to operate water mills on the Garonne River (Gimpel 1988). It was already flourishing in the thirteenth century and the shares, called uchaus, were bought and sold freely by the public, like those on a contemporary stock exchange. This company survived into the middle of the twentieth century when it was nationalized. Joint-stock companies flourished, and failed, in England in Tudor times, but because they enjoyed no limited liability many people were reluctant to buy shares because they risked not only the money they invested but all their private assets, should the company go into liquidation. Moreover this rendered it impossible to spread investment risks by investing in a number of companies. The Industrial Revolution in the UK towards the end of the eighteenth century introduced machinery and factories and made it essential that industry be able to raise large amounts of capital. So to induce small savers to invest, the British parliament granted limited liability in 1855.

Today the joint-stock company is the normal form for large business organizations. Compared with partnerships it has the advantages of limited liability, continuity, availability of capital by selling shares on the stock markets and ease of expansion within its own organization or by buying up other companies. The reader will find further valuable discussion of the role of the firm and the structure of industry in Harvey (1985). An idea of how a mining company spends its income can be gained from Fig. 1.13.

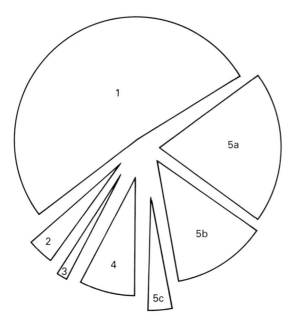

DISTRIBUTION OF INCOME 1984

1 Wages, salaries and benefits to employees
2 Taxes and imposts to governments
3 Interest to providers of loan capital
4 Dividends to shareholders
5 Reinvested in R.G.C. Group
   a) depreciation
   b) exploration and evaluation
   c) retained earnings

Fig. 1.13 Distribution of income in 1984 by Renison Goldfields Consolidated Ltd. (After *Consolidated Gold Fields PLC Annual Report* 1985.)

Important factors in the economic recovery of minerals

Principal steps in the establishment and operation of a mine

These may be summarized briefly as follows:
1 mineral exploration—to discover an orebody;
2 feasibility study—to prove its commercial viability;
3 mine development—establishment of the entire infrastructure;
4 mining—extraction of ore from the ground;
5 ore dressing (mineral processing)—milling of the ore, separation of ore minerals from gangue, separation of the ore minerals into concentrates (e.g. copper concentrate), separation and refinement of industrial mineral products;
6 smelting—recovering metals from the mineral concentrates;
7 refining—purifying the metal;
8 marketing—shipping the product (or metal concentrate if not smelted and refined at the mine) to the buyer, e.g. custom smelter, manufacturer.

Some important factors in the evaluation of a potential orebody

Ore grade

The concentration of a metal in an orebody is called its grade, usually expressed as a percentage or in parts per million (ppm). The process of determining these concentrations is called assaying. Various economic and sometimes political considerations will determine the lowest grade of ore that can be produced from an orebody; this is termed the cut-off grade. In order to delineate the boundaries of an orebody in which the level of mineralization gradually decreases to a background value many samples will have to be collected and assayed. The boundaries thus established are called assay limits. Being entirely economically determined, they may not be marked by any particular geological feature. If the price received for the product increases, then it may be possible to lower the value of the cut-off grade and thus increase the tonnage of the ore reserves; this will have the effect of lowering the overall grade of the orebody, but for the same daily production, it will increase the life of the mine. A sophisticated discussion of cut-off grade can be found in Hall (1988).

Grades vary from orebody to orebody and, clearly, the lower the grade, the greater the tonnage of ore required to provide an economic deposit. The general tendency in metalliferous mining during this century has been to mine ores of lower and lower grade. This has led to the development of more large scale operations with outputs of 40 kt of ore per day being not unusual. The drop in the average grade of copper mined this century is illustrated in Table 1.7.

Technological advances also may transform waste into ore. For example the introduction of solvent extraction enabled Nchanga Consolidated Copper Mines in Zambia to treat 9 Mt of tailings to produce 80 kt of copper (Anon. 1979).

It also will be necessary to estimate, if possible by comparison with similar orebodies, what the head grade will be. This is the grade of the ore as deliverd to the mill (mineral dressing plant). Often the head grade is lower than the measured ore grade because of mining dilution—the inadvertent or unavoidable incorporation of barren wall rock into the ore during mining.

The grade of an industrial mineral deposit is not always as critical as that for a metal deposit. The

Table 1.7 Average grade (%) of copper mined, 1900–70. (From Gentilhomme 1983)

1900	~ 2.7	1930	1.4
1910	2.0	1950	0.93
1920	1.8	1970	0.70

important criteria for assessing the usefulness of non-metallic deposits include both chemical and physical properties, and many types of deposit are used *en masse*. This means that deposit homogeneity is important; patches with different properties must either be discarded or blended to form a uniform product. For example, in an aggregate to be used for roadstone the properties that matter are the aggregate crushing, impact and abrasion values (ACV, AIV and AAV), the 10% fines value, the polished stone value (PSV), the size grading possible from the plant, and the petrography of the pebbles. As another example, limestone has a wide variety of uses, depending on such properties as the chemical purity (for making soda ash or sea water magnesia), the colour, grain-size distribution and brightness of a powder (paper and other filler applications) or its oil absorption (putty manufacture).

For a new industrial mineral deposit to be worked at a profit, it is essential firstly that the properties of the material either before or after processing match the specification for intended use, and secondly that there are adequate reserves to meet the expected demand. From many deposits a number of products with different properties can be made; a variety of different markets may therefore be required to achieve the most economical exploitation of the deposit.

By-products

In some ores several metals are present and the sale of one may help finance the mining of another. For example, silver and cadmium can be by-products of the mining of lead–zinc ores and uranium is an important by-product of many South African gold ores. Among industrial minerals the recovery of by-product baryte and lead from fluorspar operations can be cited.

Commodity prices

The price of the product to be marketed is a vital factor and this subject has been discussed above (pp. 12–15). The mineral economists of a mining company must try to forecast the future demand for, and hence the price of, the mine product(s), well in advance of mine development.

Mineralogical form

The properties of a mineral govern the ease with

which existing technology can extract and refine certain metals and this may affect the cut-off grade. Thus nickel is recovered far more readily from sulphide than from silicate ores, and sulphide ores can be worked down to about 0.5% whereas silicate ores must assay about 1.5% in order to be economic.

Tin may occur in a variety of silicate minerals, such as andradite and axinite, from which it is not recoverable, as well as in its main ore mineral form, cassiterite. Aluminium is of course abundant in many silicate rocks, but normally it must be in the form of hydrated aluminium oxides, the rock called bauxite, for economic recovery. The mineralogical nature of the ore will also place limits on the maximum possible grade of the concentrate. For example, in an ore containing native copper it is theoretically possible to produce a concentrate containing 100% Cu but, if the ore mineral was chalcopyrite ($CuFeS_2$), the principal source of copper, then the best concentrate would contain only 34.5% Cu.

Industrial mineral deposits present different problems. For example, for a silica sand deposit to be utilized for high quality glass making the Fe_2O_3 content should be less than 0.035%. Some brown-looking sands with much more Fe_2O_3 can be upgraded if most of the iron is present as a coating on the grains, which can be removed either by scrubbing or by acid-leaching. If the iron is present as inclusions within the quartz grains then upgrading may be impossible.

Grain size and shape

The recovery is the percentage of the total metal or industrial mineral contained in the ore that is recovered in the concentrate; a recovery of 90% means that 90% of the metal in the ore is recovered in the concentrate and 10% is lost in the tailings. It might be thought that if one were to grind ores to a sufficiently fine grain size then complete separation of mineral phases might occur and make 100% recovery possible. With present technology this is not the case, as most mineral processing techniques fail in the ultra-fine size range. Small mineral grains and grains finely intergrown with other minerals are difficult or impossible to recover in the processing plant, and recovery may be poor. Recoveries from primary (bedrock) tin deposits are traditionally poor, ranging over 40–80% with an average around 65%, whereas recoveries from copper ores usually lie in the range 80–90%. Sometimes fine grain size

and/or complex intergrowths may preclude a mining operation. The McArthur River deposit in the Northern Territory of Australia contains 200 Mt grading 10% zinc, 4% lead, 0.2% copper and 45 ppm silver with high grade sections running up to 24% zinc and 12% lead. This enormous deposit of base metals has remained unworked since its discovery in 1956 because of the ultra-fine grain size and despite years of mineral processing research on the 'ore'. The basic elements of a lead–zinc mineral processing plant are shown in Fig. 1.14.

As mentioned above, the grain size distribution is critical in the use of a number of different industrial rocks and minerals. Aggregate in concrete is used in specified size ranges, depending on the end use. Each different mineral filler application (paper, rubber, plastics) requires different carefully specified, often narrow, ranges. Grain shape also may be important. For example, relatively long fibres of asbestos are required to weave asbestos cloth.

Undesirable substances

Deleterious substances may be present in both ore and gangue minerals. For example, tennantite ($Cu_{12}As_4S_{13}$) in copper ores can introduce unwanted arsenic and sometimes mercury into copper concentrates. These, like phosphorus in iron concentrates and arsenic in nickel concentrates, will lead to custom smelters imposing financial penalties. The ways in which gangue minerals may lower the value of an ore are very varied. For example, an acid leach is normally used to extract uranium from the crushed ore, but if calcite is present, there will be excessive acid consumption and the less effective alkali leach method may have to be used. Some primary tin deposits contain appreciable amounts of topaz which, because of its hardness, increases the abrasion of crushing and grinding equipment, thus raising the operating costs.

Size and shape of deposits

The size, shape and nature of ore deposits also affects the workable grade. Large, low grade deposits that occur at, or near, the surface can be worked by cheap open pit methods (Fig. 1.15) whilst thin tabular vein deposits will necessitate more expensive underground methods of extraction, although generally they can be worked in much smaller volumes so that a relatively small initial capital outlay is required. Although the initial capital outlay for larger deposits may be higher, open pitting, aided

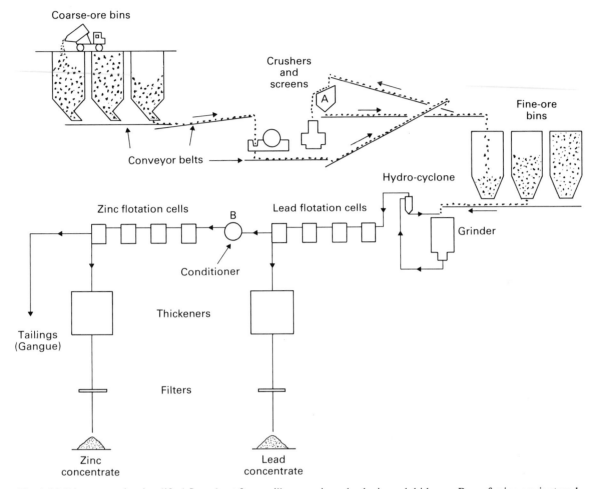

Fig. 1.14 Diagrammatic, simplified flow sheet for a mill processing a lead–zinc sulphide ore. Run-of-mine-ore is stored in the coarse ore bins to provide a continuous feed to the mill in the event of an interruption to mining operations. The ore is first crushed and then screened, with the coarse fraction being recrushed, so that it passes through the screen (A) to the fine ore bins. Material from these is then ground to the correct size for froth flotation (5–500 µm is the usual general range over which this separation process gives its best results). Sorting of the ground product is carried out in a hydrocyclone, which passes the coarse material back for further grinding. The froth flotation tanks use the surface properties of the lead and zinc minerals to float these, whilst all other minerals in the tanks sink to the bottom. The conditioner (B) is used for the storage of fine-particle suspensions. The thickeners start the process of dewatering the flotation products and the filters finish it.

Fig. 1.15 Development of an open pit mine. During the early stages (a–a′) more ore (black) is removed than waste rock; then, as the pit becomes deeper, the ratio of waste to ore mined becomes greater until at stage b–b′ it is about 1.6 to 1. (After Barnes, 1988, *Ores and Minerals*, Open University Press, with permission.)

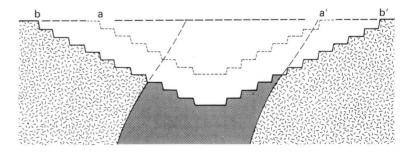

by the savings from bulk handling of large daily tonnages of ore (say > 30 kt), has led to a trend towards the large scale mining of low grade orebodies. As far as shape is concerned, orebodies of regular shape can generally be mined more cheaply than those of irregular shape particularly when they include barren zones. For an open pit mine the shape and attitude of the orebody will also determine how much waste has to be removed during mining, which is quoted as the waste-to-ore or stripping ratio. The waste will often include not only overburden (waste rock above the orebody) but waste rock around and in the orebody, which has to be cut back to maintain a safe overall slope to the sides of the pit.

As can be seen from Fig. 1.15, a time comes during exploitation when the waste-to-ore ratio becomes too high for profitable working; for low grade ores this is around 2 : 1 and the mine then must be abandoned or converted into an underground operation. Many small mines start as small, cheaply worked open pits in supergene-enriched ore (Chapter 19), and then develop into underground operations (Fig. 1.16). Haulage always used to be by narrow gauge, electrically operated railways, but now, if the orebody size permits, rubber-tyred equipment is used to produce larger tonnages more economically and shafts are then gentle spiral declines up which ore can be hauled out by diesel trucks (trackless mining). Various factors limit the depth to which under-

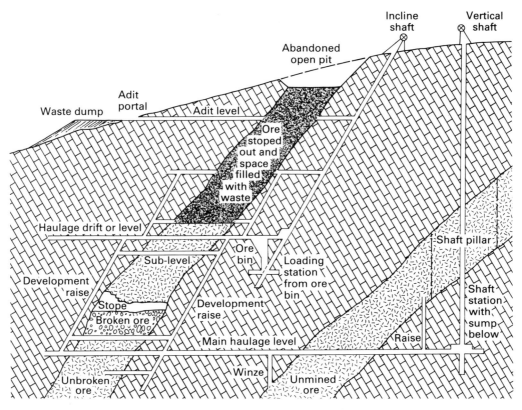

Fig. 1.16 Mining terminology. Ore was first mined at the outcrop from an open pit; then an adit was driven into the hillside to intersect and mine the ore at a lower level. An inclined shaft was sunk later to mine at even deeper levels and, eventually, a vertical shaft was sunk to serve operations to two orebodies more efficiently. Ore is mined by driving two haulage drifts at different levels and connecting them by raises which are then connected by sublevels. Ore is mined upwards from the lower sublevel to form a stope. Broken ore can be left in the stope to form a working platform and to support its walls (shrinkage stoping), or withdrawn and waste from the mill pumped in (cut-and-fill stoping). Ore between haulage and sublevel is left as supporting pillars until the level is abandoned. A shaft pillar is also left unmined. (After Barnes, 1988, *Ores and Minerals*, Open University Press, with permission).

ground mining can penetrate and the present record (1989), about 3810 m below surface, is held by the Western Deep Levels Mine, RSA.

Ore character

A loose unconsolidated beach sand deposit can be mined cheaply by dredging and does not require crushing. Hard compact ore must be drilled, blasted and crushed. In hard-rock mining operations a related aspect is the strength of the country rocks. If these are badly sheared or fractured they will be weak and require roof supports in underground working, and in open pitting a gentler slope to the pit sides will be required, which in turn will affect the waste-to-ore ratio adversely.

Cost of capital

Big mining operations have now reached the stage, thanks to inflation, where they require enormous initial capital investments. For example, to develop the 450+ Mt Cu–U–Au Roxby Downs Project in South Australia, Western Mining Corporation and British Petroleum have estimated that a capital investment of A$1200 million will be necessary, and for the 77 Mt Ag–Pb–Zn deposit of Red Dog, in northern Alaska, US$300–500 million will be required; grades there are 17% Zn, 5% Pb, $61.7\,\text{g}\,\text{t}^{-1}$ Ag. This means that the stage has been reached where few companies can afford to develop a mine with their own financial resources. They must borrow the capital from banks and elsewhere, capital which has to be repaid with interest. Thus the revenue from the mining operation must cover the running costs, the payment of taxes, royalties, the repayment of capital plus interest on it, and provide a profit to shareholders who have risked their capital to set up or invest in the company (Fig. 1.13). The order of magnitude of capital costs for industrial mineral operations in the USA is given in Table 1.8.

The models quoted in the table represent shallow underground mining in competent rock and open-cast mines in hard rock with moderate stripping ratios and short to medium haulage distances. They are thus typical for a variety of industrial mineral operations. The investment for small-scale sand and gravel operations will be much lower and, by contrast, investment costs for industrial minerals produced on a large scale, such as bauxite, phosphate and soda ash, will be several hundred million US dollars. The cost of infrastructural installations discussed in the next section are not included in the above table.

Location

Geographical factors may determine whether or not an orebody is economically viable. In a remote location there may be no electric power supply, roads, railways, houses, schools, hospitals, etc. All or some of these infrastructural elements will have to be built, the cost of transporting the mine product to its markets may be very high and wages will have to be high to attract skilled workers.

Table 1.8 Order of magnitude capital costs for model mechanized mines extracting industrial minerals in the USA. (Most data from Noetstaller, 1988)

Production capacity (t p.d.)	Type of operation	Capital cost range (10^6 1984 US$)	
		Mining operation	Model flotation mill
Undergroud			
100	Adit access or shallow shaft, shrinkage stope	2.0–4.0	2.5–3.2
1000	Adit access or shallow shaft, cut and fill stoping	10.0–12.0	9.7–10.7
5000	Adit access, room and pillar mining	18.0–20.0	25.2–27.2
Open pit			
500	Stripping ratio 1 : 1 to 2 : 1, 400 m hauls, hard rock	4.0–5.0	6.5–7.5
5000	Stripping ratio 1 : 1 to 2 : 1, 750 m hauls, hard rock	9.0–12.0	25.2–27.2
10 000	Stripping ratio 1 : 1 to 2 : 1, 2000 m hauls, hard rock	16.0–22.0	41.0–43.5

Environmental considerations

New mines bring prosperity to the areas in which they are established but they are bound to have an environmental impact. The new mine at Neves-Corvo in southern Portugal will raise that country's copper output by 93 000% and tin production by 9900%! The total labour force will be about 900. When it is remembered that one mine job creates about three indirect jobs in the community in service and construction industries, the impact clearly is considerable. Impacts of this and even much smaller size have led to conflicts over land use and opposition to the exploitation of mineral deposits by environmentalists, particularly in the more populous of the developed countries. The resolution of such conflicts may involve the payment of compensation and the eventual cost of rehabilitating mined out areas, or the abandonment of projects; '. . . whilst political risk has been cited as a barrier to investment in some countries, environmental risk is as much of a barrier, if not a greater in others.' (Select Committee 1982). Opposition by environmentalists to exploration and mining was partially responsible for the abandonment of a major copper mining project in the UK in 1973.

In its report in 1987, *Our Common Future*, the United Nations World Commission on Environment and Development, headed by Mrs Brundtland, Norway's Prime Minister, pointed out that the world manufactures seven times more goods today than it did in 1950. The Commission proposed 'sustainable development', a marriage of economy and ecology, as the only practical solution, i.e. growth without damage to the environment. White (1989) quoted James Stevenson of RTZ Corporation as admitting that sustainable growth is an awkward concept for the extractive industry. 'How does mining fit in? How can you regard a copper mine as a sustainable development?' remembering that all mines have a finite life, some of 20 years or even less. White wrote that mine operators must now take a longer term view of their operations. Feasibility studies must look at the closure costs as well as the opening and running costs. The running and closure costs must put something back into the community. The question 'What will be left behind, in physical and human terms?' must be faced squarely and adequately responded to. A number of mining companies are already engaged in environmental impact analyses, but for many companies 'the idea of predicting the effects of closure twenty to forty years ahead is still fairly novel'. However, much thought has been given to the matter and a useful reference on environmental protection during mining operations is Arndt & Luttig (1987); whilst Smith (1989) gives a good summary of the legislative controls in a number of developed and developing countries. Noetstaller (1988) has pointed out that whereas industrial mineral operations have the same general environmental impacts on land and ground water disturbance as metalliferous or coal mining, the impact is generally less marked since the mines are usually smaller and shallower, and normally less waste is produced because in most cases ore grades are higher than in metal mining. Pollution hazards owing to heavy metals or acid waters are low or non-existent and atmospheric pollution, caused by the burning of coal or the smelting of metallic ores, is much less serious or absent. The excavations created by industrial mineral operations are often close to conurbations, in which case these holes in the ground may be of great value as landfill sites for city waste. A British brick company recently sold such a site for £30 million!

Taxation

Greedy governments may demand so much tax that mining companies cannot make a reasonable profit. On the other hand, some governments have encouraged mineral development with taxation incentives, such as a waiver on tax during the early years of a mining operation. This proved to be a great attraction to mining companies in the Irish Republic in the 1960s and brought considerable economic gains to that country.

When a company only operates one mine, then it is particularly true that dividends to shareholders should represent in part a return of capital, for once an orebody is under exploitation it has become a *wasting asset* and one day there will be no ore, no mine and no further cash flow. The company will be wound up and its shares will have no value. In other words, all mines have a limited life and for this reason should not be taxed in the same manner as other commercial undertakings. When this fact is allowed for in the taxation structure of a country, it can be seen to be an important incentive to investment in mining in that country.

Political factors

Many large mining houses will not now invest in

politically unstable countries. Fear of nationalization with perhaps very inadequate or even no compensation is perhaps the main factor. Nations with a history of nationalization generally have poorly developed mining industries. Possible political turmoil, civil strife and currency controls may all combine to increase greatly the financial risks of investing in certain countries. Periodical reviews of political risks in various countries are prepared and published by specialized companies and references to these can be found in Noetstaller (1988). Useful articles on the subject are Anon. (1985c) and Anon. (1985d).

Ore reserve classification

In delineating and working an orebody the mining geologist often has to classify his ore reserves into three classes: proved, probable and possible; frequently used synonyms are: measured, indicated and inferred. Proved ore has been sampled so thoroughly that we can be certain of its outline, tonnage and average grade, within certain limits. Elsewhere in the orebody, sampling from drilling and development workings may not have been so thorough, but there may be enough information to be reasonably sure of its tonnage and grade; this is

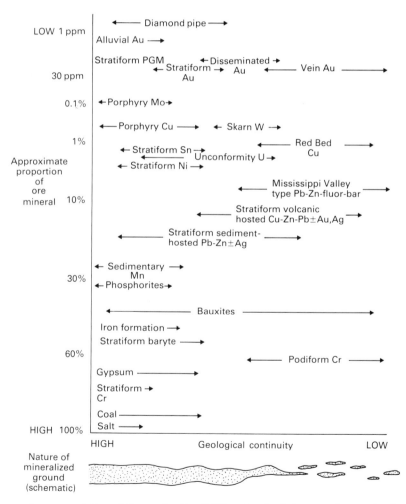

Fig. 1.17 Diagrammatic representation of the continuity of different orebody types and approximate grades. (Source King *et al.* 1982.)

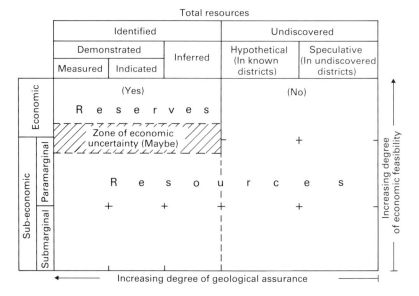

Fig. 1.18 McKelvey Box (McKelvey 1973) showing scheme of categories of reserves and resources with a modification by Taylor (1989) and indicating areas of immediate economic interest.

probable ore. On the fringes of our exploratory workings we may have enough information to infer that ore extends for some way into only partially explored ground and that it may amount to a certain volume and grade of possible ore. In most countries, these, or equivalent, words have nationally recognized definitions and legal connotations. The practising geologist must therefore know the local definitions thoroughly and make sure that he uses them correctly.

Much of the difficulty in deciding which category a particular section of an ore deposit should be assigned to arises from the differences in continuity of various types of mineralization, as indicated in Fig. 1.17. Another of many difficulties discussed by Taylor (1989) is that the term ore reserve seems to change its meaning even for a single deposit during the exploration, evaluation and mining stages, because it appears to mean different things to geologists, financiers, miners and bureaucrats, who each tend to define it in their own way! It has been argued that the ore reserve figure should itself be a forecast of the results of production—a laudable but often unattainable idea according to Taylor. Production forecasts covering early years of operation are a mandatory supplement to ore reserve statements in applications for finance (e.g. share flotations) and 1 year versions are normally included in the annual reports of working mines, but reserve estimates cannot measure mining efficiency and production is better considered as a third stage of the sequence concerning an orebody: geological exploration, reserve estimation and production.

Mineral resources

These represent the total amount of a particular commodity (e.g. tin) and usually they are estimated for a nation as a whole and not for a company. They consist of ore reserves; known but uneconomic deposits; and hypothetical deposits not yet discovered. The estimation of the undiscovered potential of a region can be made by comparison with well explored areas of similar geology.

Theoretically, world resources of most metals are enormous. Taking copper as an example, there are large amounts of rock running 0.1–0.3% and enormous volumes containing about 0.01%. The total quantity of copper in such deposits probably exceeds that in proven reserves by a factor of from 10^3 to 10^4. Nevertheless, the enormous amount of such material does not at present imply a virtually endless resource of metals. As grades approach low values, a concentration (the mineralogical limit) is reached, below which an element no longer forms a distinct physically recoverable mineral phase.

An interesting and provocative discussion of mineral resources and some of the elements of mineral economics can be found in Wolfe (1984), and the relationship between resources and reserves is shown in Fig. 1.18.

Geochemical considerations

It is traditional in the mining industry to divide metals into groups with special names. These are as follows:

1 *precious metals*—gold, silver, platinum group (PGM);

2 *non-ferrous metals*—copper, lead, zinc, tin, aluminium (the first four being commonly known as *base metals*);

3 *iron and ferroalloy metals*—iron, manganese, nickel, chromium, molybdenum, tungsten, vanadium, cobalt;

4 *minor metals and related non-metals*—antimony, arsenic, beryllium, bismuth, cadmium, magnesium, mercury, REE, selenium, tantalum, tellurium, titanium, zirconium, etc.;

5 *fissionable metals*—uranium, thorium (radium).

Table 1.9 Concentration factors

	Average crustal abundance (%)	Average minimum exploitable grade (%)	Concentration factor
Aluminium	8	30	3.75
Iron	5	25	5
Copper	0.005	0.4	80
Nickel	0.007	0.5	71
Zinc	0.007	4	571
Manganese	0.09	35	389
Tin	0.0002	0.5	2500
Chromium	0.01	30	3000
Lead	0.001	4	4000
Gold	0.000 000 4	0.0001[a]	250

[a] 1 ppm.

For the formation of an orebody the element or elements concerned must be enriched to a considerably higher level than their normal crustal abundance. The degree of enrichment is termed the concentration factor and typical values are shown in Table 1.9.

2 / The nature and morphology of the principal types of ore deposit

A good way to start an argument among mining geologists is to suggest that a deposit held by common consensus to be syngenetic is in fact epigenetic! These words are clearly concerned with the manner in which deposits have come into being and, like all matters of genesis, they are fraught with meaning and

are frequently heard on the lips of mining geologists wherever they gather. What do these magic words mean? A syngenetic deposit is one that has formed at the same time as the rocks in which it occurs and it is sometimes part of a stratigraphical succession, such as an iron-rich sedimentary horizon. An epigenetic deposit, on the other hand, is one believed to have come into being after the host rocks in which it occurs. A good igneous analogy is a dyke; an example among ore deposits is a vein. Before discussing their nature we must learn some of the terms used in describing orebodies.

If an orebody viewed in plan is longer in one direction than the other we can designate this long dimension as its strike (Fig. 2.1). The inclination of the orebody perpendicular to the strike will be its dip and the longest dimension of the orebody its axis. The plunge of the axis is measured in the vertical plane ABC, but its pitch or rake can be measured in any other plane, the usual choice being the plane containing the strike, although if the orebody is fault controlled then the pitch may be measured in the fault plane. The meanings of other terms are self-evident from the figure.

It is possible to classify orebodies in the same way as we divide up igneous intrusions according to whether they are discordant or concordant with the lithological banding (often bedding) in the enclosing rocks. Considering discordant orebodies first, this large class can be subdivided into those orebodies which have an approximately regular shape and those which are thoroughly irregular in their outlines.

Discordant orebodies

Regularly shaped bodies

Tabular orebodies

These bodies are extensive in two dimensions, but have a restricted development in their third dimension. In this class we have veins (sometimes called fissure-veins) and lodes (Fig. 2.2). In the past, some workers have made a genetic distinction between these terms; veins were considered to have resulted

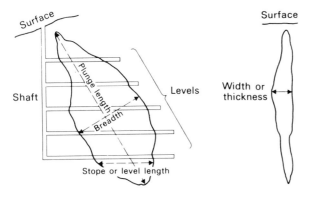

AB and CB lie in the same vertical plane.
DB, AB and EB are in the same horizontal plane and EB is perpendicular to DB

Fig. 2.1 Diagrams illustrating terms used in the description of orebodies.

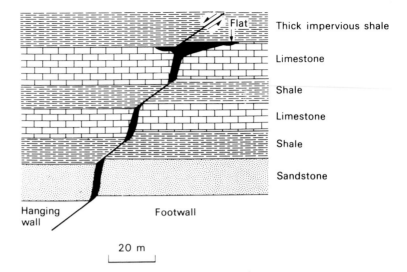

Flat

Thick impervious shale

Limestone

Shale

Limestone

Shale

Sandstone

Hanging wall   Footwall

20 m

Fig. 2.2 Vein occupying a normal fault and exhibiting pinch-and-swell structure, giving rise to ribbon ore shoots. The development of a flat beneath impervious cover is shown also.

mainly from the infilling of pre-existing open spaces, whilst the formation of lodes was held to involve the extensive replacement of pre-existing host rock. Such a genetic distinction has often proved to be unworkable and the writer advises that all such orebodies be called veins and the term lode be dropped.

Veins are often inclined, and in such cases, as with faults, we can speak of the hanging wall and the footwall. Veins frequently pinch and swell out as they are followed up or down a stratigraphical sequence (Fig. 2.2). This pinch-and-swell structure can create difficulties during both exploration and mining often because only the swells are workable and, if these are imagined in a section at right angles to that in Fig. 2.2, it can be seen that they form ribbon ore shoots. The origin of pinch-and-swell structure is shown in Fig. 2.3. An initial fracture in rocks changes its attitude as it crosses them according to the changes in physical properties of the rocks and these properties are in turn governed by changes

in lithology (Fig. 2.3a). When movement occurs producing a normal fault then the less steeply dipping sections are held against each other to become bearing surfaces, and open spaces (dilatant zones) form in the more steeply dipping sections. Then, should minerals be deposited in these cavities, a vein will be formed. If the reader carries out the experiment of reversing the movement on the initial fracture, he will find that the steeper parts of the fault now act as bearing surfaces and the dilatant zones are formed in the less steeply dipping sections. Veins are usually developed in fracture systems and therefore show regularities in their orientation (Figs 2.4, 16.2, 16.5). At the Sigma Mine, Quebec, subvertical gold–quartz veins have formed within ductile shear zones during reverse movements caused by N–S compression (Robert & Brown 1986a,b). Where the shear zones crossed subhorizontal fractures, tension veins were formed and then displaced by the reverse movement (Fig. 2.5). Shear zones are commonly important locations of vein

Bearing surfaces

Shale

Sandstone

Shale

Sandstone

Fig. 2.3 Formation of pinch-and-swell structure in veins.

(a)   (b)

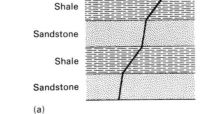

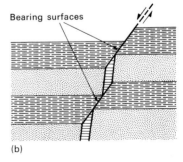

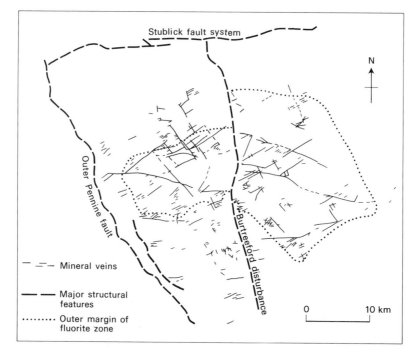

Fig. 2.4 Vein system of the Alston block of the Northern Pennine Orefield, England. Note the three dominant vein directions. (Modified from Dunham 1959.)

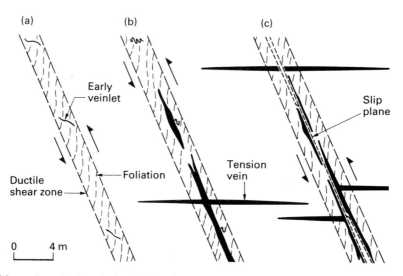

Fig. 2.5 Stages of formation of subvertical and subhorizontal veins at the Sigma Mine, Quebec. (a) Steeply dipping, ductile shear zone developed during reverse movement with formation of early sigmoidal veinlets which, with more shearing, are severely deformed. (b) With the foliation in the central part becoming parallel to the shear zone walls, mineralizing fluids penetrated the shear zone forming subvertical veins. Where the shear zones intersected subhorizontal fractures the first tension veins were formed. (c) Further mineralization as movement continued with displacement of earlier formed subhorizontal veins. (Simplified from a diagram in Robert & Brown 1986a,b.)

deposits and their development and complexities are comprehensively dealt with in Bursnall (1989).

In some deposits hundreds of thin (< 2 mm thick), parallel veins occur in densities > 30 veins per metre; these are known as sheeted veins and they probably form the most common type of primary tin deposit in the world, but they are generally high tonnage-low grade resources (Plimer 1987a).

The infilling of veins may consist of one mineral but more usually it consists of an intergrowth of ore and gangue minerals. The boundaries of vein orebodies may be the vein walls or they can be assay boundaries within the veins.

Tubular orebodies

These bodies are relatively short in two dimensions but extensive in the third. When vertical or sub-vertical they are called pipes or chimneys, when horizontal or subhorizontal, 'mantos'. The Spanish word manto is inappropriate in this context for its literal translation is blanket: it is, however, firmly entrenched in the English geological literature. The word has been and is used by some workers for flat-lying tabular bodies, but the perfectly acceptable word 'flat' is available for these; therefore the reader must look carefully at the context when he encounters the term 'manto'.

In eastern Australia, along a 2400 km belt from Queensland to New South Wales, there are hundreds of pipes in and close to granite intrusions. Most have quartz fillings and some are mineralized with bismuth, molybdenum, tungsten and tin; an example is shown in Fig. 2.6. Pipes may be of various types and origins (Mitcham 1974), but many are formed by the partial dissolution of the host rock, e.g. the Maggs Pipe in granite at the Zaaiplaats Tin Mine, RSA (Pollard *et al.* 1989). Infillings of mineralized breccia are particularly common, a good example being the copper-bearing breccia pipes of Messina in South Africa (Jacobsen & McCarthy 1976).

Mantos and pipes may branch and anastomose; an example of a branching manto is given in Fig. 2.7. Mantos and pipes are often found in association, the pipes frequently acting as feeders to the mantos. Sometimes mantos pass upwards from bed to bed by way of pipe connections, often branching as they go, an example being the Providencia Mine in Mexico where a single pipe at depth feeds into 20 mantos nearer the surface.

In some tubular deposits formed by the sub-horizontal flow of mineralizing fluid, ore grade

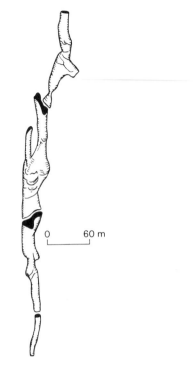

Fig. 2.6 Diagram of the Vulcan tin pipe, Herberton, Queensland. The average grade was 4.5% tin. (After Mason 1953.)

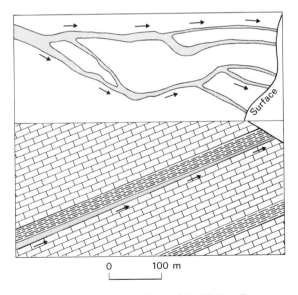

Fig. 2.7 Plan and section of part of the Hidden Treasure manto, Ophir mining district, Utah. (After Gilluly 1932.)

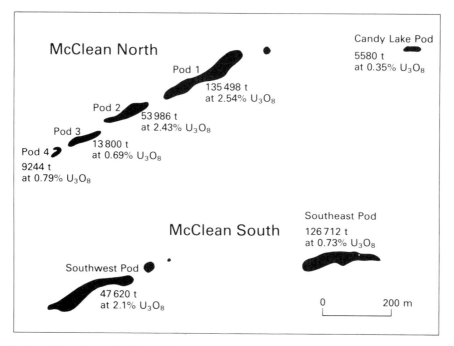

Fig. 2.8 Distribution of uranium orebodies at the McClean deposits, northern Saskatchewan. (After Wallis *et al.* 1984.)

mineralization may be discontinuous, thus creating pod-shaped orebodies (Figs 2.8, 2.9) as with the McClean deposits, Saskatchewan. These pods undulate along the unconformity between a regolith and the overlying Proterozoic Athabasca Supergroup sediments, and their position appears to have been controlled by a vertically disposed fracture system (Wallis *et al.* 1984). The mineralizing fluid removed much quartz from both the regolith and the overlying sediments and deposited new minerals including a considerable amount of pitchblende in its place.

Irregularly shaped bodies

Disseminated deposits

In these deposits, ore minerals are peppered throughout the body of the host rock in the same way as accessory minerals are disseminated through an igneous rock; in fact, they often *are* accessory minerals. A good example is that of diamonds in kimberlites; another is that of some orthomagmatic nickel–copper deposits, such as the La Perouse Layered Gabbro, Alaska (Czamanske *et al.* 1981), which contains disseminated sulphide mineraliz-

ation throughout its entire thickness of about 6 km. This deposit has over 100 Mt grading about 0.5% nickel and 0.3% copper. Given the economic climate of the 1970s and a suitable geographical location, such a body might have constituted ore. However, in its remote position and at an elevation of over 1000 m, where much of it is covered by up to 200 m of ice, it must remain for the moment an item to be grouped among mineral resources and not ore reserves (see Chapter 1), despite the 1.2–1.5 ppm of PGM in flotation concentrates which one day could form a lucrative by-product. In other deposits, the disseminations may be wholly or mainly along close-spaced veinlets cutting the host rock and forming an interlacing network called a stockwork (Fig. 14.1) or the economic minerals may be disseminated through the host rock and along veinlets (Fig. 14.2). Whatever the mode of occurrence, mineralization of this type generally fades gradually outwards into subeconomic mineralization and the boundaries of the orebody are assay limits. They are, therefore, often irregular in form and may cut across geological boundaries. The overall shapes of some are cylindrical (Fig. 14.5) and others are caplike (Fig. 14.13). The mercury-bearing

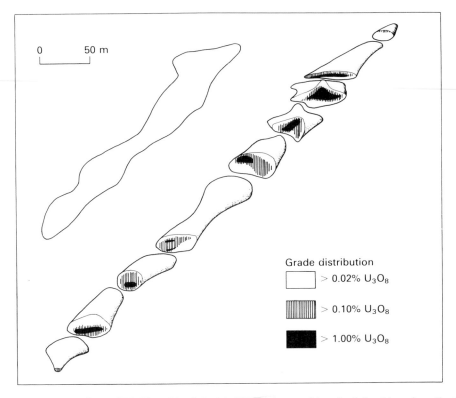

0   50 m

Grade distribution

> 0.02% U_3O_8

> 0.10% U_3O_8

> 1.00% U_3O_8

Fig. 2.9 Plan and expanded views of McClean North Pod 1. The pods are arbitrarily defined by mineralization, which averages a minimum of 0.15 m%, i.e. 0.1% U_3O_8 over a thickness of at least 1.5 m. (After Wallis *et al.* 1984.)

stockworks of Dubník in Slovakia are sometimes pear-shaped.

Stockworks occur most commonly in acid to intermediate plutonic igneous intrusions, but they may cut across the contact (Fig. 2.10) into the country rocks, and a few are wholly or mainly in the country rocks. Disseminated deposits produce most of the world's copper and molybdenum and they are also of some importance in the production of tin, gold, silver, mercury and uranium.

Irregular replacement deposits

Many ore deposits have been formed by the replacement of pre-existing rocks at low to medium temperatures ($< 400°C$), e.g. magnesite deposits in carbonate-rich sediments (Morteani 1989), pyrophyllite orebodies in altered pyroclastics (Stuckey 1967) and siderite deposits in limestones (Pohl *et al.* 1986). Other replacement processes occurred at high temperatures, at contacts with medium-sized to large igneous intrusions. Such deposits have therefore been called contact metamorphic or pyrometasomatic; however, *skarn* is now the preferred and more popular term. The orebodies are characterized by the development of calc-silicate minerals such as diopside, wollastonite, andradite garnet and actinolite. These deposits are extremely irregular in shape (Fig. 2.11); tongues of ore may project along any available planar structure—bedding, joints, faults, etc.—and the distribution within the contact aureole is often apparently capricious. Structural changes may cause abrupt termination of the orebodies. The principal materials produced from skarn deposits are: iron, copper, tungsten, graphite, zinc, lead, molybdenum, tin, uranium, garnet, talc and wollastonite.

Other replacement deposits occur which do not belong to the skarn class; examples of these include flats. These are horizontal or subhorizontal bodies of ore which commonly branch out from veins and lie in carbonate host rocks beneath an impervious cover such as shale (Figs 2.2, 2.12).

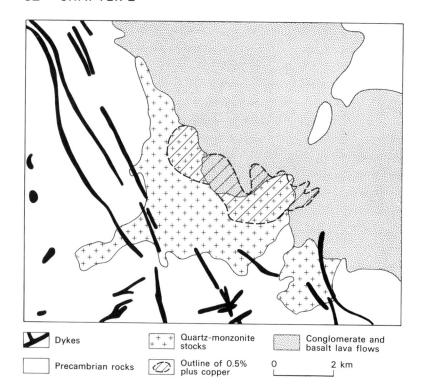

Fig. 2.10 Generalized geological map of the Bagdad mine area, Arizona. (Modified from Anderson 1948.)

	Dykes		Quartz-monzonite stocks		Conglomerate and basalt lava flows
	Precambrian rocks		Outline of 0.5% plus copper		

0   2 km

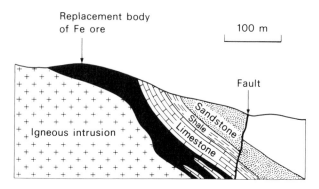

Replacement body of Fe ore

100 m

Fault

Sandstone
Shale
Limestone

Igneous intrusion

Fig. 2.11 Skarn deposit at Iron Springs, Utah. (After Gilluly *et al.* 1959.)

Concordant orebodies

Sedimentary host rocks

Concordant orebodies in sediments are very important producers of many different metals, being particularly important for base metals and iron, and are of course concordant with the bedding. They may be an integral part of the stratigraphical sequence, as is the case with Phanerozoic ironstones (syngenetic ores formed by sedimentary processes), or they may be epigenetic infillings of pore spaces or replacement orebodies. Usually these orebodies show a considerable development in two dimensions, i.e. *parallel* to the bedding and a limited development *perpendicular* to it (Figs 2.13, 2.14, 2.16) and for this reason such deposits are referred to as stratiform. This term must not be confused with strata-bound, which refers to any type or types of orebody, concordant or discordant, which are restricted to a particular part of the stratigraphical column. Thus the veins, pipes and flats of the Southern Pennine Orefield of England can be designated as strata-bound, as they are virtually re-

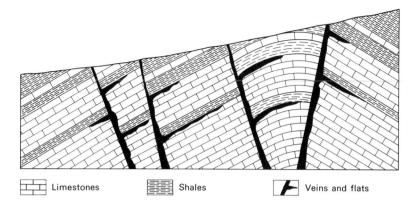

Fig. 2.12 Lead–zinc orebodies in a faulted anticline in Devonian rocks, Gyumushlug, Transcaucasia, USSR. Note the development of veins along the faults with flats branching off them beneath impervious beds of shale. (After Malyutin & Sitkovskiy 1968.)

| | Limestones | | Shales | | Veins and flats |

stricted to the Carboniferous limestone of that region. A number of examples of concordant deposits that occur in different types of sedimentary rocks will be considered.

Limestone hosts

Limestones are very common host rocks for base metal sulphide deposits. In a dominantly carbonate sequence, ore is often developed in a small number of preferred beds or at certain sedimentary interfaces. These are often zones in which the permeability has been increased by dolomitization or fracturing. When they form only a minor part of the stratigraphical succession, limestones, because of their solubility and reactivity, can become favourable horizons for mineralization. For example, the lead–zinc ores of Bingham, Utah, occur in lime-

stones, which make up 10% of a 2300 m succession mainly composed of quartzites.

At Silvermines in Ireland, lead–zinc mineralization occurred (mining ceased in 1982) as syngenetic stratiform orebodies in a limestone sequence (Fig. 2.13), as fault-bounded epigenetic stratabound orebodies in the basal Carboniferous, and as structurally controlled vein or breccia zones in the Upper Devonian sandstones (Taylor & Andrew 1978, Taylor 1984). The larger stratiform orebody shown in Fig. 2.13 occurred in massive, partly brecciated pyrite at the base of a thick sequence of dolomite breccias. The maximum thickness of the ore was 30 m, and, at the base, there was usually an abrupt change to massive pyrite from a footwall of nodular micrite, shale biomicrite and other limestone units of the Mudbank Limestone, although sometimes the contact was gradational. The upper

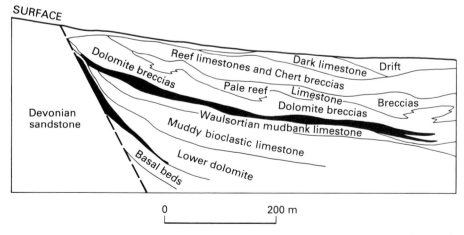

Fig. 2.13 Vertical section through the G zone at Silvermines, Co. Tipperary, Ireland. The orebodies are shown in black. (After Taylor & Andrew 1978.)

contact was always sharp. Pyrite and marcasite made up 75% of the ore, sphalerite formed 20% and galena 4%.

Argillaceous hosts

Shales, mudstones, argillites and slates are important host rocks for concordant orebodies, which are often remarkably continuous and extensive. In Germany, the Kupferschiefer of the Upper Permian is a prime example. This is a copper-bearing shale a metre or so thick which, at Mansfeld, occurred in orebodies that had plan dimensions of 8, 16, 36 and 130 km². Mineralization occurs at exactly the same horizon in Poland, where it is being worked extensively, and across the North Sea in north-eastern England, where it is subeconomic.

The world's largest, single lead–zinc orebody occurs at Sullivan, British Columbia. The host rocks are late Precambrian argillites. Above the main orebody (Fig. 2.14) there are a number of other mineralized horizons with concordant mineralization. This deposit appears to be syngenetic, and the lead, zinc and other metal sulphides form an integral part of the rocks in which they occur. They are affected by sedimentary deformation, such as slumping, pull-apart structures, load casting, etc., in a manner identical to that in which poorly consolidated sand and mud respond (Fig. 2.15).

The orebody occurs in a single generally conformable zone between 60 and 90 m thick and runs 6.6% lead and 5.9% zinc. Other metals recovered are silver, tin, cadmium, antimony, bismuth, copper and gold. Before mining commenced in 1900, the orebody contained at least 155 million tonnes of ore, and at the current rate of mining (2 Mt p.a.), the mine has a remaining life of about 20 years. The footwall rocks consist of graded impure quartzites and argillites and, in places, conglomerate. The hanging wall rocks are more thickly bedded and arenaceous. The ore zone is a mineralized argillite in which the principal sulphide–oxide minerals are pyrrhotite, sphalerite, galena, pyrite and magnetite, with minor chalcopyrite, arsenopyrite and cassiterite. Beneath the central part of the orebody there are extensive zones of brecciation and tourmalinization, which extend downwards for at least 100 m. In places, the matrix of the breccias is heavily mineralized with pyrrhotite and occasionally with galena, sphalerite, chalcopyrite and arsenopyrite. This zone

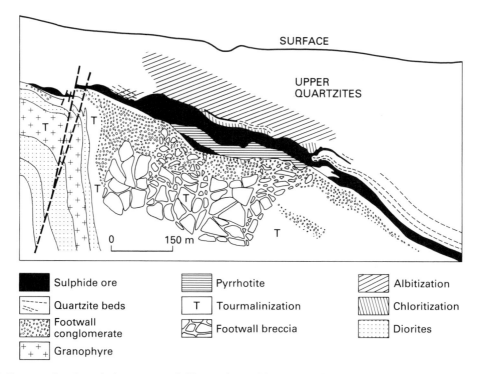

Fig. 2.14 Cross section through the ore zone, Sullivan Mine, British Columbia. (After Sangster & Scott 1976.)

Fig. 2.15 Sulphiditic siltstone, Sullivan Mine, British Columbia. The lighter grey material is sulphide. Note the pull-apart and load structures which affect both the sulphide and silicate layers.

may have been a channelway up which solutions moved to debouch on to the sea floor, to precipitate the ore minerals among the accumulating sediment; if this was the case, then Sullivan could be called a sedimentary-exhalative deposit (Hamilton *et al.* 1983).

Other good examples of concordant deposits in argillaceous rocks, or slightly metamorphosed equivalents, are the lead–zinc deposits of Mount Isa, Queensland, many of the Zambian Copperbelt deposits and the copper shales of the White Pine Mine, Michigan.

Arenaceous hosts

Not all the Zambian Copperbelt deposits occur in shales and metashales. Some orebodies occur in altered feldspathic sandstones (Fig. 2.16). The Mufulira copper deposit occurs in Proterozoic rocks on the eastern side of an anticline (Fleischer *et al.* 1976) and lies just above the unconformity with an older, strongly metamorphosed Precambrian basement. The gross ore reserves in 1974 stood at 282 Mt assaying 3.47% copper and the largest orebody stretches for 5.8 km along the strike and for several kilometres down dip. Chalcopyrite is the principal sulphide mineral, sometimes being accompanied by

significant amounts of bornite. Fluviatile and aeolian arenites form the footwall rocks. The ore zone consists of feldspathic sandstones which, in places, contain carbon-rich lenses with much sericite. The basal portion is coarse-grained and characterized by festoon cross-bedding in which bornite is concentrated along the cross-bedding together with well rounded, obviously detrital zircon; whilst in other parts of the orebody, concentrations of sulphides occur in the hollows of ripplemarks and in desiccation cracks. These features suggest that some of the sulphides are detrital in origin. Mineralization ends abruptly at the hanging wall, suggesting a regression and at this sharp cut-off the facies changes from an arenaceous one to dolomites and shallow-water muds.

Conformable deposits of copper occur in some sandstones that were laid down under desert conditions. As these rocks are frequently red, the deposits are known as red bed coppers. Dune sands are frequently porous and permeable and the copper minerals are generally developed in pore spaces. Examples of such deposits occur in the Permian of the Urals and the Don Basin in the USSR in the form of sandstone layers 10–40 cm thick running 1.5–1.9% copper. They are also found in the Trias of central England, in Nova Scotia, in Germany and in

SW   NE

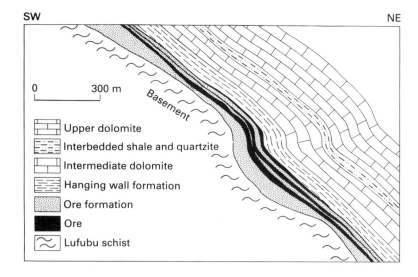

Fig. 2.16 Cross section through the Mufulira orebodies, Zambia. (After Fleischer *et al.* 1976.)

the south-western USA. At the Nacimiento Mine in New Mexico a deposit of 11 Mt averaging 0.65% copper is being worked by open pit methods. Like other red bed coppers, this deposit has a high metal/sulphur ratio as the principal mineral is chalcocite. This yields a copper concentrate low in sulphur, which is very acceptable to present day custom smelters faced with stringent anti-pollution legislation. Similar deposits are very important in China where they make up nearly 21% of that country's total stratiform copper reserves (Chen 1988).

Copper is not the only base metal that occurs in such deposits. Similar lead ores are known in Germany and silver deposits in Utah. Another important class of pore-filling deposits are the uranium–vanadium deposits of Colorado Plateau or Western States-type, which occur mainly in sandstones of continental origin but also in some siltstones and conglomerates. The orebodies are very variable in form, and pods and irregularly shaped deposits occur, although large concordant sheets up to 3 m thick are also present. The orebodies follow sedimentary structures and depositional features.

Many mechanical accumulations of high density minerals, such as magnetite, ilmenite, rutile and zircon, occur in arenaceous hosts, usually taking the form of layers rich in heavy minerals in Pleistocene and Holocene sands. As the sands are usually unlithified, the deposits are easily worked and no costly crushing of the ore is required. These orebodies belong to the group called placer deposits—

beach sand placers are a good example (Fig. 2.17). Beach placers supply much of the world's titanium, zirconium, thorium, cerium and yttrium. They occur along present day beaches or ancient beaches where longshore drift is well developed and frequent storms occur. Economic grades can be very low and sands running as little as 0.6% heavy minerals are worked along Australia's eastern coast. The deposits usually show a topographical control, the shapes of bays and the position of headlands often being very important; thus in exploring for buried orebodies a reconstruction of the palaeogeography is invaluable.

Rudaceous hosts

Alluvial gravels and conglomerates also form important recent and ancient placer deposits. Alluvial gold deposits are often marked by 'white runs' of vein quartz pebbles, as in the White Channels of the Yukon, the White Bars of California and the White Leads of Australia. Such deposits form one of the few types of economic placer deposits in fully lithified rocks, and indeed the majority of the world's gold is won from Precambrian deposits of this type in South Africa. Figure 2.18 shows the distribution of the gold orebodies in the East Rand Basin where the vein quartz pebble conglomerates occur in quartzites of the Upper Witwatersrand System. Their fan-shaped distribution strongly suggests that they occupy distributary channels. Uranium is recovered as a by-product of the working of the Witwatersrand goldfields. In the very similar Blind River area of Ontario uranium is the only

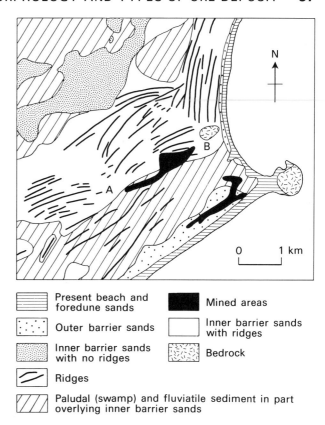

Fig. 2.17 Geology and mining areas of the beach sand deposits of Crowdy Head, New South Wales, Australia. (After Winward 1975.)

metal produced. In this field the conglomeratic orebodies lie in elongate south-easterly trending sheets (Fig. 2.19) that are composed of layers of braided, interfingering channels and beds. These ore sheets, carrying the individual orebodies, have dimensions measured in kilometres (Robertson 1962, Theis 1979). Similar mineralized conglomerates occur elsewhere in the Precambrian.

Chemical sediments

Sedimentary iron, manganese, evaporite and phosphorite formations occur scattered throughout the stratigraphical column, forming very extensive beds conformable with the stratigraphy. They are described in Chapters 18 & 20.

Igneous host rocks

Volcanic hosts

There are two principal types of deposit to be found in volcanic rocks, vesicular filling deposits and volcanic-associated massive sulphide deposits. The first deposit type is not very important but the second type is a widespread and important producer of base metals often with silver and gold as by-products.

The first type forms in the permeable vesicular tops of basic lava flows in which permeability may have been increased by autobrecciation. The mineralization is normally in the form of native copper and the best examples occurred in late Precambrian basalts of the Keweenaw Peninsula of northern Michigan. Mining commenced in 1845 and the orebodies, which are now virtually worked out, were very large and were mined down to nearly 2750 m. There were six main producing horizons, five in the tops of lava flows and one in a conglomerate. The orebodies averaged 4 m in thickness and 0.8% copper. Occasionally, so-called veins of massive copper were found, one such mass weighing 500 t. Similar deposits occur around the Coppermine River in northern Canada where 3 Mt running 3.48% copper has been found but, although the copper content was worth some £100 million at

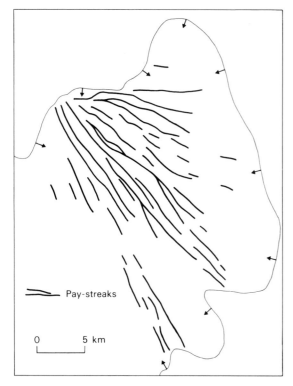

Fig. 2.18 Distribution of pay-streaks (gold orebodies) in the Main Leader Reef of the East Rand Basin of the Witwatersrand Goldfield of South Africa. The arrows indicate the direction of dip at the outcrop or suboutcrop. For the location of this ore district see Fig. 18.10. (After Du Toit 1954.)

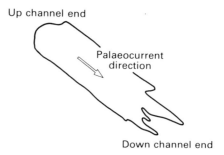

Fig. 2.19 Plan of an ore sheet, Blind River, Ontario, showing blunt up-channel end and fingering down-channel end. (After Robertson 1962.)

1979 prices, the deposit is unlikely to be worked because of its remote location. Other uneconomic deposits of this type are known in many other countries.

Volcanic-associated massive sulphide deposits often consist of over 90% iron sulphide usually as pyrite, although pyrrhotite is well developed in some deposits. They are generally stratiform bodies, lenticular to sheet-like (Fig. 2.20), developed at the interfaces between volcanic units or at volcanic–sedimentary interfaces. With increasing magnetite content, these ores grade to massive oxide ores of magnetite and/or hematite, such as at Savage River in Tasmania, Fosdalen in Norway and Kiruna in Sweden (Solomon 1976). They can be divided into three classes of deposit: (a) zinc–lead–copper, (b) zinc–copper, and (c) copper. Typical tonnages and copper grades are given in Fig. 14.3.

The most important host rock is rhyolite, and lead-bearing ores are normally only associated with this rock-type. The copper class is usually, but not invariably, associated with mafic volcanics. Massive sulphide deposits commonly occur in groups and in any one area they are found at one or a restricted number of horizons within the succession. These horizons may represent changes in composition of the volcanic rocks, a change from volcanism to sedimentation, or simply a pause in volcanism. There is a close association with volcaniclastic rocks and many orebodies overlie the explosive products of rhyolite domes. These ore deposits are usually underlain by a stockwork that may itself be ore grade and which appears to have been the feeder channel up which mineralizing fluids penetrated to form the overlying massive sulphide deposit.

Plutonic hosts

Many plutonic igneous intrusions possess rhythmic layering and this is particularly well developed in some basic intrusions. Usually the layering takes the form of alternating bands of mafic and felsic minerals, but sometimes minerals of economic interest, such as chromite, magnetite and ilmenite, may form discrete mineable seams within such layered complexes (Fig. 10.2). These seams are naturally stratiform and may extend over many kilometres, as is the case with the chromite seams in the Bushveld Complex of South Africa (Fig. 10.1).

Another form of orthomagmatic deposit is the nickel–copper sulphide orebody formed by the sinking of an immiscible sulphide liquid to the bottom of a magma chamber containing ultrabasic or basic magma. These are known to the learned as liquation deposits and they may be formed at the base of lava flows as well as in plutonic intrusions.

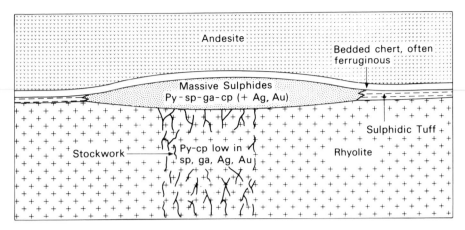

Fig. 2.20 Schematic cross section through an idealized volcanic-associated massive sulphide deposit showing the underlying feeder stockwork and typical mineralogy. Py = pyrite, sp = sphalerite, ga = galena, cp = chalcopyrite.

The sulphide usually accumulates in hollows in the base of the igneous body and generally forms sheets or irregular lenses conformable with the overlying silicate rock. From the base upwards, massive sulphide gives way through disseminated sulphides in a silicate gangue to lightly mineralized and then barren rock (Figs 11.3, 11.6, 11.9).

Metamorphic host rocks

Apart from some deposits of metamorphic origin, such as the irregular replacement deposits already described and deposits generated in contact metamorphic aureoles, e.g. of wollastonite, andalusite, garnet and graphite, metamorphic rocks are mainly important for the metamorphosed equivalents of deposits that originated in sedimentary and igneous rocks and which have been discussed above.

Residual deposits

These are deposits formed by the removal of non-ore material from protore. For example, the leaching of silica and alkalis from a nepheline-syenite may leave behind a surface capping of hydrous aluminium oxides (bauxite). Some residual bauxites occur at the present surface, others have been buried under younger sediments to which they form conformable basal beds. The weathering of feldspathic rocks (granites, arkoses, etc.) can produce important kaolin deposits which, in the Cornish granites of England, form funnel or trough-shaped bodies extending downwards from the surface for as much as 230 m.

Other examples of residual deposits include some laterites sufficiently high in iron to be worked and nickeliferous laterites formed by the weathering of peridotites (Figs 19.2–19.4).

Supergene enrichment

This is a process which may affect most orebodies to some degree. After a deposit has been formed, uplift and erosion may bring it within reach of circulating ground waters, which may leach some of the metals out of that section of the orebody above the water table. These dissolved metals may be redeposited in that part of the orebody lying beneath the water table and this can lead to a considerable enrichment in metal values. Supergene processes are discussed in Chapter 19.

3 / Textures and structures of ore and gangue minerals. Fluid inclusions. Wall rock alteration

The study of textures can tell us much about the genesis and subsequent history of orebodies. The textures of orebodies vary according to whether their constituent minerals were formed by deposition in an open space from a silicate or aqueous solution, or by replacement of pre-existing rock or ore minerals. Subsequent metamorphism may drastically alter primary textures. The interpretation of mineral textures is a very large and difficult subject and only a few important points can be touched on here; for further information the reader should consult Edwards (1952, 1960), Ramdohr (1969), Stanton (1972), Craig & Vaughan (1981) and Ineson (1989).

Open space filling

Precipitation from silicate melts

Critical factors in this situation are the time of crystallization and the presence or absence of simultaneously crystallizing silicates. Oxide ore minerals, such as chromite, often crystallize out early and thus may form good euhedral crystals, although these may be subsequently modified in various ways. Chromites deposited with interstitial silicate liquid may suffer corrosion and partial resorption to produce atoll textures (Fig. 3.1) and rounded grains, whereas those developed in monomineralic bands (Figs 3.2, 10.2) may, during the cooling of a large parent intrusion, suffer auto-annealing and develop foam texture (see Chapter 22).

When oxide and silicate minerals crystallize simultaneously, anhedral to subhedral granular textures similar to those of granitic rocks develop, owing to mutual interference during the growth of the grains of all the minerals. A minor development of micrographic textures involving oxide ore minerals may also occur at this stage.

Sulphides, because of their lower melting points, crystallize after associated silicates and, if they have not segregated from the silicates, they will be present either as rounded grain aggregates representing frozen globules of immiscible sulphide liquid (Fig. 4.1), or as anhedral grains or grain aggregates which have crystallized interstitially to the silicates

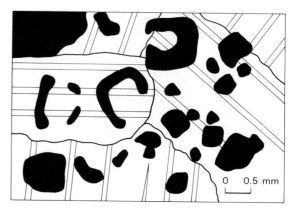

Fig. 3.1 Chromite grains in anorthosite, Bushveld Complex, RSA. The chromites are euhedral crystals which have undergone partial resorption, producing rounded grains of various shapes, including atoll texture.

and whose shapes are governed by those of the enclosing silicate grains.

Precipitation from aqueous solutions

Open spaces, such as dilatant zones along faults, solution channels in areas of karst topography, etc., may be permeated by mineralizing solutions. If the prevailing physico-chemical conditions induce precipitation then crystals will form. These will grow as the result of spontaneous nucleation within the solution, or, more commonly, by nucleation on the enclosing surface. This leads to the precipitation and outward growth of the first formed minerals on vein walls. If the solutions change in composition, there may be a change in mineralogy and crusts of minerals of different composition may give the vein filling a banded appearance (Fig. 3.3), called crustiform banding. Its development in some veins demonstrates that mineralizing solutions may change in composition with time and shows us the order in which the minerals were precipitated, this order being called the paragenetic sequence.

In an example of simple opening and filling of a fissure, as depicted in Fig. 3.3, the banding is symmetrical about the centre of the vein. With

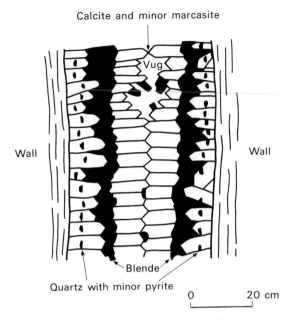

Fig. 3.3 Section across a vein showing crustiform banding.

Fig. 3.2 Chromite bands in anorthosite, Dwars River Bridge, Bushveld Complex, RSA. The central band is 1.3 cm thick at the right-hand end.

repeated opening and mineralization this symmetry will of course be disturbed, but it may still be present among the constituents deposited after the last phase of opening.

Open space deposition also occurs at the surface at sediment–water or rock–water interfaces during, for example, the formation of volcanic-associated massive sulphide deposits. Under such situations rapid flocculation of material occurs and a common primary texture that results is colloform banding. This is a very fine scale banding involving one or more sulphides, very like the banding in agate. Some workers believe that it results from colloidal deposition and that the banding is analogous to the Liesegang rings formed in some gels. This banding is very susceptible to destruction by recrystallization and may be partially or wholly destroyed by diagenesis or low grade metamorphism, producing a granular ore. The textures of

sedimentary iron and manganese ores are discussed in Chapter 18.

Replacement

Edwards (1952) defined replacement as 'the dissolving of one mineral and the simultaneous deposition of another mineral in its place, without the intervening development of appreciable open spaces, and commonly without a change of volume'. Replacement has been an important process in the formation of many ore deposits, particularly the skarn class. This process involves not only the minerals of the country rocks, but also the ore and gangue minerals. Nearly all ores, including those developed in open spaces, show some evidence of the occurrence of replacement processes.

The most compelling evidence of replacement is pseudomorphism. Pseudomorphs of cassiterite after orthoclase have been recorded from Cornwall, England, and of pyrrhotite after hornblende from Sullivan, British Columbia. Numerous other examples are known. The preservation of delicate plant cells by marcasite is well known and crinoid ossicles replaced by cassiterite occur in New South Wales. An overall view of ore deposits suggests that there is no limit to the direction of metasomatism. Given the right conditions, any mineral may replace any other mineral, although natural processes often make for unilateral reactions. Secondary (super-gene) replacement processes, leading to sulphide enrichment by downward percolating meteoric waters, are sometimes most dramatic and fraught with economic importance. They can be every bit as important as primary (hypogene) replacement brought about by solutions emanating from crustal or deeper sources.

Fluid inclusions

The growth of crystals is never perfect and as a result samples of the fluid in which the crystals grew may be trapped in tiny cavities usually < 100 μm in size; these are called fluid inclusions. Their study has proved to be useful in deciphering the formational history of many rock-types and particularly valuable in developing our understanding of ore genesis, especially the subject of ore transport and deposition (Roedder 1984). Fluid inclusions are divided into various types.

Primary inclusions, formed during the growth of crystals, provide us with samples of the ore-forming fluid. They also yield crucial geothermometric data and tell us something about the physical state of the fluid, e.g. whether it was boiling at the time of entrapment. They are common in most rocks and mineral deposits. The vast bulk of fluid inclusion work has been carried out on transparent minerals and the 10 minerals from which inclusions are most commonly reported have been listed by Shepherd *et al.* (1985) (Table 3.1).

The principal matter in most fluid inclusions is water. Second in abundance is carbon dioxide. The commonest inclusions in ore deposits fall into four groups (Nash 1976). Type I, moderate salinity inclusions, are generally two phase, consisting prin-cipally of water and a small bubble of water vapour, which forms 10–40% of the inclusion (Fig. 3.4). The presence of the bubble indicates trapping at an

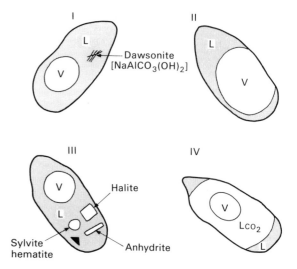

Fig. 3.4 Sketches of four important types of fluid inclusions. L = aqueous liquid, V = vapour, L_{CO_2} = liquid CO_2. For explanation see text. (After Nash 1976.)

elevated temperature with formation of the bubble on cooling. Heating on a microscope stage causes rehomogenization to one liquid phase and the homogenization temperature indicates the temper-ature of growth of that part of the containing crystal, provided the necessary pressure correction can be made. If a pressure correction cannot be made then the rehomogenization temperature can be assumed to be a minimum only, and not the actual temper-ature of trapping. Sodium, potassium, calcium and chlorine occur in solution and salinities range from 0 to 23 wt % NaCl equivalent. In some of these inclusions small amounts of daughter salts have been precipitated during cooling, among them car-bonates and anhydrite.

Type II, gas-rich inclusions, generally contain more than 60% vapour. Again they are dominantly aqueous but CO_2 may be present in small amounts. They often appear to represent trapped steam. The simultaneous presence of gas-rich and gas-poor aqueous inclusions is good evidence that the fluids were boiling at the time of trapping. Type III, halite-bearing inclusions, have salinities ranging up to more than 50%. They contain well-formed, cubic halite crystals and generally several other daughter minerals, particularly sylvite and anhy-drite. Clearly the more numerous and varied the daughter minerals the more complex was the ore fluid. Type IV, CO_2-rich inclusions, have $CO_2 : H_2O$ ratios ranging from 3 to over 30 mol %. They grade

Table 3.1 The top 10 minerals from which fluid inclusions have been reported

1. Quartz	6. Dolomite
2. Fluorite	7. Sphalerite
3. Halite	8. Baryte
4. Calcite	9. Topaz
5. Apatite	10. Cassiterite

into type II inclusions and indeed there is a general gradation in many situations, e.g. porphyry copper deposits (Chivas & Wilkins 1977), between the common types of fluid inclusion.

Perhaps one of the most surprising results of fluid inclusion studies is the evidence of the common occurrence of exceedingly strong brines in nature, brines more concentrated than any now found at the surface (Roedder 1972). These are present not only in mineral deposits but are also common in igneous and metamorphic rocks. Many, but not all, strong brine inclusions are secondary features connected with late stage magmatic and metamorphic phenomena, such as the genesis of greisens, pegmatites and ore deposits, as well as wall rock alteration processes such as sericitization and chloritization. They are compelling evidence that many ore-forming fluids are hot, saline aqueous solutions. They form a link between laboratory and field studies, and it should be noted that there is strong experimental and thermodynamic evidence which shows that chloride in hydrothermal solutions is a potent solvent for metals through the formation of metal–chloride complex ions, and indeed inclusions that carry more than 1 wt % of precipitated sulphides are known.

Secondary inclusions are those formed by any process after the crystallization of the host mineral.

One common mode of formation is during the healing of fractures and this leads to the development of planar arrays of numerous small inclusions. They are particularly common in porphyry copper deposits because most of these have suffered repeated brecciation (pp. 175, 181). *Pseudosecondary* inclusions are frequently of a dominantly primary nature but they also form a spectrum that overlaps both primary and secondary inclusions.

Kelly & Turneaure (1970) presented a detailed study of the mineralogy, paragenetic sequence (order of mineral formation) and geothermometry of tin and tungsten veins in Bolivia. These ores are plutonic to subvolcanic deposits formed at depths of 350–4000 m and over a temperature range of 530–70°C. The ore solutions of the early vein stage (Fig. 3.5) were highly saline brines, up to 46 wt % NaCl equivalent but low CO_2, and the presence of type I and II inclusions in quartz and cassiterite indicate that boiling occurred. Fluid inclusions in the later formed minerals show no evidence of boiling and the trapped fluids are of markedly lower salinity, 2–10% for both fluorites and the siderite.

Fluid inclusions can now be used to characterize ore deposit types (Roedder 1984) and this has led a number of commercial enterprises to set up fluid inclusion laboratories for use in mineral exploration programmes. Some of the important fluid inclusion

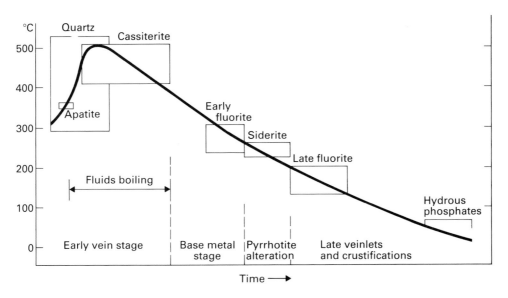

Fig. 3.5 Diagram summarizing the stages of formation and the general trend of temperature variation during the deposition of tin and tungsten ores in Bolivia. The height of the time–temperature squares indicates the minimum range of corrected homogenization temperatures. (Modified from Kelly and Turneaure 1970.)

characteristics of certain mineral deposit types are mentioned in the relevant chapters.

Wall rock alteration

Frequently alongside veins or around irregularly shaped orebodies of hydrothermal origin we find alteration of the country rocks. This may take the form of colour, textural, mineralogical or chemical changes, or any combination of these. Alteration is not always present, but when it is, it may vary from minor colour changes to extensive mineralogical transformations and complete recrystallization. Generally speaking, the higher the temperature of deposition of the ore minerals the more intense is the alteration, but it is not necessarily more widespread. This alteration, which shows a spatial and usually a close temporal relationship to ore deposits, is called wall rock alteration.

The areal extent of the alteration can vary considerably, sometimes being limited to a few centimetres on either side of a vein, at other times forming a thick halo around an orebody and then, since it widens the drilling target, it may be of considerable exploration value. Hoeve (1984) estimated that the drilling targets in the uranium field of the Athabasca Basin in Saskatchewan are enlarged by a factor of 10 to 20 times by the wall rock alteration (see Fig. 16.12), and Williams-Jones (1986) has shown how a study of the distribution of illite crystallinity indices can reveal a series of concentric zones of different crystallinity values several kilometres across and centred on an orebody. The spatial and temporal relationships suggest that wall rock alteration is due to reactions caused by the mineralizing fluid permeating parts of the wall rocks. Many alteration haloes show a zonation of mineral assemblages resulting from the changing nature of the hydro-thermal solution as it passed through the wall rocks. The associated orebodies in certain deposits (e.g. porphyry coppers, Chapter 14) may show a special spatial relationship to the zoning, knowledge of which may be invaluable in probing for the deposit with a diamond drill.

The different wall rock alteration mineral assemblages can be compared with metamorphic facies as, like these, they are formed in response to various pressure, temperature and compositional changes. They are not, however, generally referred to as alteration facies but as types of wall rock alteration. It can be very difficult in some areas to distinguish wall rock alteration, such as chloritization, from the effects of low grade regional or contact metamor-phism, but it is, however, essential that the two effects are separately identified, otherwise a consid-erable degree of exploration effort may be expended in vain.

There are two main divisions of wall rock alter-ation: hypogene and supergene. Hypogene alter-ation is caused by ascending hydrothermal solu-tions, and supergene alteration by descending meteoric water reacting with previously mineralized ground. A third mechanism giving rise to the formation of wall rock alteration is the metamor-phism of sulphide orebodies. In this chapter we will be concerned mainly with hypogene alteration. Studies of this alteration are important because they (a) contribute to our knowledge of the nature and evolution of ore-forming solutions, (b) are often valuable in exploration, and (c) produce minerals such as phyllosilicates that can be used to obtain radiometric dates on the wall rock alteration and, by inference, on the associated mineralization.

The study of hydrothermal fluids has shown that they are commonly weakly acidic, but may become neutral or slightly alkaline by reaction with wall rocks (or by mixing with other waters, e.g. ground water). The solutions contain dissolved ions that are important in ion exchange reactions, and the com-position of a particular hydrothermal solution will have an important bearing on the nature of the wall rock alteration it may give rise to. Since the chemistry of wall rocks can also vary greatly accord-ing to their petrography, it is clear that predictions as to the course of wall rock alteration reactions are fraught with difficulties. Nevertheless, there is, despite a variety of controls, a considerable unifor-mity in the types of wall rock alteration, which facilitates their study and classification. The controls of wall rock alteration fall into two groups governed respectively by the nature of the host rocks and the nature of the ore-forming solutions. Besides the chemistry of the host rocks, other factors of impor-tance are their grain size, physical state (e.g. sheared or unsheared) and permeability, and for the hydro-thermal solution important properties are its chem-istry, pH, Eh, pressure and temperature.

At the actual site of alteration, pressure gradients will probably be insignificant and unlikely to influ-ence mineral zoning. Temperature gradients will have more importance but there is little evidence as yet that steep gradients, having a direct control on alteration assemblages, have ever been developed on the small scale of an orebody, e.g. alongside veins.

Zonal changes on this scale are more likely to have resulted in response to compositional gradients. In and around large orebodies, such as porphyry coppers and molybdenums, gentle thermal gradients (several degrees per metre) obtaining over long periods probably do exert a control on the alteration assemblages. The importance of compositional changes in the mineralizing fluid as it penetrates the wall rocks (producing chemical gradients) has been shown by Hemley (1959), Hemley & Jones (1964), Meyer & Hemley (1967) and other workers. For example in Fig. 3.6 it can be seen that the stability fields of kaolinite, muscovite and K-feldspar as well as being temperature dependent also vary according to the activity ratio of postassium and hydrogen ions in the solution.

Some of the processes occurring during hydrothermal alteration can be illustrated by reference to Fig. 3.6. Curve 1 represents the alteration of K-feldspar to muscovite (sericite)

$$3KAlSi_3O_8 + 2H^+(aq) \rightleftharpoons KAl_3Si_3O_{10}(OH)_2 + 2K^+(aq) + 6SiO_2$$

K-feldspar   muscovite   quartz
                (sericite)

in for example a granite undergoing alteration. Hydrogen ion is consumed (hydrolysis) and potassium released thus altering the activity ratio. This will affect the pH of the ore fluid and the degree of dissociation of dissolved HCl which, in turn, affects the degree of combination of metal–chlorine complexes and therefore the solubility of metals in the solution. This is one of many illustrations of the interdependency of alteration and mineralization. Note also that silica is released by this reaction. This

may be precipitated as secondary quartz, so that some silicification accompanies the sericitization, or it may be carried away in solution perhaps to form a quartz vein or veinlet in the vicinity. By a similar reaction albite can be hydrolysed to paragonite and in this case sodium and silica are removed from the rock. Curve 2 represents another step in the removal of alkalis from a rock by hydroysis:

$$2KAl_3Si_3O_{10}(OH)_2 + 2H^+ + 3H_2O \rightleftharpoons 3Al_2Si_2O_5(OH)_4 + 2K^+$$

muscovite   kaolinite

which allows the growth of a clay mineral typical of intermediate argillic alteration. On the other hand if this reaction took place at a higher temperature, pyrophyllite, a key mineral of advanced argillic alteration, would be formed.

It is clear from the above considerations and many wall rock alteration studies that extensive metasomatism often accompanies the alteration. In some cases the nature of the original rock may be uncertain, but examination of the field relations and petrology of an area may allow the determination of a 'least altered equivalent'. The relative chemical gains and losses then can be calculated after making assumptions of constant volume or constant alumina content during the alteration. This is Gresens' approach, which was further developed by Grant (1986). During wall rock alteration, although most rock-forming minerals are susceptible to attack by acid solution, carbonates, zeolites, feldspathoids and calcic plagioclase are least resistant; pyroxenes, amphiboles and biotite are moderately resistant, and sodic plagioclase, potash feldspar and muscovite are strongly resistant. Quartz is often entirely

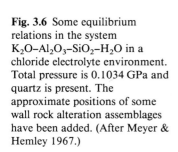

Fig. 3.6 Some equilibrium relations in the system $K_2O–Al_2O_3–SiO_2–H_2O$ in a chloride electrolyte environment. Total pressure is 0.1034 GPa and quartz is present. The approximate positions of some wall rock alteration assemblages have been added. (After Meyer & Hemley 1967.)

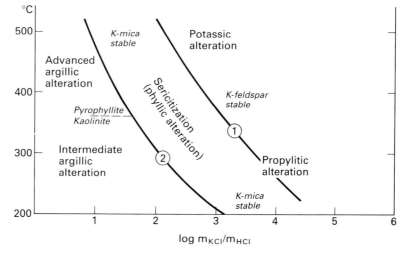

unaffected. As wall rock alteration is in many cases a relatively rapid and incomplete geological process, compared say with regional and contact metamorphism, some of these more resistant minerals may persist as unstable relics in new assemblages more commonly than is the case during metamorphism. Nevertheless the application of the metamorphic facies approach in the study of wall rock alteration has proved fruitful, and the common wall rock alteration assemblages are displayed in Fig. 3.7 using ACF and AKF diagrams whose use is described in Best (1982), Winkler (1979) and other textbooks on metamorphic petrology. These assemblages and others are described in the next section.

Types of wall rock alteration

These have been described by Meyer & Hemley (1967) and by Rose & Burt (1979) from whose work much of the following is drawn.

Advanced argillic alteration

This alteration is characterized by dickite, kaolinite (both $Al_2Si_2O_5(OH)_4$), pyrophyllite ($Al_2Si_4O_{10}(OH)_2$) and quartz. Sericite is usually present and frequently alunite, pyrite, tourmaline, topaz and zunyite. At high temperatures andalusite may be present. This is one of the more intense forms of alteration, often present as an inner zone adjoining many base metal vein or pipe deposits associated with acid plutonic stocks, as at Butte, Montana, and Cerro de Pasco in Peru. It is found also in hot spring environments and in telescoped shallow precious metal deposits. The associated sulphides of the orebodies are generally sulphur-rich; covellite, digenite, pyrite and enargite are most common.

This alteration involves extreme leaching of bases (alkalis and calcium) from all aluminous phases such as feldspars and micas, but is present only if aluminium is not appreciably mobilized. When aluminium is also removed it grades into silicification and, with increasing sericite, it grades outwards into sericitization. The generation of advanced argillic alteration can be very important in developing the high permeability necessary for the circulation of enormous quantities of hydrothermal fluids and vein growth (Brimhall & Ghiorso 1983).

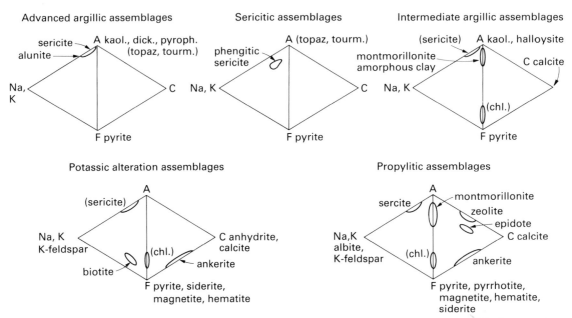

Fig. 3.7 Common wall rock alteration assemblages in aluminosilicate rocks plotted on ACF and AKF diagrams. A represents Al_2O_3 and other components that behave chemically in the same way, C is CaO plus similar components, K is $K_2O + Na_2O$ and F is $FeO + MgO + MnO$. Minerals in parentheses may be present as unstable relics. Quartz will usually be present in all these assemblages. kaol. = kaolinite, dick. = dickite, pyroph. = pyrophyllite, tourm. = tourmaline, chl. = chlorite. (After Meyer & Hemley 1967.)

Sericitization

In orefields the world over this is one of the commonest types of alteration in aluminium-rich rocks such as slates, granites, etc. The dominant minerals are sericite and quartz, pyrite often accompanying them. Care should be taken to ensure that the sericite is muscovite because illite, paragonite, phlogopite, talc and pyrophyllite can be mistaken for sericite. Muscovite is stable over a wide pressure–temperature field and this accounts for its common occurrence as an alteration mineral. If potassium is introduced into the wall rocks then rocks low in this element, such as diorites, can be sericitized. The reader must not assume that during this and other wall rock alteration processes the wall rocks necessarily become solid sericite or clay minerals, as the case may be. What we see is the appearance in significant amounts, or an increase in quantity, of the mineral or minerals concerned. Sometimes the new mineral(s) will develop, to the exclusion of all other minerals, but this is not necessarily the case. During the sericitization of granite, the feldspars and micas may be transformed to sericite, with secondary quartz as a reaction by-product, but the primary quartz may be largely unaffected except for the development of secondary fluid inclusions. Wall rock alteration is progressive, with some minerals reacting and being altered more rapidly than others. If shearing accompanies this, or any of the other types of alteration carrying phyllosilicates, then a schistose rock may result, otherwise hornfelsic-like textures develop.

With the appearance of secondary potash feldspar and/or secondary biotite, sericitization grades into potassic alteration, which is very common in the central deeper portions of porphyry copper deposits as at Bingham Canyon, Utah. In fluorine-rich environments, topaz together with zunyite and quartz may accompany the sericite to form greisen. Outside the sericitization zone lower grade intermediate argillic alteration may occur. Thus sericitization may grade into three types of higher grade alteration and one of lower grade.

Intermediate argillic alteration

Formerly known as argillic alteration and given its present name by Hemley & Jones (1964) to avoid any confusion with advanced argillic alteration. The principal minerals are now kaolin- and montmorillonite-group minerals occurring mainly as alteration products of plagioclase. These may be accompanied by 'amorphous clays' (clays sensibly amorphous to X-rays and commonly called allophane). The intermediate argillic zone may itself be zoned with montmorillonite minerals dominant near the outer fringe of alteration and kaolin minerals nearer the sericitic zone. Sulphides are generally unimportant. Outwards from the intermediate argillic alteration zone propylitic alteration may be present before fresh rock is reached.

Propylitic alteration

This is a complex alteration generally characterized by chlorite, epidote, albite and carbonate (calcite, dolomite or ankerite). Minor sericite, pyrite and magnetite may be present, less commonly zeolites and montmorillonites. The term propylitic alteration was first used by Becker in 1882 for the alteration of diorite and andesite beside the Comstock Lode, Nevada (a big gold–silver producer in the boom days of the last century). Here the main alteration products are epidote, chlorite and albite. The propylitic alteration zone is often very wide and therefore, when present, is a useful guide in mineral exploration. For example at Telluride, Colorado, narrow sericite zones along the veins are succeeded outwards by a wide zone of propylitization.

With the intense development of one of the main propylitic minerals we have what are sometimes considered as subdivisions of propylitization: chloritization, albitization and carbonatization. Albitization will be dealt with under feldspathization.

Chloritization

Chlorite may be present alone or with quartz or tourmaline in very simple assemblages; however, other propylitic minerals are usually present, and anhydrite also may be in evidence. Hydrothermal chlorites often show a change in their Fe : Mg ratio with distance from the orebody, usually being richer in iron adjacent to the sulphides, although the reverse has been reported. The change in this ratio can be recorded by simple refractive index measurements so that this offers the possibility of a cheap exploration tool.

The development of secondary chlorite may result from the alteration of mafic minerals already present in the country rocks or from the introduction of magnesium and iron, though of course both processes can occur together, as at Ajo, Arizona.

Chloritization is common in alteration zones alongside tin veins in Cornwall where progressive alteration of the country rocks occurs as shown in the diagram at the bottom of this page.

Carbonatization

Dolomitization is a common accompaniment of low to medium temperature ore deposition in limestones, and dolomite is probably the commonest of the carbonates formed by hydrothermal activity. Dolomitization is associated most commonly with low temperature, lead–zinc deposits of 'Mississippi Valley-type'. These can have wall rocks of pure dolomite; usually this rock is coarser and lighter in colour than the surrounding limestone. Dolomitization generally appears to have preceded sulphide deposition, but the relationship and timing have been much debated in the various fields where dolomitization occurs. Nevertheless, dolomitization often appears to have preceded mineralization, to have increased the permeability of the host limestones and thus to have prepared them for mineralization. Of course not all dolomitized limestone contains ore and sometimes in such fields ore may occur in unaltered limestone. Often the limestone is then recrystallized to a coarse white calcite rock, as at Bisbee, Arizona.

Other carbonates may be developed in silicate rocks, especially where iron is available and ankerite may then be common, particularly in the calcium–iron environment of carbonatized basic igneous rocks and volcaniclastics. This is particularly the case with many Precambrian and Phanerozoic vein gold deposits, for example the Mother Lode in California where ankerite, sericite, albite, quartz, pyrite and arsenopyrite are well developed in the altered wall rocks. At Larder Lake in Ontario, dolomite and ankerite have replaced large masses of greenstone. As with the chlorites, there may be a chemical variation in the Fe : Mg ratio with proximity to ore.

Most of the alteration types we have dealt with so far have involved hydrolysis, i.e. the introduction of hydrogen ion for the formation of hydroxyl-bearing minerals such as micas and chlorite, often accompanied by the removal of bases (K^+, Na^+, Ca^{2+}). Thus the $H^+ : OH^-$ ratio of the mineralizing solutions will decrease and concomitantly they will be enriched in bases. The chemical significance of carbonatization may take two different forms. Dolomitization of limestone involves very extensive magnesium metasomatism, but only as a base exchange process. This is called cation metasomatism and in this case the mineralizing (or pre-mineralization) solutions must have carried abundant magnesium. Carbonatization of silicate rocks involves *inter alia* anion metasomatism with the introduction of CO_3^{2-} rather than bases.

Potassic alteration

Secondary potash feldspar and/or biotite are the essential minerals of this alteration. Clay minerals are absent but minor chlorite may be present. Anhydrite is often important especially in prophyry copper deposits, e.g. El Salvador, Chile, and it can form up to 15% of the altered rock. It is, however, easily hydrated and removed in solution, even at depths of up to 1000 m, so that it has probably been removed from many deposits by ground water solution. Magnetite and hematite may be present and the common sulphides are pyrite, molybdenite and chalcopyrite, i.e. there is an intermediate sulphur : metal ratio.

Silicification

This involves an increase in the proportion of quartz or crypto-crystalline silica (i.e. cherty or opaline silica) in the altered rock. The silica may be introduced from the hydrothermal solutions, as in the case of chertified limestones associated with lead–zinc–fluorite–baryte deposits, or it may be the by-product of the alteration of feldspars and other minerals during the leaching of bases. Silicification is often a good guide to ore, e.g. the Black Hills, Dakota. At the Climax porphyry molybdenum

Unaltered porphyritic granite	Pinking of feldspars	Chlorite in groundmass developed from biotite and small feldspars	Rim, core and cleavage replacement of phenocrysts	Quartz–chlorite rock carrying tin values	Quartz vein with tin

deposit in Colorado, intensive and widespread silicification accompanied the mineralization.

The silification of carbonate rocks leading to the development of skarn is dealt with in Chapter 13.

Feldspathization

This is the term often applied when potassium or sodium metasomatism has produced new potash feldspar or albite unaccompanied by the other alteration products characteristic of potassic alteration. Albitization is found adjacent to some gold deposits, often replacing potash feldspar, e.g. Treadwell, Alaska.

Tourmalinization

This is associated with medium to high temperature deposits, e.g. many tin and some gold veins have a strong development of tourmaline in the wall rocks and often in the veins as well. The Sigma Gold Mine in Quebec has veins that in places are massive tourmaline and this mineral is well developed in the adjacent wall rock. This is also the case in the granodiorite wall rock of the Siscoe Mine, Quebec. At Llallagua, Bolivia, the world's largest primary tin mine, the porphyry host is altered to a quartz–sericite–tourmaline rock.

If the altered country rocks are lime-rich then axinite rather than tourmaline may be formed.

Other alteration types

There are many other types of alteration; among these may be mentioned *alunitization*, which may be of either hypogene or supergene origin; *pyritization*, due to the introduction of sulphur which may attack both iron oxides and mafic minerals; *hematitization*, an alteration type often associated with uranium (particularly pitchblende deposits); *bleaching*, due in many cases to the reduction of hematite; *greisenization*, a frequent form of alteration alongside tin–tungsten and beryllium deposits in granitic rocks or gneisses—see 'sericitization' above and Chapter 12; *fenitization*, which is associated with carbonatite hosted deposits and is characterized by the development of nepheline, aegirine, sodic amphiboles and alkali feldspars in the aureoles of the carbonatite masses; *serpentinization*, and the allied development of talc, can occur in both ultrabasic rocks and limestones; it is associated with some gold and nickel deposits but where serpentine and talc are developed in limestones there is generally an introduction of SiO_2 and H_2O and frequently some Mg; finally *zeolitization* is marked by the development of stilbite, natrolite, heulandite, etc., and often accompanies native copper mineralization in amygdaloidal basalts—calcite, prehnite, pectolite, apophyllite and datolite are generally also present.

Influence of original rock types

A survey of world literature on the subject shows that wall rock alteration exhibits a certain regularity with respect to the nature of the host rock (Boyle 1970). There are of course exceptions, but certain generalizations can be made. Thus, for example, the most prevalent types of alteration in acidic rocks are sericitization, argillization, silicification and pyritization. Intermediate and basic rocks generally show chloritization, carbonatization, sericitization, pyritization and propylitization. In carbonate rocks the principal high temperature alteration is skarnification, whereas normal shales, slates and schists are frequently characterized by tourmalinization, especially when hosting tin and tungsten deposits. A more detailed discussion of this subject can be found in Boyle's work.

Correlation with type of mineralization

Boyle has also shown that certain types of mineralization tend to be accompanied by characteristic types of alteration but space permits only the mention here of a few examples. Red bed uranium, vanadium, copper, lead and silver deposits are generally accompanied by bleaching. Vein deposits of native silver are usually characterized by carbonatization and chloritization and molybdenum-bearing veins by silicification and sericitization. Other examples have been given above.

Timing of wall rock alteration

The problem of disentangling age relationships between different assemblages of wall rock alteration minerals in a given deposit can be extremely difficult. Their presence, together with evidence such as crustiform banding, has suggested to many workers that the mineralizing solutions came in pulses of different compositions (polyascendant solutions). In some deposits, however, such as Butte, Montana, there is a considerable body of data suggesting a single long-lasting phase of mineralization and accompanying alteration during the formation of the main stage veins. Both mechanisms have probably taken place and some deposits may have been formed from monoascendant solutions whilst others are the result of polyascendant mineralization. This important aspect is discussed at length by Meyer & Hemley (1967).

The nature of ore-forming solutions as deduced from wall rock alteration

Studies of wall rock alteration indicate that aqueous solutions played a large part in the formation of epigenetic deposits. Clearly, in some cases these solutions carried other volatiles such as CO_2, S, B and F. The pH of ore-forming solutions is difficult to assess from wall rock alteration studies, but in cases of hydrogen metasomatism the pH value must have been low; it would have increased in value during reactions with the wall rocks, so that in some cases solutions may have become neutral or even slightly alkaline. A very instructive study of the use of wall rock alteration studies in throwing light on the nature of ore-forming solutions is to be found in Fournier (1967).

An example of wall rock alteration

Piroshco & Hodgson (1988) investigated the alteration beside the auriferous quartz veins of the Coniaurum Mine in the Porcupine Camp of eastern Ontario (Fig. 3.8). These authors recognized in the mafic metavolcanics that form the wall rocks: (a) a least altered facies consisting of the regional greenschist metamorphic mineral assemblage which, on approaching a vein, gave way to (b) a chlorite facies in which the actinolite is replaced by chlorite, albite is progressively replaced by calcite + quartz + chlorite and clinozoisite disappears as sericite develops; (c) an ankerite facies zone, in which the fresh rock takes on a bleached appearance in contrast to the green colour of the chlorite facies rocks; this colour

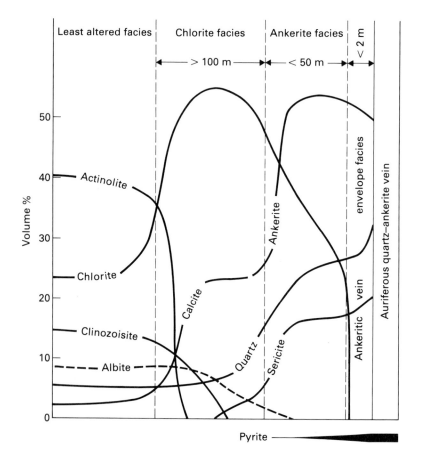

Fig. 3.8 Wall rock alteration zonation at the Coniaurum Mine, Ontario showing the volume percentage of minerals in the different alteration assemblages.

change correlates with the appearance of ankerite as the carbonate mineral and its considerable volumetric increase; there are concomitant increases in quartz and sericite as the vein is approached and chlorite decreases; and (d) a narrow zone occupied by the vein envelope facies. In the last named, chlorite is virtually absent, sericite and quartz are prominent, the ankerite appears to have a higher Mg : Fe ratio and pyrite is significantly developed.

The mineral assemblages of the various zones can be considered as metamorphic greenschist facies assemblages, which grew under the influence of different CO_2 concentrations in the interstitial fluid. The mineral assemblage of the least altered facies had the lowest CO_2 concentration and the other assemblages grew in environments having successively higher CO_2 concentrations. Piroshco and Hodgson therefore concluded that the zonation is the result of lateral gradients in X_{CO_2}, rather than temperature gradients.

4 / Some major theories of ore genesis

Some indication has been given in Chapter 2 of the very varied nature of ore deposits and their occurrence. This variety of form has given rise over the last 100 years or more to an equally great variety of hypotheses of ore genesis. The history of the evolution of these ideas is an interesting study in itself and one which has been well documented by Stanton (1972). There is no room for such a discussion here and the reader is referred to Stanton's work and to references given by him. This chapter will, therefore, be concerned only with major theories of ore genesis current at the present time and these will be divided for the sake of convenience into internal and surface processes. The reader should be warned, however, that very often several processes contribute to the formation of an orebody. Thus, where we have rising hot aqueous solutions forming an epigenetic stockwork deposit just below the surface and passing on upwards through it to form a contiguous syngenetic deposit under, say, marine conditions, even the above simple classification is in difficulties. This is the reason why ore geologists, besides producing a plethora of ore genesis theories, have also created a plethora of orebody classifications! A summary of the principal theories of ore genesis is given in Table 4.1.

Origin due to internal processes

Magmatic crystallization

This covers the ordinary processes of crystallization that provide us with volcanic and plutonic rocks. Some of these, such as granites and basalts, we may exploit as bulk materials, others may be important for their possession of one or more economically important minerals, e.g. diamonds in kimberlites, feldspar in pegmatites. Further attention will be given to these and other examples later in this book.

Magmatic segregation

The terms magmatic segregation deposit or ortho-magmatic deposit are used for those ore deposits that have crystallized direct from a magma. Those formed by fractional crystallization are usually found in plutonic igneous rocks; those produced by liquation (separation into immiscible liquids) may be found associated with both plutonic and volcanic rocks. Magmatic segregation deposits may consist of layers within or beneath the rock mass (chromite layers, subjacent copper–nickel sulphide ores).

Fractional crystallization

This includes any process by which early formed crystals are prevented from equilibrating with the melt from which they grew. The important processes until recently were thought to be gravity fraction-ation, flowage differentiation, filter pressing and dilatation (Carmichael *et al.* 1974), but the simple hypotheses of separation and mechanical sorting by magmatic currents invoked by various workers as being at least partly responsible for stratiform chromite accumulations (e.g. Cameron & Emerson 1959, Irvine & Smith 1969) were questioned as long ago as 1961 by E.D. Jackson, following his work on layering in the Stillwater Complex. The critical evidence and the new explanations of deposition from density currents and *in situ* bottom crystalli-zation were succinctly summarized by Best (1982). The second of these two processes is favoured by Eales & Reynolds (1985) to explain evidence from the Bushveld Complex.

Whatever the formative processes may be, their products are the rocks called cumulates, which often display conspicuous lithological alternations called rhythmic layering, owing to their frequent repetition in vertical sections of the plutonic bodies in which they occur. Most of these intrusions appear to be funnel shaped, but the diameter of the funnel relative to its height varies greatly. The layering is generally discordant to the walls of the funnel. Usually, olivine-, pyroxene-, or plagioclase-rich lay-ers are formed. However, when oxides such as chromite are precipitated, layers of this mineral may develop, as in the Bushveld Complex of South Africa. This enormous layered intrusion is charac-terized by cumulus magnetite in the upper zone. The chromite layers have been mined for decades, the

Table 4.1 Simple classification of the theories of mineral deposit genesis

Theory	Nature of process	Typical deposits
Origin due to internal processes		
Magmatic crystallization	Precipitation of ore minerals as major or minor constituents of igneous rocks in the form of disseminated grains or segregations	Diamonds disseminated in kimberlites, REE minerals in carbonatites. Lithium–tin–caesium pegmatites of Bikita, Zimbabwe. Uranium pegmatites of Bancroft, Canada and Rössing, Namibia. Bulk material deposits of granite, basalt, dunite, nepheline-syenite
Magmatic segregation	Separation of ore minerals by fractional crystallization and related processes during magmatic differentiation	Chromite layers in the Great Dyke of Zimbabwe and the Bushveld Complex, RSA
	Liquation, liquid immiscibility. Settling out from magmas of sulphide, sulphide–oxide or oxide melts that accumulated beneath the silicates or were injected into wall rocks or in rare cases erupted on the surface	Copper–nickel orebodies of Sudbury, Canada; Pechenga, USSR and the Yilgarn Block, Western Australia. Allard Lake titanium deposits, Quebec, Canada
Hydrothermal	Deposition from hot aqueous solutions, which may have had a magmatic, metamorphic, surface or other source	Tin–tungsten–copper veins and stockworks of Cornwall, UK. Molybdenum stockworks of Climax, USA. Porphyry copper deposits of Panguna, PNG and Bingham, USA. Fluorspar veins of Derbyshire, UK
Lateral secretion	Diffusion of ore- and gangue-forming materials from the country rocks into faults and other structures	Yellowknife gold deposits, Canada. Mother Lode gold deposit, USA
Metamorphic processes	Contact and regional metamorphism producing industrial mineral deposits	Andalusite deposits, Transvaal, RSA. Garnet deposits, NY, USA
	Pyrometasomatic (skarn) deposits formed by replacement of wall rocks adjacent to an intrusion	Copper deposits of Mackay, USA and Craigmont, Canada, Magnetite bodies of Iron Springs, USA. Talc deposits, Luzenac, France
	Initial or further concentration of ore elements by metamorphic processes, e.g. granitization, alteration processes	Some gold veins, and disseminated nickel deposits in ultramafic bodies
Origin due to surface processes		
Mechanical accumulation	Concentration of heavy, durable minerals into placer deposits	Rutile–zircon sands of New South Wales, Australia, and Trail Ridge, USA. Tin placers of Malaysia. Gold placers of the Yukon, Canada. Industrial sands and gravels. Kaolin deposits, Georgia, USA. Bauxites of Guyana
Sedimentary precipitates	Precipitation of particular elements in suitable sedimentary environments, with or without the intervention of biological organisms	Banded iron formations of the Precambrian shields. Manganese deposits of Chiaturi, USSR. Zechstein evaporite deposits of Europe. Floridan phosphate deposits, USA
Residual processes	Leaching from rocks of soluble elements leaving concentrations of insoluble elements in the remaining material	Nickel laterites of New Caledonia. Bauxites of Hungary, France, Jamaica and Arkansas, USA. Kaolin deposits, Nigeria

Table 4.1 *Continued*

Theory	Nature of process	Typical deposits
Secondary or supergene enrichment	Leaching of valuable elements from the upper parts of mineral deposits and their precipitation at depth to produce higher concentrations	Many gold and silver bonanzas. The upper parts of a number of porphyry copper deposits
Volcanic exhalative (= sedimentary exhalative)	Exhalations of hydrothermal solutions at the surface, usually under marine conditions and generally producing stratiform orebodies	Base metal deposits of Meggan, Germany; Sullivan, Canada; Mount Isa, Australia; Rio Tinto, Spain; Kuroko deposits of Japan; black smoker deposits of modern oceans. Mercury of Almaden, Spain. Solfatara deposits (kaolin + alunite), Sicily

magnetite now being exploited for its high vanadium content. Detailed studies of the rhythmic layering suggests that each unit results from the influx of a new magma pulse which forms a layer at the base of the magma chamber where it cools and precipitates a mineral or mineral phases until its reduced density permits mixing with the overlying magma. The precipitated crystals are thought to be nucleated and to grow *in situ* on the floor and walls of the chamber (Wilson 1989). Another mineral which may be concentrated in this way is ilmenite. Whilst chromite accumulations are nearly all in ultrabasic rocks and to a lesser extent in gabbroic or noritic rocks, ilmenite accumulations show an association with anorthosites or anorthositic gabbros. These striking rock associations are strong evidence for the magmatic origin of the minerals.

Liquation

A different form of segregation results from liquid immiscibility. In exactly the same way that oil and water will not mix but form immiscibile globules of one within the other, so in a mixed sulphide–silicate magma the two liquids will tend to segregate. Sulphide droplets separate out and coalesce to form globules which, being denser than the magma, sink through it to accumulate at the base of the intrusion or lava flow (Fig. 4.1). Iron sulphide is the principal constituent of these droplets, which are associated with basic and ultrabasic rocks because sulphur and iron are both more abundant in these rocks than in acid or intermediate rocks. Chalcophile elements, such as copper and nickel, also enter ('partition into' is the pundits' phrase) these droplets and sometimes the platinum group metals. Groves *et al.* (1986)

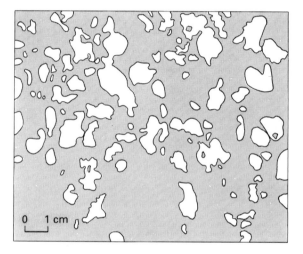

Fig. 4.1 Tracing of an ore specimen from Sudbury, Ontario. Sulphides, mainly pyrrhotite with minor pentlandite and chalcopyrite, are shown white; surrounding silicates are shown grey. Note the rounded discontinuous nature of the sulphide globules. They appear to have formed as a result of liquid immiscibility from a silicate–sulphide melt. Note especially the rounded silicate blebs within the sulphide bodies, and that many of the sulphide globules appear to have formed from the coalescence of smaller bodies of sulphide liquid.

provided strong evidence for the coexistence of sulphide and silicate melts in nature.

A basic or ultrabasic magma is generated by partial melting in the mantle and it may acquire its sulphur at this time, or later by assimilation in the crust. For significant sulphide segregation to occur the magma must be sulphide saturated. If immiscible sulphides form, equilibrate with the silicate magma, settle and accumulate on the substrate in a single stage, after which the magma begins to crystallize olivine, then we have the process known as batch equilibrium. If the proportion of sulphide formed is large, much of the available Ni in the magma is removed, and if the ratio of mass of magma to mass of sulphide is about 1000 or below, then there will be obvious Ni depletion in the magma and the consolidated silicate rocks will carry a geochemical mark that can be used both to identify favourable exploration targets and give an indication of the tenor of Ni in the possible, associated mineralization.

Contrasting with batch equilibrium we may have fractional segregation. This is a more continual process in which small amounts of sulphide become immiscible and then settle. The silicate melt is still sulphide saturated and crystallization of just a small amount of olivine will increase the sulphide concentration in the remaining liquid, thereby forcing some more dissolved sulphide out of solution (Naldrett 1989).

For the formation of an ore deposit the timing of the liquation is critical (Naldrett *et al.* 1984). If it is too early then the sulphides may settle out in the mantle or the lower crust; if it is late then crystallization of silicates may be in full swing and they will dilute any sulphide accumulations. The processes that can promote sulphide immiscibility are: cooling, silication (increase of silica content by assimilation), sulphur assimilation and magma mixing.

The accumulation of Fe–Ni–Cu sulphide droplets beneath the silicate fraction can produce massive sulphide orebodies. These are overlain by a zone with subordinate silicates enclosed in a network of sulphides—net-textured ore, sometimes called disseminated ore. This zone is, in turn, overlain by one of weak mineralization which grades up into overlying peridotite, gabbro or komatiite, depending on the nature of the associated silicate fraction (see Fig. 11.3). To explain the mechanism of formation of these zones Naldrett (1973) proposed his 'billiard ball' model (Fig. 4.2).

Imagine a large beaker partly filled with billiard balls and water (Fig. 4.2a). These represent olivine grains and interstitial silicate liquid. Then consider the effect of adding mercury to represent the immiscible sulphide liquid. This will sink to the bottom and the balls will tend to float on it with the lower balls being forced down into the mercury by the weight of the overlying ones. If the contents of

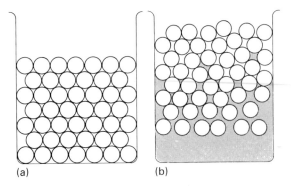

(a)   (b)

Fig. 4.2 The billiard ball model illustrating the formation of Fe–Ni–Cu sulphide deposits by liquation. For explanation see text. (After Naldrett 1973.)

the beaker are frozen before all the mercury has had a chance to percolate to the bottom then the situation shown in Fig. 4.2b will exist. There is an obvious analogy between the massive mercury and the zone of massive sulphides, between the overlying zone of balls immersed in mercury and net-textured ore, and between the zone of scattered globules of mercury and the zone of weak mineralization. Orthomagmatic titanium deposits are also considered to be of liquation origin (see Chapter 10).

Hydrothermal processes

Hot aqueous solutions have played a part in the formation of many different types of mineral and ore deposit, for example veins, stockworks of various types, volcanic-exhalative deposits and others. Such fluids are usually called hydrothermal solutions and many lines of evidence attest to their important role as mineralizers. The evidence from wall rock alteration and fluid inclusions has been discussed in Chapter 3. Homogenization of fluid inclusions in minerals from hydrothermal deposits and other geothermometers have shown that the depositional range for all types of deposit is approximately 50–650°C. Analysis of the fluid has shown water to be the common phase and usually it has salinities far higher than that of sea water. Hydrothermal solutions are believed to be capable of carrying a wide variety of materials and of depositing these to form minerals as diverse as gold and muscovite, showing that the physical chemistry of such solutions is complex and very difficult to imitate in the laboratory. Our knowledge of their properties and behaviour is still somewhat hazy and there are many ideas about the origin of such solutions and the materials they carry (see relevant chapters in Barnes 1979a).

It always should be remembered that hydrothermal ore deposits are small compared with most geological features—the largest are only a few km^3 in volume. At first sight they may appear to be random accidents with little or no control over the position or geological environment in which they occur within the crust. Nevertheless deposits can be classified into families and individual family members occur more frequently in some areas of the crust than others. Moreover, despite their variation in occurrence and the large number of minerals known in nature, hydrothermal deposits display a chemical consistency that is best expressed by the limited and repetitive ranges of minerals, mostly sulphides and oxides, that are found concentrated within them. This chemical consistency suggests that relatively few chemical processes are important in their genesis (Skinner 1979).

Ore geology, like most branches of geology, is what might be termed a forensic science. We study the evidence collected in the field and attempt to reconstruct the crime, but many decades of study have revealed that the evidence is always partial and open to various interpretations, which implies that the best investigated deposits are still worthy of further research and our ideas concerning their modes of genesis will be modified continually. The principal problems are the source and nature of the solutions, the sources of the metals and sulphur in them and the driving force that moved the solutions through the crust, the means of transport of these substances by the solutions, and the mechanisms of deposition.

Sources of the solutions and their contents

As explained in the section on fluid inclusions in Chapter 3, there is much evidence that saline hydrothermal solutions are, and have been, very active and widespread in the crust. In some present day geothermal systems the circulation of hydrothermal solutions is under intensive study. Whence the water of these solutions? Data from water in mines, tunnels, drill holes, hot springs, fluid inclusions, minerals and rocks suggest that there are five sources of subsurface hydrothermal waters:

1 surface water, including ground water, commonly referred to by geologists as meteoric water;
2 ocean (sea) water;
3 formation and deeply penetrating meteoric water;
4 metamorphic water;
5 magmatic water.

Most formation water may have been meteoric water originally, but long burial in sediments and reactions with the rock minerals give it a different character.

Measurements of the relative abundances of oxygen and hydrogen isotopes give us information on the sources of water (see Fig. 4.15), but there are problems of interpretation of the data that we obtain. Both formation and metamorphic water (produced by dehydration of minerals during metamorphism) may once have been meteoric water, but subsurface rock–water reactions may change the isotopic compositions and, if these reactions are incomplete, then a range of isotopic compositions will result. Another mechanism that may produce intermediate isotopic compositions is the mixing of waters, e.g. magmatic and meteoric.

Examples of some hydrogen–oxygen isotopic studies are shown in Fig. 4.3 and in examining these please refer to Fig. 4.15. From the values obtained Rye & Sawkins (1974) inferred that the main-stage mineralizing fluids at the Ag–Pb–Zn–Cu Casapalca Mine, Peru were of magmatic origin, whereas for the Pasto Buena tungsten–base metal deposit of northern Peru the data according to Landis & Rye (1974) suggest mixing of some meteoric (or metamorphic or formation) water with magmatic water during mineralization. By contrast Richardson *et al.* (1988) interpreted fluids in the Cave-in-Rock Fluorspar District, Illinois (the mines also produce zinc and lead) as being dominantly meteoric-recharged formation water, and Ohmoto (1986) ascribed the main stage sulphide mineralization period in the Upper Mississippi Valley Pb–Zn Field to the action of somewhat modified meteoric water.

Present evidence thus appears to show that similar deposits can be formed from detectably different types of water and, additionally, that waters of at least two parentages have played an important role in the formation of some orebodies. As White (1974) has warned us, we must not try to make our mineralization models too simple but be prepared to accept diversity of sources for the water in hydrothermal solutions. We must also keep in mind the possibility of overprinting, as for example when a post-mineralization flow of meteoric water, through rock–water reactions (and the formation of new fluid inclusions) partially or wholly changes the isotopic character of the orebody and its immediate wall rocks. In many cases where an intrusion, acting

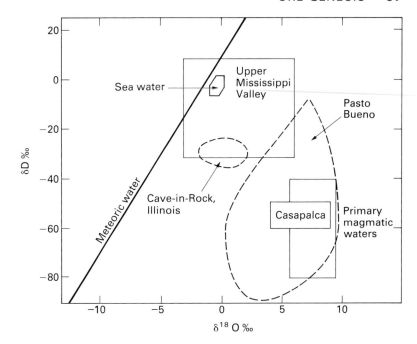

Fig. 4.3 Hydrogen versus oxygen isotopic plot showing observed values of hydrothermal fluids involved in the main stages of mineralization in four mineralized areas. For further discussion and sources of data see text.

as a heat engine, provided the driving force for the movement of solutions, such an overprinting may be expected to have taken place, and evidence for this should be sought (e.g. Scratch *et al.* 1984). Finally it must be said that in putting forward a mineralization model for a particular deposit or class of deposit, thought should be given to the amount of water that will be required for its formation. Can the postulated source provide a sufficient volume?

Let us now turn our thoughts to the source of the dissolved contents. Firstly, what are they? Secondly, where do we look for this information? The answer to the second question is fluid inclusions (which provide most of our data), modern geothermal systems, hot springs, and waters encountered during deep drilling operations in oilfields—oilfield brines. Some data from these sources are given in Table 4.2. They indicate that the major constituents are sodium, potassium, calcium and chlorine. Other elements and radicles are usually present in amounts less than 1000 ppm. Little can be said about the source of any of these except helium, lead and strontium.

Few helium isotopic studies have been published yet, but an interesting one was that of Simmons *et al.* (1987) on the well-known Casapalca and Pasto Bueno silver–base metal and tungsten–base metal deposits of Peru. These both have $^3He/^4He$ values indicative of mantle origin, in keeping with the hydrogen and oxygen isotopic ratios which indicate the presence of magmatic fluids during mineralization.

Lead always contains a number of stable isotopes developed by the radioactive decay of ^{232}Th, ^{235}U and ^{238}U plus non-radiogenic lead. As with hydrogen and oxygen isotopes, these have been used as tracers to seek the source of the lead and these studies have yielded some interesting results. For example, hydrothermal fluids responsible for depositing lead in the Old Lead Belt of south-east Missouri appear to have leached it from a sandstone underlying the orefield (Doe & Delavaux 1972); but a significant amount of the lead in the famous stratabound lead–zinc deposits of the eastern and southern Alps, which are hosted by Triassic carbonates, appears to have been derived from feldspars in the crystalline basement (Köppel & Schroll 1988). Stacey *et al.* (1968) showed that much of the lead in some leadfields of Utah was derived from associated igneous intrusions, and MacFarlane & Petersen (1990) investigated the lead isotopes of the important silver veins and mantos of the Hualgayoc district, Peru; their investigations revealed that (a) the ore solutions of the entire district had essentially a single lead source, (b) all the ores probably formed during a single event or in a series of closely related

Table 4.2 Concentrations in ppm of *some* of the elements in some modern and ancient hydrothermal solutions. The data are from Skiner (1979) in which the original literature sources are to be found. (1) Salton Sea geothermal brine (California); (2) Cheleken geothermal brine (USSR); (3) oilfield brine, 3350 m depth (Michigan); (4) fluid inclusion in fluorite (Illinois); (5) fluid inclusion in sphalerite, Creede (Colorado); (6) fluid inclusions, Core Zone, Bingham (Utah)

Element	Modern solutions			Ancient solutions		
	1	2	3	4	5	6
Cl	155 000	157 000	158 200	87 000	46 500	295 000
Na	50 400	76 140	59 500	40 400	19 700	152 000
Ca	28 000	19 708	36 400	8 600	7 500	4 400
K	17 500	409	538	3 500	3 700	67 000
Mg	54	3 080	1 730	5 600	570	—
B	390	—	—	< 100	185	—
Br	120	526.5	870	—	—	—
F	15	—	—	—	—	—
NH_4	409	—	39	—	—	—
HCO_3^-	> 150	31.9	—	—	—	—
H_2S	16[a]	0	—	—	—	—
SO_4^{2-}	5	309	310	1 200	1 600	11 000
Fe	2 290	14.0	298	—	—	8 000
Mn	1 400	46.5	—	450	690	—
Zn	540	3.0	300	10 900	1 330	—
Pb	102	9.2	80	—	—	—
Cu	8	1.4	—	9 100	140	—

[a] Sulphide present; all S reported as H_2S.

events, and (c) that the ore lead isotopic compositions match those of the Miocene intrusions, some of which act as partial hosts to the veins, but are dissimilar to those of all the other possible source rocks of the district. Thus the plutons were the only important source of lead in the ore-forming solutions. Marcoux (1982) demonstrated that lead in the Pentivy district of France was derived direct from the mantle, and Plimer (1985), citing lead isotopic and other evidence, also postulated a mantle source for the lead and other metals in the huge Broken Hill orebodies of New South Wales. The evidence from lead isotopic studies thus suggests that ore fluids may collect their metals from a magma, if that was their source. Alternatively, they may collect more metals from the rocks they pass through, or obtain all their metallic content from the rocks they traverse, which can contain (in trace amounts) all the metals required to form an orebody. We must not, however, assume that other metallic elements in a mineral deposit necessarily come from the same source as the lead. A gold-bearing quartz vein at the Dome Mine in Ontario is a case in point. Here gold-bearing solutions from a deep source, appear, from isotopic investigations, to have obtained the lead of the galena in this vein from porhyries at and

close to the site of deposition (Moritz *et al.* 1990). In this case the deep source of the gold could well be basalts or komatiites, which indeed appear to be the source of the lead (and by inference, the gold) in the gold mineralization of the Archaean Yilgarn Block, Western Australia (Browning *et al.* 1987).

Strontium tends to be concentrated in wall rock alteration zones and perhaps for this reason has not received anything like the attention that has been given to lead. Two studies on very different deposit types are worthy of note. Spooner *et al.* (1977) working on the volcanic-associated massive sulphide deposits of Cyprus suggested that the mineralizing fluid was probably circulating Cretaceous sea water. Their investigation of the mineralized zones showed that these are enriched in [87]Sr relative to the initial magmatic [87]Sr/[86]Sr of the igneous rocks hosting the deposits. The values obtained range up to the value for Upper Cretaceous sea water but no higher. Dickin *et al.* (1986) working on the molybdenum and precious metal mineralization near Central City, Colorado found low initial ratios in sericite, pyrite and fluorite (associated with the molybdenite mineralization) that indicate a large content of magmatically derived strontium, whereas the radiogenic strontium of the uraninite, base metal and late tellurides stages was probably derived, along (they suggest) with other cations, from the country rocks.

Our present knowledge indicates that most rocks can act as a source of geochemically scarce elements, which can be leached out under suitable conditions by hydrothermal solutions. In laboratory experiments, Bischoff *et al.* (1981) have shown that heavy metals present in trace quantities will fractionate from greywacke into sea water or natural brine at 350°C. Similar experiments have revealed that basalt will also yield up much of its heavy metal content under similar conditions, and both these experiments produced sufficiently high concentrations for the solutions to be ranked as ore-forming fluids. High trace concentrations of most heavy metals in source rocks do not seem to be a *sine qua non* for the formation of ore fluids. The greywacke of the above experiments contained only 15 ppm of lead. Differences in source rocks and differences in leaching conditions will of course produce different concentrations of economic metals in the reacting solutions. Very scarce elements, such as tin, mercury and silver, may require pre-enrichment.

Because of the spatial relationship that exists between many hydrothermal deposits and igneous rocks, a strong school of thought holds that consolidating magmas are the source of many, if not all, hydrothermal solutions. The solutions are considered to be low temperature residual fluids left over after pegmatite crystallization, and containing the base metals and other incompatible elements that could not be accommodated in the crystal lattices of the silicate minerals precipitated by the freezing magma (e.g. highly charged cations, such as W^{6+}, Ta^{5+}, U^{4+} and Mo^{6+}, very large cations, such as Cs^+ and Rb^+, and small variably charged cations, such as Li^+, Be^{2+}, B^{3+} and P^{5+}, are incompatible with the major silicate phases and are enriched in residual liquids). This model derives not only the water, the metals and other elements from a hot body of igneous rock, but also the heat to drive the mineralizing system. The solutions are assumed to move upwards along fractures and other channelways to cooler parts of the crust where deposition of minerals occurs. Laboratory (e.g. Khitarov *et al.* 1982, Manning 1984) and field studies (Strong, 1981), backed by thermodynamic considerations, show that this is a perfectly feasible process of mineralization, and it has been discussed at length by Burnham (1979).

How much water do magmas contain? Burnham (1979) estimated that water concentrations in felsic magmas vary from 2.5 to 6.5 wt % with a median close to 3%. The data in Whitney (1975) show that a rising monzogranitic magma with 3% water will begin to exsolve copious quantities at about 3.5 km depth (equivalent to a load pressure of 0.1 GPa). The same magma with 4% water will behave similarly at above 4.5 km depth. Below these depths water remains in solution in melts because of the high containing pressure. As these magmas crystallize to produce a largely anhydrous mineral assemblage an absolutely enormous volume of water can be given off by a cooling magma—1 km^3 of felsic magma with 3% water could exsolve approximately 100 Mt (10^{11} l) (Brimhall & Crerar 1987).

Now water is only one of the volatile components of late stage magmatic fluids; these also contain H_2S, HCl, HF, CO_2, SO_2 and H_2. Of these H_2S and HCl may be of particular importance, as will be shown later, and experimental data quoted by Burnham (1979) suggest that the solubilities of H_2O, H_2S and HCl in granitic magmas are comparable and high. H_2S and HCl would be expected to fractionate strongly into an exsolving aqueous phase.

All base metals and many others that have been investigated so far can be extracted efficiently from a melt into an exsolving aqueous phase provided that sufficient water has exsolved. Candela & Holland (1986) have shown that with a 3 wt % water content of the melt about 95% of the copper in a felsic magma would be extracted, and Urabe (1987) has shown experimentally that magmatic fluids should contain a high concentration of base metals when they are released from water-saturated magmas at high levels within the crust.

Whether *any* granite magma can produce economic mineral deposits under favourable conditions, or whether the ability to develop significantly rich mineralizing fluids is dependent on the source region of magma generation itself having above normal concentrations of base and precious metals, is at present under active debate. Hannah & Stein (1990) favoured the first possibility and wrote as follows. 'The delicate relations between a granite magma, its crystallizing phases, volatile content and species, and oxygen fugacity, plus the timing and mechanism of fluid release and the efficiency of metal extraction, ultimately control the formation of an ore deposit.' They quote the work of a number of authors in support of this contention, e.g. Newberry *et al.* (1990) who found no differences between Sn-related and W-related plutons in the Fairbanks–Circle area of Alaska, which have similar crystallization histories, initial Sr and Pb isotopic ratios, etc. Newberry *et al.* contended that a major factor controlling metallogeny is the depth of emplacement, which exerts a control over the timing and volume of volatile release. However, the Fairbanks example and others cited by Hannah & Stein are of very minor economic significance, and in areas where economically large granite-associated base metal deposits occur the granites appear to be exclusively of a particular type, e.g. the S-type granites of the South-west England Orefield, which are significantly richer in trace element tin than average granite (Bromley 1989). This suggests that partial melting of tin-rich protoliths was involved in their generation, and indeed this may be a *sine qua non* for the formation of important tinfields (Shaw & Guilbert 1990). There are many recent papers in which the authors, who studied particular deposits, favoured a magmatic origin for the hydrothermal solutions that formed them; see for example Eadington (1983), Kay & Strong (1983), Norman & Trangcotchasan (1982), Lehmann (1985), Wilton (1985) and Simmons *et al.* (1988). Recently Meeker (1988), as recorded in Henley (1991), has shown that in December 1986, Mount Erebus, Antarctica discharged daily about 0.1 kg Au and 0.2 kg Cu—equivalent to 360 t Au in 10 000 years. This, and similar evidence from other volcanoes, demonstrates the ability of degassing magmas to supply metals.

In many orefields, however, such as the Northern Pennine Orefield of England, there are no acid or intermediate plutonic intrusions which might be the source of the ores. Some workers have therefore postulated a more remote magmatic source, such as the lower crust or, more frequently, magmatic processes in the mantle, whilst an important body of opinion has favoured deposition from formation solutions—that is water which was trapped in sediments during deposition and that has been driven up dip by the rise in temperature and pressure caused by deep burial. Such burial might occur in sedimentary basins, and solutions from this source are often called basinal brines. With a geothermal gradient of 1°C per 30 m, temperatures around 300°C would be reached at a depth of 9 km. Hot solutions from this source are believed to leach metals, but not necessarily sulphur, from the rocks through which they pass, ultimately precipitating them near the surface in shelf facies carbonates on the fringe of the basin, and far from any igneous intrusion. This, too, is a model favoured at present by many workers, partic-

ularly as an explanation for the genesis of low temperature, carbonate-hosted lead–zinc–fluorite–baryte deposits (Mississippi Valley-type). Useful discussions are Hanor (1979), Cathles & Smith (1983) and Sverjensky (1984). It has been suggested, however, that the available volumes of formation water are insufficient to carry the amount of metal that is present in such deposits (Duhovnik 1967). If this is a serious objection to the model as depicted in the simple terms above, then there are various hypotheses with which it may be refuted. Hsü (1984) has suggested one and applied it in particular to deposits of this nature in the southern Alps. Other workers favour a comparable flow of water, under a hydrostatic head, passing through sedimentary basins to produce the ore fluid. There is no problem of a lack of water for this model and recent statements in support of it are to be found in Garven (1985) and Bethke (1986)—see Fig 4.7.

An important paper by White in 1955 engendered a fertile field of research into geothermal systems as possible generators of orebodies and a number of epigenetic deposits are now considered by many researchers to have formed in ancient geothermal systems. Useful reading on this subject includes Ellis (1979), Weissberg *et al.* (1979), White (1981), Henley & Ellis (1983), Henley *et al.* (1984), Höll (1985) and Henley (1986). Geothermal systems form where a heat engine (usually magmatic) at depths of a few kilometres sets deep ground waters in motion (Fig. 4.4a). These waters are usually meteoric in origin but in some systems formation or other saline waters (Salton Sea) may be present. Systems near the coast may be fed by sea water or both sea water and meteoric water (Svartsengi, Iceland). Magmatic water may be added by the heat engine, and some ancient systems appear to have been dominated by magmatic water, at least in their early stages, e.g. porphyry copper and molybdenum deposits. Dissolved constituents (see Table 4.2, columns 1 and 2) may be derived by the circulating waters from a magmatic body at depth, or from the country rocks which contain the system. These may be altered by the solutions to mineral assemblages identical with those found in some wall rock alteration zones associated with orebodies. Common sulphides, such as galena, sphalerite, chalcopyrite and chalcocite, occur in a number of modern systems, e.g. Salton Sea (McKibben *et al.* 1988).

At Broadlands, New Zealand, an amorphous Sb–As–Hg–Tl sulphide precipitate enriched to ore grade in gold and silver has been formed, and in the

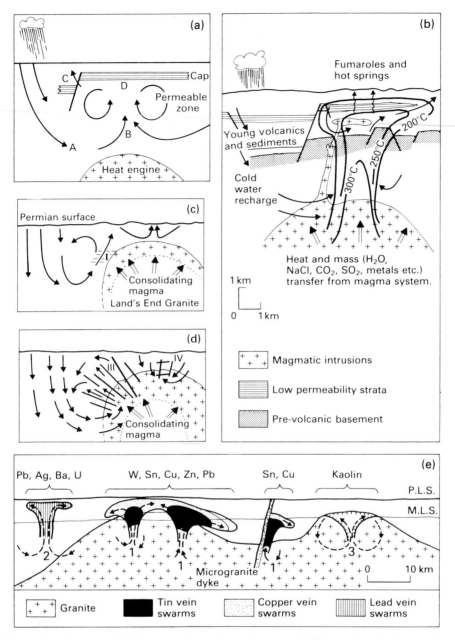

Fig. 4.4 (a) Schema showing some of the features of a geothermal system. (b) Schema showing the structure of a geothermal system like that of the Taupo Volcanic Zone, NZ. (After Henley & Ellis 1983.) (c and d) Schemata illustrating the evolution of some of the mineralization in a flank of the Land's End Granite; in detail these show: (c) initial emplacement of the pluton with the development of an H_2O-saturated carapace enclosing still-consolidating magma. It also shows formation of tin- and magnetite-bearing skarns (I) in aureole rocks, by aqueous solutions of a dominantly magmatic origin—at about 290–270 Ma ago. (d) Further crystallization of the pluton has taken place, and joints and fractures have formed in the crystallized carapace. With the formation of a water-rich phase that has separated from the H_2O-saturated melt, an extensive geothermal system has come into being. This has produced the main stage mineralization (III and IV) of tin- and copper-bearing quartz veins—at about 270 Ma (type II mineralization is that of pegmatites). (After Jackson *et al.* 1982.) (e) Schema (after Moore 1982) of possible fossil geothermal systems associated with the granite batholith of south-west England, illustrating the different types and settings of mineralization in that region, and the district zoning developed there. (1) Dines (1956) type emanative centres. (2) Cross course mineralization (succinctly described by Alderton 1978). (3) Kaolin deposits (weathering may have played a part in their formation).

nearby Rotokawa Geothermal System base metal precipitates are accompanied by argentite. Since the formation 6060 years ago of the large hydrothermal explosion crater in which Lake Rotokawa is now situated, about 370 Mt of gold may have been transported into the rocks beneath the crater (Krupp & Seward 1987).

The principal features of a geothermal system are shown in Fig. 4.4a. Meteoric water sinking to several kilometres depth (A) enters a zone of high heat flow, absorbs heat and rises into one or a succession of permeable zones (BD). There may be outflow at an appreciable rate along a path such as (C) or much slower outflow by permeation of the cap rock (mudstone, tuff, etc.). Outflow through (C) depends on the permeability of the rocks and the pressure at the top of the zone (BD). If the outflow rate does not exceed that of the inflow, an all liquid system will prevail. With a higher outflow rate, a steam phase will form in (BD) and the steam pressure will decrease until the mass outflow through (C) is reduced to equal the mass inflow. A dynamic balance then obtains with a lowered water level in the permeable horizon, boiling water and, most important from our point of view, the development of convection currents in the water.

Figure 4.4b illustrates the structure of a geothermal system in a volcanic terrane like that of the Taupo Volcanic Zone, New Zealand. Note that the hot waters are circulating through, reacting with and probably obtaining dissolved constituents from both the magmatic intrusion and the country rocks. In Figs 4.4c and d geothermal systems are postulated to explain vein tin and copper mineralization in and adjacent to the Land's End Granite in south-west England. In Fig. 4.4e we have a broader picture, with geothermal systems being invoked to explain some of the different types of mineralization in south-west England and the zoning of metals that is one of the well-known features of this orefield. Recent geothermal work (Hall 1990) indicates that the boron, lithium and tin in Cornish deposits was probably derived from the granites, whereas the copper and sulphur was apparently leached from the country rocks, particularly shales. A similar model, to explain the epigenetic uranium mineralization of the Variscan Metallogenic Province of western Europe, has been proposed by Cathelineau (1982). An important and detailed application of this model to epithermal, precious metal deposits is to be found in Hayba et al. (1986) and other recent papers—see Fig. 16.6b.

Although hydrothermal deposits are only small geological features, fossil geothermal systems can be very large. One of the largest so far documented is the Casto Ring Zone in central Idaho, which occupies an area of 4500 km^2 (Criss et al. 1984).

Means of transport

Sulphides and other minerals have such low solubilities in pure water that it is now generally believed that the metals were transported as complex ions. A few simple figures will illustrate this. The amount of zinc in a saturated zinc sulphide solution at a pH of 5 and a temperature of 100°C (possible mineralizing conditions) is about 1×10^{-5} g l^{-1}. A small orebody containing 1 Mt of zinc could have been formed from a solution of this strength (assuming all the zinc was precipitated) provided that 10^{17} l of solution passed through the orebody. This is equal to the volume of a tank having an area of 10 000 km^2 and sides 10 km high—an impossible quantity of solution. This difficulty is further illustrated in Fig. 4.5 where the calculated lead ion (line AB) and H$_2$S concentrations in water in equilibrium with galena at 80°C are shown. This indicates that (in the absence of ion complexing) concentrations of lead

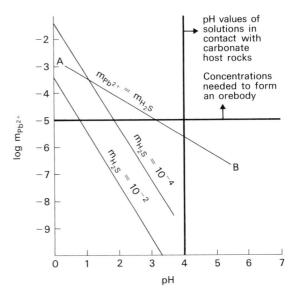

Fig. 4.5 Calculated lead ion concentrations in water in equilibrium with galena at 80°C: line AB. Two lines showing H$_2$S concentrations of 10^{-2} and 10^{-4} m are plotted. (After Anderson 1977.)

and H_2S high enough to form an orebody can be achieved only in very acid solutions, about pH = 0–3 (Anderson 1977). These pH values are most unlikely to hold for hydrothermal solutions except those in contact with a relatively insoluble rock such as quartzite. With other rocks, hydrogen ion would be consumed by wall rock alteration reactions until a pH nearer neutral was achieved. Probable values of pH for hydrothermal solutions lie to the right of the thick vertical line, particularly when the solutions are passing through limestones. So we can rule out the possibility that pure acidified water could be the transporting medium. What we require is a mechanism of transportation that will operate in the upper right-hand portion of such a diagram.

Laboratory, thermodynamic studies and the examination of modern geothermal systems have led geochemists to conclude that metals are transported in hydrothermal solutions as complex ions, i.e. the metals are joined to complexing groups (ligands). The most important are HS^- or H_2S, Cl^- and OH^-; other ligands, including organic ligands, may also contribute to complexing in ore fluids (Barnes (1979b). Bisulphide complexes can exist stably in near-neutral solutions containing abundant H_2S. Ions such as $PbS(HS)^-$ are formed and these have much higher solubilities than pure ionic solutions. The main objection to what is a very useful and promising hypothesis is the high concentration of H_2S and HS^- required to keep the complexes stable, a concentration much higher than that usually found in hot springs, fluid inclusions and geothermal systems (Table 4.2). For this and other reasons many workers favour the idea of metal transport in chloride complexes such as $AgCl_2^-$ and $PbCl^{3-}$. It is likely, however, that both, and other complexes, play a part in metal transportation. Barnes (1979b) has pointed out that available data show that the geologically improbable requirement of very alkaline solutions is necessary above approximately 300°C for HS^- to become a dominant species. For this reason bisulphide complexes may only be important for ore transportation at temperatures below about 350°C. Henley et al. (1984) suggested that gold probably travels as a bisulphide complex up to about 300°C but that chloride complexing may be important at higher temperatures. However, Seward (1991) has pointed out that, although $Au(HS)_2^-$ will play an increasingly diminished role in gold transport at temperatures above 300°C, sulphide complexes as a whole may still be important carriers of gold, for example the very stable $AuHS^0$

complex may be important at temperatures above 300°C. The solubility of gold as $Au(HS)_2^-$ in nature is by no means a simple matter, for Shenberger & Barnes (1989) showed that it varies appreciably according to which mineral pairs are buffering the oxidation state of the system. For example whilst the gold solubility at 300°C with a sulphate–sulphide buffer is decreasing, at the same total activity of aqueous sulphur and with a pyrite–pyrrhotite buffer the solubility is still increasing steadily at 350°C. It is interesting to note that Weissberg et al. (1979) indicated that in the Broadlands (NZ) geothermal system, gold is probably in solution as a bisulphide complex at 260°C, whilst lead is probably travelling as a chloride complex, and Krupp & Seward (1987) found the concentration of $Au(HS)_2^-$ at 311°C in the Rotokawa (NZ) geothermal system to be about six orders of magnitude larger than that of $AuCl_2^-$. One particular advantage of the bisulphide hypothesis is that it can be used to account for the zonal distribution of minerals in epigenetic deposits, as will be discussed later. A succcessful explanation for zoning using the chloride complex hypothesis has not yet been produced. The work of Wood et al. (1987), however, cautions us against taking any remotely simple view of element transport in complex ions. Their laboratory experiments on Fe, Zn, Pb, Au, Sb, Bi, Ag and Mo in H_2O–Nacl–CO_2 solutions at 200–300°C showed that 'No single complex or species can be expected to predominate completely for any metal over reasonable ranges of solution composition and temperature; many different ligands may be significant in any given solution, different metals are likely to be transported by quite different mechanisms ... and mixed ligand and perhaps polynuclear species may be expected.'

The origin of the sulphur at the site of deposition is also a problem; did it originate at this site or was it carried there with the metals in solution? Some mineralization situations, e.g. structurally deep sulphide veins in granites and quartzites, seem to demand that sulphur and the metals travelled together. Isotopic evidence, evidence from modern geothermal systems (Weissberg et al. 1979), ocean floor fissures venting hydrothermal solutions, and evidence from some fluid inclusions, many of which carry daughter sulphides in amounts often implying quite high concentrations of metals and sulphur in the ore fluid (Sawkins & Scherkenbach 1981), also favour this interpretation for many deposits. It must be emphasized that the apparent absence of H_2S from many fluid inclusions may be due to analytical difficulties of detection. The advent of Raman spectroscopy as an analytical tool may result in many more reports of the presence of H_2S in fluid inclusions (Touray & Guilhaumou 1984).

If the ore metals are transported as bisulphide complexes then abundant sulphur will be available for the precipitation of sulphides at the site of deposition. On the other hand, the postulate that chloride complexes are the metal transporters, which is the most favoured hypothesis of those studying carbonate-hosted base metal deposits, creates difficulties as far as sulphur supply and metal transport are concerned. These difficulties may be explained in this way: Anderson (1977) has shown that, if the molality of H_2S is 10^{-5}, a 3 molal NaCl solution will transport almost 10^{-5} m (about 2 ppm) Pb at a pH of 4 at 80°C. At temperatures up to 150°C and otherwise similar conditions a little more than 10^{-5} m Pb would dissolve. This means that ore transport and deposition in equilibrium with carbonate rocks could occur at these temperatures. pH 4 is at the acid end of the possible pH range. With pH values in the more probable range 5–6, chloride solutions at up to 150°C cannot carry more than 10^{-5} m of Pb and H_2S at the same time. This implies that transportation and precipitation

conditions would be in the very restricted field labelled *Galena precipitation* (Fig. 4.6), which is too acid to co-exist with limestone or dolomite. Thus, as Anderson (1983) remarks, we have a transport problem. Lead (and zinc) solubilities appear to be too low, even at 100–150°C, to allow metal transport in realistic amounts at realistic pH values to form an orebody. Two of the possible ways round this problem for low temperature carbonate-hosted deposits are as follows. The first way is to accept the hypothesis of Beales & Jackson (1966) that sulphur is added from another solution to the ore fluid at the site of deposition—the mixing model. We can then propose transport under relatively oxidizing and alkaline conditions well above the line AB in Fig. 4.6, which will permit the transport of significantly greater amounts of metal, but of almost no H_2S. In support of this possibility it can be recorded that Ghazban *et al.* (1990) have published isotopic evidence that marine sulphate reduction produced sulphur at the site of formation of a lead–zinc deposit in Baffin Island, together with isotopic data showing that the reductant was organic matter. From another carbonate-hosted lead–zinc deposit, this time in Missouri, Leventhal (1990) has produced evidence of the degradation of organic matter

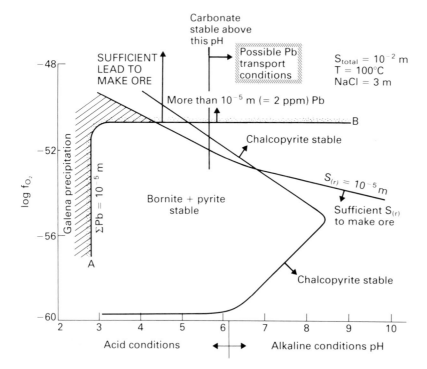

Fig. 4.6 Some mineral stabilities as a function of oxygen fugacities and pH. Other conditions as indicated. The stippled line is the 10^{-5} contour for total lead in equilibrium with galena. Larger solubilities lie to the left and above this line. $S_{(r)}$ = reduced sulphur (i.e. H_2S). For further discussion see text. (Modified from Anderson 1975.)

in the vicinity of ore. He proposed than non-biological sulphate reduction by organic matter has supplied at least some of the sulphur in the ore deposit. A second way out of the problem, and one that allows us to keep a single solution model with metal and sulphur travelling together, has been put forward by Giordano & Barnes (1981), Barnes, H.L. (1983) and Giordano (1985), who have proposed that organometallic complexes were the metal carriers. Such complexes could carry many thousand parts per million lead and zinc. The presence of these complexing agents would have the effect of shifting the solubility contours to lower f_{O_2} values, possibly permitting metals and H_2S to travel together.

Gize & Barnes (1987), however, from their investigation of organic matter in a carbonate-hosted, Nova Scotian, lead–zinc deposit concluded that organometallic complexing in the mineralizing solutions was not important during the formation of this deposit, because only minor amounts of introduced organic matter are present. Crocetti & Holland (1989), also, did not feel that organic complexing models are promising ideas as there are no known organic complexing agents of sufficient strength present in sufficient abundance in natural brines to act as efficient ligands for lead. A third possibility is the co-transport of metals with sulphur as sulphate, as proposed by Barton (1967) and tentatively favoured by Crocetti & Holland (1989) for the formation of the important deposits of the Viburnum Trend, Missouri for which their isotopic results indicated co-transport of lead and sulphur to the site of deposition. The sulphate would then be reduced by reaction with organic compounds to precipitate metal sulphides.

In connexion with the idea of oilfield brines as potential ore fluids, particularly in the formation of carbonate-hosted lead–zinc deposits (Mississippi Valley-type), it must be pointed out that there are both single solution models as postulated by Sverjensky (1984), which may be criticized as having too dilute a solution to form an orebody and as pushing the chemical parameters of the ore fluid to the fringes of what might be expected under natural conditions (Giordano 1985), and the mixing model as proposed by Beales & Jackson (1966).

Whatever finer details may be agreed upon eventually, Mississippi Valley-type deposits are now widely recognized as the products of basin dewatering, though uncertainty continues with regard to the mechanisms and timing of dewatering and the relation between specific deposits and potential source basins (Kesler *et al.* 1989, Sangster 1990). The presence of exotic oil in the probable aquifer, followed by mineralizing fluids, at Gays River, Nova Scotia (Gize & Barnes 1987) and oil-like hydrocarbons trapped in sphalerite and gangue minerals in deposits in the Canning Basin, Australia (Etminan & Hoffman 1989), suggest the action of metal-rich oilfield brines, such as those described by Carpenter *et al.* (1974), or similar formation waters derived from organic-rich, basinal shales. Hydrogen and oxygen isotopic values on fluid inclusion fluids support this hypothesis (Ravenhurst *et al.* 1989) and, using Na : K and Na : Li ratios in fluid inclusions from the central fluorite zone of the Northern Pennine Orefield, UK, Rankin & Graham (1988) have employed alkali geothermometers to estimate the temperatures in the source region of the mineralizing fluids as 150–250°C, which is entirely in keeping with derivation of the fluids from deeply buried basinal shales, but not incidentally with a magmatic source! Two of the many models that have been put forward to explain the genesis of Mississippi Valley-type deposits are shown in Fig. 4.7.

Sulphide deposition

There is a useful concise illustration of various factors involved in ore mineral deposition in Henley *et al.* (1984) which we will follow here. If $PbCl_2$ is the dominant lead complex in solution then we may write:

$$PbS + 2H^+ + 2Cl^- \rightleftharpoons PbCl_2 + H_2S.$$

From this equation we can see that *dilution or addition of H_2S* would precipitate galena from an initially saturated solution. Dilution can occur in various ways, the most important being when a rising ore fluid enters a zone saturated with ground water. Addition of H_2S may take place when another solution carrying H_2S is encountered or if sulphate *in the ore fluid* is reduced by organic material. (It should be stressed that it is only reduced sulphur in ore solutions that causes precipitation and Anderson's 'transport problem' discussed above. Much sulphate ion is usually present in geothermal systems and sulphate values in fluid inclusions are usually comparable with base metal concentrations—see Table 4.2.) An *increase in pH* due to *boiling* can induce precipitation and *cooling* of course also reduces solubility. What is the relative effectiveness of these factors? We can illustrate this by reference to

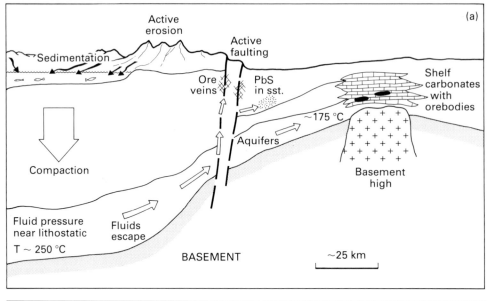

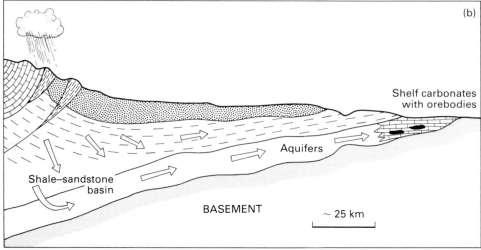

Fig. 4.7 Schemata for the formation of Mississippi Valley-type deposits. (a) Overpressured, hot pore fluids escape from a shale basin (perhaps aided by hydraulic fracturing) and move up aquifers to form deposits in cooler strata, filling fractures or forming other types of orebody. (b) Gravity-driven fluids flowing from a hydraulic head in a highland area flush through a basin driving out and replenishing the formation waters. These diagrams draw on the work of Garven (1985), Bethke (1986), Ohmoto (1986) and Ravenhurst *et al.* (1989).

the above solution, initially saturated with respect to galena, when it is cooled from 300°C to 280°C. A temperature change of this amount causes the solubility to drop to less than one-fifth of that at 300°C, whereas the solubility change due to dilution is much less. On the other hand, if boiling occurs (how could this come about?), accompanied by a temperature drop to 280°C, then the solubility is one hundred times less than that of the original solution. Of the three processes leading to galena precipitation, dilution, cooling and boiling, boiling is clearly the most efficient. If the initial solution was considerably undersaturated then dilution alone might lead to saturation and nothing more.

If, after reading the above sections, the reader feels baffled, and uncertain of his own opinion on the difficult subject of hydrothermal ore genesis, then let him take consolation from the recent statement of Giordano & Barnes (1981) concerning the vast number of geological and geochemical studies of Mississippi Valley-type deposits, 'In spite of this wealth of information, the mechanisms of base metal transport and deposition remain poorly understood'. If that is not consolation enough let him turn to Henley *et al.* (1984, pp. 119–200) who demonstrate that we could use our present state of geochemical knowledge to prove that it would be impossible for nature to form Mississippi Valley-type deposits. To quote from them, 'It is a good thing that someone had already found these deposits because geochemists might have proven that they couldn't exist!'

In the foregoing, some ideas concerning the genesis and nature of hydrothermal solutions have been dealt with briefly by particular reference to lead–zinc deposits in limestone host rocks. The subject is, however, vast and the above discussion must be considered only as a short introduction to some of the principles involved. The reader who wishes to further his study of the chemistry of hydrothermal solutions should turn to Henley *et al.* (1984) and then progress onwards from that excellent work, which is well endowed with pertinent references. An excellent recent summary of the hydrothermal behaviour of gold is to be found in Seward (1991).

Lateral secretion

It has been accepted for many years that quartz lenses and veins in metamorphic rocks commonly result from the infilling of dilatational zones and open fractures by silica which has migrated out of the enclosing rocks, and that this silica may be accompanied by other constituents of the wall rocks including metallic components and sulphur. This derivation of materials *from the immediate neighbourhood* of the vein is called lateral secretion. A very interesting example of deposits formed in this way has been described by Boyle (1959) from the Yellowknife Goldfield of the Northwest Territories of Canada; but before discussing it we must consider the probable behaviour of element levels in rocks adjacent to veins forming under different conditions. In Fig. 4.8a we have a vein forming from an uprising hydrothermal solution supersaturated in silica. Some of this diffuses into the wall rocks and causes some silicification. The curve showing the level of silica decreases away from the source (i.e. the vein). In Fig. 4.8b we have the opposite situation where silica is being supplied to the vein from the wall rocks. The curve now climbs as it leaves the vein, indicating a zone of silica depletion in the rocks next to the vein. Clearly silica has been abstracted from the wall rocks and presumably has accumulated in the vein.

The principal economic deposits of the Yellowknife Field occur in quartz–carbonate lenses in extensive chloritic shear zones cutting amphibolites (metabasites). The deposits represent concentrations of silica, carbon dioxide, sulphur, water, gold, silver and other metallic elements. The principal minerals are quartz, carbonates, sericite, pyrite, arsenopyrite, stibnite, chalcopyrite, sphalerite, pyrrhotite, various sulphosalts, galena, scheelite, gold and aurostibnite. The regional metamorphism of the host rocks varies from amphibolite to greenschist facies. Alteration haloes of carbonate–sericite-schist and chlorite–carbonate-schist occur in the host rocks adjoining the deposits.

It is very instructive to remember that the dominant mineral of the veins is quartz. The profile of silica alongside the lenses is shown in Fig. 4.9. This demonstrates that a very substantial amount of silica has been subtracted from both the alteration zones

Fig. 4.8 Comparison of hypothetical profiles of silica. In (a) silica is added to the wall rocks from the hydrothermal solution, which is depositing quartz in the vein. In (b) silica is abstracted from the wall rocks and deposited as quartz in the vein. C indicates the normal level of silica in the country rocks.

(a)

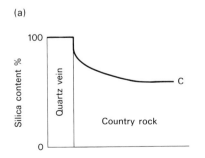

(b)

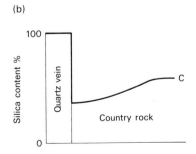

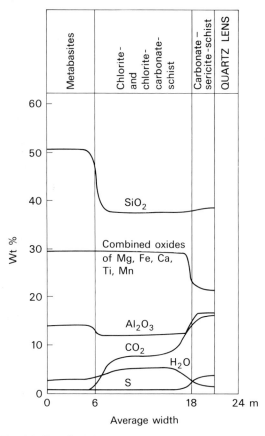

Fig. 4.9 Chemical changes produced by alteration of metabasites. Yellowknife gold deposits, Canada. (Modified from Boyle 1959.)

develops to a much higher level than in the unaltered country rocks. Water shows a similar but not identical behaviour. Boyle produced good evidence that these two oxides were passing through the rocks in considerable quantity, being mobilized by regional metamorphism and migrating down the metamorphic facies. They passed into the shear zones to form the chlorite, carbonate and sericite.

It appears highly probable that the major constituents of the shear zones resulted from rearrangement and introduction of material from the country rocks; the remaining question is whether the metabasites could have been the source of the sulphur and metallic elements in the deposits. The metabasites consist of metamorphosed basic volcanic lavas and tuffs. These rocks are richer in elements such as gold, silver, arsenic, copper, etc., than other igneous rocks, and for the unsheared metabasites of the Yellowknife area Boyle obtained the following values (all in ppm): S = 1500; As = 12; Sb = 1; Cu = 50; Zn = 50; Au = 0.01; Ag = 1. For purposes of calculation the rock system was taken to be: length = 16 km; width = 152 m; depth = 4.8 km. The amount of ore in the system was assumed to be 6×10^6 t with average grades of S = 2.34%; As = 1.35%; Sb = 0.15%; Cu = 0.07%; Zn = 0.28%; Au = 0.654 oz ton^{-1} and Ag = 0.139 oz ton^{-1}. The total contents of these elements in the shear system prior to shearing and alteration, and in the deposits is shown in Table 4.3. It is apparent from these figures that all the elements considered could have been derived solely from the sheared rock of the shear zone and there is no need to postulate another source. Indeed, there is such a difference between the values in columns two and three that it may well be that significant quantities of chalcophile elements ac-

and this has occurred of course *on both sides* of the vein. Clearly more silica has been subtracted from the wall rocks than is present in the lenses and the problem is not, where has the silica in the lenses come from, but where has the surplus silica gone to? Some subtraction of magnesia, iron oxides, lime, titania and manganese oxide has also occurred and doubtless this is the source of iron in such minerals as pyrite, pyrrhotite and chalcopyrite in the lenses. Alumina shows a depletion in the outer zone of alteration and a concentration in the inner zone where it has collected for the formation of sericite. A similar depletion in these elements has been reported from wall rock alteration zones beside the gold veins of the Sigma Mine, Quebec (Robert & Brown 1986 a,b).

As a result of the extensive development of carbonates in the alteration zones carbon dioxide

Table 4.3 Contents of chalcophile elements in shear zones and deposits, Yellowknife gold deposit, Canada

Element	Total content in shear system before shearing and alteration (millions of tons)	Total content in deposits (millions of tons)
S	62	0.14
As	0.5	0.081
Sb	0.04	0.009
Cu	2.0	0.004
Zn	2.0	0.017
Au	12.2×10^6 oz	3.9×10^6 oz
Ag	1219×10^6 oz	0.834×10^6 oz

companied the surplus silica to higher zones in the crust to form deposits which have now been eroded away.

Other applications of the lateral secretion theory in very different geological settings include Brimhall (1979) and Dejonghe & de Walque (1981). Brimhall has proposed that the main veins at Butte were formed by the concentration of ore-forming material already present in the host granodiorite intrusion in the form of protore (see Chapter 16). Dejonghe & de Walque (1981) have discussed the formation of a lead–zinc–copper-bearing vein in Carboniferous sediments in Belgium in terms of this theory. Finally it can be noted that even low grade metamorphism (lower greenschist facies) permits the subtraction of 50% of the gold content of andesites (Dostal & Dupuy 1987). Other references and a review of secretion theories can be found in Boyle (1991).

Metamorphic processes

Contact and regional metamorphism

Isochemical metamorphism of many rocks can produce materials having an industrial use. An obvious example long used by mankind is marble, which may be produced by either contact or regional metamorphism of pure and impure limestones and dolomites. Another much used metamorphic rock is slate. Other important industrial materials of metamorphic origin are asbestos, corundum and emery, garnet, some gemstones, graphite, magnesite, pyrophyllite, sillimanite minerals, talc and wollastonite.

Allochemical metamorphism (metasomatism) may accompany contact or regional metamorphism. In the former case in particular it may lead to the formation of skarn deposits carrying economic amounts of metals or industrial minerals. Their general morphology and nature have been summarized in Chapter 2. Their genesis will be further discussed in Chapter 13.

The role of other metamorphic processes in ore formation

Some examples of lateral secretion are clearly the result of metamorphism. This subject has, however, already been covered above and will not be discussed further. In this section we are concerned with those metamorphic changes that involve recrystallization and redistribution of materials by ionic diffusion in the solid state or through the medium of volatiles, especially water. Under such conditions relatively mobile ore constituents may be transported to sites of lower pressure, such as shear zones, fractures or the crests of folds. In this way, the occurrrence of quartz–chalcopyrite–pyrite veins in amphibolites and schists and many gold veins in greenstone belts (Saager et al. 1982) may have come about.

The behaviour of trace amounts of ore minerals in large volumes of rock undergoing regional metamorphism is uncertain and is a field for more extensive research. It might be thought that with the progressive explusion of large volumes of water and other volatiles during prograde metamorphism, natural hydrothermal systems might evolve that would carry away elements such as copper, zinc or uranium, which are enriched in trace amounts in pelites. Shaw (1954), however, in a study of the progressive metamorphism of pelitic sediments from clays through to gneisses showed that such changes are but slight. Taylor (1955) in a study of greywackes reported similar results. On the other hand, De Vore (1955) calculated that during the transformation of one cubic mile of epidote–amphibolite facies hornblendite into the granulite facies there may be a release of 9 Mt of Cr_2O_3, 4.5 Mt of NiO and 900 000 t of CuO. Similarly, retrograde metamorphism can release large quantities of zinc, lead and manganese.

However, recent studies of mass balance changes accompanying the development of foliation in metamorphic rocks have shown that regional metamorphic terranes are large hydrothermal systems analogous to the smaller scale systems in young oceanic crust (Cox et al. 1987). These systems have the capacity to leach a wide range of components including ore-forming materials from a very large volume of crust. For flow to take place in such systems regional permeability must be developed. It is now known that major roles in enhancing rock permeability are played by the development of grain-scale dilatancy (Fischer & Paterson 1985) and mineral filled fractures (Yardley 1983). Both features result from rock fracture, but on very different scales. If this fluid flow is channelled through a small volume of rock in which dilatant zones develop then mineral deposits may be formed. Fyfe & Henley (1973) considered just such a mechanism.

They envisaged a situation where a volcanic–sedimentary pile is being metamorphosed under amphibolite facies conditions. It would be losing about 2% water, and if salt is present and oxygen buffered by magnetite–ferrous silicate assemblages, then gold solubilities of the order of 0.1 ppm at 500°C would be achieved. This gold would either be dispersed through greenschist facies rocks or concentrated into a favourable structure. This could happen if the solution flow was focused into a large vein or fault system where seismic pumping could have forced the ore fluid upwards (Sibson *et al.* 1975) (Fig. 4.10).

Fyfe & Henley showed that a source region of 30 km³ could have provided all the gold, silica and water required to form a deposit as large as Morro Velho, Brazil. Their figures are as follows. The orebody of auriferous quartz occupies 0.01 km³ and contains about 3×10^8 g of gold. With an average crustal gold content of about 3 ppb, approximately 30 km³ of volcanics or sediments are required to form the source region. About 2×10^{15} g of water would be released, which at 0.1 ppm could transport 2×10^8 g of gold. At 0.5 GPa and 500°C this volume of water could also transport 2×10^{13} g silica (solubility under these conditions 10 g kg^{-1}). This silica is about 0.01 km³ in volume, i.e. that of the orebody. This model of gold mineralization is similar to that proposed by Boyle (see earlier), but now the major

constituents are derived from a deep source region and not by lateral secretion.

This model, and variations of it, have been used to explain the formation of a number of gold deposits—see for example Kerrich & Fryer (1979), Phillips *et al.* (1984), Shepherd & Allen (1985), Annels & Roberts (1989), Williams (1990). Some of these and other workers have advanced the results of isotopic studies in support of this model.

In developing this model for exploration purposes the comments of Phillips & Groves (1983, p. 36) should be carefully considered. A recent counter-blast to this metamorphic model for the genesis of hydrothermal, gold-depositing solutions has come from Burrows *et al.* (1986) who have used a great deal of carbon isotope data from the world's two largest Archaean gold vein and shear zone systems, Hollinger-McIntyre in Canada and the Golden Mile in Western Australia, to support a return to the once orthodox hypothesis of a magmatic-hydrothermal origin of the mineralizing solutions! They have support from a number of workers: for example Siddaiah & Rajamani (1989) have used the evidence from REE patterns, oxygen and lead isotopes to support the application of this hypothesis to the

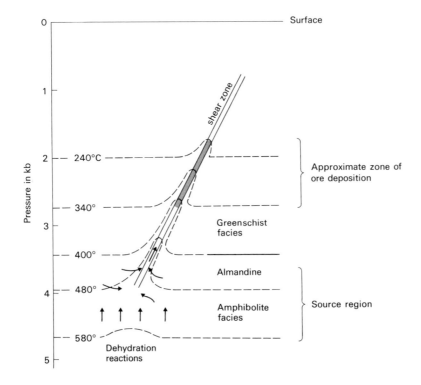

Fig. 4.10 Diagram of a shear zone where metamorphic water from a large volume of rock is rising to higher levels. (After Fyfe & Henley 1973.)

Kolar gold deposits of India; and Cameron & Hattori (1987) produced evidence for the presence of oxidized fluids during the formation of a number of Canadian and Australian Archaean gold deposits— such fluids, they contended, can only be of magmatic origin. Nesbitt *et al.* (1989) and Madu *et al.* (1990), however, argued cogently and produced compelling isotopic evidence for regarding these mineralizing fluids as evolved meteoric water that had gained hydrogen and oxygen isotopic values close to those of magmatic and metamorphic water by interaction with sedimentary rocks. We have a three cornered fight to referee!

Examples of other deposits considered to have formed from metamorphic fluids include uranium deposits in Sweden (Adamek & Wilson 1979), cobalt– tungsten mineralization in Queensland (Nisbet *et al.* 1983), the 27 Mt Elura Zn–Pb–Ag massive sulphide orebody, New South Wales (Seccombe 1990), the copper orebodies at the famous Mount Isa Mine, Queensland (Heinrich *et al.* 1989) and the important Sb–Au orebodies along the Antimony Line, a 40 km long shear zone, in RSA (Vearncombe *et al.* 1988).

Origin due to surface processes

Processes involving mechanical and chemical sedimentation will not be dealt with here. The reader is referred to texts on sedimentology for the general principles involved, e.g. Blatt *et al.* (1980), Reading (1986) and Tucker (1988). However, certain aspects will be touched upon in Chapter 18. Residual processes and supergene enrichment will be dealt with in Chapter 19. The remaining space in this chapter will be devoted to a consideration of exhalative processes. These, it should be noted, are a surface expression of the activity of hydrothermal solutions.

Volcanic-exhalative (sedimentary-exhalative) processes

We are concerned here with a group of deposits often referred to as exhalites and including massive sulphide ores. Some of the characteristics of these ores have been dealt with on pp. 38–39. They frequently show a close spatial relationship to volcanic rocks, but this is not the case with all the deposits, e.g. Sullivan, Canada (Fig. 2.14) which is sediment-hosted and this and similar examples are referred to commonly as sedex (sedimentary-exhalative) deposits. They are conformable and frequently banded; and in the volcanic-associated types the principal constituent is usually pyrite with varying amounts of copper, lead, zinc and baryte; precious metals together with other minerals may be present. For many decades they were considered to be epigenetic hydrothermal replacement orebodies (Bateman 1950). In the 1950s, however, they were recognized as being syngenetic, submarine-exhalative, sedimentary orebodies, and deposits of this type have been observed in the process of formation from hydrothermal vents (black smokers) at a large number of places along sea-floor spreading centres (Rona 1988). These deposits are now often referred to by one or other of the two terms in this section heading, or as volcanic-associated (or volcanogenic) massive sulphide deposits.

The ores with a volcanic affiliation show a progression of types. Associated with basic volcanics, usually in the form of ophiolites and presumably formed at oceanic or back-arc spreading ridges, we find the Cyprus types. These are essentially cupriferous pyrite bodies. They are exemplified by the deposits of the Troodos Massif in Cyprus and the Ordovician Bay of Islands Complex in Newfoundland. Associated with the early part of the main calc-alkaline stage of island arc formation are the Besshi-type deposits. These occur in successions of mafic volcanics in complex structural settings characterized by thick greywacke sequences. They commonly carry zinc as well as copper and are exemplified by the Palaeozoic Sanbagawa deposits in Japan, and the Ordovician deposits of Folldal in Norway. The more felsic volcanics, developed at a later stage in island arc evolution, have a more varied metal association. They are copper–zinc–lead ores often carrying gold and silver. Large amounts of baryte, quartz and gypsum may be associated with them. They are called Kuroko deposits after the Miocene ores of that name in Japan, but similar deposits in the Precambrian are known as Primitive-type. All these different types are normally underlain in part by a stockwork up which the generating hydrothermal solutions appear to have passed (Fig. 2.20).

There is today wide agreement that these deposits are submarine-hydrothermal in origin, but there is a divergence of opinion as to whether the solutions responsible for their formation are magmatic in origin or whether they represent circulating sea water. Here, there is only room to deal briefly with this controversy. We will start by looking at the evidence from hydrothermal activity on the ocean floors.

Hydrothermal mineralization at sea-floor spreading centres was first discovered in the Red Sea in the mid 1960s (see p. 324), but the resultant deposits do not appear to be true modern analogues of volcanic- and sediment-hosted massive sulphide deposits. Since then various forms of hydrothermal mineralization have been found at many sites along spreading centres with the black smoker type producing obvious analogues of ancient massive sulphide deposits (Rona 1988).

Black smokers were discovered in the late 1970s during ocean floor investigations using a submersible. They are plumes of hot, black, or sometimes white, hydrothermal fluid issuing from chimney-like vents that connect with fractures in the sea floor. The black smoke is so coloured by a high content of fine-grained metallic sulphide particles and the white by calcium and barium sulphates. The chimneys are generally less than 6 m high and are about 2 m across. They stand on mounds of massive ore-grade sulphides (Fig. 4.11) that occur within the grabens and on the flanks of oceanic ridges. Ten of the largest mounds in the eastern axial valley of the southern Explorer Ridge (about 350 km west of Vancouver Island) average 150 m across and 5 m

thick and are estimated to contain a total of 3–5 Mt of sulphides (Hannington & Scott 1988). The largest mound-chimney deposit so far found is the TAG mound on the Mid-Atlantic Ridge at 26°N, which is estimated to contain 4.5 Mt (Rona *et al.* 1986). The mineralogy of the mounds is similar to that of massive sulphide deposits on land with high temperature copper–iron sulphides beneath lower temperature zinc- and iron-rich sulphides, baryte and amorphous silica. Silver-bearing sulphosalts with minor galena occur in the lower temperature (< 300°C) Zn–Fe assemblages rather than the higher temperature (> 300°C) Cu assemblages (Hannington & Scott 1988). Gold values ranging up to 16.4 ppm have been found.

Growth of a sulphide chimney commences with the precipitation of anhydrite from the cold sea water around the hot ascending plume, forming a porous wall that continues to grow upward during the life of the plume. Most of the hydrothermal fluid flows up the chimney to discharge as a plume into the surrounding sea water, but a small proportion flows through the porous anhydrite wall. In doing so the fluid moves rapidly from high temperature (> 300°C), acidic (pH~3.5) and reduced ($H_2S \gg SO_4$) conditions to

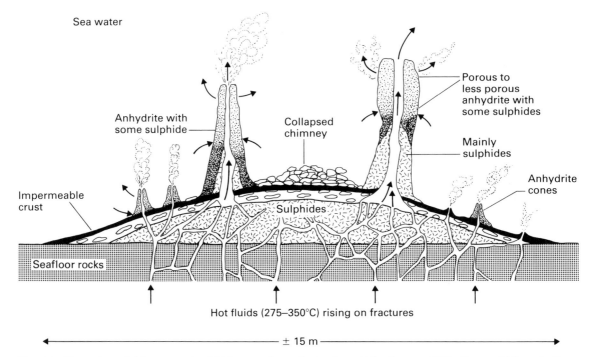

Fig. 4.11 Formation of chimneys and sulphide mounds on the sea floor. (After Barnes, 1988, *Ores and Minerals*, Open University Press, with permission.)

conditions close to those of normal sea water at oceanic depths (2°C, alkaline, pH ~7.8, oxidized, $SO_4 \gg H_2S$) as it meets sea water penetrating the chimney from the outside. Thus, whilst the anhydrite chimney top grows upwards, the walls of the lower part thicken by the precipitation of sulphides in the interior portions and anhydrite on the outside (Fig. 4.11). This process leads to a concentric zoning with chalcopyrite ($\pm$ isocubanite $\pm$ pyrrhotite) on the inside, an intermediate zone of pyrite, sphalerite, wurzite and anhydrite, and an outer zone of anhydrite with minor sulphides, amorphous silica, baryte and other minerals. The zoning is probably caused by the temperature decrease across the wall rather than other factors (Lydon 1989).

After growing upwards at a rate of perhaps 8–30 cm a day, chimneys eventually become unstable and collapse forming a mound of chimney debris mixed with anhydrite and sulphides upon which further chimney growth and collapse occur. Once a mound has developed it grows both by accumulation of chimney debris on its upper surface *and* by precipitation of sulphides within the mound. The covering of chimney debris performs the same role as the vertical porous anhydrite chimney walls producing sulphide and silica precipitation in the outer part of the mound. This decreases the permeability of the mound and forms a crust that constrains fluid escape and leads to considerable circulation of high temperature solutions within the mound. The isotherms within the mound then rise, leading to the replacement of lower temperature mineral assemblages by higher temperature ones, thus producing a similar zoning to that in the chimneys and that of volcanic-associated massive sulphide deposits found on land (pp. 202–203).

Volcanic-associated massive sulphide (VMS) deposits may have the mound shape of modern massive sulphide deposits or they may be bowl-shaped. The latter type probably developed when hydrothermal solutions, more saline (denser) than the surrounding sea water, vented into a submarine depression (Fig. 4.12). Many Cyprus-type deposits appear to have developed in this way and the available fluid inclusion data is consistent with this model, as explained by Rona (1988).

A most surprising feature of modern submarine hydrothermal vents is the associated prolific biota and their food chain based on chemosynthetic bacteria. The megafauna is varied and characteristic and some individuals, particularly tube casts of vestimentifera have been preserved in chimney

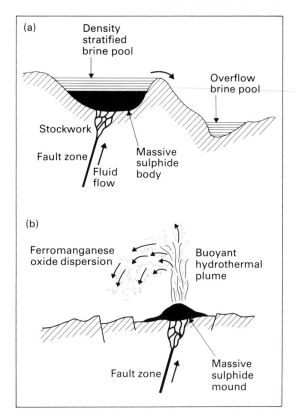

Fig. 4.12 Sketches showing the development of massive sulphide deposits on the sea floor. In (a) a ponded hydrothermal solution whose density is greater than that of the surrounding sea water collects in a depression to form a bowl-shaped deposit. In (b) a solution less dense than sea water forms a sulphide mound and rises buoyantly above it to form a hydrothermal plume. From this, oxide and sulphide particles and silica rain down on the surrounding rocks to deposit ferromanganese oxide rocks, cherts with or without pyrite and other hydrothermal sedimentary accumulations that are called exhalites. (After Rona 1988.)

sulphides. The remains of ancient vents have been found in the Cretaceous deposits of Cyprus, in similar deposits in the Oman where traces of this characteristic fauna are also present (Oudin & Constantinou 1984, Haymon *et al.* 1984) and in the Philippines (Boirat & Fouquet 1986).

This brief summary of the formation of black smoker deposits, and there is much more important supporting evidence in the literature (e.g. see *Canadian Mineralogist*, **26**, 429–888 and references therein), shows that our knowledge of their mode of formation confirms the model put forward for the genesis of volcanic-associated massive sulphide deposits by Eldridge *et al.* (1983) and Pisutha-Arnond & Ohmoto (1983) and termed by Huston & Large (1989) the Kuroko model. The principal stages of development of this model are as follows (Fig. 4.13).

1 Precipitation of fine-grained sphalerite, galena, pyrite, tetrahedrite, baryte with minor chalcopyrite (black ore) by the mixing of relatively cool (~200°C) hydrothermal solutions with cold sea water.

2 Recrystallization and grain growth of these minerals at the base of the evolving mound by hotter (~250°C) solutions, together with deposition of more sphalerite, etc.

3 Influx of hotter (~300°C) copper-rich solutions which replace the earlier deposited minerals with chalcopyrite in the lower part of the deposit (yellow ore). Redeposition of these replaced minerals at a higher level.

4 Still hotter, copper undersaturated solutions then dissolve some chalcopyrite to form pyrite-rich bases in the deposits.

5 Deposition of chert–hematite exhalites above and around the sulphide deposit. Similar exhalites will also have formed during previous stages. Silica is slow to precipitate, it needs silicate minerals to nucleate on and so, although much may be deposited

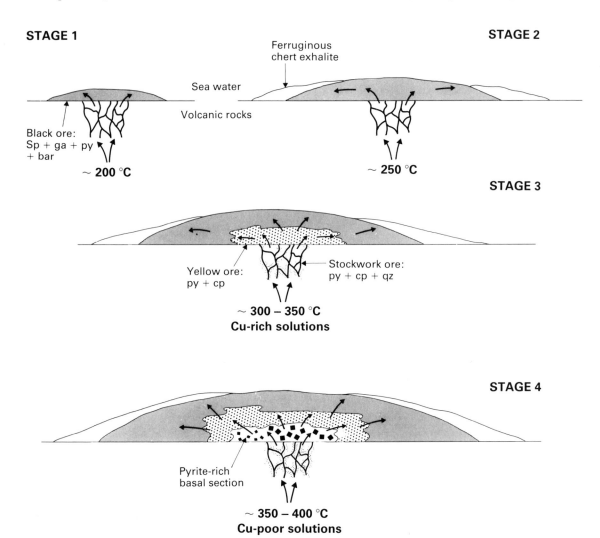

Fig. 4.13 Diagrams to illustrate the first four stages during the formation of volcanic-associated massive sulphide deposits as described in the text; bar = baryte, cp = chalcopyrite, ga = galena, py = pyrite, qz = quartz and sp = sphalerite.

in the underlying stockwork, the rest is mainly carried through the sulphide body to form exhalites above it.

Note that an evolution with time of the hydrothermal solutions is postulated from cooler to hotter and back to cooler and that the solutions of stages (2–4) cool as they move upwards and outwards through the sulphide mound to produce stage 1 type mineralization at the top of the deposit; see Fig. 4.13. The most important evidence on which this model is based, mineral replacement textures and fluid inclusion filling temperature trends, comes from young unaltered deposits and generally is difficult or impossible to find in older (Palaeozoic and Precambrian) deposits, which usually have suffered some degree of deformation and metamorphism.

Most workers in this field agree at least on the broad outline of the above mechanisms for the formation of VMS deposits, but they disagree on the origin of the formative solutions, with the majority, at the time of writing, favouring the hypothesis that they are composed of sea water which has circulated deep into the crust, where it has been heated, become more saline and dissolved base and precious metals from the volcanic and other rocks it has passed through, and then risen along permeable zones to be exhaled on the sea floor as sketched in Fig. 4.17. These workers point, in particular, to the evidence from isotopic studies, which is not unequivocal as we shall see. Before discussing the isotopic data the increase in salinity must be accounted for. Rock hydration reactions by consuming water will concentrate chloride in the pore solutions, phase separation of the $NaCl–H_2O$ fluid under high PT conditions can produce a brine with salinities two to three times that of sea water and other processes, including dissolution of evaporites, can produce the same result.

As was mentioned earlier in this chapter, different types of water have characteristic hydrogen (D/H) and oxygen ($^{18}O/^{16}O$) isotopic ratios (Sheppard 1977). Using these ratios it has been shown that various types of water were involved in general mineralization processes. Magmatic fluids were dominant in some cases, and in others initiated the mineralization and wall rock alteration only to be swamped by convective meteoric water set in motion by hot intrusions or other heat sources. In massive sulphide deposits, much isotopic evidence favours sea water as the principal or only fluid. The possibility must be borne in mind, however, that

the sea water was a late addition to a magmatic hydrothermal system and that it has overprinted the pre-existing magmatic values (Sato 1977, Stanton 1986). It is worthwhile examining the evidence in some depth as isotopic studies are now being applied to all types of mineral deposits and they will be mentioned again in later chapters. The use of stable isotopes in research into high temperature geological processes, including ore deposits, is well covered in Valley *et al.* (1986).

Variations in the isotopic ratios of hydrogen and oxygen are given in the δ notation in parts per thousand (per mil, ‰) where:

$$\delta_x = \left(\frac{R_{\text{sample}}}{R_{\text{standard}}} - 1 \right) \times 1000.$$

In the above formula for hydrogen, $\delta_x = \delta D$ and $R = D/H$; for oxygen, $\delta_x = \delta^{18}O$ and $R = {}^{18}O/{}^{16}O$.

The standard for both hydrogen and oxygen is standard mean ocean water or SMOW. In nature, D/H is about 1/7000 and $^{18}O/^{16}O$ is about 1/500. These values are measured directly on natural substances, such as thermal waters, formation waters in sediments and fluid inclusions, or they are determined indirectly using minerals after removal of all the absorbed water. In the latter case the isotopic composition of the mineral is not that of the fluid with which it was in contact at the time of crystallization or recrystallization. The δ values for the fluid have to be calculated from the mineral values using equilibrium fractionation factors determined by laboratory experiment or from studies of active geothermal systems. A temperature fractionation effect also occurs, so the temperature must be known (from fluid inclusion studies, etc.) in order to determine the isotopic composition of the water in equilibrium with the mineral. For example, in Fig. 4.14, raising the temperature from 10 to 200°C gives rise to isotope exchange and re-equilibrium such that the isotopic composition of the water changes from Y_1 to X_1 whilst that of the coexisting kaolinite changes from Y_2 to X_2. In other words rock–water reactions cause a shift in the δ values of both the circulating meteoric water and the rock with which it is in contact, with the result that the water is enriched in ^{18}O as the temperature rises.

The isotopic compositions of the various types of water show useful differences. Sea water in general plots very close to SMOW (Fig. 4.15) and shows very little variation. Meteoric water varies fairly systematically with latitude along the line shown in

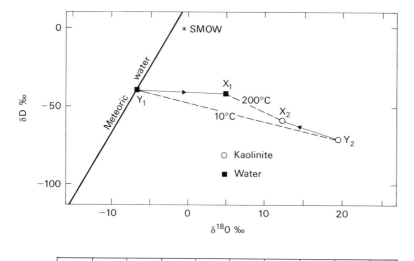

Fig. 4.14 Hydrogen and oxygen isotopic fractionations between water and kaolinite at two different temperatures. From a knowledge of the system kaolinite–water, and given the values X_2 and 200°C, the value of X_1 can be calculated. (Modified from Sheppard 1977.)

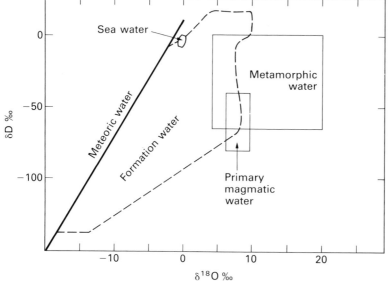

Fig. 4.15 Fields of isotopic composition of sea water, formation water, metamorphic water and magmatic water. (Modified from Sheppard 1986.)

Fig. 4.15. Values for metamorphic and magmatic waters have been deduced from measurements on minerals. Formation water (interstitial water in sediments) may have been trapped during sedimentation or may have entered the interstices at a later time, when it may be of any origin or age. It can be measured directly and plots as shown. Since many of the formation waters are richer in ^{18}O than SMOW they cannot have resulted from simple mixing of meteoric and sea water. There must have been isotopic exchange with the sediments at elevated temperatures (shown by Fig. 4.14 to result in ^{18}O enrichment of the water, e.g. the change from Y_1 to

X_1), addition of rising metamorphic water or some other process. We are now in a position to look at some results of studies on volcanic-associated massive sulphide deposits. The data of course comes from associated gangue and wall rock alteration minerals and not from the ore minerals themselves.

These results are plotted in Fig. 4.16. Those for the Cretaceous Cyprus stockwork deposits coincide exactly with the values for sea water, and Heaton & Sheppard (1977) suggest a model involving deeply circulating sea water as sketched in Fig. 4.17. They also present evidence that the associated country rocks were thoroughly permeated by sea water

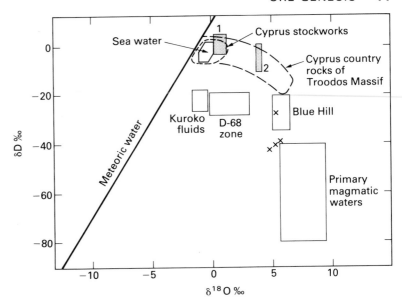

Fig. 4.16 Plot of δD against δ^{18}O for ore-forming fluids in the Cyprus stockworks, associated country rocks and Kuroko deposits. Crosses mark values for sericites from the Kosaka deposit, Japan. (Modified from Sheppard 1977.) Data for Kosaka are from Hattori & Muehlenbachs (1980), for the Iberian Pyrite Belt (1 = Rio Tinto, 2 = Salgadhino) and Blue Hill, Maine from Munha *et al.* (1986) and for the D-68 Zone, Quebec from Ikingura *et al.* (1989).

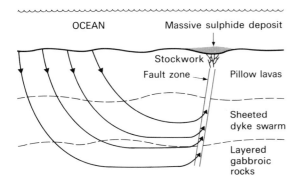

Fig. 4.17 Diagram showing how sea water circulation through oceanic crust might give rise to the formation of an exhalative volcanic-associated massive sulphide deposit.

during their metamorphism into the greenschist and zeolitic facies. The Kuroko fluids show δ^{18}O values commensurate with a sea water origin but δD is depleted by 11–26‰ relative to sea water. Ohmoto & Rye (1974) concluded that sea water was the dominant source of the hydrothermal fluid but that it contained a small meteoric and/or magmatic contribution. As Kuroko deposits belong to the Island Arc environment, meteoric water could be involved. This would suggest a model, similar to that in Fig. 4.17, of circulating sea water becoming a concentrated brine at depth, dissolving copper and other metals from the rocks it traversed and carrying these up to the surface where they were

precipitated as sulphides with sulphur derived from the sea water and/or the rocks through which it passed. A large number of other deposits give similar plots to that of the Cyprus stockworks (Fig. 4.16) and the positions of two, Rio Tinto (the world's largest, ~500 Mt) and Salgadhino from the Iberian Pyrite Belt (Carboniferous) are shown. The D-68 Zone deposit, Noranda, Quebec (Archaean), plots closer to the magmatic water box than the Kuroko solutions and Ikingura *et al.* (1989) suggested that 25–30% of magmatic water may have been present in the mineralizing fluids. The Blue Hill, Maine (Siluro-Devonian) solutions and values for the Kosaka deposit, Japan plot even closer. In discussing the Blue Hill deposit Munha *et al.* (1986) suggested a 40% magmatic component.

Did the solutions responsible for the bulk of the mineralization in these bodies really vary as much in isotopic composition as these results suggest, or was there massive but variable overprinting by sea water in convection cells set in motion by nearby igneous intrusions that may have been the source of the magmatic water? Solomon *et al.* (1987) have shown that during the cooling cycle, after deposit formation, sea water convection cells would remain in motion for a considerable time, long enough indeed to produce wall rock alteration on the later formed hanging wall rocks above the deposits! Perhaps, as Stanton (1986) has suggested, the process is a hybrid one, a magmatic source producing concentrated ore solutions that move up mixing with large quantities of sea water at higher levels so acquiring hydrogen, oxygen and sulphur isotope characteristics near to those of sea water. The evidence is equivocal so let us pass on to look at the sulphur isotopic data for these deposits.

Sulphur isotopic variation is reported in terms of $\delta^{34}S$ representing the changes in $^{34}S/^{32}S$ (pp. 87–88). It is sometimes possible to differentiate between crustal and mantle sources of sulphur on the assumption that mantle values of $\delta^{34}S$ are zero or close to zero, and Ohmoto (1986) concluded that, unless assimilation of crustal sulphur has occurred, magmas generated in the mantle will have $\delta^{34}S$ values between -3 and $3‰$. This may be an oversimplification and evidence is appearing which suggests that contamination of the upper mantle by subduction is producing compositional heterogeneities, e.g. Chaussidon *et al.* (1987) obtained values up to $+9.5‰$ from sulphide inclusions in diamonds. However magmatic rocks from the mantle do, in general, have values close to zero. On the other hand much crustal sulphide has undergone biogenic fractionation giving $\delta^{34}S$ values as high as $+30‰$ and ocean water has $\delta^{34}S \sim 20‰$.

$\delta^{34}S$ values of modern sulphide mounds vary from about 1.5–4‰ (Lydon 1989), indicating a mainly mantle source with a smaller component derived from reduced sea water sulphur. This is interpreted by those favouring a circulating sea water hypothesis as indicating sulphur leaching from the consolidated volcanic substrate, but it does not rule out derivation of this sulphur from a consolidating magma that yielded a concentrated ore solution at some stage during its crystallization. The same argument applies to ancient deposits. Thus sulphur isotopic data for the Cyprus ores was reviewed by Spooner (1977) who pointed out that $\delta^{34}S$ for the pyrite is somewhat higher than that of the basaltic country rocks from which the sulphur may have been partly derived. This suggests an additional source of isotopically heavy sulphur. Spooner felt that this was probably the circulating Cretaceous sea water which would have had a value of $\delta^{34}S = +16‰$. Later studies on other deposits have produced parallel findings and similar interpretations have been advanced (Lydon 1989). However Aye (1982) postulated a mantle or magmatic source for the ore solutions that formed some Brittany deposits in which $\delta^{34}S$ values are close to zero and, in their paper on Palaeozoic deposits of this type in northern Queensland, Gregory & Robinson (1984) used sulphur isotope analyses to infer that a magmatic ore fluid was responsible for the formation of one deposit, but that in a nearby one the magmatic ore fluid was diluted with sea water. The reader must beware of taking such studies too much at their face value; sulphur isotope values in sea water passing through mantle-derived rocks may, through water–rock reactions, become close to or virtually the same as magmatic (mantle) values. For further discussion on this point see Skirrow & Coleman (1982).

Other isotopic tracers, including strontium and lead, also allow for a circulating sea water hypothesis or a magmatic-hydrothermal interpretation, e.g. Sato (1977) who suggested a direct magmatic source for the lead in Kuroko deposits based on his work on lead isotopes.

One of the difficulties concerning the isotopic studies outlined above is that they have been confined largely to mineralized areas. One non-mineralized area where such studies have been carried out in depth is Mull in Scotland. Another is the Bushveld Complex, RSA where Schiffries & Skinner (1987) demonstrated that an enormous palaeohydrothermal system had produced veins over a crustal thickness of 9 km along a strike length of 100 km without producing any significant mineralization. The isotopic evidence from these districts shows without doubt that vast quantities of meteoric water can be flushed through igneous intrusive and extrusive rocks without the genesis of any significant mineralization. This fact should be remembered when considering the evidence of the circulation of meteoric water through some porphyry copper deposits (Chapter 14).

As mentioned above there is wide acceptance of the hypothesis that the mineralizing fluids represent heated sea water that has leached metals from subjacent igneous rocks. It is therefore of value to note some of the arguments, not so far discussed, that the minority have advanced in favour of a magmatic-hydrothermal origin.

Most persuasive and perhaps influential has been Stanton (1978, 1986 and many other papers). He has argued that the ore–rock relations are the opposite of what might be expected from the circulating sea water hypothesis. He pointed out firstly that Cyprus-type deposits are small (usually < 5 Mt) and associated with basaltic successions of crustal dimensions whilst Kuroko (and Primitive) types range up to immense sizes (> 200 Mt) and are often associated with thin successions of andesitic to rhyolitic rocks, e.g. at Rio Tinto the thickness of the subjacent volcanics is in places less than 100 m (Williams *et al.* 1975) and indeed the volcanic-sedimentary complex underlying the many enormous orebodies of the 250 km long Iberian Pyrite Belt is only 50–800 m thick with much of the succession being sedimentary; secondly that basalts contain appproximately twice the amount of Cu–Zn–Pb as dacites; and thirdly that the basalts of the oceanic crust are accessible to leaching solutions over long periods compared with the andesites, dacites and rhyolites, which often are erupted in shallow water just before volcanism becomes subaerial.

Surely then, Stanton argues, the basalts, enormously larger in bulk, containing more base metal and exposed to leaching to a far greater degree than the andesite–rhyolite series, should have far larger and more numerous associated orebodies; but exactly the opposite is the case. He further argues that if the leaching hypothesis were correct, then Ni and Co should be as plentiful as Cu and Zn in basalt-associated orebodies as these elements occur in the same abundance in basalts and are leached out just as easily. The Sherlock Bay deposits of Western Australia (Keays *et al.* 1982) and the ores of Saskatchewan (Chapter 16) are impressive evidence that Ni can be mobilized within a hydrothermal medium and precipitated as a sulphide. Drawing attention to the fact that massive sulphide deposits occur in clusters rather than randomly and are normally confined to just one or two horizons, Stanton suggested that the ore-forming exhalations result from some relatively brief and clearly defined event in an underlying magma chamber, such as the attainment of a particular stage in crystallization, a sudden decrease in confining pressure with consequent degassing, abrupt ingress of external water or a combination of two or more of these.

Sawkins (1986) considered that the isotopic evidence supports the involvement of magmatic water in the formation of these deposits and he feels that we must have a hypothesis of origin that accounts for the narrow time–stratigraphical interval during which such deposits form. To satisfy all the evidence so far available he suggests a genetic model having many long-lived, magmatically driven convection cells of sea water, similar to that of Fig. 4.17, which are responsible for most of the fluid producing the hydrothermal alteration and to which additions of metal-rich magmatic solutions from the heat source occurred during a short time interval.

The most important factor in controlling the behaviour of exhalative hydrothermal solutions when they reach the sea floor is probably the density difference between the ore solution and sea water. Solomon & Walshe (1979) deduced that the ore solutions were buoyant on entering sea water (all the evidence from black smokers bears out their conclusion) and sometimes rose from the sea floor as conical plumes, especially in shallow water, distributing their load of sulphides over a wide area. As a result, massive ores may not necessarily be underlain by a feeder vent or stockwork. The location of ore deposition is, however, mainly controlled by the site of discharge of the hydrothermal solutions and this may lead to deposition in gravitationally unstable positions. The unconsolidated ores may therefore be reworked by submarine sliding, leading perhaps to turbidity current transportation and deposition. The common occurrence of graded fragmental ore is a characteristic feature of many Kuroko deposits (Sato 1977), and Thurlow (1977) and Badham (1978) have described the results of transportation by slumping and sliding at Buchans, Newfoundland, and Avoca, Ireland.

Hydraulic fracturing

All the hydrothermal processes that have been described so far in this chapter require the development of sufficient rock permeability to allow the mineralizing solutions to flow from their source to the site of mineral deposition. This permeability was achieved by rock fracturing on scales from the

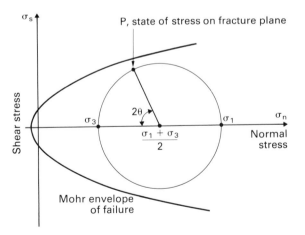

Fig. 4.18 Main features of a Mohr circle of stress and failure envelope. The point P represents the values of shear stress and normal stress on a plane making an angle θ with σ_1.

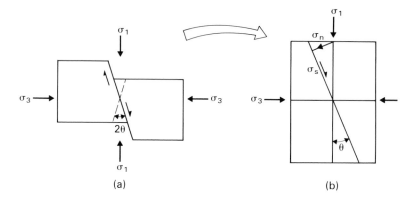

Fig. 4.19 (a) Section through a normal fault showing the angular relationship of the fault to the causal stresses. (b) The same fault illustrating the concepts of normal stress (σ_n) and shear stress (σ_s).

microscopic to that of major crustal faults. Much of the fracturing is due to the solutions themselves for it is now well known that fluids under high pressure can influence the mechanical behaviour of the crust dramatically and develop sites for precipitation of material held in solution. The fracturing of rocks by water under high pressure is known as hydrofracturing or hydraulic fracturing. The process is used artificially to increase the permeability in oil, gas and geothermal reservoirs. Important papers dealing with this process in connection with the formation of mineralized faults and breccia zones are those by Phillips (1972, 1986).

The Mohr envelope of failure

The state of stress on differently oriented planes in the same stress field can be depicted using Mohr diagrams, which are well explained in Davis (1984), Rowland (1986), Ramsay & Huber (1987) and other textbooks on structural geology to which the reader is referred for revision purposes! For recall purposes a Mohr diagram is shown in Fig. 4.18 for the normal fault depicted in Fig. 4.19a. σ_1 is the maximum stress and σ_3 the minimum. The stress acting on the fault plane can be resolved into the normal stress (σ_n) acting perpendicular to the plane and the shear stress (σ_s) acting parallel to it (Fig. 4.19b).

On the Mohr diagram we plot values of σ_n against σ_s. Thus if $2\theta = 0°$, then σ_s is clearly zero and when $2\theta = 90°$ it is at a maximum. Actual values fall on a circle whose centre is $(\sigma_1 - \sigma_3)/2$. For a plane of any given orientation, θ, the coordinates of the point P on this circle give the values of σ_n and σ_s. In the diagram as drawn failure *would not occur* as the values of σ_n and σ_s are too small for the circle to

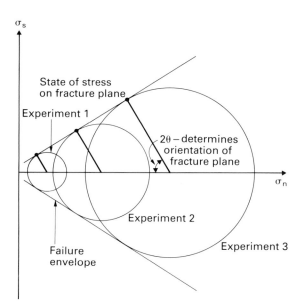

Fig. 4.20 Construction of a failure envelope using the data from three experiments in which fracturing is induced in identical rock samples by subjecting it to axial compression.

touch the failure envelope. It is when this happens that fracturing takes place. Failure envelopes for rocks can be determined by subjecting a cylindrical specimen to axial compression (σ_a) at different confining pressures (σ_c) and plotting the Mohr circles for the values of σ_a and σ_c at which failure occurs. As σ_c increases so does σ_a and Mohr circles become larger (Fig. 4.20). The tangents to these circles give us the Mohr envelope.

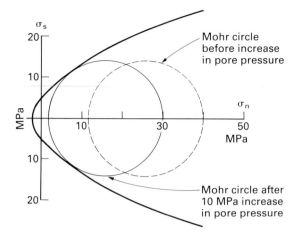

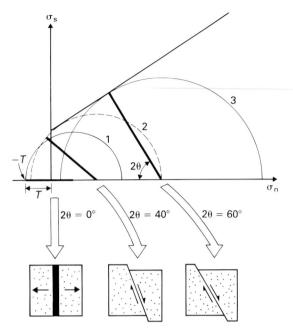

Fig. 4.21 Effect of pore pressure on brittle failure. An increase in pore pressure translates the Mohr circle to the left, it intersects the failure envelope and fracturing results.

Fig. 4.22 Mohr diagram with modified Griffith envelope and stress circles representing the following stress states: 1 = a stress circle with a stress difference of $4T$ producing a vertical fissure or group of tension gashes, 2 = a larger stress difference with a compressional shear fracture dipping at 70°, 3 = a much greater stress difference resulting in a shear fracture dipping at 60°. Note: the thick black lines are the circle's radii, the dip of each fracture is 90° – θ and T is the tensile strength of the rock. This diagram is drawn for a normal fault situation where σ_1, maximum compressive stress, is vertical.

The effect of pore pressure

Fluids in the pore spaces of rocks support some of the load that would otherwise be supported by the rock matrix. If we have a porous sandstone as depicted in Fig. 4.21 with $\sigma_1 = 40$ MPa and $\sigma_3 = 13$ MPa, and we add 10 MPa of pore pressure (P_f) to the rock, then the principal stresses σ_1 and σ_3 are lowered but the circle remains the same size; it simply moves to the left for a distance of 10 MPa. This causes it to intersect the failure envelope and fracturing will occur. Thus the effect of the pore fluid pressure is to decrease the *effective stress* acting on the rock. This is the phenomenon known as hydraulic fracturing, which is used routinely in the oil industry to create fractures and enhance permeability in reservoir rocks. Increased pore pressure can trigger earthquakes, and Rowland (1986) has summarized an experiment by the US Geological Survey in which a fault in Weber Sandstone was activated when water under pressure was injected into the ground to raise the pore pressure to 27 MPa. The earthquakes along the fault ceased when the pore pressure was reduced by 3.5 MPa.

Development of hydraulic fracturing

In 1920 Griffith suggested that the strength of any brittle material was controlled by the presence of small cracks. Viewed under the microscope rocks are seen to be full of such flaws. Griffith indicated that

when such an array of cracks is stressed, very high local stresses can build up which can lead to propagation of the microcracks, interconnections between cracks and eventually to the development of large discontinuous shear fractures. A mathematical analysis of the Griffith crack theory indicates that the Mohr envelope is parabolic, where T is the tensile strength of the material concerned (Fig. 4.22). As this diagram shows, when stress differences are very small then tensile fractures occur, 2θ decreases, the dip of the fracture(s) increases and at 2θ = 0° becomes vertical (Fig. 4.22). Tensile failure may only occur at relatively small stress differences. At stress differences greater than about $4T$ the stress circle is too large to intercept the failure envelope at $-T$ and compressional shear or shear extension fractures result. The frequent occurrence of mineral-filled extension fractures in regionally metamorphosed terranes suggests that high fluid pressures during prograde metamorphism continually produce fracturing in the tensile field (Cox *et al.* 1987).

Normal faults may be important loci of hydrothermal mineralization and frequently there are extensive volumes of mineralized breccia in the fault and the bordering rock, especially the hanging wall. Although it was often assumed that the fault movements created the breccias, their volume is often out of all proportion to the amount of movement on the fault zone and they usually consist of very angular fragments which are generally not in contact and do not appear to have been formed by grinding. Phillips (1972) suggested that these features indicate that the breccias are the result of hydraulic fracturing.

To understand the mechanism at work let us consider the situation at the top of a fault (Fig. 4.23a). If hydrothermal fluid flowed up the fault then, when the fault plane was fully permeated, fluid pressure at the top of the fracture would increase and set up a pressure gradient between this fluid and the pore water in the adjacent rocks. Hydrothermal fluid would then flow into the pore spaces of the adjacent rock, particularly the hanging wall. When the pore

pressure had built up to an amount that caused the stress circle to intersect the failure envelope as in Fig. 4.21, then abrupt fracturing would occur. The fault would be extended upwards and this new fracture would momentarily have such low fluid pressure on it that the wall rocks above, into which the hydrothermal solution had flowed at high pressure, would burst apart forming angular breccias (Fig. 4.23b). The effects of hydraulic fracturing and fault propagation are probably accompanied by seismic pumping (Sibson *et al.* 1975), which is certainly in operation along the underlying section of the fault zone. This is a process of dilatation and then collapse of multitudes of cracks in the wall rocks—a process which pumps fluid up faults every time that shear failure occurs.

If the stress difference in the above sequence of events dropped below $4T$ the fault would be extended as a vertical tension fracture or breccia zone (Fig. 4.23c). Later increases in stress might reestablish fault propagation so that a mineralized fault zone with important vertical branches of mineralization evolves.

Increases in pore pressure leading to hydraulic fracturing may well play a part in increasing permeability and allowing formation water to move up dip in sedimentary basins. One of the simplest situations in which high pore pressure regimes may develop is underneath low permeability zones, such as thick shale or evaporite sequences (Fig. 4.24). High pore pressure in such regions is often encountered during oil drilling in sedimentary basins. With such a situation upward extension of fractures might be very limited or negligible if the overlying impermeable barrier deformed by flow because it was more

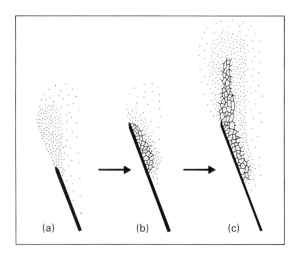

Fig. 4.23 Propagation of normal faults and accompanying development of breccia zones. (a) Hydrothermal solution saturates fault zone and permeates into the hanging wall. (b) Sudden fracturing extends fault upwards and a breccia zone forms in the hanging wall, with further similar extensions and brecciation a large breccia zone may form. (c) Vertical hydraulic fracturing occurs when stress difference is $< 4T$. (After Phillips 1972.)

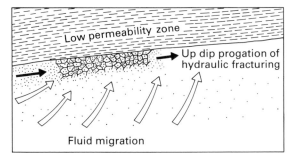

Fig. 4.24 Development of hydraulic fracturing in a sedimentary basin. Fluids accumulate beneath an impermeable zone, e.g. shale or evaporite sequence, pore pressure builds up and hydraulic fracturing is initiated and propagated up dip.

ductile than the underlying sandstones and limestones. If the latter rocks fractured in brittle fashion then a zone of hydraulic fracturing might be steadily propagated up dip. Fluid concentration and hence pore pressure may increase within sedimentary basins as the result of a number of mechanisms, some or all of which might be coeval. Rapid burial can trap abnormal amounts of water in the sediments; closure of pore spaces in one site by tectonic deformation may cause pore fluid to migrate along the tectonic gradient to produce high pore pressure in another site; aquathermal pressuring, which results when a fluid containing dissolved material is heated, the rise in temperature in this case being due to burial; and lastly mineral phase changes, such as the regional transformation of montmorillonite to illite and gypsum to anhydrite, which can release large volumes of water.

The effects of hydraulic fracturing have been reported from a number of sedimentary basins where they have given rise to horizontal bands of breccia with shale fragments in carbonate and other diagenetic cements (Roehl 1981, Lindgren 1985).

In the Kupferschiefer of central Europe (see Chapter 15), i.e. the copper–silver-bearing ore shales of south-western Poland, crowds of horizontal and vertical gypsum, calcite and base metal veinlets are present. These appear to be of late diagenetic origin and Jowett (1987) suggested that they were formed by hydraulic fracturing resulting from high pore pressures. These pressures were built up by aquathermal pressuring, generation of water, CO_2 and CH_4 from the coaly material in the Kupferschiefer and over-pressuring of pore waters trapped during sedimentation. These fluids were not able to escape in the course of normal dewatering, because the Kupferschiefer and the underlying rocks are effectively capped by the thick Zechstein evaporites—the same caprock that led to the formation of the extensive gas deposits of the southern North Sea, Holland and Germany.

As might be predicted from the above discussions hydraulic fracturing plays an important role in the escape of the end-magmatic fluids from consolidating intrusions. This leads to the formation of veins including sheeted vein systems, which may host economic Sn–W deposits, and other deposit types. A succinct description of these and their modes of formation can be found in Plimer (1987a).

5 / Geothermometry, geobarometry, paragenetic sequence, zoning and dating of ore deposits

Geothermometry and geobarometry

Ores are deposited at temperatures and pressures ranging from very high, at deep crustal levels, to atmospheric, at the surface. Some pegmatites and magmatic segregation deposits have formed at temperatures around 1000°C and under many kilometres of overlying rock, whilst placer deposits and sedimentary ores have formed under surface conditions. Most orebodies were deposited between these two extremes. Clearly, knowledge of the temperatures and pressures obtaining during the precipitation of the various minerals will be invaluable in assessing their probable mode of genesis and such knowledge also will be of great value in formulating exploration programmes. In this small volume it is only possible to touch on a few of the methods that can be used.

Fluid inclusions

The nature of fluid inclusions and the principle of this method have been outlined in Chapter 3. Clearly, it is the primary inclusions that must be examined. Secondary inclusions produced after the mineral was deposited and commonly formed by the healing of fractures will not give us data on the mineral depositional conditions, but their study can be very important in the investigation of certain deposits, e.g. porphyry coppers (Chapter 14).

The filling (homogenization) temperatures of aqueous inclusions (i.e. the temperature at which the inclusion becomes a single phase fluid) indicate the depositional temperature of the enclosing mineral if a correction can be made for the confining pressure and salinity of the fluid. The salinity, in terms of equivalent weight per cent NaCl, can be determined by studying the depression of the freezing point using a freezing stage. Frequently, confining pressures have to be estimated by reconstructing the stratigraphical and structural succession above the point of mineral deposition. This clearly leads to a degree of uncertainty. Pressure corrections increase the homogenization temperatures obtained in the laboratory, so these are still of great value even when

uncorrected as they record minimum temperatures of deposition and can be used to map temperature gradients along mineral deposits. Where CO_2-rich and H_2O-rich inclusions coexist, the pressure can be estimated from the filling temperature of the aqueous inclusions and the density of the CO_2 inclusions (Groves & Solomon 1969). If the fluid inclusion assemblage indicates boiling at the time of trapping then the depth of formation can be estimated (Haas 1971). However, if boiling did not occur, then the confining pressure must have exceeded the vapour pressure of the fluid, which can be calculated from the salinity and temperature data, giving us a minimum pressure value. Studies on the use of fluid inclusions in geobarometry were reviewed by Roedder & Bodnar (1980).

Inversion points

Some natural substances exist in various mineral forms (polymorphs) and some use of their inversion temperatures can be made in geothermometry. For example, β-quartz inverts to α-quartz with falling temperature at 573°C. We can determine whether quartz originally crystallized as β-quartz by etching with hydrofluoric acid and thus decide whether it was deposited above or below 573°C. Examples among ore minerals include:

$$177°C$$
acanthite (monoclinic) $\rightleftharpoons$ argentite (cubic),

$$104°C$$
orthorhombic chalcocite $\rightleftharpoons$ hexagonal chalcocite.

This second low inversion temperature is important in distinguishing hypogene chalcocite from low temperature, near-surface supergene chalcocite, a distinction which can be of great economic importance in evaluating many copper deposits, for if the near surface ore is only just of economic grade and much of the mineralization is supergene, then the ore below the zone of supergene enrichment may be uneconomic. Fortunately, the distinction can be made, for relict cleavage remains after chalcocite has inverted from the high temperature hexagonal form, and this cleavage is revealed by etching polished sections with nitric acid. In addition transformation (inversion) lamellae may be seen using oil immersion and crossed polars. A comprehensive list of invariant points (including inversion points) is included in Barton & Skinner (1979).

Exsolution textures

As a result of restricted solid solution at lower temperatures between various pairs of oxide and sulphide minerals, exsolution bodies of the minor phase segregate from the host solid solution on cooling (Fig. 5.1). Their presence indicates the former existence of a solid solution of the two minerals which was deposited at an elevated temperature. An idea of that temperature can be obtained by reference to laboratory work on the sulphide system concerned (Edwards 1960), or by resolution of natural exsolution bodies in the host grain by heating samples in the laboratory. An example of the latter approach was reported by Edwards & Lyon (1957) who performed resolution experiments on samples from the Aberfoyle tin mine, Tasmania. After the geologically short annealing time of one week they obtained the following temperatures for the onset of resolution:

chalcopyrite and stannite in sphalerite	550°C
sphalerite in chalcopyrite	400°C
stannite in chalcopyrite	475°C
sphalerite in stannite	325°C
chalcopyrite in stannite	400–475°C

These results, they suggested, indicated a temperature of about 600°C for the formation of the original solid solutions. There are much data in the literature suggesting that with longer annealing times lower temperatures would have been obtained and the original solid solutions might have been deposited at temperatures in the range 400–500°C. This range would agree better with the homogenization temperatures of around 400°C obtained by Groves *et al.* (1970) on fluid inclusions in cassiterites from this deposit.

Sulphide systems

The initial studies of these systems were carried out to discover those which would provide data on the temperature and pressure of ore deposition. The first results, e.g. the sphalerite geothermometer, were very promising. Further study, however, revealed that the formation of mineral assemblages and variations in mineral composition depend on many

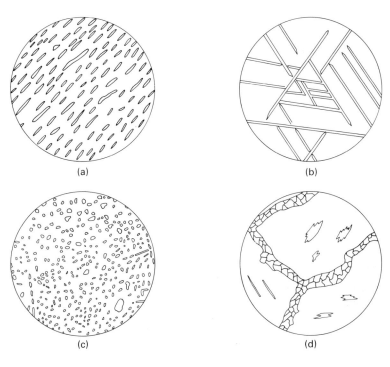

Fig. 5.1 Exsolution textures in oxide and sulphide mineral systems. (a) Seriate exsolution bodies of hematite-rich material in an ilmenite-rich base, × 263, Tellnes, Norway. (b) Exsolution lamellae of ilmenite in magnetite, × 117, Sudbury, Ontario. (c) Emulsoid exsolution bodies of chalcopyrite in sphalerite, * × 70, Geevor mine, Cornwall. (d) Rim or net exsolution texture formed by the exsolution of pentlandite from pyrrhotite, × 470, Sudbury, Ontario. *Most workers would now agree that this density of small chalcopyrite bodies is too high to have resulted entirely from exsolution. A minor amount of chalcopyrite may have been in solid solution above ~500°C. The texture is of economic importance as it illustrates how fine-grained chalcopyrite, in bodies too small to be liberated during grinding, can be locked into the sphalerite, thus giving rise to a copper loss which passes into the zinc concentrate.

(a)   (b)   (c)   (d)

factors. Thus, Kullerud (1953) suggested that the FeS content of sphalerite gave a direct measurement of its temperature of deposition. Later, Barton & Toulmin (1963) showed that the fugacity of sulphur was also an important control but unfortunately it is often an unknown quantity.

More research into this system, however, has produced some rewards. Although the work of Scott & Barnes (1971) indicated that the composition of sphalerite coexisting with pyrrhotite and pyrite over the range 525–250°C is essentially constant (Fig. 5.2) their microprobe studies did reveal a possible geothermometer. They discovered metastable iron-rich patches within sphalerites whose composition, compared with their matrix, appeared to be constant and temperature dependent. This applies only to coexisting sphalerite–pyrrhotite–pyrite assemblages. Iron-rich patches have been found in natural

sphalerites from veins but they do not appear to be present in metamorphosed sphalerite because of annealing and re-equilibration.

Scott & Barnes (1971) and Scott (1973) indicated that the iron content of sphalerite *in equilibrium* with pyrite and pyrrhotite, although not temperature dependent below 525°C (Fig. 5.2), is strongly controlled by pressure above 300°C (Fig. 5.3). Thus, sphalerite can be used as a geobarometer provided equilibrium has been reached. This is most likely to be the case in metamorphosed ores, and Lusk *et al.* (1975) have described its use in just such a case. Further refinement of this geobarometer can be found in Hutchinson & Scott (1981) and Bryndzia *et al.* (1990), who used it to show that the ores at Broken Hill, NSW were metamorphosed at a pressure of 0.58 GPa

This brief discussion must suffice to illustrate the use of sulphide systems in geothermometry and geobarometry. Excellent summaries of experimental techniques and problems are provided by Barton & Skinner (1979) and Scott (1974). A list of the sulphide systems studied experimentally up to 1974 is given by Craig & Scott (1974) and the use of arsenopyrite as a geothermometer is discussed by

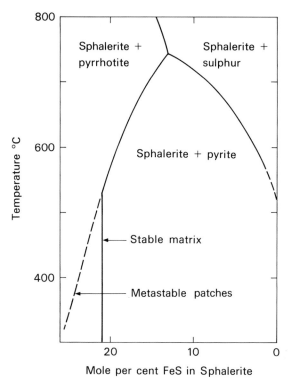

Fig. 5.2 Temperature–composition projection of a portion of the system ZnS–FeS–S showing the constant composition of sphalerite coexisting with pyrite and pyrrhotite over the range 525–300°C (vertical line), and the composition of metastable patches of iron-rich sphalerite and their variation in composition with temperature. (After Scott & Barnes 1971.)

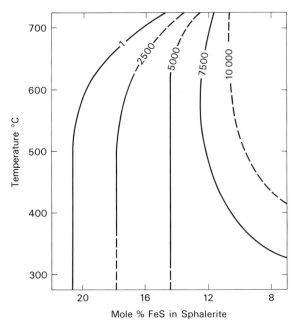

Fig. 5.3 Temperature–composition projection of the sphalerite + pyrite + hexagonal pyrrhotite solvus isobars. Pressures are in bars. Solid lines are measured isobars, pecked lines are extrapolations. (After Scott 1973.)

Kretschmar & Scott (1976) but Sharp *et al.* (1985) considered that its applications are very limited. However, Touray *et al.* (1989) used this geo-thermometer on the Le Bourneix gold deposit of central France and obtained depositional temperatures of $400° \pm 30°C$ for the arsenopyrite, which is in very good agreement with the temperatures they obtained using fluid inclusions in associated minerals. There is also considerable experimental work on non-sulphide systems which is applicable to orebodies, particularly their gangue minerals (Levin *et al.* 1969). For carbonate composition variation with temperature see Goldsmith & Newton (1969).

A very interesting recent development is the Ga/Ge-geothermometer using sphalerite (Möller 1985). This can be used to determine temperatures in the source regions of ore solutions and to estimate the degree of mixing of hot parental ore fluids with cool, near surface waters. Möller's first applications of this method appear to be very promising and have shown that for the vein mineralization of Bad Grund and Andreasberg the ore fluid at its time of genesis had a temperature above $220°C$, but must have mixed with waters of less than $130°C$ at the level of mineral deposition. However, at Ramsbeck there is no difference between the temperature deduced by Ga/Ge-geothermometry for the source region and the fluid inclusion homogenization temperatures, indicating that no significant fluid mixing occurred.

For the sediment-hosted ore deposits of various types investigated by Möller, source fluid temperatures of $180–270°C$ were obtained and, on the assumption that these fluids were generated by sediment dewatering in sedimentary basins, this indicates a source depth of about 6–11 km, if geothermal gradients of $25–30°C \ km^{-1}$ are assumed.

Several empirical chemical geothermometers based on the Na, K and Ca contents of modern geothermal brines, which can be used to determine the temperatures in the fluid source region, have been reviewed by Truesdell (1984). Variations in alkali ratios in these brines appear to be controlled by only a small number of fluid–mineral equilibria, particularly those involving feldspars. Graphs relating alkali ratios and temperature are given by Truesdell. Rankin & Graham (1988) extended the use of this method from geothermal waters to the hydrothermal solutions in fluid inclusions. They worked on specimens of fluorite, calcite and quartz from the Mississippi Valley-type deposits of the Alston Block of the Northern Pennine Orefield of England and obtained temperatures for the source region, considered to be a nearby sedimentary basin, of $150–250°C$. These temperatures, as might be expected, are about $50°C$ higher than the temperatures of deposition of the fluorite. The source region temperatures are interestingly very similar to those obtained by Möller.

Stable isotope studies

In Chapter 4, when discussing the uses of hydrogen and oxygen isotopes to determine the origin of waters that had reacted with minerals, it was pointed out that if we know the temperature of reaction we can determine the isotopic composition of the water. Similarly, given the isotopic compositions of cogenetic minerals and water, we can determine the temperature. This is the principle of the use of oxygen isotopes on gangue minerals as a geo-thermometer. A method used much more extensively for ore deposits has been the sulphur isotope geothermometer.

Sulphur has four isotopes, but the major geological interest is in the variation $^{34}S/^{32}S$ because this varies significantly in the minerals of hydrothermal deposits for a number of reasons:

1 the fractionation of the isotopes between individual sulphide and sulphate minerals in the ore varies with the temperature of deposition;

2 the initial isotopic ratio is controlled by the source of the sulphur, e.g. mantle, crust, sea water, etc;

3 the variable proportions of oxidized and reduced sulphur species in solution—in its simplest form the H_2S/SO_4^{2-} ratio. As this ratio depends on temperature, pH and f_{O_2}, we can also evaluate the variation of sulphur isotopic values in terms of T, pH and f_{O_2} (Rye & Ohmoto 1974).

Variations in $^{34}S/^{32}S$ are expressed in delta notation ($\delta^{34}S$) where:

$$\delta^{34}S = \left(\frac{R_{34 \ (sample)}}{R_{34 \ (standard)}} - 1 \right) \times 1000.$$

Point (1) above implies that sulphides and sulphates in hydrothermal deposits will show a variation (fractionation) in $\delta^{34}S$ values from mineral to mineral which is dependent on their temperature of deposition, provided the minerals crystallized in equilibrium with each other. This fractionation is found in nature and its extent is illustrated in Fig. 5.4. The curves in this figure are based on experimental and theoretical data. They show that sulphur

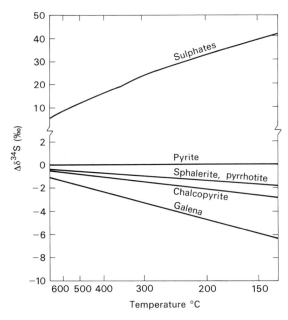

Fig. 5.4 Sulphur isotope fractionation between some hydrothermal minerals plotted with respect to pyrite. (Modified from Rye & Ohmoto 1974.)

isotopic fractionation at 200°C between pyrite and galena is about 4.6‰ and between sphalerite and galena is a little over 3‰. Thus, these and other mineral pairs can be used in sulphur isotope geothermometry.

Depositional temperatures obtained by fluid inclusion methods on transparent minerals, sulphur isotope methods on sulphides intergrown with them, and oxygen isotopic compositions of coexisting oxides, generally give results in close agreement, as in the work on the Echo Bay uranium–nickel–cobalt–silver–copper deposits by Robinson & Ohmoto (1973). This paper is a good illustration of the use of these methods in obtaining temperatures of mineral deposition.

Paragenetic sequence and zoning

The time sequence of deposition of minerals in a rock or mineral deposit is known as its paragenetic sequence. If the minerals show a spatial distribution then this is known as zoning. The paragenetic sequence is determined from studying such structures in deposits as crustiform banding and from the microscopic observation of textures in polished sections.

Paragenetic sequence

Abundant evidence has been accumulated from world-wide studies of epigenetic hydrothermal deposits indicating that there is a general order of deposition of minerals in these deposits. Exceptions and reversals are known but not in sufficient number to suggest that anything other than a common order of deposition is generally the case. A simplified, general paragenetic sequence is as follows:
1 silicates;
2 magnetite, ilmenite, hematite;
3 cassiterite, wolframite, molybdenite;
4 pyrrhotite, löllingite, arsenopyrite, pyrite, cobalt and nickel arsenides;
5 chalcopyrite, bornite, sphalerite;
6 galena, tetrahedrite, lead sulphosalts, tellurides, cinnabar.
Of course, not all these minerals are necessarily present in any one deposit and the above list has been drawn up from evidence from a great number of orebodies.

Zoning

Zones may be defined by changes in the mineralogy of ore or gangue minerals or both, by changes in the percentage of metals present, or by more subtle changes from place to place in an orebody or mineralized district of the ratios between certain elements or even the isotopic ratios within one element. Zoning was first described from epigenetic vein deposits but it is present also in other types of deposit. For example, syngenetic deposits may show zoning parallel to a former shore line, as is the case with the iron ores of the Mesabi Range, Minnesota; alluvial deposits may show zoning along the course of a river leading from the source area; some exhalative syngenetic sulphide deposits show a marked zonation of their metals and skarn deposits often show a zoning running parallel to the igneous–sedimentary contact. In this discussion, attention will be focused on the zoning of epigenetic hydrothermal, exhalative syngenetic and sedimentary syngenetic sulphide deposits.

Epigenetic hydrothermal zoning

Zoning of this type can be divided into three intergradational classes; regional, district and orebody zoning (Park & MacDiarmid 1975). Regional zoning occurs on a very large scale, often corre-

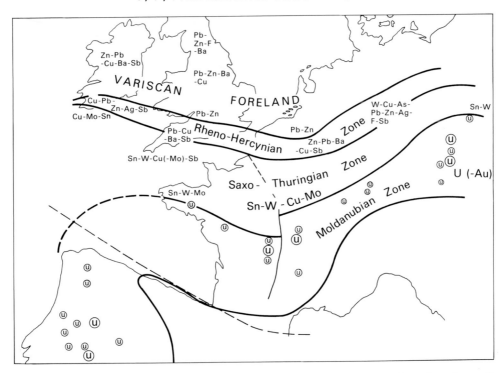

Fig. 5.5 Regional metal zonation of epigenetic deposits in the Variscan Metallogenic Province of north-west Europe. (After Evans 1976a and Cuney 1978.) Sizes of symbols in Moldanubian Zone and Spain indicate relative sizes of uranium deposits. Note: the zonal boundaries in this figure are broadly in agreement with Pouba & Ilavský (1986) and Chaloupský (1989), but other authors, e.g. Derré (1982) and Holder & Leveridge (1986), place the southern boundary of the Rheno-Hercynian Zone immediately south of Cornwall. This would place the Cornish granites just outside the zone containing the Cu–Mo–Sn-bearing granites stretching from the Erzgebirge in Bohemia to Brittany (north-west France). These Carboniferous S-type granitoids have similar compositions to the S-type granitoids of the Moldanubian Zone (Finger & Steyrer 1990).

sponding to large sections of orogenic belts and their foreland (Fig. 5.5). A number of examples of zoning on this scale were described from the circum-Pacific orogenic belts by Radkevich (1972). Some regional zoning of this type, e.g. the Andes, appears to be related to the depth of the underlying Benioff Zone, which suggests a deep level origin for the metals as well as the associated magmas (Chapter 23). District zoning is the zoning seen in individual orefields such as Cornwall, England (Fig. 5.6 and Table 5.1), Flat River, Missouri (Fig. 5.7) and the Copper Canyon area, Nevada (Fig. 13.10). Zoning of this type is most clearly displayed where the mineralization is of considerable vertical extent and was formed at depth where changes in the pressure and temperature gradients were very gradual. If deposition took place near to the surface, then steep temperature gradients may have caused superimposition of what

would, at deeper levels, be distinct zones, thus giving rise to the effect known as telescoping. Some geothermal systems also show metal zoning, e.g. Broadlands, NZ where there is Sb, Au and Tl enrichment near the surface and Pb, Zn, Ag, Cu, etc. at depth (Ewers & Keays 1977, Krupp & Seward 1990).

In Cornwall, Dines (1956) showed that the zonal distribution is by no means always as simply related to the granite outcrops as is depicted in Fig. 5.6. The centres of some districts with zoned mineralization occur within granite outcrops but others lie within the country rocks. Dines called all these foci 'emanative centres'—each centre is elliptical in plan with a long axis of up to about 4 km and a tendency for the Pb–Zn zone, being associated with the cross-courses (Table 5.1), to be elongated at right angles to the other zones. More recent work has shown that

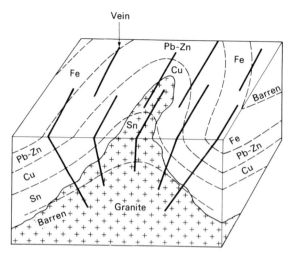

Fig. 5.6 Diagrammatic illustration of district zoning in the orefield of south-west England showing the relationship of the zonal boundaries to the granite–metasediment contact. (After Hosking 1951.)

although these relationships are true in many areas, in some this zoning is not evident and in others it has been created by different phases of mineralization. Sometimes there is a barren or uneconomic zone of tens of metres between the copper and tin zones and

at other times they overlap, as at the Dolcoath Mine (p. 214). In some deposits reactivation of veins has led to a complex assemblage of economic amounts of Sn, Cu and Zn, as at Wheal Jane. (Wheal is the Cornish, Celtic, word for mine.) Despite these exceptions, district zoning in Cornwall and many other epigenetic orefields appears to be a valid concept.

Orebody zoning takes the form of changes in the mineralization within a single orebody. A good example occurs in the Emperor Gold Mine, Fiji, where vertical zoning of gold–silver tellurides in one of the main ore shoots gives rise to an increase in the Ag : Au ratio with depth (Forsythe 1971). A comprehensive discussion of zoning with many examples is given in Routhier (1963).

Syngenetic hydrothermal zoning

This is the zoning found in stratiform sulphide bodies principally of volcanic affiliation (Chapter 2). These deposits are frequently underlain by stockwork deposits and many of them appear to have been formed from hydrothermal solutions that reached the sea floor. As Barnes (1975) has pointed out, the zonal sequence is not always clearly seen in this type of ore deposit, but where it has been

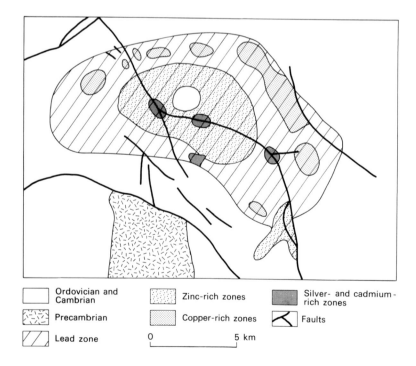

	Ordovician and Cambrian		Zinc-rich zones		Silver- and cadmium-rich zones
	Precambrian		Copper-rich zones		Faults
	Lead zone				

0   5 km

Fig. 5.7 District zoning in the Flat River area of the Old Lead Belt of south-east Missouri. (After Heyl 1969.)

Table 5.1 A generalized mineral paragenesis of the mineral deposits of south-west England. (After El Shazly *et al.* 1957 and Stone & Exley 1986.)

Wall rock alteration	Gangue minerals	Zone	Ore minerals[a]		Economically important elements
		7	Barren (pyrite) Hematite Stibnite. Jamesonite	*Mesothermal and epithermal veins* Generally at right-angles to granite ridges (cross courses)	Fe Sb
		6	Tetrahedrite Bournonite. Pyrargyrite? Siderite. Sphalerite Pyrite (marcasite)		
		5	Argenite. Galena Sphalerite Pitchblende. Niccolite Smaltite Cobaltite (Native bismuth—bismuthinite?)		Ag Pb Zn U Ni Co Bi
		4	Chalcopyrite Sphalerite Wolframite, scheelite Arsenopyrite Pyrite Sphalerite Chalcopyrite, stannite	*Hypothermal veins* Generally parallel to granite ridges and dykes	Cu
		3	Wolframite, scheelite Arsenopyrite, pyrite Cassiterite (wood tin)		
		2	Wolframite (scheelite) Arsenopyrite Molybdenite Cassiterite		Sn W As
		1	Cassiterite. Specularite Wolframite		
		b	Arsenopyrite. Stannite Wolframite, Cassiterite Molybdenite	Veins often in granite cusps	
		c	Arsenopyrite. Wolframite Cassiterite Molybdenite		

Wall rock alteration (vertical ranges): Greisenization — Tourmalinization — Silicification — Feldspathization — Chloritization — Hematitization

Gangue minerals (vertical ranges): Quartz — Felspar — Mica — Tourmaline — Chlorite — Hematite — Flourite — Chalcedony — Dolomite — Baryte — Calcite

[a] Latest minerals at top through to earliest minerals at base of column.
[b] Greisen bordered veins, sheeted veins, some stockworks.
[c] Pegmatites.

established, it is identical to that of epigenetic hydrothermal ores. This zonal sequence is well exemplified by the Precambrian deposits of the Canadian Shield, by the Devonian Rammelsberg and Meggen deposits of Germany and the Kuroko deposits of Japan. In these deposits there is a general sequence upwards and outwards through the ore-bodies of Fe → (Sn) → Cu → Zn → Pb → Ag → Ba, which should be compared with the similar zoning shown in Table 5.1. As in vein deposits, the zonal boundaries are normally gradational with frequent overlap of zones. The cause and evolution of this zoning are discussed on pp. 73–75.

Sedimentary syngenetic sulphide zoning

This zoning is found in stratiform sulphide deposits of sedimentary affiliation usually of wide regional development such as the Permian Kupferschiefer of Germany and Poland, and the Zambian Copperbelt deposits. Underlying stockwork feeder channelways are rarely known beneath these deposits and usually they have formed in euxinic environments. The zoning appears to show a relationship to the palaeogeography, and proceeding basinwards through a deposit it takes the form of Cu + Ag → Pb → Zn. In the Zambian Copperbelt, however, the zoning is principally one of copper minerals and pyrite, as lead, zinc and silver are virtually absent.

Different patterns of *orebody* zoning are seen in the lead–zinc sediment-hosted deposits (Chapter 15), which Plimer (1986) has divided into the Broken Hill and Mount Isa types. At Broken Hill the stack of orebodies shows a vertical zonation from a copper zone, in the lowest orebody, upwards to Pb–Zn orebodies, which themselves have an upward and outward decrease in Cu and Zn and an increase in Pb, Ag, Mn and F. A barium-rich horizon overlies these orebodies. Similar zoning is found in other deposits of this type. Plimer points out that this zoning strongly resembles that of volcanic-hosted massive sulphide deposits described in the foregoing section, and, like the zoning in these, it shows a spatial relationship to the feeder stockwork zone(s). By contrast, Plimer contended that Mount Isa type deposits have their highest copper content in the middle or upper lenses of sedimentary sequences that contain numerous horizons of Pb–Zn mineralization over hundreds of metres of stratigraphy. On the other hand the work of McGoldrick & Keays (1990) suggested that a very similar spatial relationship obtains between the Cu and Pb–Zn mineralization at Mount Isa to that at Broken Hill.

Sulphide precipitation, paragenetic sequence and zoning

In Chapter 4 it was pointed out that there are two principal schools of thought concerning the transport of metal ions in hydrothermal solutions. The first favours transportation as bisulphide complexes, and the second as complex chloride ions in chloride-rich, sulphide-poor brines. Barnes (1975) has shown how the first hypothesis can account for the zoning seen in epigenetic and syngenetic hydrothermal deposits. Clearly, the relative stabilities of complex metal bisulphide ions will control their relative times of precipitation and hence both the resulting paragenetic sequence and any zoning which may be developed. Barnes has calculated these stabilities, which are shown in Table 5.2. The data in this table suggest that iron and tin would be precipitated early in the paragenetic sequence and would be present in the lowest zone of a zoned deposit, whilst silver and mercury would be late precipitates which would travel furthest from the source of the mineralizing solutions (cf. Table 5.1). Precipitation itself would be occasioned by the interplay of a number of factors including changes in pH due to wall rock alteration reactions, decrease in temperature with distance from source, reaction with carbonaceous materials in wall rocks, mixing with meteoric waters, and so on.

Syngenetic sedimentary sulphide deposits may have been precipitated from chloride-rich brines, such as those found near the Salton Sea (Chapter 4), if these brines encountered H_2S or HS^- supplied by a large volume of marine, euxinic water. On the other hand, they may have been introduced into a basin of deposition in ionic solution by rivers. It is, therefore, pertinent to look at the relative stabilities of metal sulphides that show a zonal distribution in such deposits in the presence and absence of chloride ions. Some relative solubilities are shown in Tables 5.3 and 5.4 (drawn from data in Barnes 1975).

Table 5.2 Predicted sequence of stabilities of bisulphide complexes (in kilocalories)

Least soluble..Most soluble							
Fe	Ni	Sn	Zn	Cu	Pb	Ag	Hg
79	84	126	132	135	153	157	226

Table 5.3 Relative stabilities of sulphides in chloride solutions (as expressed by the equilibrium constant for the reaction: $MeCl_2(aq) + HS^- \rightarrow MeS(s) + H^+ + Cl^-$)

Least soluble		Most soluble
CuS	PbS	ZnS
At 25°C 38.4	28.2	27.7

Table 5.4 Relative stabilities of sulphides in ionic solutions (as expressed by the equilibrium constant for the reaction: $Me^{2+} + HS^- \rightarrow MeS(s) + H^+$)

Least soluble		Most soluble
$CuFeS_2$	PbS	ZnS
23.7	20.7	19.2

From these tables it can be seen that there is excellent agreement between the zonal arrangement of metals in syngenetic sedimentary deposits and the relative solubilities shown. Barnes felt that this suggests that the metals travelled in ionic solution or as chloride complexes. It must be noted, however, that studies of the transport of base metals in surface streams at the present day show that only a small proportion of the base metal contents are carried in ionic solution, which means that the above picture is a somewhat simplified version of the true situation. In addition, faunal (including bacterial) and facies factors can influence zonation significantly. Zoning, however, cannot be used to distinguish between the products of these two types of solution, because chloride complexing up to at least 100°C is not sufficiently strong to change significantly the normal order of precipitation of sulphides from that found for simple metallic ions. It might be thought from the work of Barnes that the nature of the zoning in hydrothermal deposits points to deposition from sulphide complexes, and that chloride complexes or ionic solutions may play a significant role as metal-transporting agents in the formation of sedimentary sulphide deposits. Unfortunately, the situation is not as simple as this. Different metals may be travelling as chloride and bisulphide complexes in the same solution and above 300°C chloride complexing may be dominant (Chapter 4). This leaves us with an unresolved problem—probably attributable to inadequate experimental data on ore mineral solvation reactions and the high temperature molecular properties of aqueous metal complexes (Susak & Crerar 1982).

The dating of ore deposits

When orebodies form part of a stratigraphical succession, such as the Mesozoic ironstones of north-western Europe, their age is not in dispute. Similarly the ages of orthomagmatic deposits may be fixed almost as certainly if their parent pluton can be well dated. On the other hand, epigenetic deposits may be very difficult to date, especially as there is now abundant evidence that many of them may have resulted from polyphase mineralization, with epochs of mineralization being separated by intervals in excess of 100 Ma. There are three main lines of evidence that can be used: the field data, radiometric and palaeomagnetic age determinations. The field evidence may not be very exact. It may take the form of an unconformity cutting a vein and thus giving a minimum age for the mineralization. Similarly, if mineralization occurs in rocks reaching up to Cretaceous in age we know that some of it is at least as young as the Cretaceous, but we cannot tell when the first mineralization occurred or how young some of it may be.

Radiometric dating

Sometimes a mineral clearly related to a mineralization episode will permit a direct dating, e.g. uraninite in a vein, but more commonly it is necessary to use wall rock alteration products such as micas, feldspar and clay minerals. In the latter case, the underlying assumption is that the wall rock alteration is coeval with some of or all the mineralization. Brownlow (1979) gives a simple summary of the theory of how ages are deduced from the radioactive decay of an unstable parent isotope to a more stable isotope of a different element (daughter) and it is assumed here that the reader is familiar with the principles of the theory. Time is but part of the story. Almost as a by-product, the isochron method also reveals the isotopic composition of the lead and strontium that were present when a rock or orebody formed; information which may point to a specific source for the element concerned. Table 5.5 summarizes the parent–daughter relationships that are of major interest for dating rocks and ores. Perusal of this table will show that in dating mineralization by these methods we are heavily dependent on the products of wall rock alteration and the assumption of its contemporaneity with the associated mineralization.

It also is important to note that our radiometric

Table 5.5 Major isotopic dating methods

Method	Generalized decay scheme	Half-life of radioactive isotope (years)	Materials that can be dated
U–Pb	$^{238}U \rightarrow {}^{206}Pb$	4.5×10^9	Uraninite, pitchblende, zircon, sphene, apatite, epidote, whole rock
Rb–Sr	$^{87}Rb \rightarrow {}^{87}Sr$	5.0×10^{10}	Mica, feldspar, amphibole, whole rock
K–Ar	$^{40}K \rightarrow {}^{40}Ar$	1.31×10^9	Mica, feldspar, amphibole, pyroxene, whole rock

clocks can be reset. Broken Hill, New South Wales, gives us a dramatic example of this phenomenon. The lead–zinc–silver orebodies of Broken Hill are generally held to be regionally metamorphosed syngenetic deposits. Rb–Sr and K–Ar ages of biotites (450–500 Ma), however, markedly post-date the 1700 Ma of the dominant metamorphism in the country rocks (Shaw 1968, Richards & Pidgeon 1983). The heating event reflected by the mica ages may have been associated with formation of the nearby Thackeringa-type orebodies.

Porphyry coppers in many parts of the world have been dated using Rb–Sr and K–Ar methods. The results of this work have shown (a) that it is usually the youngest intrusions in any particular area that carry mineralization, and (b) that there is a close temporal relationship between porphyry copper mineralization and associated calc-alkaline magmatism. Thus at Panguna, Papua New Guinea, the mineralization dates at 3.4 ± 0.3 Ma, compared with intrusives 4–5 Ma old. At Ok Tedi, north-west Papua, alteration as young as 1.1–1.2 Ma has been found (Page & McDougall 1972). This contrasts with the oldest mineralization yet found, a banded iron formation with associated stratabound copper sulphides at Isua, West Greenland, dated at 3760 ± 70 Ma (Moorbath et al. 1973, Appel 1979).

Another interesting but very different example of the use of Rb–Sr dating is that of the work of Bornhorst et al. (1988) on the famous native copper mineralization in late Proterozoic basalt flows in the Keweenaw Peninsula of Michigan (see p. 37). Using the Rb–Sr method these workers showed that the amygdule-filling minerals (microcline, calcite, epidote, chlorite) with which the native copper is intergrown have ages lying between 1060 and 1047 Ma (± 20 Ma). These dates are supported by a fission track age on epidote of 1044 ± 169 Ma and are appreciably younger than the peak of eruptive activity of the host basalts—1086–1098 Ma (a recent refined age is 1095 ± 1.3 Ma). There is thus an age

difference of about 15–70 Ma between the end of the volcanism and the hydrothermal mineralization, which these authors feel militates against a hypothesis of magmatically derived hydrothermal fluids being the mineralizers.

The use of $^{40}Ar/^{39}Ar$ age spectrum data (a dating technique developed from the K–Ar method) is also making important contributions in dating epochs and phases of mineralization. For example there was for many years a controversy as to whether the age of the Au–Ag veins of the Buffalo Hump District of central Idaho is Cretaceous or Tertiary. As other deposits in Central Idaho are hosted by Tertiary igneous rocks some workers believed all these deposits to be Tertiary. Using $^{40}Ar/^{39}Ar$ age spectrum data, Lund et al. (1986) demonstrated the Cretaceous age of the veins in the Buffalo Hump District and the occurrence of at least two precious-metal mineralizing episodes in central Idaho. Snee et al. (1988) showed what a powerful tool this technique can be in the study of an individual mine. Using it to investigate the tin–tungsten orebodies of Panasqueira, Portugal, they demonstrated that muscovite samples from different types of mineralization and alteration in this c. 300 Ma old deposit can be distinguished from each other even when their age difference is as little as 0.9 Ma. Many distinct pulses of mineralization within a time span of less than 5 Ma were recognized. Clearly this high precision technique is going to provide us in the future with much valuable evidence of the age and evolution of mineral deposits.

Determining the age of Mississippi Valley-type deposits using radiometric methods has proved to be about the most difficult task in ore deposit geochemistry. This knotty problem apparently arises in part from their mode of genesis; ore materials quite probably being derived from a number of source rocks, which implies isotopic heterogeneity in the minerals of those deposits. Ore fluid mixing, reactions with wall rocks, etc. lead to further difficulties.

The reader will find it rewarding to read the short note by Ruiz *et al.* (1985) on this subject.

Finally, mention should be made of the model lead ages obtained on lead mineralization (normally from galenas). These, the reader should remember, record the time when the radiometric clock stopped, not the time when it was started or restarted as with the U–Pb, Rb–Sr and K–Ar methods. The Pb–Pb method is dependent on assumptions concerning the evolution of the lead isotopes in the source material, and for this mathematical models have to be deduced. The method is regarded with suspicion by many workers but if the basis of the method is borne in mind, its results can be of considerable value, not only for dating purposes but also in throwing light on the origin of the lead.

Palaeomagnetic dating

This method deserves to be used more widely than it has been. Krs & Stovičková (1966) applied it to veins of the Jáchymov (Joachimsthal) region of the Czechoslovakian section of the Erzgebirge, where they demonstrated the presence of Hercynian (late Carboniferous to early Permian) and Saxonian (middle Triassic to Jurassic) mineralization. Further, they showed that the Ag–Co–Ni–U–Bi mineralization was not associated temporally with the Hercynian granites but with the later Saxonian to Tertiary basaltic magmatism. Similar work by Krs in the Freiberg region of Germany showed an excellent correlation with radiometric data on the veins he investigated (Baumann & Krs 1967).

The successful palaeomagnetic dating of ore deposits depends on a number of factors (Evans & Evans 1977). Some of the most important are:
1 the development of magnetic minerals in a deposit or its wall rocks during one of the principal phases of mineralization;
2 a lack of complete oxidation or alteration, which may be accompanied by overprinting with a later period of magnetization;
3 the availability of an accurate polar wandering curve for the continent or plate in which the deposit occurs.

As is well known, magnetite and hematite are the two principal carriers of magnetization in rocks; this is true also for ore deposits. In general, magnetization carried wholly or in the main by magnetite is, with present palaeomagnetic techniques, measured and interpreted more easily. This mineral is, however, by no means common in epigenetic ore deposits and hematite-mineralized specimens often have to be used. Using specimens mineralized with hematite, Evans & Evans (1977) dated as Saxonian the primary hematite mineralization in the base metal veins of the Mendip Orefield, England, and Evans & El-Nikhely (1982) obtained a Permian age for epigenetic hematite deposits at two mines in Cumbria, England, contrary to the Triassic or post-Triassic ages proposed by previous workers.

An important development in this method of dating was the perfecting of the cryogenic magnetometer in the early 1970s. This instrument is beginning to provide us with new data on mineralization ages, particularly for Mississippi Valley-type deposits, which, as mentioned above, present many problems to the geochronologist using radiometric dating methods. Wu & Beales (1981) used a cryogenic magnetometer to date this type of mineralization in south-east Missouri as late Pennsylvanian to early Permian and Symons & Sangster (1991) have extended this work with important results—see pp. 238–239.

Part 2
Examples of the more important types of ore deposit

6 / Classification of ore deposits

The classification of objects in the natural world has been always one of the prime interests of the physical and biological sciences. In geology and biology it is particularly important as it provides a shorthand method of referring to groups of objects having common properties. Without the use of classifications, the comparison of fossils from an evolutionary or palaeogeographical point of view would be practically impossible. Similarly, for the discussion of magmatic processes we must classify igneous rocks. The best classifications are generally those that have no reference to the origin of geological material. Once genetic factors are brought into a classification, difficulties arise. The student

Table 6.1 General characteristics of hypothermal deposits. (After Lindgren 1933)

Depth of formation	3000–15 000 m
Temperature of formation	≈ 300–$600\,°C$
Occurrence	In or near deep-seated acid plutonic rocks in deeply eroded areas. Usually found in Precambrian terranes, rarely in young rocks. Often found in reverse faults
Nature of ore zones	Fracture-filling and replacement bodies, with the latter phenomenon often more prevalent, leading to irregular shaped orebodies; nevertheless these are frequently broadly tabular. Sheeted zones common, also bedding plane deposits and short, irregular veins. Boundaries usually sharp with limited amount of ore disseminated in walls. Good persistence in depth
Ores of	Au, Sn, Mo, W, Cu, Pb, Zn, As
Ore minerals	*Magnetite*, specularite, *pyrrhotite*, *cassiterite*, *arsenopyrite*, molybdenite, bornite, chalcopyrite, Ag-poor gold, *wolframite*, *scheelite*, pyrite, galena, *Fe-rich sphalerite* (marmatite)
Gangue minerals	*Garnet*, plagioclase, *biotite*, *muscovite*, *topaz*, *tourmaline*, epidote, quartz (often originally *high quartz*), *chlorite* (high Fe variety), carbonates
Wall rock alteration	*Albitization*, *tourmalinization*, *rutile development*, sericitization in siliceous rocks, chloritization. Wall rocks are often crisp and sparkling
Textures and structures	Often very coarse-grained, frequently banded, fluid inclusions present in quartz
Zoning	Textural and mineralogical changes with increasing depth are very gradual over thousands of metres. Gold tellurides may give rise to spectacular bonanzas
Examples	Au of Kirkland Lake, Ontario; Kolar, Mysore; Kalgoorlie, W. Australia; Homestake, Dakota. Cu–Au of Rouyn area, Quebec. Sn of Cornwall

may have noted already that most igneous rock classifications in use today are essentially free of genetic parameters, whereas classifications of pyroclastic (volcaniclastic) rocks have many genetic overtones. These can affect the usefulness of the classification as our ideas on genesis evolve.

In older classifications of ore deposits much emphasis was placed on the mode of origins of deposits, with the result that as ideas concerning these changed, many classifications became obsolete. A good example of this occurs with the volcanic-associated massive sulphide deposits which, 30 years ago, were generally held to be formed by replacement at considerable depths within the crust, but are now thought to be the product of deposition in open spaces at the volcanic or sediment–sea water interface. They have, therefore, moved from the class of hydrothermal-replacement deposits to that of volcanic-exhalative deposits, and who will be so intrepid as to suggest that future generations of geologists may not postulate new theories concerning the genesis of this and other classes of deposit! If a classification is to be of any value it must be capable of including all known ore deposits so that it will provide a framework and a terminology for discussion and so be of use to the mining geologist, the prospector and the exploration geologist.

Classifications of ore deposits have been based upon commodity (copper deposits, iron deposits, etc.), morphology, environment and origin. Commodity and morphological classifications may be of value to economists and mining engineers, but they

Table 6.2 General characteristics of mesothermal deposits. (After Lindgren 1933)

Depth of formation	1200–4500 m
Temperature of formation	200–300 °C
Occurrence	Generally in or near intrusive igneous rocks. May be associated with regional tectonic fractures. Common in both normal and reversed faults
Nature of ore zones	Extensive replacement deposits or fracture-fillings. Boundaries of orebodies often gradational from massive to disseminated ore. Tabular bodies, sheeted zones, stockworks, pipes, saddle-reefs, bedding-surface deposits. Fissures fairly regular in dip and strike
Ores of	Au, Ag, Cu, As, Pb, Zn, Ni, Co, W, Mo, U, etc.
Ore minerals	*Native Au, chalcopyrite, bornite,* pyrite, *sphalerite, galena, enargite, chalcocite, bournonite,* argentite, *pitchblende, niccolite, cobaltite, tetrahedrite,* sulphosalts
Gangue minerals	Lack high temperature minerals (garnet, tourmaline, topaz, etc.), albite, *quartz, sericite, chlorite, carbonates, siderite,* epidote, montmorillonite
Wall rock alteration	Intense chloritization, carbonization or sericitization. Walls often dull
Textures and structures	Less coarse than hypothermal ores, pyrite, when present, is often very fine-grained. Veins are often banded, large lenses usually massive
Zoning	Gradual but definite change in mineralization with depth, e.g. Butte. Good vertical range, many deposits not bottomed after 1800 m of mining
Examples	Au of Bendigo, Australia. Ag of Cobalt, Ontario. Ag–Pb of Coeur d'Alene, Idaho; Leadville, Colorado. Cu of Butte, Montana

lump too many fundamentally different deposit types together to be of much use to geologists. In the past, ore geologists have been inclined to favour genetic classifications, but in recent years there has been a swing away from such ideas towards environmental–rock association classifications. Good examples of this trend are to be found in Stanton (1972) and Dixon (1979) and a comprehensive discussion of classification schemes can be found in Wolf (1981).

Looking at some selected classifications we may start with Lindgren's, which was first put forward in 1913 and has been the most influential classification ever proposed. A *brief summary* of the modified version of 1933 is as follows.

1 Deposits formed by mechanical processes.

2 Deposits formed by chemical processes of concentration.

(a) In surface waters: (i) by reactions (between solutions), and (ii) by evaporation (of solutions).

(b) In bodies of rock: (i) by concentration of substances contained within rocks by (a) weathering, (b) ground water and (c) metamorphism; (ii) by introduced substances—(a) without igneous activity, and (b) related to igneous activity by (I) ascending hydrothermal waters, i.e. epithermal, mesothermal and hypothermal deposits, and (II)

Table 6.3 General characteristics of epithermal deposits. (After Lindgren 1933)

Depth of formation	Near surface to 1500 m
Temperature of formation	50–200°C
Occurrence	In sedimentary or igneous rocks, especially in or associated with extrusive or near surface intrusive rocks, usually in post-Precambrian rocks not deeply eroded since ore formation. Often occupy normal fault systems, joints, etc.
Nature of ore zones	Simple veins—some irregular with development of ore chambers—also commonly in pipes and stockworks. Rarely formed along bedding surfaces. Little replacement phenomena
Ores of	Pb, Zn, Au, Ag, Hg, Sb, Cu, Se, Bi, U
Ore minerals	*Native Au now often Ag-rich*, native Ag, *Cu*, Bi. Pyrite, *marcasite, sphalerite, galena*, chalcopyrite, *cinnabar*, jamesonite, *stibnite, realgar, orpiment, ruby silvers, argentite, selenides*, tellurides
Gangue minerals	SiO_2 as *chert, chalcedony* or crystalline quartz—often amethystine, (sericite), low Fe chlorite, epidote, carbonates, fluorite, baryte, *andularia, alunite, dickite*, rhodochrosite, *zeolites*
Wall rock alteration	Often lacking, otherwise chertification, kaolinization, pyritization, dolomitization, chloritization
Textures and structures	Crustification (banding) very common, often with development of fine banding, cockade ore, vugs and brecciation of veins. Grain size very variable
Zoning	This type of mineralization may vary abruptly with depth, often having only a small vertical range (telescoping), mostly bottoms at 300–900 m. Grade variable with occurrence of bonanzas within low grade ore
Examples	Au of Cripple Creek, Colorado; Comstock, Nevada. Keweenawan Coppers. Sb of China

by direct igneous emanations, i.e. pyrometa-somatic and sublimate deposits.

(c) In magmas by differentiation: (i) magmatic deposits, and (ii) pegmatites.

This classification is still used by many geologists, particularly when discussing epigenetic hydro-thermal deposits, and all ore geologists should be aware of the descriptive terms used by Lindgren and added to by L.C. Graton in later modifications. In this part of Lindgren's classification, epigenetic hydrothermal deposits are classified according to their depth and temperature of formation—hypothermal deposits being deep-seated, high temperature deposits; mesothermal deposits those formed at low temperatures and medium depths, and epithermal deposits being near surface. Later terms include leptothermal to cover deposits gradational from mesothermal to epithermal, and tele-thermal for very low temperature deposits formed at great distances from the source of the hydrothermal solutions that gave rise to them. In the field, these various deposit types have to be recognized from their mineral assemblages, type of wall rock alter-ation, etc. The features used for this purpose and general features of these deposits are given in Tables 6.1–6.4. Excellent recent reviews of mesothermal gold deposits can be found in Nesbitt (1991) and of epithermal ones in Berger & Henley (1989) and Henley (1991). One of the uses of these and other classifications is that if we can classify a particular deposit then we can compare it with others of the same class and make predictions as to its behaviour in depth. Thus, recognition that a deposit is hypo-thermal suggests that it will have great continuity in depth, whereas epithermal and telethermal deposits may bottom very quickly and be of limited vertical development. Again, recognition that a deposit is of massive volcanic-exhalative type should stimulate the mining geologist into searching for a possible underlying feeder stockwork that may be exploit-able.

As Stanton (1972) pointed out the simplest classifications recognize ores as rocks, which is just what H. Schneiderhöhn did in 1932. Unfortunately Schneiderhöhn later elaborated his classification into an extremely detailed tabular arrangement of little use to the exploration geologist and which has attracted scant attention from the bulk of academic geologists. Stanton's was the first modern textbook in which mineral deposits were treated as common rocks. He recognized seven important worldwide associations, which can be condensed as follows.

1 Ores in igneous rocks:
  (a) ores of mafic and ultramafic association;
  (b) ores of felsic association.
2 Iron concentrations of sedimentary affiliation.
3 Manganese concentrations of sedimentary affili-ation.

Table 6.4 General characteristics of telethermal deposits. These are believed to have formed a long way from the parent magma at low temperatures and high in the crust. Temperatures are still higher than in ground waters

Depth of formation	Near surface
Temperature	$\pm 100°C$
Occurrence	In sedimentary rocks or lava flows, often in areas where plutonic rocks are apparently absent
Nature of ore zones	In open fractures, cavities, joints, fissures, caverns, etc. No replacement phenomena
Ores of	Pb, Zn, Cd, Ge
Ore minerals	*Galena* (poor in Ag), *sphalerite* (poor in Fe, may be rich in Cd), *marcasite* in excess over pyrite, cinnabar, etc. Viz: similar to epithermal mineralogy (Table 6.3)
Gangue minerals	Calcite, low-Fe dolomite, etc.
Wall rock alteration	Dolomitization and chertification
Textures and structures	As epithermal
Examples	Tri-State, USA, Pb–Zn ores, etc. Many Hg deposits

4 Stratiform sulphides of marine and marine-volcanic association.
5 Strata-bound ores of sedimentary affiliation.
6 Ores of vein association.
7 Ore deposits of metamorphic affiliation.

Smirnov (1976) also recognized mineral deposits as part of the spectrum of crustal rocks but his classification is not organized as clearly as Stanton's. A more elaborate classification of mineral deposits based on host rock lithology is that of Aleva & Dijkstra (1986). Like Stanton's, which it resembles, it is basically simple, although the complete classification has many divisions and subdivisions, and has a useful column giving the known geological age range of particular deposit types. It is undoubtedly of value for the mineral explorationist at whom it is aimed.

A glance at the contents pages of this book will show that an environmental–rock association classification is favoured although a whiff of genesis and morphology is included! The various types of industrial mineral deposits dealt with in Chapters 20 and 21 could have been assigned to the other chapters of Part 2 of this book in order to conform with the rock association classification, but for the sake of convenience they have been grouped together for reasons that are discussed in the introduction to Chapter 20.

7 / Diamond deposits in kimberlites and lamproites

Long known as the hardest of naturally occurring minerals, the exciting recent work of Richardson *et al.* (1984) has shown diamonds to be among the oldest minerals in the earth, capable of being picked up as exotic fragments by magmas generated at great depths and then surviving both this ordeal and subsequent violent volcanic activity at and near the earth's surface. The endurance of this mineral is thus truly epitomized by the etymology of its name, which is derived from the Greek *adamas*—unconquerable. Besides being exotic in their genesis and their properties, diamonds are still weighed in the old units known as carats. This was formerly defined as 3.17 grains (avoirdupois or troy) but the international (metric) carat is now standardized as 0.2 g. (This carat should not be confused with that used to state the number of parts of gold in 24 parts of an alloy (usual American spelling 'karat').)

The ultimate bedrock sources of diamonds are the igneous rocks kimberlite and lamproite and it is these occurrences that are dealt with in this chapter. Important amounts of diamond are recovered from beach and alluvial placer deposits, which are described in Chapter 18. Not all kimberlites and lamproites contain diamond and, in those that do, it is present only in minute concentrations. For example, in the famous Kimberley Mine (RSA), 24 Mt of kimberlite yielded only 3 t of diamond, or one part in eight million. From the revenue point of view it is not necessarily the highest grade mines that provide the greatest return—what matters is the percentage of gem quality diamonds. Each mining region and even each mine has a different percentage, and some may be 90% or more while others may rely on only an occasional gem stone to boost their income. Total world production of natural diamonds is about 97 Mct p.a., of which about half are of industrial grade. In addition, 60–80 Mct p.a. of synthetic industrial diamonds are manufactured. The leading world producers are shown in Table 7.1 and world bedrock diamond fields and the general occurrence of kimberlites in Fig. 7.1. In 1984 world production was only about 64 Mct, most of the increased production since then has resulted from the coming on stream in

Table 7.1 World production of natural diamonds (Mct) in 1989 (bedrock and placer deposits)

Australia	37.0
Zaïre	20.0
Botswana	15.2
USSR	12.0
RSA	9.0
Angola	1.2
Namibia	0.9
South America	0.9
Sierra Leone	0.6
CAR	0.6
Others	1.1
Total	97.4

1985 of the Argyle Mine's AK1 Pipe in Western Australia.

Morphology and nature of diamond pipes

Many near surface, diamond-bearing kimberlites and lamproites occur in pipe-like diatremes (often just called pipes), which are small, generally less than 1 km² in horizontal area. They are often grouped in clusters, of which the Lesotho occurrences are a good example (Nixon 1973). Lesotho probably has more kimberlites per unit area than anywhere else in the world. Seventeen pipes, 21 dyke enlargements or 'blows' and over 200 dykes are listed by Nixon. In northern Lesotho over 180 kimberlites are known, i.e. about one per 25 km², but a high proportion are barren. Some diatremes are known to coalesce at depth with dykes of non-fragmental kimberlite (Fig. 7.2). These dykes are thin, usually less than 10 m in diameter but may be as much as 14 km long. Diatremes and dykes may be mutually cross-cutting. Their morphology and internal structures have been succinctly reviewed by Nixon (1980a) and the interested reader is strongly recommended to look at the entire contents of the book in which this paper occurs.

Some recently formed diatremes terminate at the surface in maars; these volcanic craters may be filled with lacustrine sediments to a depth of over 300 m.

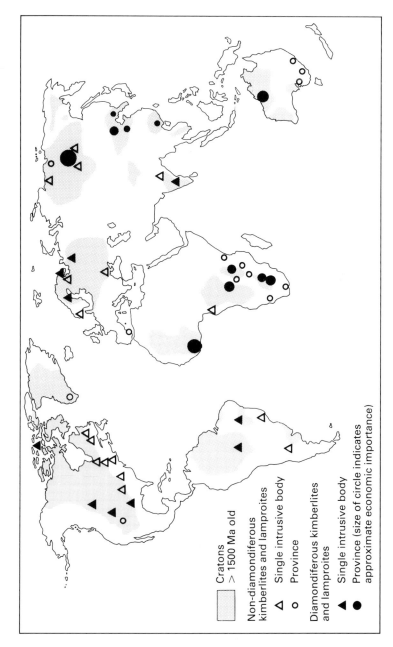

Fig. 7.1 Distribution of diamondiferous and non-diamondiferous kimberlites and lamproites. (After Ferguson 1980a, Groves *et al.* 1987 and other sources.)

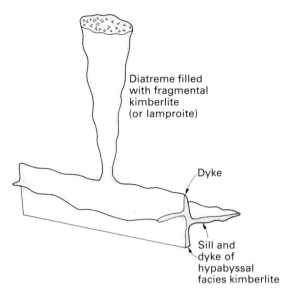

Diatreme filled
with fragmental
kimberlite
(or lamproite)

Dyke

Sill and
dyke of
hypabyssal
facies kimberlite

Fig. 7.2 Schematic drawing showing the relationship
between an explosive kimberlite diatreme and
deep-seated sills and a feeder dyke filled with
non-fragmental kimberlite magma that solidified *in situ*.
(After Dawson 1971.)

The sediments, which may be diamondiferous with
a heavier concentration of more and bigger diamonds near the crater rim shoreline, are sometimes
affected by subsidence, perhaps of a cauldron
nature. The kimberlite beneath the sediments may
be relatively barren. Below the flared crater area is
the vertical or near vertical pipe itself which typically has walls dipping inwards at about 82° and a
fairly regular outline producing the classic, carrot-
shaped diatreme (Fig. 7.3) that may exceed 2 km in
depth. Smaller, more eroded intrusions often have
more irregular outlines (Fig. 7.4). Lower down in the
'zone of pipe generation', pipes may no longer have
vertical axes and may separate into discrete root-like
bodies. Upward terminating bodies ('blind pipes')
are also known.

At the surface kimberlite may be weathered and
oxidized to a hydrated 'yellow ground', which gives
way at depth to fresher 'blue ground', and 'harde-
bank' (resistant kimberlite that often crops out and
does not disintegrate easily upon exposure). In the
upper levels of pipes the kimberlite is usually in the
form of so-called agglomerate (really a tuffisitic
breccia with many rounded and embayed fragments
in a finer grained matrix) and tuff. The rounded
fragments are often xenoliths of metamorphic rocks

from deeper crust, or garnet-peridotite or eclogite
from the upper mantle. Their rounded nature is
attributed to a gas-fluidized origin (Dawson 1971).
Magmatic kimberlite bodies are confined mainly to
the root zones of pipes, where they are often in the
form of intrusion breccias which grade up with
gradational boundaries into the agglomerate, or to
sills and dykes.

Kimberlites and lamproites and their emplacement

As Scott Smith & Skinner (1984) have pointed out,
kimberlite has traditionally been considered to be
the only important primary source of diamond.
However, both these authors and others, e.g. Jaques
& Ferguson (1983) and Jaques *et al.* (1984), have
pointed out that lamproites in Arkansas and Western Australia (including the AK1 pipe mentioned
above) carry significant amounts of diamond, a
point of such exploration importance that it is of
value to look briefly at these two rock-types.

Kimberlite may be defined (Best 1982; Clement *et
al.* 1984) as a potassic ultrabasic hybrid igneous rock
containing large crystals (megacrysts) of olivine,
enstatite, Cr-rich diopside, phlogopite, pyrope-
almandine and Mg-rich ilmenite in a fine-grained
matrix containing several of the following minerals
as prominent constituents: olivine, phlogopite, calcite, serpentine, diopside, monticellite, apatite,
spinel, perovskite and ilmenite. Some of these
minerals are used as indicator minerals in stream
sediment and soil samples in the search for kimberlites, e.g. red-brown pyrope, purple-red chromium
pyrope, Mg-rich ilmenite, chromium diopside
(Nixon 1980b) and a study of their morphology can
be used to indicate the proximity of their source
(Mosig 1980).

Lamproites are defined by Jaques *et al.* (1984),
Scott Smith & Skinner (1984) and Rock (1991) as
potash- and magnesia-rich lamprophyric rocks of
volcanic or hypabyssal origin comprising mineral
assemblages containing one or more of the following
primary phenocrystal and/or groundmass phases:
leucite, Ti-rich phlogopite, clinopyroxene, amphib-
ole (typically Ti-rich, potassic richterite), olivine
and sanidine. Accessories may include priderite,
apatite, nepheline, spinel, pervoskite, wadeite and
ilmenite. Glass may be an important constituent of
rapidly chilled lamproites. Xenoliths and xenocrysts, including olivine, pyroxene, garnet and spinel
of upper mantle origin, may be present and diamond

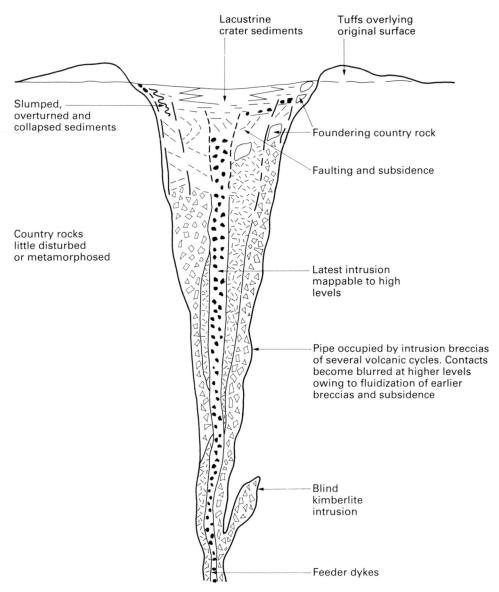

Lacustrine
crater sediments

Tuffs overlying
original surface

Slumped,
overturned and
collapsed sediments

Foundering country rock

Faulting and subsidence

Country rocks
little disturbed
or metamorphosed

Latest intrusion
mappable to high
levels

Pipe occupied by intrusion breccias
of several volcanic cycles. Contacts
become blurred at higher levels
owing to fluidization of earlier
breccias and subsidence

Blind
kimberlite
intrusion

Feeder dykes

Fig.7.3 Schematic diagram of a kimberlite diatreme (pipe) and maar (volcanic crater below ground level and surrounded by a low tuff rim). The maar can be up to 2 km across. (After Nixon 1980a.)

as a rare accessory. Lamproites have greater mineralogical and textural variations than kimberlites and although they do contain some minerals characteristic of kimberlites, the chemical compositions of these minerals are often distinctly different. Lamproitic craters are generally wider and shallower than those of kimberlites, and in the crater, magmatic lamproite, analogous to the magmatic kimberlite found only in the deeper levels of kimberlite

pipes, may be present as intrusions into the crater sediments and in the form of ponded lava lakes overlying the pyroclastic rocks and sediments. Fertile lamproites appear to be the silica-saturated orendites and madupites which carry sanidine rather than leucite (Gold 1984). Diamondiferous lamproites and other lamprophyres with diamonds have now been discovered in Western Australia, Quebec, India, Ivory Coast, RSA, Sweden, USA,

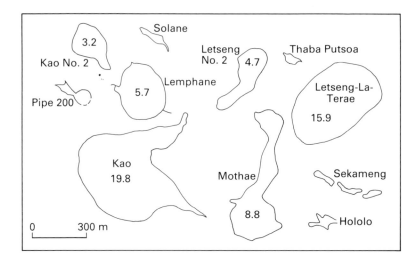

Fig. 7.4 Surface outlines and areas (in hectares) of some Lesotho kimberlite pipes from various localities. (After Nixon 1980a.)

USSR and Zambia. Certain diamondiferous rocks classified as kimberlites, simply because of their diamond content, have now been recognized to be lamproites (Rock 1991).

Kimberlites and lamproites are generally regarded as having been intruded upwards through a series of deep-seated tension fractures, often in areas of regional doming and rifting, in which the magmas started to consolidate as dykes. Then highly gas-charged magma broke through explosively to the surface at points of weakness, such as cross-cutting fractures, to form the explosion vent, which was filled with fluidized fragmented kimberlite or lam-

proite and xenoliths of country rock. Quietly intruded magma was then sometimes emplaced as described above.

Some examples

Western Australia

The very recent (1977) find of diamonds in Western Australia in and around the Kimberley Craton (Fig. 7.5) has led to the discovery of three diamondiferous provinces with over 100 kimberlites and lamproites. A radiometric age of 20 Ma (Jaques & Ferguson

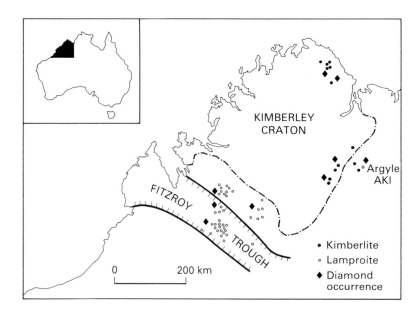

Fig. 7.5 Kimberlites, lamproites and diamond occurrences in the Kimberley region of Western Australia. (After Atkinson *et al.* 1984.)

1983) makes some of them the youngest known diamondiferous pipes, and the lamproitic Argyle AK1 pipe is richer in diamond than any known kimberlite. The AG1 pipe is, however, Proterozoic— 1126 ± 9 Ma with diamonds dated at 1580 Ma (Skinner *et al.* 1985, Richardson 1986). Proven reserves in 1985 were 61 Mt grading 6.8 ct t^{-1}, and probable reserves were 14 Mt at 6.1 ct t^{-1} with a gem content of 5%, cheap gem 40% and industrial diamonds 55%. A review of the long term mining plans and the definition of ore reserves for the AK1 Mine has led to 1990 figures for proven reserves of 94.2 Mt at 4 ct t^{-1} and for probable reserves of 5.9 Mt at 2.8 ct t^{-1}. In 1990 the AK1 plant processed 5.1 Mt which implies a mine life extending well into the next century. Before ore could be produced, some 20 Mt of waste had to be removed from the southern end of the deposit (Anon. 1983a,b).

The AK1 deposit is emplaced in Proterozoic rocks aged about 1200 Ma. The crater zone is 1600 m long and 150–600 m wide (Fig. 7.6), has a surface area of nearly 50 ha and is filled with sandy lapilli-tuff (about 40% rounded quartz grains with lamproite clasts) which contains some interbedded mudstone

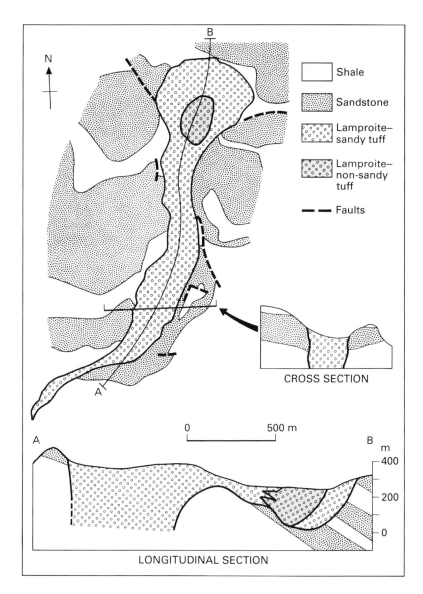

Fig. 7.6 Map and sections of the Argyle AK1 Lamproite. (After Anon. 1983b and Atkinson *et al.* 1984.)

and quartzite. Soft rock deformation structures (slumping, microfaulting) are common. In the centre of the northern end of the crater is a body of non-sandy tuff devoid of detrital quartz which is thought to be younger than the sandy tuff. Occasional late, magmatic kimberlite dykes 1–2 m thick and kimberlitic tuffisite dykes cut the sandy tuff. Diamond grades are considerably lower in the non-sandy tuff and the intrusives appear to be barren.

Jaques & Ferguson (1983) and Jaques *et al.* (1984) have suggested that the diamondiferous lamproites of the Fitzroy area formed by the partial melting of metasomatized phlogopite-peridotite under H_2O- and F-rich and CO_2-poor conditions. They suggested that low degrees of melting under these conditions produced silica-saturated, olivine-lamproite melts rather than the undersaturated kimberlitic-carbonatitic melts that form under high P_{CO_2}.

Lesotho

The deposits of this small nation, already touched on above, make an interesting contrast with those of Western Australia. Among the many non-productive kimberlite pipes the Letseng-La-Terae and neighbouring Kao pipes are notable exceptions. The main pipe is almost 776×366 m at the surface and its tadpole shaped satellite pipe is almost 426×131 m. The mine, owned by De Beers (75%) and the Government (25%) was opened in 1977 at a capital cost of £23 million. Although production is small, just over 50 000 ct p.a. and the grade at 0.309 ct t^{-1} is low, 13% of the output is 10 ct and larger stones. 'Profitability', to quote Harry Oppenheimer, De Beers' former Chairman, 'depends on a small number of exceptionally beautiful stones.'

USSR

The first diamondiferous kimberlite pipe, named Zarnitsa (Dawn), was found in Siberia in 1954. The grade of this pipe was too low for exploitation but since then exploitable pipes have been discovered, among them the rich Mir (Peace) pipe. Despite the arrival of glasnost, production figures for the USSR are still a closely guarded secret and the 12 Mct of Table 7.1 does not reflect the exact amount mined nor do we know the exact proportions of gem, cheap gem and industrial diamonds. Some 25% of USSR production is of gem quality.

The Siberian kimberlites lie within the Siberian Platform (Fig. 7.7), forming groups of different ages (Dawson 1980). As in South Africa, there appears to be a regional zoning of the kimberlites. A central zone of diamondiferous kimberlites is surrounded by a zone with pyrope and low diamond values, then by a zone from which diamond is absent but pyrope present, and finally by an outer zone in which neither of these high pressure minerals is present in the kimberlites. Elsewhere in the USSR feasibility studies are being carried out on a possible mine in the Kola Peninsula and exploration work is in progress along the Finnish border.

Origin of diamonds and some geological guides to exploration

Extremely high temperature and pressure are required to form diamond rather than graphite from pure carbon—of the order of $1000°K$ and 3.5 GPa, equivalent, in areas of 60 km thick continental crust, to a depth of about 117 km. As coesite rather than stishovite is found as inclusions in diamonds, and the inversion curve for these two silica polymorphs is equivalent to a depth of about 300 km, the approximate range for diamond genesis is 100–300 km.

For many decades there has been a very active debate as to whether diamonds crystallized from the magmas which cooled to form the igneous rocks in which they are now found (phenocrysts), or whether they were picked up by these magmas as exotic fragments derived from the diamond stability field within the upper mantle (xenocrysts). The much greater abundance of diamonds in eclogite xenoliths than in the surrounding kimberlite suggests that they have been derived from disaggregated eclogite (Robinson 1978). The former hypothesis infers that successful exploration consists of finding kimberlites and lamproites but the latter implies a more sophisticated approach (Rogers & Hawkesworth 1984). Furthermore, in southern Africa, kimberlites erupted within the confines of the Archaean craton are diamondiferous, while those in adjacent younger orogenic belts are barren (Nixon *et al.* 1983); a general point that might be deduced on a global scale from Fig. 7.1. These data have led Gurney (1990) to remark that the most prospective areas for diamond exploration are on cratons stabilized by 3.2 Ga and that the diamond-carrying potential of kimberlites decreases as one moves away from such core zones.

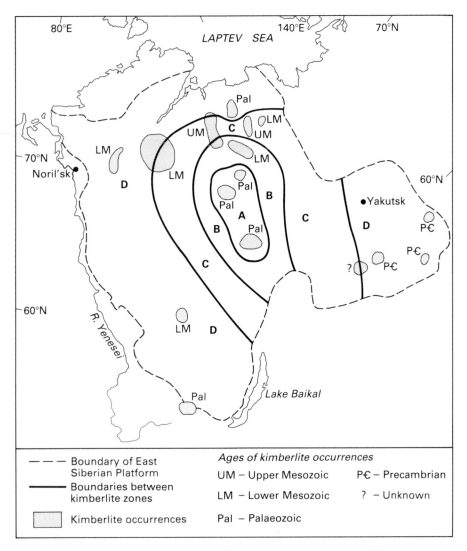

Fig. 7.7 Distribution of kimberlites in the East Siberian Platform showing the zones of different mineralogical subfacies: A = diamond, B = diamond-pyrope, C = pyrope, D = other kimberlites. (Modified from Dawson 1980.)

Convincing dating evidence that many, perhaps most, diamonds are xenocrysts and often much older than the magmas that transported them to the surface has been accumulating since 1984, when Richardson *et al.* (1984) obtained dates on garnet inclusions in diamonds from two southern African kimberlites (Kimberley and Finsch, 160 km apart and about 90 Ma old) of greater than 3000 Ma. It seems highly probable that the diamond hosts are of a similar great age. At the time of writing the available radiometric ages on diamond inclusions suggest two main periods of diamond formation; the

first about 3.3 Ga ago and the second about 1580–990 Ma ago. The dated inclusions in the first group of diamonds are peridotitic minerals and in the second eclogitic. Diamonds with these different inclusion assemblages are referred to as peridotitic or eclogitic. Whereas peridotitic diamonds are all associated with much younger hosts some eclogitic diamonds occur in kimberlites of essentially the same age (Phillips *et al.* 1989) suggesting a link between eclogitic diamond formation and kimberlitic or lamproitic magmatism. This now appears to be unlikely as a result of the discovery of Proterozoic eclogitic diamonds in host kimberlites of about 100 Ma (Richardson, S.H. *et al.* 1990).

This and other evidence is leading workers to the opinion that diamonds grew stably within the upper mantle in eclogite and ultramafic rocks. The estimated conditions of equilibration for ultramafic suite minerals coexisting with diamonds (Meyer 1985, Nickel & Green 1985) suggest that this growth takes, or took, place in a layer between about 132 and 208 km in depth beneath continents and 121–197 km beneath oceans, at temperatures of 1200–1600°K, provided that carbon is (or was) present. However, a very deep origin, perhaps 300 km, for a small proportion of eclogitic diamonds has been suggested by recent work (see Gurney 1990). Thus any magma that samples a diamondiferous zone may bring diamonds to the surface if it moves swiftly enough. The speed of ascent of such magmas has been calculated to be around 70 km h^{-1} (Meyer 1985). Slow ascent could allow time for the resorption of diamonds by transporting magmas as the pressure decreased. That resorption is important appears from the fact that although some pipes produce good crystals (cubes, octahedra or cubo-octahedra), good crystals are rare in many kimberlites and only rounded forms are present. Such rounded forms are rare or absent in diamonds included in xenoliths and thus protected from corrosive magma.

Eclogitic and peridotitic diamonds have statistically different although overlapping δ^{13}C values. Values for peridotitic diamonds are restricted but eclogitic diamonds have a wide range of values which matches that seen in crustal carbon (Gurney 1990). This suggests a recycling of crustal carbon back into the mantle by subduction, an idea supported by the wide spread of oxygen isotope ratios in the Roberts Victor eclogitic xenoliths, which Jagoutz *et al.* (1984) (quoted in Gurney 1990) interpreted as evidence for the hydrothermal alteration of sea floor basalt prior to ancient subduction. Gurney (1990) suggested that this subducted material was underplated during the Proterozoic as eclogite on to the existing diamondiferous Archaean peridotitic keels of cratonic areas in which peridotitic diamonds had formed by 3.2 Ga ago.

The diamondiferous layers in the upper mantle are probably discontinuous, because their formation requires the existence (and preservation) of a thick 'cold' crust above, otherwise diamond generating (and destroying) temperatures may be present at too shallow a depth. Absence of the layers may account for the absence of diamonds from many kimberlites, but another important reason is that many kimberlitic magmas were developed at too shallow a level to be able to sample them. This was probably the case in New South Wales where *most* of the kimberlites were generated at 60–70 km depth (Ferguson 1980b). A mining company exploring a particular pipe could now save itself considerable expenditure by having a study made of the equilibrium conditions of crystallization of ultramafic or eclogitic xenoliths and xenocrysts. This would determine whether they crystallized within the diamond generating zone and therefore whether the pipe was likely to be fertile or barren. (Another good example of the industrial application of what might once have been described as useless, esoteric scientific research.) This now important tool of applied science also may be used to identify other igneous rocks that had sampled diamond zones because, as Gold (1984) has pointed out, we now have reports of diamonds in alkali basalts, ophiolites and andesites, and we should keep open minds on the subject of host rocks for diamonds. Gurney (1985) and McGee (1988) have suggested the use of garnet compositions to differentiate between diamondiferous and non-diamondiferous intrusions.

Are all diamonds of great age? The answer will come only from further work. Eclogitic diamonds may have been formed from recycled 'crustal' carbon (Milledge *et al.* 1983, Gurney 1990) and some might yield Phanerozoic ages. Why are diamondiferous intrusions virtually confined to old crustal areas? Is it because thicker lithosphere was stabilized during the Archaean than more recently, or because the upper mantle in Archaean times had a much higher carbon content (Rogers & Hawkesworth 1984)? Should we confine diamond exploration to continental areas underlain by old Precambrian cratons, as present experience suggests we should? Or should we keep a more open mind and pay some heed to the occurrence of diamondiferous breccia pipes in south Borneo, New South Wales, the Kamchatka Peninsula and the Solomon Islands (Hutchinson, 1983); and to the diamondiferous ophiolites listed from many orogenic belts by Pearson *et al.* (1989)? (It should be noted that the Kalimanton alluvial diamond field in Borneo, at present being exploited, is on the margin of the Sundaland Craton and Bergman *et al.* (1988) have suggested derivation of these diamonds from a yet undiscovered lamproitic pipe.)

The West Australian occurrences in cratonized

Proterozoic mobile belts will surely lead diamond explorationists to investigate other environments, and the report of a diamondiferous kimberlite in British Columbia (Anon. 1985a), appears to confirm that already some are ranging far afield in the diamond hunt. Unfortunately they now have to look even more keenly over their shoulders at the back-room boys who already have taken several steps forward in the manufacture of bigger and better synthetic diamonds (Anon. 1985b, Jones 1990).

Jones wrote that in 5 years synthetic diamonds could be commonplace in jewelry. The threat is so real that De Beers itself (see Chapter 1) is researching in the field of synthetic diamonds and has produced an 11 ct stone. Sumitomo Electric of Japan has produced a 10 ct stone. Synthetic stones are now of such high quality that experts sometimes cannot distinguish them from natural varieties. The managing director of Sumitomo Electric in Europe has stated that the only important factor stopping them from making larger gem stones is the cost. When this obstacle is overcome the market could receive a significant supply of synthetic gem quality diamonds. 'Beware of substitutes!'—as a well-known toothpaste producer used to print on its products!

8 / The carbonatite-alkaline igneous ore environment

Carbonatites

Carbonatite complexes consist of intrusive magmatic carbonates and associated alkaline igneous rocks, mainly ranging in age from Proterozoic to Recent; only a few Archaean examples are known. They belong to alkaline igneous provinces and are generally found in stable cratonic regions sometimes with major rift faulting, such as the East African Rift Valley and the St Lawrence River Graben. There are, however, exceptions where carbonatite complexes are not associated directly with any alkalic rocks, e.g. Sangu Complex, Tanzania, and Kaluwe, Zambia; moreover, not all alkalic rock provinces and complexes have associated carbonatites. Woolley (1989) remarked that, of the then known worldwide total of about 330 carbonatites, probably less than half occur in rift valleys, the remainder being spatially related to major faults cutting cratonic regions and in some cases, as in the Kola Peninsula, USSR, located at major fault intersections. In some regions, particularly Africa, carbonatites and their related alkalic rocks are associated with regional domes of various sizes to which much of the faulting, including rifting, is related. The carbonatites may be linked genetically in some way to these structures. Carbonatitic activity appears to have increased with time, but this increase was episodic and seems to have been related temporally and spatially to orogenic events. Carbonatites often form clusters or provinces within which there may have been several episodes of activity, e.g. the East African Rift Valleys, where late Proterozoic, early Palaeozoic and Cretaceous carbonatites occur. Carbonatites occur as volcanic and plutonic bodies, in dilatant fractures where they have been precipitated from hydrothermal solutions, and as replacements of earlier carbonatites and silicate rocks. Carbonatites are normally emplaced later than the bulk of alkalic silicate rocks with which they are associated (Barker 1989).

Carbonatites together with ijolites and other alkalic rocks commonly form the plutonic complex underlying volcanoes that erupted nephelinitic lavas and pyroclastics. Surrounding such complexes there is a zone of fenitization (mainly of a potassium nature and producing orthoclasites) which has a variable width. The emplacement of carbonatites was effected in stages (Le Bas 1977), with the dominant rock often made up of early C_1 (sövite) intrusions, usually emplaced in an envelope of explosively brecciated rocks. C_1 carbonatite is principally calcite with apatite, pyrochlore, magnetite, biotite and aegirine–augite. C_2 carbonatite (alvikite) usually shows marked flow banding and is medium- to fine-grained compared with the coarse-grained sövite. C_3 carbonatite (ferrocarbonatite) contains essential iron-bearing carbonate minerals and commonly some rare earth and radioactive minerals. C_4 carbonatites (late-stage alvikites) are usually barren. Sövites form penetrative stock-like intrusions and C_2 and C_3 carbonatites form cone-sheets and dykes. The intrusion of C_1 and most C_2 carbonatites is preceded by intense fenitization; other carbonatites show little or no fenitization.

Economic aspects

The most important products of this environment are phosphorus (from apatite), magnetite, niobium (pyrochlore), zirconia and rare earth elements (monazite, bastnäsite). To date, only one carbonatite complex is a major producer of copper (Palabora, South Africa), but others are known to contain traces of copper mineralization, e.g. Glenover, South Africa; Callandar Bay, Canada and Sulphide Queen, Mountain Pass, USA (Jacobsen 1975). Other economic minerals in carbonatites include fluorite, baryte and strontianite, and the carbonatites themselves are useful as a source of lime in areas devoid of good limestones. About 22 mines are now operating on 19 different carbonatites in 14 countries. Their combined production of minerals and mineral products in 1988 was worth about US$1250 million. Apatite produced around 39% of this, copper 23%, magnetite 12% and pyrochlore 11% (Notholt *et al.* 1990). Carbonatites and associated alkaline rocks carry an impressive resource of economic minerals and a growing number are under development. They supply most of the world's niobium (Brazil, Canada), REE (USA, China) and vermiculite (RSA). The other important source of REE is monazite in beach placers (Highley *et al.* 1988). The largest reserves of REE appear to be in China, which has several important deposits; one of these, an enormous proven reserve, with 36 Mt of REE, is hosted by an iron ore deposit in Inner Mongolia (O'Driscoll 1988a). The occurrence, exploration for, processing and applications of many of the products mentioned in this paragraph are discussed in Mariano (1989) and Möller *et al.* (1989). Carbonatites are highly susceptible to weathering and therefore may have residual, eluvial and alluvial placers associated with them (see Chapters 18, 19).

The Mountain Pass occurrences, California

The Mountain Pass Carbonatite of California occurs within the Rocky Mountain area, but it is a Precambrian intrusion (1400 Ma) into Precambrian gneisses so that its present tectonic environment is a function of considerably later deformation (Moore 1973), and this may be the reason why it is not now in an obviously alkaline igneous province. The deposits lie in a belt about 10 km long and 2.5 km wide. One of the deposits, the Sulphide Queen carbonate body, carries a great concentration of rare earth minerals (Fig. 8.1). The metamorphic country rocks have been intruded by potash-rich igneous rocks and the REE-bearing carbonate rocks are related spatially, and probably genetically, to these granites, syenites and shonkinites (Olson *et al.* 1954). The rare earth elements are carried by bastnäsite and parisite, these minerals being in veins that are most abundant in and near the largest shonkinite–syenite body. Most of the 200 veins that have been mapped are less than 2 m thick. One mass of carbonate rock, however, is about 200 m in maximum width and about 730 m long and is the largest orebody of rare earth minerals in the world at present under exploitation. It is called the Sulphide Queen Mine, not because of a high sulphide content, but because it is situated in Sulphide Queen Hill.

Carbonate minerals make up about 60% of the veins and the large carbonate body; they are chiefly calcite, dolomite, ankerite and siderite. The other constituents are baryte, bastnäsite, parisite, quartz and variable small quantities of 23 other minerals. The REE content of much of the orebody is 5–15%.

The Palabora Igneous Complex

This Proterozoic Complex (*c.* 2047 Ma) lies in the Archaean of the north-eastern Transvaal. It resulted from alkaline intrusive activity in which there were emplaced in successive stages pyroxenite, syenite and ultrabasic pegmatoids (Palabora Mining Co. Staff 1976). The first intrusion was that of apatite-rich, phlogopite pyroxenite, kidney-shaped in outcrop (but forming a pipe in depth) and about 6×2.5 km (Fig. 8.2). Ultrabasic pegmatoids were then developed at three centres within the pyroxenite pipe. In the central one, phoskorite (magnetite–olivine–apatite rock) and banded carbonatite were emplaced to form the Loolekop carbonatite–phoskorite pipe, which is about 1.4×0.8 km (Fig. 8.3). Fracturing of this pipe led to the intrusion of a dyke-like body of transgressive carbonatite and the development of a stockwork of carbonatite veinlets. The zone along which the main body of transgressive carbonatite was emplaced suffered repeated fracturing, and mineralizing fluids migrated along it depositing copper sulphides which healed the fine discontinuous fractures. These near vertical veinlets occur in parallel-trending zones up to 10 m wide, although individually the veinlets are usually less than 1 cm wide and do not continue for more than 1 m. Chalcopyrite with minor cubanite is the principal copper mineral in the core, especially in the transgressive carbonatite, which runs 1% Cu, while bornite is dominant in the lower grade banded carbonatite and phoskorite. Vallerite occurs in both the phoskorite and the transgressive carbonatite and presents recovery problems, because (a) recovery of vallerite in the flotation cells is less than 20%, and (b) it interferes with the flotation of the other sulphides. Magnetite is zoned in both quantity and quality (Eriksson 1989). Magnetite with up to 4 wt % TiO_2 makes up 25–50% of the phoskorite. Better quality magnetite (lower TiO_2) occurs in smaller quantities in the transgressive carbonatite. Apatite is present in economic amounts in the phoskorite. Baddeleyite is recovered from both phoskorite and carbonatite.

Earlier workers and the Palabora Mining Co. Staff (1976) suggested that the sulphide mineralization was a late stage, hydrothermal event perhaps not related to the magmatism. Sulphur isotopic and fluid inclusion compositions, however, show that copper sulphide liquid was present very early in the crystallization sequence and before olivine crystallization. These points led Eriksson (1989) to surmise

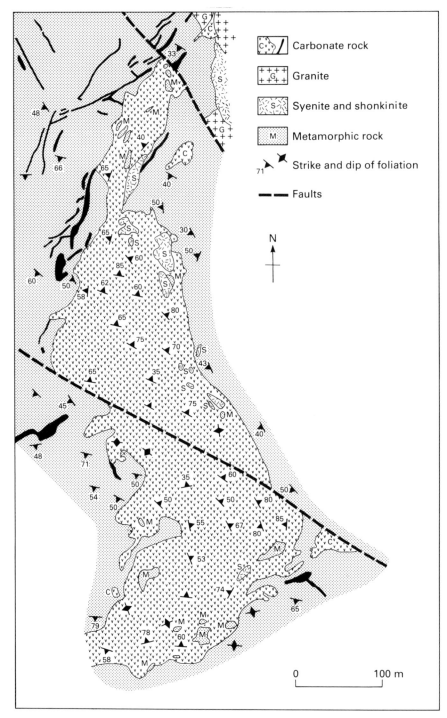

Fig. 8.1 Simplified map of the Sulphide Queen Carbonate Body, Mountain Pass District, California. (After Olson & Pray 1954.)

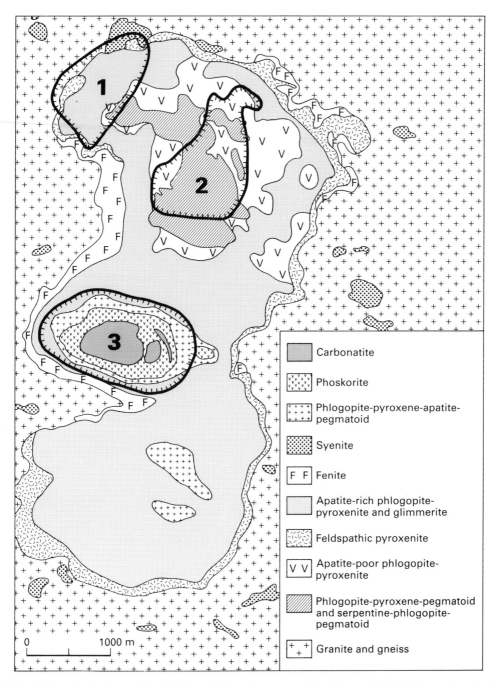

Fig. 8.2 Simplified geological map of the Palabora alkaline igneous complex, Transvaal, RSA. The north-westerly trending dolerite dyke swarm has been totally omitted and some of the satellite intrusions of syenite. (1) Open pit operation of Foskor which produces phosphate. (2) The vermiculite open pit. (3) The Palabora open pit at Loolekop, which produces copper, magnetite, etc. Both (2) and (3) are operated by the Palabora Mining Co., who sell their apatite concentrate to Foskor. North is parallel to the vertical map margins.

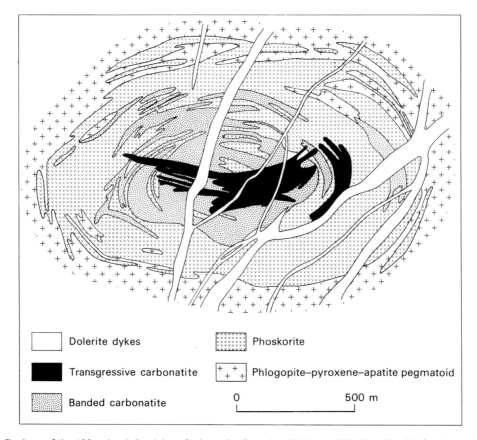

Fig. 8.3 Geology of the 122 m level, Loolekop Carbonatite Complex, Palabora, RSA. See Fig. 8.2 for the location of this complex. (Modified from Jacobsen 1975.)

Legend:
- Dolerite dykes
- Transgressive carbonatite
- Banded carbonatite
- Phoskorite
- Phlogopite–pyroxene–apatite pegmatoid

0   500 m

that sulphide mineralization may be an intrinsic part of carbonatite magmatism and, from his studies on the Guide Copper Mine, 5 km north-west of the Palabora Complex, Aldous (1986) produced evidence suggesting that the copper-rich carbonatites at Palabora could have arisen from the pyroxenites by liquid immiscibility.

Diamond drilling has shown that the orebody continues beyond 1000 m below the surface. Ore reserves are about 300 Mt grading 0.69% Cu. This is the world's second largest open pit mine and the size of the operation can be judged from the fact that in the first 24 years of operation to 1989 the total of material mined was 1680 Mt; the ore milled 558 Mt, and 2.7 Mt of copper was produced. With a cut-off grade of 0.15% Cu and an average mill grade of 0.48–0.54% Palabora is one of the most efficient, low grade copper mines in the world. No other mine in the world can match Palabora for the variety of its

products and potentially extractable metals (Th, Al, K and REE) (Verwoerd 1986). It set a world record in 1980 when over 521 kt of ore and waste were loaded and hauled in one 24-hour period.

There are three separate open pits at Palabora. The carbonatite and phoskorite are worked for copper with by-product magnetite, apatite, gold, silver, PGM, baddeleyite (ZrO_2), uranium, nickel sulphate and sulphuric acid. About 2 km away in the same alkaline complex is the Foskor Open Pit in an apatite-rich pyroxenite that forms the world's largest igneous phosphate deposit. Reserves in the pit area alone, which only covers part of the pyroxenite, are 3 Gt of apatite *concentrates* (36.5% P_2O_5). In 1984 Foskor produced 2.6 Mt of phosphate concentrate, nearly 9 kt of baddeleyite concentrate and 43 kt of copper concentrate carrying 35% Cu. In 1990 Foskor increased its mill capacity to 3.7 Mt p.a.

In a neigbouring pit, vermiculite (Fig. 8.2), a weathering product of phlogopite, is worked. This is the second largest vermiculite mine in the world; it started production in 1946 and in 1990 produced nearly 218 kt of concentrate (90% vermiculite). Evaluation of the adjoining PP&V deposit in 1990 indicated the presence of 46 Mt of vermiculite, which makes it the world's largest orebody of this type.

Vermiculite sells at US$134–204 (June 1991) a short ton (2000 lb) depending on the particle size. Nearly all crude vermiculite is transported in the unexfoliated (i.e. unexpanded) form because of the far higher bulk density of the raw material. Exfoliation plants are sited close to final markets and are thus in most cases thousands of kilometres from the mines. Vermiculite is a form of mica which loses water and expands up to 20 times in size when it is heated. The light, porous product finds a wide range of uses—as an insulating material, particularly in the building industry, for fire-proofing, for acoustic damping, as a carrier for fertilizers, and in the horticultural industry. It has extreme lightness, useful cation exchange properties and is free from health hazards. World production in 1989 was about 650 kt of which about 195 kt (30%) was in RSA. About 50% came from the USA and much of the rest from the Kovdor deposit, USSR and Brazil. Palabora strongly resembles Kovdor, which contains traces of copper, except that Palabora is devoid of nepheline- and melilite-bearing rocks.

The Kola Peninsula—Northern Karelia Alkaline Province

Alkaline igneous complexes and their associated carbonatites show a broad spatial relationship to areas of hot spot activity, which may be accompanied by doming and fracturing, the Kola Peninsula (Fig. 8.4) being an excellent example. The Upper Palaeozoic alkaline igneous and carbonatite complexes of this region host a number of extremely large orebodies, of which the most important are Khibina, Kovdor and Sokli.

The USSR is the world's second largest producer of phosphate rock with much of its production coming from Khibina (Notholt 1979). This is a ring complex about 40 km across with inward dipping, layered intrusions of various alkaline rock types. One apatite–nepheline orebody forms an arcuate, irregular lens-shaped mass with a strike length of 11 km and a proved dip extension of 2 km. The thickness ranges from 10 to 200 m and averages

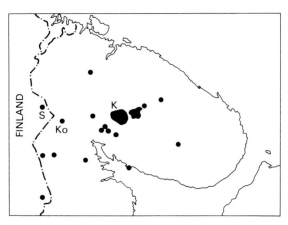

Fig. 8.4 Alkaline igneous and carbonatite complexes in the Kola Peninsula–Northern Karelia Alkaline Province. K = Khibina, Ko = Kovdor, S = Sokli. (After Vartiainen & Paarma 1979.)

100 m and at least 2.7 Gt of ore averaging 18% P_2O_5 are present. The apatite concentrates contain significant SrO and Re_2O_3 values. Annual production is 18 Mt from four mines. Nepheline concentrates are also produced for the manufacture of alumina.

At Kovdor (Fig. 8.5) there is a late stage development of apatite–forsterite rocks and magnetite ores. The orebody is mined by open pit methods. Ore reserves are about 708 Mt grading 36% Fe and 6.6% P_2O_5. Baddeleyite forms a by-product and vermiculite is produced from a separate mining operation. In its mineralization and rock types this complex has affinities with Palabora. A drawback at Kovdor is the high magnesium content of the apatite concentrate, which presents processing problems at fertilizer plants.

Across the border in Finland is the Sokli Complex. With an areal extent of about 18 km^2 it is one of the world's largest carbonatites and it is remarkable for its partial cover of residual ferruginous phosphate rock, forming an apatite–francolite regolith of a type hitherto regarded as being of tropical origin. Proved reserves are over 50 Mt averaging 19% P_2O_5, with possible lower grade reserves of 50–100 Mt. The deposits vary from a few metres to 70 m in thickness. Pyrochlores with both U–Ta and Th enrichment occur in the carbonatites and apatite–magnetite mineralization in metaphoskorites which runs 11–22% Fe (Notholt 1979, Vartiainen & Paarma 1979). Kimberlitic dykes are also present but no diamonds have been found. Seismic evidence suggests that the intrusion tapers downwards from a

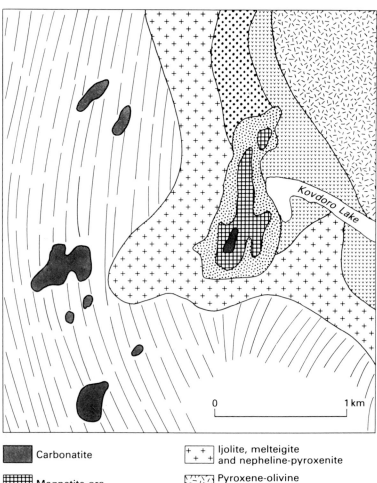

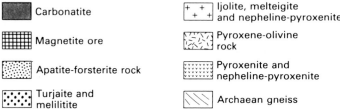

Carbonatite	Ijolite, melteigite and nepheline-pyroxenite
Magnetite ore	Pyroxene-olivine rock
Apatite-forsterite rock	Pyroxenite and nepheline-pyroxenite
Turjaite and melilitite	Archaean gneiss

Fig. 8.5 Sketch map of part of the western section of the Kovdor Complex showing the position of the magnetite orebody. (After Rimskaya-Korsakova 1964.)

diameter of about 6 km at the surface to about 1 km at a depth of 5 km. Other carbonatites, e.g. Palabora, appear to taper downwards, a point of some economic significance. This evidence also suggests that at great depths carbonatites, like kimberlites, become dyke-like bodies within deep fracture zones that have tapped the levels in the mantle from which these alkaline rocks and their rare earth and associated elements have probably been derived (Samoilov & Plyusnin 1982).

Accounts of other important carbonatite deposits that the reader may like to study are as follows: Gold *et al.* (1967) on Oka, Quebec (Nb); Melcher (1966) on Jacupiranga, Brazil; and Reedman (1984) on Sukulu, Uganda (phosphate with possible Nb, zircon, baddeleyite and magnetite by-products).

9 / The pegmatitic environment

Some general points

Pegmatites are very coarse-grained igneous or metamorphic rocks, generally of granitic composition. Those of granulite and some amphibolite facies terranes are frequently indistinguishable mineralogically from the migmatitic leucosomes associated with them, but those developed at higher structural levels and often spatially related to intrusive, late tectonic granite plutons, are often marked by minerals with volatile components (OH, F, B) and a wide range of accessory minerals containing rare lithophile elements. These include Be, Li, Sn, W, Rb, Cs, Nb, Ta, REE and U, for which pegmatites are mined (rare element or (better) rare metal pegmatites). Chemically, the *bulk* composition of most pegmatites is close to that of granite, but components such as Li_2O, Rb_2O, B_2O_3, F and rarely Cs_2O may range up to or just over 1%.

Pegmatite bodies vary greatly in size and shape. They range from pegmatitic schlieren and patches in parent granites, through thick dykes many kilometres long and wholly divorced in space from any possible parent intrusion, to pegmatitic granite plutons many kilometres squared in area. They form simple to complicated fracture-filling bodies in competent country rocks, or ellipsoidal, lenticular, turnip-shaped or amoeboid forms in incompetent hosts. Pegmatites are often classed as simple or complex. Simple pegmatites have simple mineralogy and no well developed internal zoning while complex pegmatites may have a complex mineralogy with many rare minerals, such as pollucite and amblygonite, but their marked feature is the arrangement of their minerals in a zonal sequence from the contact inwards. An example of a complex, zoned pegmatite from Zimbabwe is given in Fig. 9.1 and Table 9.1. Contacts between different zones may be sharp or gradational. Inner zones may cut across or replace outer zones, but not vice versa, so that inside the wall zones at Bikita no two cross-cuts expose the same zonal sequence. The crystals in complex pegmatites can be very large and at Bikita, for example, the spodumene crystals are commonly 3 m long. The Bikita Pegmatite, which is about 2360–2650 Ma old—it is notoriously difficult to obtain concordant radiometric ages for zoned pegmatites (Clark 1982)—is emplaced in the Archaean

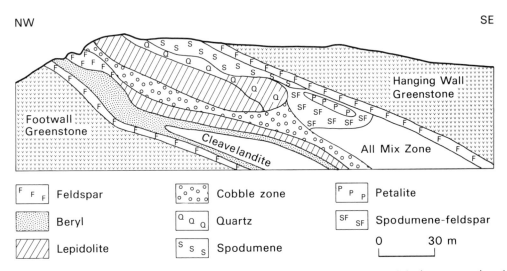

Fig. 9.1 Section through the Bikita Pegmatite showing the generalized zonal structure and the important minerals of each zone. Cleavelandite is a lamellar variety of white albite. (After Symons 1961.)

Table 9.1 Zoning in the Bikita Pegmatite, Zimbabwe. (From Symons 1961)

	Hanging wall greenstone
Border zone	Selvage of fine-grained albite, quartz, muscovite
Wall zones	Mica band. Coarse muscovite, some quartz. Hanging wall feldspar zone. Large microcline crystals

Intermediate zones	Petalite-feldspar zone
	Spodumene zone
	  (a) massive
	  (b) mixed spodumene, quartz, plagioclase and lepidolite
	Pollucite zone. Massive pollucite with 40% quartz
	Feldspar-quartz zone. Virtually devoid of lithium minerals
	'All mix' zone. Microline, lepidolite, quartz

Core zones	Massive lepidolite
	  (a) high grade core, nearly pure lepidolite
	  (b) lepidolite-quartz subzone
	Lepidolite-quartz shell

Intermediate zones	'Cobble' zone. Rounded masses of lepidolite in an albite matrix Feldspathic lepidolite zone
Wall zone	Beryl zone. Albite, lepidolite, beryl Footwall feldspar. Albite, muscovite, quartz
	Footwall greenstone

Fort Victoria Greenstone Belt. It is one of the world's largest Li–Cs–Be deposits; the main pegmatite is about 2 km long and 45–60 m thick, and the minerals of major economic importance were petalite, lepidolite, spodumene, pollucite, beryl, eucryptite and amblygonite. Cassiterite, tantalite and microlite were disseminated through quartz-rich zones in lepidolite-greisen, but these had been mined out by 1950. Mining ceased in 1986 but the company will continue production for many years using stockpile material to produce petalite concentrates (Russell 1988c). Broadly speaking there is a worldwide similarity of zonal sequences in pegmatites, an important point which, among many others characteristic of complex pegmatites, must be explained by any proposed genetic theory. Useful discussions of internal zoning, terminology used, etc., are to be found in Cameron *et al.* (1949), Černý (1982a), Jahns (1955, 1982) and Norton (1983).

Pegmatites have been classified in numerous fashions using as many as nine major criteria (Černý

1982a). A simple classification into four pegmatite formations which has exploration implications is that of Ginsburg *et al.* (1979) and Černý (1989).

1 Pegmatite formation of shallow depths (miarolitic pegmatites, 1.5–3.5 km); pegmatite pods in the upper parts of epizonal granites. Cavities may supply piezoelectric rock-quartz, optical fluorite, gemstones.

2 Pegmatite formation of intermediate depths (rare metal pegmatites, 3.5–7 km); filling fractures in and around possible parent granites or quite isolated from intrusive granites; emplaced in low pressure (Abukumu Series), amphibolite facies metamorphites. Mainly or wholly magmatic differentiates.

3 Pegmatite formation of great depths (mica-bearing pegmatites, 7–8 to 10–11 km); only minor rare metal mineralization, emplaced in intermediate pressure (Barrovian Series), upper amphibolite facies metamorphites. Largely the products of anatexis.

4 Pegmatite formation of maximal depth (>11 km);

in granulite facies terranes, usually no economic mineralization. Commonly grade into migmatites.

Reference to texts on metamorphic petrology suggests that the depths given by Ginsburg *et al.* (1979) err very much on the shallow side. For example, the pegmatites of the Bancroft field, Ontario, belonging to formation 2, are emplaced in pelitic schists carrying cordierite and almandine, which suggests a depth of emplacement of about 20 km or so (Winkler 1979).

Pegmatites may occur singly or in swarms forming pegmatite fields, and these in turn may be strung out in a linear fashion to form pegmatite belts, one of the largest being the rare metal pegmatite belt of the Mongolian Altai, about 450 km long and 20–70 km broad, which contains over 20 pegmatite fields. Pegmatites in a particular field may show a regional mineralogical–chemical zonation. When a swarm of pegmatites is associated with a parent granite pluton then the more highly fractionated pegmatites, enriched in rare metals and volatile components, are found at greater distances from the pluton. The degree of internal zoning also increases with an increase in rare metals and volatiles. The regional zoning may reflect the fact that melts enriched in Li, P, B and F have considerably lower solidi than H_2O-only saturated magma and can penetrate further from their source. This regional zoning is of course of great importance to the exploration geologist (Trueman & Černý 1982).

Parental (fertile) granites of rare metal pegmatites are found in compressional regimes of orogenic chains within back-arc type sedimentary–volcanic sequences or their analogues in flysch-filled cratonic margins or rifts. They are forceful intrusions that post-date the peak of regional metamorphism and granite emplacement, and most are leucocratic alkali granites that display heterogeneity in ranging from biotite granite at depth to two-mica and muscovite–garnet facies in their cupolas. In contrast to their surrounding haloes of mineralized pegmatites, fertile granites are typically barren. Rare earth element abundances are low and disturbed, and stable and radiogenic isotopic ratios also show disturbances. The above and much more useful information on parental granites can be found in Černý & Meintzer (1988).

Most pegmatites, whether igneous or metamorphic in origin, have similar *bulk* compositions and these correspond closely to low temperature melts near the minima of Ab–An–Or–Q–H_2O systems (Černý 1982b). This is to be expected for melts developed by extreme magmatic differentiation or anatexis. Many rare metal pegmatites appear to be of magmatic origin; their unusual composition appears to be a consequence of retrograde boiling and the manner in which elements are partitioned between crystals, melt and volatile phases during the cooling of the magma. Bonding factors such as ionic size and charge largely prevent many constituents, originally present in minor or trace concentrations in the magma, from being incorporated into precipitating crystals. They thus become more highly concentrated in the residual melt which, in the case of granites, is also enriched in water, since quartz and feldspars are anhydrous minerals. The list of residual elements is long but includes Li, Be, B, C, P, F, Nb, Ta, Sn and W. Tin and tungsten are exploited more commonly in hydrothermal deposits, but there is no doubt of their orthomagmatic origin in pegmatite occurrences since there is never any evidence of hydrothermal activity (Gouanvic & Gagny 1983). A steady increase in water content, as anhydrous phases are precipitated, will also mark the crystallization of pegmatitic melts and a point may be reached where retrograde boiling produces a water-rich phase. Na, K, Si and some other residual elements listed above will fractionate from the melt into the water-rich phase within which atoms can diffuse much more rapidly than in a condensed silicate melt. Consequently, rates of crystal growth are greatly enhanced and this may have an important bearing on the growth of giant crystals in pegmatites. London (1984) has shown how different pressures give rise to the crystallization of different phases, e.g. whether spodumene or petalite is the primary lithium alumino-silicate phase.

Broadly speaking, three hypotheses have been put forward to account for the development of internal zoning. The first is that of fractional crystallization under non-equilibrium conditions leading to a steady change in the composition of the melt with time. The second is of deposition along open channels from solutions of changing composition. The third is a two stage model: (a) crystallization of a simple pegmatite, with (b) partial or complete replacement of the pegmatite as hot aqueous solutions pass through it. At the present time, most workers prefer the first or third hypothesis as the majority of the evidence, such as complete enclosure in many pegmatites of the interior zones, does not favour the existence of open channels during crystallization. As the third theory encounters difficulties in explaining the worldwide similarity of zonal sequences, the first theory finds most favour. An early statement of this preferred theory was given by Jahns & Burnham (1969) who, on the basis of field observations and experimental data, postulated that the internal evolution of zoned granitic pegmatites resulted from the crystallization of water-saturated melts that evolved to produce systems with a melt and a separate aqueous fluid. A useful discussion of this theory appears in Thomas *et al.* (1988) who, following a comprehensive study of fluid inclusions from the Tanco Pegmatite, south-eastern Manitoba, were able to confirm the Jahns–Burnham theory. They traced the thermal evolution of this pegmatite from the initial intrusion of a mixture of an alumino-silicate melt plus an H_2O–CO_2–dissolved salt fluid at $\sim 720°C$ to final crystallization of the quartz zone at $\sim 262°C$. Initial crystallization was from the wall rock inwards with the wall zone crystallization commencing at about $600°C$ and the intermediate zone temperatures being around $475°C$. Useful and more general discussions may be found in London (1987, 1990) and London *et al.* (1989).

Some economic aspects

Pegmatitic deposits of spodumene, petalite, lepidolite and other Li minerals are exploited throughout the world for use in glass, ceramics, fluxes in aluminium reduction cells and the manufacture of numerous lithium compounds. These deposits often yield by-product Be, Rb, Cs, Nb, Ta and Sn. They may be internally zoned, or unzoned as are the highly productive lithium pegmatites of King's Mountain, North Carolina. The largest known pegmatitic lithium resources are in Zaïre in two laccoliths each about 5 km long and 0.4 km wide. Reserves have been put at 300 Mt and Ta, Nb, Zr and Ti values have been reported (Harben & Bates 1984). Political insecurity, however, is a major drawback as far as overseas investors are concerned and these projects are in abeyance. Another big resource, mainly in granite, is the Echassières deposit in France, which contains some 50 Mt grading 0.71% Li, 0.022% Nb, 0.13% Sn and 0.023% Ta.

The reader should note that lithium has already been produced from brines in the USA, and a brine producer in Chile is now in production (Crozier 1986). In May 1989 the Japanese Industrial Research Institute announced the development of an efficient method of removing lithium from sea water. These new sources will present an economic challenge to the hard rock producers and the market price may suffer, as by 1984 lithium production had already exceeded demand.

The traditional source of beryllium (beryl), has now been overtaken by bertrandite, which occurs on a commercial scale in hydrothermal deposits (Farr 1984). The main source of beryl is pegmatites, with the USSR (2 kt p.a.) being the world's largest producer, followed by Brazil with less than 400 t p.a.

Pegmatites are also important as a source of tantalum, but it must be noted that the largest reserves—about 7.25 Gt of Ta—are in slags (running about 12% Ta_2O_5) produced during the smelting of tin ores in Thailand. Significant reserves are present in the Greenbushes Pegmatite, Western Australia (Hatcher & Bolitho 1982), and in the Tanco Pegmatite at Bernic Lake, Manitoba (Černý 1982c) which up to 1982 supplied about 20% of the market, but then closed down owing to the weak tantalum price. It reopened in 1988. The backbone of the Greenbushes operation is tin production and most of the tin mined in Thailand is won from pegmatites, not veins (Manning 1986). London (1986) pointed out that many economically important rare metal pegmatites are marked by the development of holmquistite in their immediate wall rocks and that a search for this mineral during pegmatite prospecting is a valuable exploration tool. Pegmatites are also important producers of feldspar and sheet mica.

Uraniferous pegmatites

Pegmatites and pegmatitic granite have been exploited in a number of localities. Among the more important deposits are those of Bancroft, Ontario and the enormous Rössing Deposit in Namibia.

Bancroft Field, Ontario

This area is part of the south-western extremity of the Grenville Province of the Canadian Shield. Granitic and related pegmatites of this province yield ages of 1100–900 Ma (Lumbers 1979), coincident with the waning stages of the high grade Grenvillian metamorphism. The Bancroft Field lies in the Central Metasedimentary Belt, which consists of a metasedimentary–metavolcanic sequence developed around a number of complex granitoid and syenitoid gneiss bodies having domal and periclinal shapes. The pegmatites occur within these gneisses, in the adjacent metasedimentary and metavolcanic sequence and in large metagabbro bodies (Ayres & Černý 1982). Most of the pegmatites are conformable with the metamorphic fabric of their host rocks and internal zoning is but poorly developed, except in the less common fracture-filling bodies that cut across the regional structures. The common presence of pyroxene, hornblende and biotite relates the pegmatitic compositions to the upper amphibolite facies grade of the enclosing rocks. Rare metal mineralization consists of a variety of U, Th, Nb–Ta, REE, Y, Ti, Zr and Be minerals. One school of thought (references in Ayres & Černý 1982) relates the origin of these pegmatites to igneous differentiation, the other to some form of ultra-

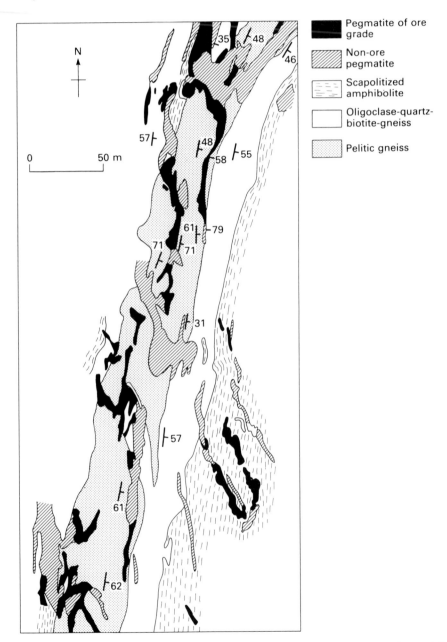

Fig. 9.2 Geology of the First Level, Bircroft Mine, Ontario. (From the work of the mine geologists and the author.)

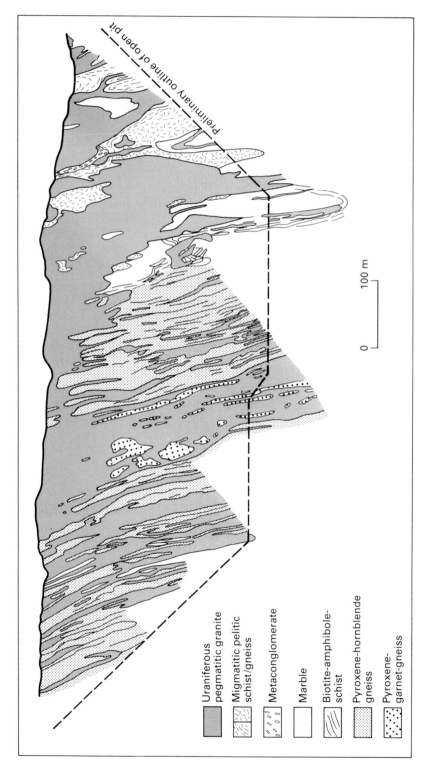

Fig. 9.3 Cross section of the Rössing uranium deposit. (After Berning *et al.* 1976.)

metamorphic activity such as anataxis (Evans 1966, Nash *et al.* 1981, and references in Ayres & Černý 1982). The large amounts of uranium and other rare metals would be derived from the country rocks, a view held by many Russian workers for certain pegmatites in the USSR, e.g. Shmakin (1983). Fowler & Doig (1983), on the basis of stable isotope data have, however, suggested a mantle source for these Grenvillian pegmatites and their incompatible elements.

Average uranium grades in the Bancroft Field were a little above 0.1% U_3O_8 with each of the five main mines working a number of orebodies usually located in swarms of dominantly granitic pegmatite bodies. The Bicroft Mine was one of the largest operations (Fig. 9.2). The pegmatite dykes of this mine area occur in a 5 km long zone only about a quarter of which was developed (Hewitt 1967). The orebodies occurred at the footwall or hanging wall contacts but sometimes occupied the entire pegmatite. The excellent vertical continuity of the orebodies contrasted markedly with their rather tortuous planform (Bryce *et al.* 1958) and two of the largest orebodies were about 90 m long by 3 m wide but extended vertically for over 400 m. The pegmatites are very variable in their lithology, frequently carry aegirine–augite, show no signs of forcible intrusion, have a metamorphic fabric, contain non-rotated enclaves and show a tendency for their lithology to be governed by changes in the nature of their host rocks. The principal uranium minerals are uraninite and uranothorite and the mineralization is best developed where the pegmatites cross a zone containing graphitic pelitic gneiss. Evans (1962) suggested that this association may indicate that the ultimate source of much of the uranium is to be found in these altered black shales.

The Rössing uranium deposit, Namibia

This is the world's largest uranium producer. The operation is a large tonnage, low grade one—15–16 Mt of ore p.a., grading 0.031% U_3O_8, being produced from an open pit and underground workings (Anon. 1982). The uranium mineralization occurs within a migmatite zone (Fig. 9.3), characterized by largely concordant relationships between uraniferous, pegmatitic granites and the country rocks. These are metasediments of similar age (*c.* 900 Ma)

to the Grenvillian rocks of the Bancroft area, although the pegmatitic granites have been dated at 950–550 Ma (Jacob *et al.* 1986). The metasediments are very similar to those of the Bancroft area and, as in that area, occupy the ground between and around granite–gneiss domes and periclines (Berning *et al.* 1976, 1986). The grade of regional metamorphism, upper amphibolite facies, is identical, with cordierite in the pelitic gneisses of both areas indicating low pressure, Abukuma type metamorphism. The pegmatitic granites have a very low colour index and are termed alaskites. About 55% of the uranium is in uraninite and 40% in secondary uranium minerals. Economic uranium mineralization is concentrated where the pegmatites are emplaced in certain metasedimentary zones, including pelitic gneisses, in a manner reminiscent of the Bicroft Mine mineralization. The arid climate of the Namib Desert played an important role in the formation of this deposit because, while giving rise to the enrichment of the primary mineralization with secondary minerals released by weathering, it prevented the leaching of this secondary mineralization by meteoric water. Berning *et al.* (1976) favoured an initial distribution of uranium within the metasedimentary sequence and its later concentration in anatectic melts of alaskitic composition, which show only minor evidence of movement from their zone of generation. The host megastructure has been interpreted as an aborted and closed aulacogen that developed 700–500 Ma ago (Burke & Dewey 1973), but 1000–500 Ma is a more probable time span.

As an example of a large tonnage, low grade, disseminated deposit, Rössing could have been included in Chapter 14 and some authors have unfortunately referred to it as a porphyry uranium deposit, but there is little justification for this description. In its geological environment, host rock type, lack of hydrothermal alteration, etc., it bears no resemblance to porphyry copper deposits.

As an appropriate ending to this chapter I refer the interested reader to the excellent volume edited by Černý (1982d) which provides a comprehensive survey of granitic pegmatites and their economic importance and to the 30 papers in the Jahns Memorial Issue of the *American Mineralogist*, **71**, 233–651.

10 / Orthomagmatic deposits of chromium, platinum, titanium and iron associated with basic and ultrabasic rocks

Orthomagmatic ores of these metals are found almost exclusively in association with basic and ultrabasic plutonic igneous rocks—some platinum being found in nickel–copper deposits associated with extrusive komatiites.

Chromium

Chromium is won only from chromite ($FeCr_2O_4$). This spinel mineral can show a considerable variation in composition, with magnesium substituting for the ferrous iron and aluminium, and/or ferric iron substituting for the chromium. It may also be so intimately intergrown with silicate minerals that these too act as an ore dilutant. Because of these variations there are three grades of chromite ore. The specifications for these are somewhat variable; typical figures are given in Table 10.1. Metallurgical grade ore was for many years marketed as ferrochrome for steel making, but RSA production is now moving towards value added steel products rather than raw ferrochrome. The non-metallurgical markets for chromite consume some 25% of the chromium ore mined and can be divided into refractory, chemical and foundry industries. The demand for refractory chromite has been reduced considerably by the demise of open-hearth steel furnaces, but some recent recovery has followed the use of chromite in steel refining ladles. Chemical grade material is enjoying a buoyant market with a firm demand for chromium chemicals for pigments, wood preservatives and leather tanning salts—three of the more important non-metallurgical uses. Sodium dichromate is also the raw material for chromium electroplating. As a foundry sand, chromite is suffering competition from zircon but does have some advantages over zircon sand (McMichael 1989).

Chromite ores are classified as lumpy or friable, depending on their cohesiveness. Hard, lumpy ore is required for the Perrin process, which produces low carbon ferrochrome (< 0.03% C). Fines from the mining of lumpy ore and low grade ores are milled and concentrated to remove silicate gangue and, when present, magnetite is removed magnetically to improve the Cr/Fe ratio. The fines may be pelletized for smelting.

Occurrence

Three-quarters of the world's chromium reserves are in the Republic of South Africa and 23% in Zimbabwe, most deposits outside these two countries being small. The only other countries with appreciable reserves are the USSR, Albania, India and Turkey. Lower grade deposits were found recently in the Fiskenæsset Complex of Greenland and there are large subeconomic deposits in Canada.

All economic deposits of chromite are in ultrabasic and anorthositic plutonic rocks. There are two major types: stratiform and podiform (often referred to respectively as Bushveld and Alpine types). Each type yields about 50% of world chromite production, which was 12.31 Mt in 1989.

Stratiform deposits

This type contains over 98% of the world's chromite resources. These deposits consist of layers (see Fig. 3.2) usually formed in the lower parts of stratified igneous complexes of either funnel-shaped intrusions (Bushveld, Great Dyke) or sill-like intrusions

Table 10.1 Chromite ore grades and specifications

	Cr_2O_3	Cr/Fe	$Cr_2O_3 + Al_2O_3$	Fe	SiO_2
Metallurgical grade	>48%	>1.5	—	—	—
Refractory grade	>30%	Not critical	>57%	⊁10%	⊁5%
Chemical grade	>45%	—	—	—	⊁8%

(Stillwater, Kemi in Finland, Selukwe in Zimbabwe, Fiskenæsset); see Duke (1983).

The immediate country rocks of the complexes are ultrabasic differentiates of the parent gabbroic magma—dunites, peridotites and pyroxenites. The layers of these rocks have great lateral extent, uniformity and consistent positions within the complexes. The layers of massive chromite (chromitite) range from a few millimetres to over 1 m in thickness and can be traced laterally for as much as tens of kilometres. Orebodies may consist of a single layer or a number of closely spaced layers. Chromite in these deposits is usually iron-rich but an outstanding exception is the Great Dyke of Zimbabwe with its high chromium ores.

Bushveld Complex

This is an enormous differentiated igneous complex in the RSA (Fig. 10.1), which is generally considered to have resulted from the repeated intrusion of two main magma types into partly overlapping conical intrusions—these eventually coalesced into three larger magma chambers corresponding to the eastern, western and northern segments (Duke 1983, van Gruenewaldt et al. 1985). The chromite occurs in the western and eastern outcrops of ultrabasic rocks, with layers a few centimetres to 2 m thick

(Fig. 10.2). They make up an enormous tonnage (van Gruenewaldt 1977). Assuming a maximum vertical mining depth of only 300 m gives a figure of 2.3×10^9 t, a figure that can be multiplied by 10 if lower grade deposits are included and the vertical mining depth is increased to 1200 m. Potentially the largest orebodies are the LG3 and LG4 chromite layers present only in the western Bushveld. In these layers the chromite grades 50% Cr_2O_3, Cr/Fe = 2.0, the strike length is 63 km, the thickness 50 cm and the resources (300 m vertical depth) are 156×10^6 t. The ore grades about 45% Cr_2O_3. At present, however, the majority of mines exploit the 90–128 cm thick LG6 (Steelpoort) chromitite in the eastern Bushveld. In general this provides friable ore with small amounts of hard lumpy material. The friable ore is processed to remove interstitial silicates and a very consistent concentrator product is derived. In the 1990s prodigious amounts of chromite may be produced as a by-product from the mining of the UG2 layer (de Villiers 1990)—see p. 135.

The Bushveld chromite zone as a whole contains as many as 29 chromite layers or groups of layers. Above this zone is the platinum-bearing Merensky Reef (Figs 10.1, 10.2), and near the top of the basic part of the complex, vanadiferous magnetite layers occur.

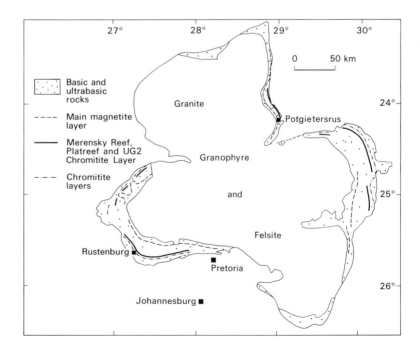

Fig. 10.1 Sketch map of the Bushveld Complex. (After van Gruenewaldt 1977.)

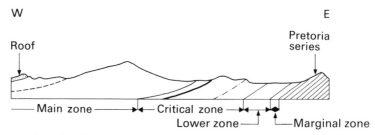

Section showing major zones in the Bushveld Complex, north of Steelport. Length of section, 30.5 km. (After Hall, 1932)

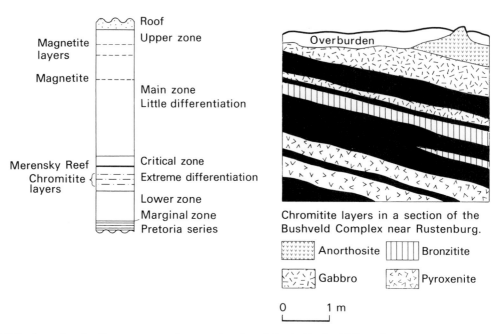

Chromitite layers in a section of the Bushveld Complex near Rustenburg.

Fig. 10.2 Sections showing the occurrence of economic minerals in the Bushveld Complex.

Great Dyke of Zimbabwe

This early Proterozoic (2470 Ma), mafic–ultramafic intrusion, 532 km long and 5–9.5 km broad, consists of four narrow, layered complexes. In cross section the mineral layers are synclinal (Fig. 10.3). Chromite layers occur along the entire length and individual layers extend across the entire width. The layers are in the range 5–100 cm and nearly all the chromite is the high chromium variety. Only layers 15 cm thick or more are mined. The Great Dyke has an estimated reserve of 10 000 Mt of chromite (Prendergast 1987) in as many as 11 persistent main seams.

Podiform deposits

The morphology of podiform chromite orebodies is irregular and unpredictable and they vary from sheet-like to pod-like in basic form, but can be very variable in their shapes—see Fig. 10.4. Their mass ranges from a few kilogrammes to several million tonnes. Throughout most of the world, production comes from bodies containing 100 000 t or so and reserves greater than 1 Mt are most uncommon, except in the Urals where some very big bodies occur, as in the Kempirsai Ultramafic Massif, a large ophiolite complex in the southern Urals (Fig. 10.5). Most pod deposits are of high chromium type, but

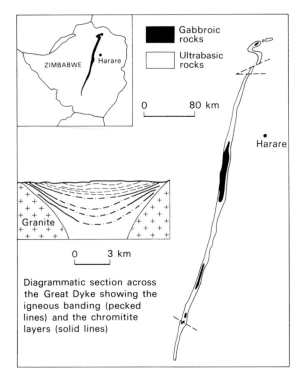

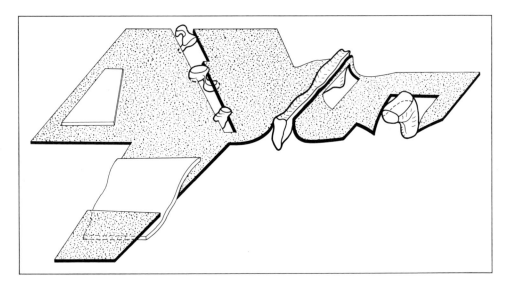

Diagrammatic section across the Great Dyke showing the igneous banding (pecked lines) and the chromitite layers (solid lines)

Fig. 10.3 Sketch diagrams illustrating the Great Dyke of Zimbabwe and the occurrence of chromitite layers in it. (In part after Bichan 1969.)

these deposits are also the only source of high aluminium chromite. The chromite layers are in the range 1–40 cm, or above.

Podiform deposits occur in irregular peridotite masses or peridotite–gabbro complexes of the Alpine type which are restricted mainly to orogenic zones, such as the Urals and the Philippine island arc. These are often ophiolites or parts of dismembered ophiolites. Most ophiolites are allochthonous and are believed to represent transported fragments of oceanic lithosphere. Within many large complexes the chromite deposits are generally near contacts between peridotite and gabbro, but where there is no gabbro the deposits appear to be distributed at random through the peridotite. The host intrusions are usually only a few tens of kilometres squared, or less, in area. Generally they are strongly elongated and lenticular in shape. Large numbers of these small intrusions occur in narrow zones (serpentinite belts) running parallel to regional thrust zones and the general trend of the orogen in which they occur. The intrusions are usually layered, but the layering does not often show the perfection nor the continuity of the stratiform deposits. They are short lenses rather than extensive sheets. Compositions range from dunite to gabbro and, whilst the average composition of stratiform hosts is close to gabbro, that of

Fig. 10.4 Diagram showing the shapes of podiform chromite deposits in New Caledonia and their relationships to the plane of the foliation in the host peridotites. (The foliation is nearly always parallel to the compositional banding.) Note the disturbance of the foliation around the two deposit types on the right. The tabular deposits are of the order of 50–100 kt. (After Cassard *et al.* 1981.)

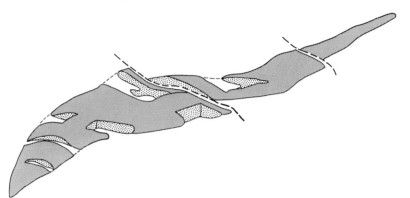

Fig. 10.5 The Molodezhnoe Orebody, southern Urals. Length 1.54 km, width 200–300 m, maximum thickness 140 m, ore reserves about 90 Mt. (After Pavlov & Grigor'eva 1977.)

Alpine intrusions is near peridotite. There are two subtypes, harzburgite and lherzolite. It is very important for exploration purposes to note that it is almost invariably the harzburgite subtype only that carries chromite deposits (Jackson & Thayer 1972); the only exceptions known to the writer are the minor and unusual occurrences in lherzolites of southern Spain and Morocco. The majority of podiform deposits occur in the uppermost 500–1000 m of olivine-rich harzburgite and dunite which make up the transition zone between normal harzburgite and cumulate crustal rocks. Most lie close to (either above or below) and parallel to the mappable harzburgite–dunite contact.

It is also important to remark that all major podiform chromite deposits are found in ophiolites formed in marginal basins and not in those formed at mid-ocean ridges (Pearce *et al.* 1984). Most podiform deposits are Palaeozoic or younger; many of them are Mesozoic or Tertiary. The largest known deposits are those of the Urals, Albania, the Philippines and Turkey. Significant deposits occur in Cuba, New Caledonia, Yugoslavia and Greece.

Whilst podiform deposits belong to mobile belts, stratiform deposits (apart perhaps from the Fiskenæsset Complex of western Greenland) were intruded into stable cratons. The Fiskenæsset Complex, being associated with what were oceanic basalts, may have been emplaced in stable oceanic crust. Thus, the two major types of chromite deposit belong to very different geological environments. Other differences are their general age—stratiform deposits with abundant chromite are Precambrian—and number of deposits. Whilst there are numerous podiform deposits, only a few stratiform deposits have produced chromite.

Disrupted stratiform deposits

Tectonic dismemberment of stratiform deposits has led to their individual parts being identified as podiform deposits. Recognition of parts of a once continuous stratiform complex can be very important in mineral exploration as it will lead to the search for missing segments, whereas a similar programme in a podiform chromite district could be financially ruinous (Thayer 1973). The chromite deposits of Campo Formoso in Brazil were formerly thought to be isolated blocks. They have been recently shown to be parts of a highly faulted, layered complex about 18 km long.

Genesis of primary chromite deposits

In most, perhaps all, stratiform complexes, emplacement of the magma and crystallization took place in a stable cratonic environment and many delicate primary igneous and sedimentary features, in particular layering, are preserved. They were thus originally emplaced in the upper crust. The host rocks of podiform deposits, on the other hand, probably originally crystallized in the mantle and were then incorporated into highly unstable tectonic environments in the crust by movement up thrusts and reverse faults. They are usually part of the ophiolite suite and were probably developed originally at mid-oceanic or back-arc spreading ridges. Despite their tectonic deformation they often preserve layering and textures due to crystal settling, some identical with those in stratiform deposits, indicating a common origin. The transport of the peridotite and chromite from the upper mantle into the upper crust, where they now are, probably

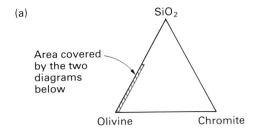

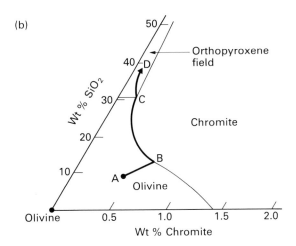

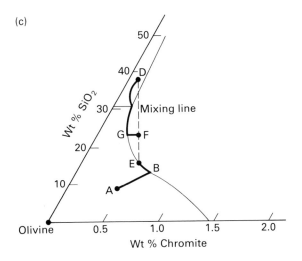

Fig. 10.6 The phase diagram for the system olivine–chromite–silica. (a) Plot of mineral compositions. (b, c) A small portion adjoining the olivine corner. The horizontal scale is greatly expanded. (Modified from Irvine 1977.)

occurred by plastic flowage at high temperatures possibly over many kilometres; this fragmented the original layering and has produced many metamorphic features in the chromite and their host rocks. These peridotites have usually suffered a high degree of serpentinization in contrast to the stratiform deposits, in which it is relatively negligible.

There is no doubt then that the chromite deposits of both stratiform and podiform deposits are orthomagmatic. The principal genetic problem is the chromitite layers in which chromite is the only cumulus mineral. During the normal course of fractional crystallization, olivine and chromite crystallize out together and chromite comprises at most a few per cent of the solid fraction. There must be some special control which 'pushes' the magma composition into the chromite field of the relevant phase diagram. Several possible mechanisms have been put forward including changes in oxygen fugacity, pressure and contamination of the magma by assimilation of roof rocks. At the present time, when much rhythmic layering in basic intrusions is considered to have resulted from the influx of many successive magma pulses (pp. 52, 54), a hypothesis of magma mixing, as proposed by Irvine (1977), gives an elegant solution. Figure 10.6b shows the normal crystallization course of a primitive basaltic fluid that plots at A. As cooling proceeds it precipitates olivine and thereby moves to B where *co-precipitation* of chromite commences and the liquid composition follows the cotectic to C where it enters the orthopyroxene field of precipitation and olivine *and* chromite are no longer formed. Suppose it has reached the point D when it is mixed with a later magma pulse that has evolved only as far as point E (Fig. 10.6c), then the resulting mixture could lie within the liquidus field of chromite at F to produce the precipitation of chromite *alone* as the magma changed in composition to G and resumed the more normal evolutionary path along the cotectic to the orthopyroxene field.

Constitution of chromite concentrations

As will have been gathered from the foregoing, there are some chemical differences between the chromites from stratiform and podiform deposits. These, together with the relationship of chromites from the Fiskenæsset Complex of western Greenland, are shown in Fig. 10.7. Apart from the differences brought out by the figure, it should be noted that the Fiskenæsset chromites are

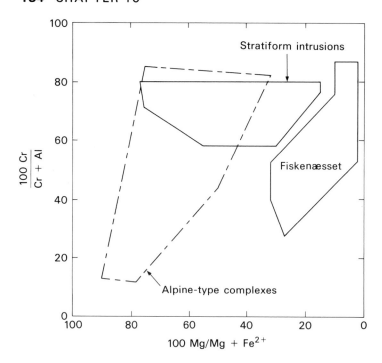

Fig. 10.7 Plot of mg against cr for spinels from stratiform and Alpine type complexes together with the Fiskenæsset Complex of western Greenland. (After Steele *et al.* 1977.)

vanadiferous (about 1.5% V_2O_5) like the magnetites of the upper basic levels of the Bushveld Complex; and, like the Bushveld magnetite layers, the Fiskenæsset chromitites occur in the upper part of the intrusion. These differences are perhaps to be attributed to crystallization from a water-rich magma in an oceanic environment, compared with the dry magmas of the Bushveld and similar lopoliths.

Useful discussions of chromite deposits can be found in Stowe (1987).

Platinum group metals (PGM)

These metals used to be produced entirely from placer deposits, but now primary deposits are more important. From 1778 to 1823 Colombia was the only producer, then in 1822 placers were also discovered in the Urals. In 1919, the recovery of by-product platinum metals from the Sudbury copper–nickel ores of Canada started. In 1924, the South African deposits in the Bushveld Complex were found and in 1956 South African production outstripped Canadian. The principal producers are now: (a) RSA, (b) USSR, and (c) Canada; for production figures see Table 10.2 and for reserves and grades see Table 10.3.

Table 10.2 Platinum and palladium mine production (kg) in 1987

USSR	115 000[a]
RSA	113 000
Canada	10 930
USA	3600[a]
Japan	2170
Colombia	699
Others	946
World total	264 000

[a] Estimated.

Primary deposits

There are two principal types of deposit exemplified firstly by those associated with the basic intrusions of Noril'sk-Talnakh, USSR and Sudbury, Ontario; and secondly by those of the basic–ultrabasic intrusions of the Bushveld Complex, RSA and Stillwater, Montana. In the former, platinum minerals occur in liquation ores at the base of the intrusions and in the latter, platinum occurs in the Merensky Reef and similar layers some 2000 m above the base of the intrusions. The platinum in these deposits occurs

Table 10.3 Principal reserves of all platinum group metals (i.e. Pt + Pd + Rh + Ru + Ir + Os *in tonnes of metal*) in major deposits and their grade. (Data from Naldrett (1989) is taken from 1979 and 1981 publications)

Deposit	Tonnage (t)	Grade (ppm)
Bushveld Complex	60 388	8.27
Stillwater Complex	1057	22.3
Great Dyke	7892	4.7
Noril'sk-Talnakh	6200	3.8
Sudbury	217	0.9

mainly as arsenides, sulphides and antimonides (Mertie 1969).

A third type of primary deposit is sometimes important. It consists of deposits of platinum metals in peridotites and perknites, commonly in dunite and serpentinite, less commonly in pyroxenic rocks containing no olivine. In these rocks the platinum occurs mainly as platinum alloys associated with lenticular masses of chromite or disseminated through the host rock in association with chromite. Such deposits have occasionally yielded small orebodies of phenomenally high grade, but are more important as source rocks for the formation of platinum placers, which may be particularly rich in Ru, Ir and Os, reflecting the high proportion of these elements compared with Pt and Pd in the source rocks.

The Sudbury copper–nickel ores will be covered in the next chapter. Attention will therefore be focused here on the Bushveld Complex on which an excellent summary of the voluminous literature is to be found in Naldrett (1989).

Bushveld Complex

In this intrusion there are three very extensive deposits: the Merensky Reef, the UG2 Chromitite Layer (both in the western and eastern Bushveld) and the Platreef in the Potgietersrus area. The first-named occurs in the Merensky Zone—an exceedingly persistent igneous zone traced for over 200 km in the western Bushveld where it has been proved to contain workable ore for over 110 km (Fig. 10.1). This zone is also well developed in two other districts. It is 0.6–11 m thick and generally consists of dark-coloured norite.

The Merensky Reef is a thin sheet of coarsely crystalline pyroxenite with a pegmatitic habit that lies near the base of the Merensky Zone. It has a thickness of 0.3–0.6 m. Chromite bands approximately 1 cm thick mark the top and bottom of the reef and are enriched with platinum metals relative to the reef. The platinum minerals are sperrylite, $PtAs_2$; braggite $(Pt,Pd,Ni)S$; stibiopalladinite, Pd_3Sb and laurite, RuS_2, together with some native platinum and gold. Grades at Rustenberg are 7.5–$11 g t^{-1}$, but grades elsewhere are lower. The average stoping width is 0.8 m and the exploitable ore contains 3.3×10^9 t. Sulphides (chalcopyrite, pyrrhotite, pentlandite, nickeliferous pyrite, cubanite, millerite and violarite) are also present, and copper (0.11%) and nickel (0.18%) are recovered.

The UG2 Chromitite Layer lies 150–300 m below the Merensky Reef. It is 60–250 cm thick, carries 3.5–19 g t^{-1} platinum together with copper and nickel values and occurs within a bronzite cumulate which overlies a zone of bronzite–plagioclase-pegmatoid. Chromium will also be produced. The chromitite consists of 60–90 vol % chromite, 5–25 vol % bronzite and 5–15 vol % plagioclase with accessory clinopyroxene, base metal sulphides, PGM minerals, ilmenite, magnetite, rutile and biotite. There is a close correlation of sulphides and PGM mineralization. The reserves are far greater than for the Merensky Reef and total about 5.42×10^9 t. The Platreef has been correlated with the Merensky Reef but it is much thicker, being up to 200 m with rich mineralization over thicknesses of 6–45 m and the mineralization is never far from the floor of the intrusion. Grades are very irregular, estimates ranging from 7 to 27 ppm total PGM. Ore reserves are about 4.08×10^9 t.

Other deposits

The South African ores are essentially platinum ores with by-product nickel and copper. A horizon, the J-M Reef, similar to the Merensky Reef has been traced within the Banded Zone of the Stillwater Complex, Montana over its entire exposed length of 39 km. The host section of the Banded Zone consists dominantly of plagioclase-rich cumulates. The J-M Reef itself is a plagioclase–olivine cumulate enriched in sulphides. Discontinuities and stacking of ore layers may have been created by the action of convection cells (Raedeke & Vian 1985). The Great Dyke of Zimbabwe mineralization is very different from that of the Merensky and J-M Reefs but again occurs in sulphide-rich zones (Prendergast 1990, Wilson & Treadoux 1990). The huge Dufek Complex in Antarctica also may have a PGM rich horizon (de Wit 1985). At Sudbury, Ontario, the platinum metals are by-products of copper–nickel ores and the grades of the platinum metals are much lower, being about 0.6–0.8 g t^{-1}. Similar ores to those at Sudbury occur at Noril'sk in Siberia and the grade is said to average 3.8 g t^{-1}.

Naldrett (1989) drew attention to another class of layered basic complexes occurring in the rifted continental environment that appear to have considerable potential for the future. A good example of these is the Penikat Layered Intrusion of northern Finland (Alapieti & Lahtinen 1986), which is part of an early Proterozoic belt of layered intrusions that extends 300 km from the Russian border right across Finland to the Swedish border. Three major mineralized zones enriched in PGM have been delineated along strike for almost 23 km in this intrusion, which is only up to 2.3 km thick.

Genesis of platinum deposits

The genesis of the platinum-rich layers in the Bushveld Complex is still very much an open question. The mystery centres on why the platinum and associated sulphides are concentrated into just a few thin layers. There is a strong correlation between PGM and sulphur contents in these layers supporting the idea that sulphides acted as collectors scavenging the PGM from the silicate magma. The sulphides of both the Bushveld and Stillwater Complexes, however, contain very much higher concentrations of PGM than the similar Fe–Ni–Cu magmatic sulphides of orthomagmatic nickel deposits (e.g. Noril'sk and Sudbury), which are believed to have acquired PGM in an identical manner. By contrast the Ni and Cu contents of the sulphide phase are similar to those found in sulphides formed in equilibrium with gabbroic rocks, such as at Noril'sk, Sudbury, etc. Many ingenious hypotheses for these high platinum concentrations have been put forward, e.g. Vermaak (1976) who suggested that the platinum layers were enriched in PGM by ascending, fractionated, magmatic fluids displaced from underlying cumulates as these compacted. Others, e.g. Ballhaus & Stumpfl (1985), have suggested deposition from hydrothermal solutions, citing the evidence of fluid inclusions, high F and Cl values and the presence of graphite in support of their arguments. Naldrett (1989) has built on his own and others' work to produce a hypothesis in which, as a result of unusual magmatic events, sulphide droplets come into contact with large volumes of turbulent convecting magma allowing them to scavenge much PGM from the magma. When cooling gives rise to laminar flow the sulphides and suspended crystals sink to form a cumulate layer such as the Merensky Reef. This model is very convincing and the interested reader is strongly recommended to study Naldrett's account.

Both the nature and origin of the platinum mineralization in the Platreef differ completely from those of the Merensky Reef and the UG2 Chromitite Layer. The mineralization in the Platreef is thought to have resulted from contamination of the magma by country rocks (Buchanan & Rouse 1984, Cawthorn et al. 1985).

Titanium

Titanium metal and TiO$_2$ pigments are the two main products of titanium mines and on them large industries depend. The pigment industry consumes over 90% of all titanium mined. Only titanium-bearing oxide minerals with more than about 25% TiO$_2$ have present or potential economic value and all titaniferous silicate minerals are valueless. The most important oxides are rutile (> 95% TiO$_2$) and ilmenite (about 52% TiO$_2$). The market price of a titanium mineral concentrate depends on its TiO$_2$ content *and* its suitability for a given industrial process. This means that different deposits obtain very different prices for their product. At present shoreline placer deposits supply over half the annual production, with nearly all the rest coming from primary magmatic sources. A fluvial placer deposit produces rutile in Sierra Leone and a residual deposit on alkalic igneous rock produces anatase in Brazil. Titanium production was one of the few sectors of the minerals industry that suffered no downturn in the 1980s and prices are climbing. The development of substitutes for titanium pigments or metal seems remote and, as suggested in Chapter 1, titanium appears to be one of the few metals with a bright future. The market situation will change considerably over the coming decades as most currently known shoreline placers will be exhausted in about 30 years. New types of deposit are already being sought. In addition to residual deposits (p. 267), these could be eclogites carrying 6% or more TiO$_2$, rutile-bearing contact-metasomatic zones of alkalic anorthosites, pervoskite-bearing pyroxenites, rutile by-products of porphyry copper deposits, placer deposits on continental shelves and other possibilities (Force 1991).

In this chapter we will consider orthomagmatic ilmenite deposits; these are always associated with anorthosite or anorthosite–gabbro complexes. Anorthositic massifs show a spectrum from bodies of labradoritic composition to those dominantly andesine. Workable titanium ores appear to be restricted to the andesine anorthosite massifs, but the major ores—ilmenite-rich rocks—cut the anorthosites, often having a dyke-like form; the anorthosite functions only as country rock.

At Allard Lake, Quebec, the ores grade 32–35% TiO_2 and occur in anorthosites. The Lac Tio deposit contains about 125 Mt ore and, at present, this single deposit supplies 19% of world production. It lies 40 km inland from Havre St Pierre on the north shore of the St Lawrence in an uninhabited area and is reached by a company railway. The ore is coarse-grained, with ilmenite grains containing exsolved hematite blebs up to 10 mm in diameter. Gangue minerals are chiefly plagioclase, pyroxene, biotite, pyrite, pyrrhotite and chalcopyrite. The exsolution hematite is too fine-grained to be sepa-rated by grinding from the ilmenite, and dilutes the ilmenite concentrate. The orebodies form irregular lenses, narrow dykes, large sill-like masses and various combinations of these forms. Some of these clearly cut the anorthosite and appear to be later in age.

The world's largest ilmenite orebody is at Tellnes in the anorthosite belt of southern Norway about 120 km south of Stavanger. The deposit is boat-shaped, elongated north-west, 2.3 km long, 400 m wide and about 350 m deep (Fig. 10.8). It occurs in the base of a noritic anorthosite. Proven reserves are 300 Mt of 18% TiO_2, 2% magnetite and 0.25% sulphides (pyrite and Cu–Ni sulphides). The ilmenite carries up to 12% hematite as exsolution lamellae (Fig. 5.1a). The ilmeno-norite intrudes the anorthosite with which it has sharp contacts. The ilmeno-norite intrusion's shape is apparently an original feature of intrusion. The annual production is 2.76 Mt of ore, and extraction of this by opencast methods means that 1.36 Mt of waste have to be removed.

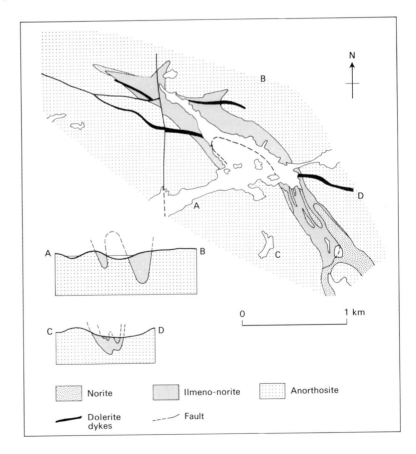

Fig. 10.8 Map and sections of the Tellnes titanium orebody. (After Dybdahl 1960.)

Field relations, textures and the results of experimental petrology support the interpretation of these ores as igneous rocks formed from magmas generated by immiscibility within parent intrusions. Titanium-rich droplets appear to have separated out and then, because of their high density, they sank to the magma floor to form iron–titanium-rich heavy liquids. These liquids then escaped through fractures to intrude structurally lower units, which are most commonly anorthosite, to form discordant ilmenite deposits. The accumulative process and the evidence supporting it are analogous to the sulphide immiscibility that gives rise to orthomagmatic nickel deposits (Chapter 4). For further discussion and relevant references the reader is referred to Force (1991).

Iron and vanadium

Many small to medium sized magnetite deposits occur in gabbroic intrusions but the really big tonnages occur in the stratiform lopoliths. Some of these deposits are worked for vanadium as well as for iron. This is the case with the vanadiferous magnetite in the Upper Zone of the Bushveld Complex (Fig. 10.1). This carries 0.3–2% V_2O_5. Values above 1.6% V_2O_5 are usually found in the Main Magnetite Layer and several thinner layers below it, but only the main layer (some 1.8 m thick) can be considered as ore. Based on a vertical limit of 30 m for opencast mining, the reserves are about 1030 Mt. Much more iron ore than this is present, but it is spoilt by a titanium content of up to 19%. About 50% of total world vanadium production ($\sim$ 33 467 t in 1988) now comes from the Bushveld Complex. Similar occurrences are present in the USSR (29%) and China (12%). Most vanadium production (about 85%) goes into steel manufacture, 10% into titanium alloys and 5% into chemicals.

The exact mechanism by which those oxide-rich layers form is still uncertain. It is generally considered that the precipitation of copious quantities of titaniferous magnetite is triggered by episodic increases in f_{O_2}, but the process giving rise to such increases is still uncertain. The problem, as far as the Bushveld Complex is concerned, has been reviewed by Reynolds (1985) who considers that a complex interplay of factors resulted in copious Ti–V–magnetite precipitation. These included concentration of Fe, Ti and V in the residual magma and large scale, *in situ* bottom crystallization of plagioclase with development of a layer of stagnant magma above, from which the magnetite crystallized, as well as changes in f_{O_2}, temperature and f_{H_2O}/f_{H_2}.

At the present time about 7% of vanadium production is from heavy crude oils rich in sulphur that may contain several hundred parts per million vanadium; such oils are exported from Venezuela and Mexico. Production from this source will probably increase in the future. A minor amount of by-product vanadium is won from some uranium deposits see (Chapter 17).

11 / Orthomagmatic copper–nickel–iron (–platinoid) deposits associated with basic and ultrabasic rocks

Introduction

World nickel metal production is currently running at about 800 000 t p.a. It is a metal that commands a high price: £4772 t^{-1} compared with £1309 t^{-1} for copper in July 1991. Nickel is produced from two principal ore types: nickeliferous laterites and nickel sulphide ores. We are concerned with the latter deposit type in this chapter. These deposits usually carry copper, often in economic amounts, and sometimes recoverable platinoids. Iron is produced in some cases from the pyrrhotite concentrates, which are a by-product of the dressing of these ores. Nickel sulphide deposits are not common and so just a few countries are important for production of nickel sulphide ores; Canada is pre-eminent, while the USSR and Australia are the only other important producers. A small production comes from Zimbabwe, RSA, Botswana and Finland.

The mineralogy of these deposits is usually simple, consisting of pyrrhotite, pentlandite ((Fe, Ni)$_9$S$_8$), chalcopyrite and magnetite. Ore grades are somewhat variable. The lowest grade of a working deposit in western countries appears to be an Outukumpu (Finnish) mine working 0.2% Ni ore. This low grade can be compared with the very high grade sections of some Western Australian deposits that run about 12% Ni, and the Kambalda ores range from 8 to 22% Ni in 100% sulphide ore (Cowden & Woolrich 1987). Of course, the overall grade for Australian deposits is less than this because lower grade ore is mined with these high grades. Therefore, although such high grades occur at Kambalda in the Western Mining Corporation mines, the ore reserves at June 1984 were 26 Mt running 3.3% Ni. In this mining camp the ore treated in 1983–84 totalled 1.373 Mt having a mill grade of 3.43% and nickel recovery of 92.4%.

All nickel sulphide deposits are associated with basic or ultrabasic igneous rocks. There is both a spatial and a geochemical relationship in that deposits associated with gabbroic igneous rocks normally have a high Cu/Ni ratio (e.g. Sudbury, Ontario; Noril'sk, USSR), and those associated with ultrabasic rocks a low Cu/Ni ratio (e.g. Thompson Belt, Manitoba; Western Australian deposits).

Classification of ultrabasic and basic bodies with special reference to nickel sulphide mineralization

There are a number of different types of basic and ultrabasic rocks; not all of these have nickel sulphide deposits associated with them and therefore, in exploring for these deposits, it is important to know which classes of basic and ultrabasic rocks are likely to have associated nickel sulphide ores. A classification is given in Table 11.1 with an indication of the tectonic settings. Particular combinations of rock type and tectonic setting have proved to be especially productive. These are:
1 noritic rocks intruded into an area that has suffered a catastrophic release of energy, e.g. an astrobleme (Sudbury);
2 intrusions associated with flood basalts in intracontinental rift zones (Noril'sk-Talnakh, Duluth);
3 komatiitic and tholeiitic flows and intrusions in greenstone belts (Kambalda, Agnew, Pechenga).

Bodies in an orogenic setting (class A)

Two broad groups of ultrabasic and basic bodies can be seen in this setting: bodies coeval with plate margin volcanism and syntectonic intrusions.

The first group occurs in the Archaean and Proterozoic greenstone belts and can be divided into the tholeiitic and komatiitic suites. The tholeiitic suite contains the picritic and anorthositic classes. The anorthositic class is important for titanium mineralization but so far no substantial nickel mineralization has been found in rocks of this class. The picritic class is an important nickel ore carrier and ultrabasic rocks in this class occur as basal accumulations in differentiated sills and lava flows, some having basal sulphide segregations—the Dundonald Sill of the Abitibi Greenstone Belt, Canada, is a good example. The tholeiitic activity in this and other areas was often contemporaneous

Table 11.1 Classification of basic and ultrabasic bodies and associated nickel mineralization. (Based on Naldrett & Cabri 1976, Besson *et al*. 1979, Naldrett 1981, 1989 and Eckstrand 1984)

Examples	Remarks	Examples of associated nickel deposits
Class A: Bodies emplaced in active orogenic areas 1. Bodies contemporaneous with plate margin volcanism, largely restricted to Archaean greenstone belts		
(i) Tholeiitic suite: (a) Picritic subtype Dundonald Sill, Ontario; Eastern Goldfields, Western Australia	Differentiated sills and stock-like intrusions, 1–10 Mt orebodies	Pechenga, USSR; Lynn Lake, Manitoba; Carr Boyd, Western Australia
(b) Anorthositic subtype Doré Lake Complex, Quebec; Bell River Complex, Ontario; Kamiscotia Complex, Ontario	Some examples of this class are conformable, others appear to be discordant	No known deposits
(ii) Komatiitic suite: (a) Lava Flows, Munro Township, Ontario; Eastern Goldfields, Western Australia	Komatiites are principal hosts, also closely underlying small sills, in some cases adjacent metasedimentary or metavolcanic rocks. 1–5 Mt orebodies, commonly several orebodies per deposit	Langmuir, Ontario; Marbridge, Quebec; Kambalda, Windarra, Western Australia; Shangani, Inyati-Damba, Zimbabwe
(b) Intrusive dunite lenses,[a] Eastern Goldfields, Western Australia; Dumont, Quebec; Thompson Nickel Belt, Manitoba	Significant development in Proterozoic as well as Archaean rocks. Highly magnesian ultramafic sills larger than those in (iia). Segregated orebodies up to tens of Mt, disseminated up to 250 Mt	Mt Keith, Western Australia; Dumont, Ungava, Quebec; Pipe, Birchtree, Manibridge, Manitoba
2. Bodies emplaced during orogenesis generally in Phanerozoic orogens		
(i) Synorogenic intrusions: Aberdeenshire Gabbros, Scotland; Rôna, Norway; Moxie, USA	Only small nickel deposits known	
(ii) Large obducted sheets: New Caledonia; Papua New Guinea	No significant mineralization	
(iii) Ophiolite complexes: Troodos, Cyprus; Bay of Islands, Newfoundland; Luzon Complex, Philippines	Only small deposits known	Acoje Mine, Philippines
(iv) Alaskan-type complexes: Duke Island and Union Bay, Alaska; Northern California; Urals; British Columbia; Venezuela	Concentrically zoned intrusions	No economic deposits known

Table 11.1 *Continued*

Examples	Remarks	Examples of associated nickel deposits
Class B: Bodies emplaced in cratons 3. Largely stratiformly layered complexes Bushveld, RSA; Stillwater, Montana; Sudbury, Ontario; Great Dyke, Zimbabwe; Jimberlana, Western Australia	Individual orebodies run up to tens of Mt. Each intrusive complex generally contains a number of orebodies	Bushveld mines produce PGM with by-product nickel. Sudbury production + reserves = 700 Mt in numerous deposits
4. Intrusions related to flood basalts and associated with rifting Duluth, Minnesota; Noril'sk-Talnakh, USSR; Dufek Complex, Antarctica; Insizwa–Ingeli Complex, RSA; Palisades Sill, New Jersey	These intrusions generally occur in areas in which flood basalts are developed. They are chemically similar to the extruded basalts. Segregated orebodies up to tens of Mt. Enormous resources of disseminated mineralization present in some complexes, e.g. Duluth	Noril'sk-Talnakh, USSR
5. Medium- and small-sized intrusions associated with rifted plate margins and ocean basins		
(i) Komatiitic: Cape Smith, Quebec; Thompson Belt, Manitoba	Often much deformed orebodies associated with serpentinized ultramafic rocks	Manitoban Ni belt
(ii) Largely gabbroic: Kemi-Koillismaa Belt, Finland; Skaergaard, Greenland; Rhum, Scotland	Some small, low grade orebodies	
6. Alkalic ultramafic rocks in ring complexes, kimberlite and lamproite pipes	No known examples	

[a] Hill & Gole (1985) have questioned the intrusive nature of bodies of this group in Western Australia and have suggested that these rocks are mainly or wholly volcanic.

with komatiitic volcanicity. The komatiitic suite is a much more important carrier of nickel mineralization. Disagreements persist regarding the definition of komatiite and the full range and authenticity of the komatiitic suite (Arndt & Nisbet 1982a, Best 1982). Many workers consider that the komatiitic suite ranges from dunite (> 40 wt % MgO) through peridotite (30–40%), pyroxene–peridotite (20–30%), pyroxenite (12–20%) and magnesian basalt to basalt, thus forming a magma series with the status of the tholeiitic or calc-alkaline series. The term komatiitic basalt is applied to those rocks with less than 18% MgO that show a definite spatial and chemical link with komatiites. Komatiites are both

extrusive and intrusive, and ultrabasic members are believed to have crystallized from liquid with up to 35 wt % MgO and carrying 20–30% of olivine phenocrysts in suspension. In some flows and near surface sills, quench textures (probably due to contact with sea water and consisting of platy and skeletal olivine and pyroxene growths) are present in the upper part. This is called spinifex texture. These flows clearly crystallized from magnesian-rich undifferentiated magma, extruded (in the case of 35 wt % MgO) at up to 1650°C. Spinifex textures resemble those of silica-poor slags, having a low viscosity and a high rate of internal diffusion—ideal conditions for the sinking of sulphide droplets to form accumulations at the flow bottom (see pp. 54–55). It is very unlikely that the sulphides were in solution when rapid crystallization started or they would have been trapped before they could sink to the flow bottom. They were therefore probably already in droplet form when the magma was erupted.

The recognition of the presence of komatiites in an exploration area is of great importance and the nickel prospector should be aware of the distinguishing characteristics, both petrographic and geochemical, of these rocks. Excellent descriptions of komatiites are to be found in Arndt & Nisbet (1982b).

The synorogenic intrusions are of little importance as carriers of nickel sulphide ores, but constitute a small resource for the future. The large obducted sheets and ophiolite complexes figure large in the preceding chapter but are of little note as hosts for nickel mineralization. Alaskan-type complexes form concentrically zoned intrusions, which are well developed along the Alaskan pan-handle. As a group they are distinguished from Alpine-type ultrabasics or stratiform intrusions by highly calcic clinopyroxene, no orthopyroxene or plagioclase, much hornblende, more iron-rich chromite, and magnetite. The last-named occurs in concentrations that are occasionally of economic interest.

Bodies emplaced in a cratonic setting (class B)

There are three main groups to be noted. The first is an important metal producer because it is that of the large stratiform complexes. In the last chapter, we noted the importance of the Bushveld Complex and its by-product nickel–copper won from the platinum-rich horizons. The Sudbury intrusion also belongs to this group and it hosts the world's greatest known concentration of nickel ores. The overall composition of this group is basic rather than ultrabasic but a lower ultrabasic zone is usually present. Sudbury is a notable exception to this rule but it possesses a sublayer rich in ultrabasic xenoliths probably derived from a hidden layered sequence.

The second group includes intrusions related to flood basalts and usually associated with the early stages of continental rifting. Very important here are the gabbroic intrusions of the Noril'sk-Talnakh Nickel Field. The Duluth Complex and the troctolite hosting the Great Lakes nickel deposit of Ontario are included in this group as they are both in a rift setting and are petrogenetically related to the Keweenawan flood basalts.

Thirdly there is a group of medium- and small-sized intrusions associated with rifted plate margins and ocean basins which can divide into two subgroups. First are a number of intrusive and extrusive, mafic and ultramafic bodies (including komatiites) emplaced within Proterozoic sedimentary sequences that appear to be parts of a rifting event along Archaean continental margins, particularly the Circum-Superior Belt of Canada. This belt includes the Thompson Belt, Manitoba and other belts carrying nickel deposits. The second subgroup contains intrusions associated with much more recent continental rifting, such as Skaergaard and Rhum, which carry nickel mineralization of only academic interest.

Relationship of nickel sulphide mineralization to classes of ultrabasic and basic rocks

Figure 11.1 indicates the relationship between known reserves plus past production of nickel in the main deposits of the world and the rock classes given in Table 11.1. Apart from the unique position of Sudbury, production from there being responsible for much of the Canadian section of classes 3, 4 and 5 komatiitic magmatism [1(ii)] is clearly the most important. Tholeiitic volcanism is much less important. Deposits near the basal contacts of the Stillwater and Duluth Complexes (classes 3 and 4) are low grade disseminated deposits which are unlikely to be producers in the near future.

The Noril'sk-Talnakh Field, like Sudbury, has many unusual features, which further emphasizes the importance of class 1(ii) as the best bet for further nickel exploration.

Genesis of sulphur-rich magmas

For a rich concentration of magmatic sulphides it is necessary that (a) the host magma is saturated in sulphur, and (b) a reasonably high proportion of sulphide droplets can settle rapidly to form an orebody. Slow settling may give rise to a disseminated, uneconomic ore. The production of a high proportion of immiscible sulphides is possible if the magma assimilates much sulphur from its country rocks, e.g. Duluth, Noril'sk, or if the magma carries excess amounts of mantle-derived sulphides, e.g. some komatiites. We can often differentiate between these two sources (crustal and mantle) by examining

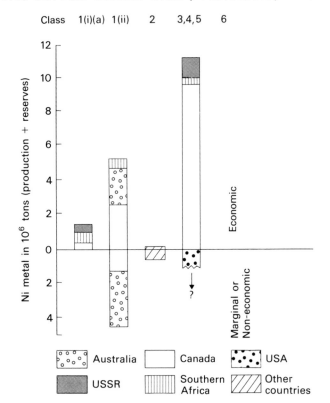

Fig. 11.1 Present reserves plus past production of sulphide nickel as a function of the class of host rock. The key to the classes is given in Table 11.1. (Modified from Naldrett 1973, 1989 and Naldrett & Cabri 1976.)

variations in $^{34}S/^{32}S$. These are reported in the delta notation (p. 87). For this work the standard is troilite of the Cañon Diablo meteorite, which is taken to be equivalent to the earth's mantle and for which $\delta^{34}S = 0‰$. Originally, all the earth's sulphur was in the mantle and much of that transferred to the crust has undergone biogenic fractionation producing an enrichment in ^{34}S and giving $\delta^{34}S$ values as high as +30‰. It is now held to be an oversimplification to claim that for all natural sulphur a narrow range of $\delta^{34}S$ close to zero indicates a mantle source, and a wide range of $\delta^{34}S$ a crustal source, but this statement is still probably true for the majority of magmatic systems.

Let us look first at two examples where there is no evidence to suggest that the sulphur came from anywhere but the mantle. At Sudbury, $\delta^{34}S$ varies from -0.2 to $+5.9‰$ with a mode of 1.7‰; this is a narrow spread of values close to $\delta^{34}S = 0‰$ and suggestive of a mantle origin. Similarly, ore associated with komatiites at the Alexo Mine, Ontario gives $\delta^{34}S = 4.4–6.2‰$; again close to a zero value but perhaps showing a little temperature fractionation. At Noril'sk, however, the values are $+7$ to

+17‰. The Noril'sk Gabbro is Triassic (an unusual and perhaps unique situation as practically all economic nickel mineralization is early to middle Precambrian), and its magma is intruded through gypsum beds, where it is thought to have picked up ^{34}S-rich sulphur. A comparable example is the Water Hen intrusion of the Duluth Complex with values of $+11$ to $+16‰$. This is believed to have gained its sulphur by assimilating sulphur-rich sediments of the Virginia Formation in which $\delta^{34}S$ ranges from $+17$ to $+19‰$ (Mainwaring & Naldrett 1977). Further evidence for the assimilation origin of the sulphur in Cu–Ni mineralization in the Duluth Complex comes from the work of Ripley (1990) who has demonstrated similar Se : S ratios in the Babbit deposit Cu–Ni sulphides and the Virginia Formation. The mineralized gabbros of northern Maine have also yielded evidence of having acquired much of their sulphur by the assimilation of country rocks (Naldrett *et al.* 1984). Clearly basic intrusions in sulphur-rich country rocks should be prospected carefully for nickel mineralization!

In the sulphides of komatiitic-related deposits of Western Australia, $\delta^{34}S$ for a number of deposits ranges from -2.6 to $+3.4$‰ with a mode of nearly zero. This is clearly sulphur of direct or recent mantle origin, but it must be pointed out that the interflow sediments are rich in sulphur of almost identical $\delta^{34}S$ values (Groves & Hudson 1981). These sediments represent a mixture of volcaniclastic detritus and chemical–sedimentary material derived largely from volcanic exhalations between times of lava extrusion. The associated komatiitic magmas may therefore have brought much of their sulphur up from the mantle but could also have acquired more by assimilation of interflow sediments as they flowed over them (Huppert & Sparks 1985). Many writers at the moment favour the hypothesis that komatiites incorporate sulphur during their genesis by partial melting deep within the mantle.

We must now consider the problem of how some mantle-derived magmas could acquire a high sulphur content. This was discussed by Naldrett & Cabri (1976) with the aid of the relationships shown in Fig. 11.2, which shows the degree of melting of the mantle and the oceanic geothermal gradients for the present day and the Archaean. Consider mantle material at A; it is sufficiently above the solidus to yield a 5% partial melt. Any slight perturbation (such as a tectonic effect, accession of H_2O or magma from the descending slab of a Benioff Zone, normal convective overturn) would cause a diapir tens of kilometres across to rise adiabatically to B. Partial melting would increase to 30% and if this melt separated from the diapir it might rise to the surface along the non-adiabatic curve BF. The continued rise of the diapir would produce further partial melting but now the melt would form from depleted mantle and it would be much more magnesian since basaltic material has been removed. Partial melting might reach 30% at C (the contours hold for undepleted mantle only) to produce a much more magnesian (komatiitic) magma, which might be intruded or extruded at E.

Figure 11.2 shows that the zone of general sulphide melting intersects the Archaean geothermal gradient at about 100 km depth. Below this depth sulphides were molten and probably percolated downwards leaving a zone depleted in sulphides and producing a deeper enriched zone—this sulphur-enriched zone will lie at the level from which komatiitic magmas are ultimately derived. In this way, we can explain the relationship between komatiites and sulphide ores and the fact that they may be associated with tholeiitic picrites, which themselves may carry sulphide ores, as in the Abitibi region of Ontario.

It must be noted, however, that another school of thought, e.g. Keays (1982), opposes this concept. Keays contended that the low Pd/Ir ratios of komatiites implies that their magmas formed during

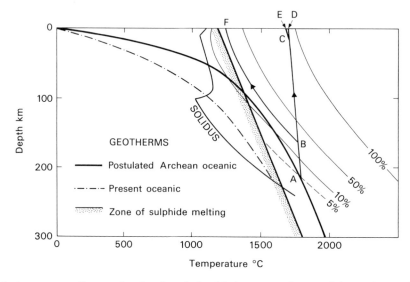

Fig. 11.2 A depth–temperature diagram showing the relationship between estimates of the modern and Archaean oceanic geotherms and melting relations of possible mantle material (pyrolite II + 0.2 wt % H_2O); drawn to illustrate the generation of komatiitic magma. (After Naldrett & Cabri 1976.)

sulphur-undersaturated conditions and that sulphur saturation within the mantle would have led to the loss of sulphides owing to pressure release during the ascent of the magma.

Nickel mineralization in time and depletion of sulphur in the mantle

The only large sulphide deposits which are younger than 1700 Ma are Duluth (1115 Ma) and Noril'sk (Triassic). As we have seen, both host intrusions probably acquired most of their sulphur by assimilation of crustal rocks. This restriction of economically significant orthomagmatic nickel deposits to the Archaean and early Proterozoic may reflect sulphur depletion of the upper mantle caused by many cycles of plate tectonic or similar processes, if one accepts Naldrett's and Cabri's hypothesis. On the other hand, the concentration of nickel sulphide mineralization early in the stratigraphical column may be related largely to the concentration of komatiitic activity early in earth history, no matter whether the necessary sulphur was acquired deep within the mantle or by assimilation at or near the surface.

Origin of the metals

There is no difficulty here. Ultrabasic and basic magmas are rich in iron and in trace amounts of copper, nickel and platinoid elements. These would be scavenged by the sulphur to form sulphide droplets. The reason why other chalcophile elements, such as lead and zinc, are not present in significant amounts in nickel sulphide ores has been explained by Shimazaki & MacLean (1976). This process of nickel scavenging leads to the development of nickel-depleted olivines in the silicate phase, a circumstance which may have great exploration significance (Naldrett *et al.* 1984, Naldrett 1989).

Examples of nickel sulphide orefields

Volcanic association

Here we are concerned mainly with class 1(ii) (Table 11.1), the komatiitic suite. In some of these, sulphides occur at or near the base of the flow or sill suggesting gravitational settling of a sulphide liquid. Typical sections through two deposits are given in Fig. 11.3. These have certain features in common:

1 massive ore at the base (the banding in the Lunnon orebody is probably the result of metamorphism);
2 a sharp contact between the massive ore and the overlying disseminated ore, which consists of net-textured sulphides in peridotite;
3 another sharp contact between the net-textured ore and the weak mineralization above it, which grades up into peridotite with a very low sulphur content.

This situation is strongly reminiscent of the billiard ball model of sulphide segregation described on p. 55.

In nearly all examples of this deposit type the orebodies are in former topographical depressions beneath the ultramafic lava flow. Some of these depressions, but by no means all, appear to be fault related while others may be thermal erosion channels formed by these exceptionally hot and very fluid lavas (Huppert & Sparks 1985). The depressions are narrow and elongate with length-to-width ratios of about 10 : 1 and are a few metres to 100 m deep. Orebodies may have formed in these depressions not just by simple sinking of sulphide droplets from a static silicate liquid but also by a riffling process as the main lava stream moved for some time over these footwall embayments (Naldrett 1982). It must be noted that Cowden (1988) has argued that these depresions never existed and that the present ore–host rock relationships are the result of the developments of thick sections of komatiite involving much turbulent flow, with the deposition of the sulphides having occurred in deep lava channels between flanking flows.

The Eastern Goldfields Province of Western Australia is a typical Archaean region having a considerable development of greenstone belts (Fig. 11.4), with nickel sulphide deposits of two main types (Groves & Lesher 1982). The first consists of segregations of massive and disseminated ores at the base of small lens-like peridotitic to dunitic flows or subvolcanic sills at the bottom of thick sequences of komatiitic flows, e.g. Kambalda, Windarra, Scotia (Fig. 11.4), which are termed volcanic-type deposits. The second type, dyke-like or sill deposits, occurs in largely concordant, but partially discordant, dunitic intrusions emplaced in narrow zones up to several hundred kilometres in length, e.g. Perseverance, Mount Keith. These deposits occur in the 800 km long Norseman–Wiluna Greenstone Belt (Fig. 11.4), which was interpreted by Groves *et al.* (1984) as a rift zone 200 km wide. There are two ultramafic to

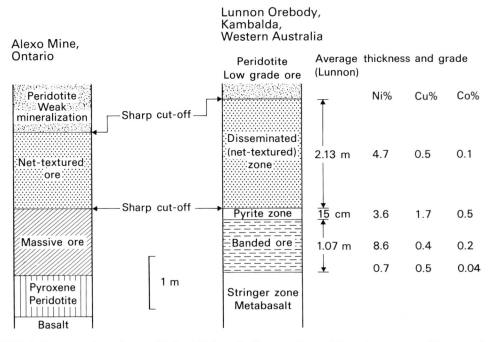

Fig. 11.3 Typical sections through two nickel sulphide orebodies associated with Archaean class 1(ii) ultrabasic bodies. The Alexo Mine is 40 km east-north-east of Timmins, Ontario. (After Naldrett 1973.)

felsic volcanic cycles with some sediments at the top of each. All the important known nickel deposits are in the ultramafic rocks of the lower cycle, which is about 2700 Ma old.

The volcanic-type deposits are commonly clustered around the periphery of granitoid periclines as at Kambalda (Fig. 11.5). These periclines trend north-north-westerly. Most ores are at, or close to, the basal contact of the mineralized ultrabasic sequence (Fig. 11.6) and they are commonly associated with, or even confined to, embayments in the footwall. Essentially the orebodies are tabular, with their greatest elongation subparallel to the penetrative linear fabrics in the enclosing rocks and/or the trend and plunge of both regional and parasitic folds. The ores have been metamorphosed and complexly deformed and are developed only in amphibolite facies metamorphic domains.

The dyke-like deposits are much larger than the volcanic-type. For example, in the Perseverance deposit at Agnew, 33 Mt averaging 2.2% nickel with a cut-off grade of 1% have been outlined and at Mount Keith 290 Mt grading 0.6%. They occur in long dunite dykes or sills especially where these bulge out to thicknesses of several hundred metres;

e.g. at Perseverance the host dyke thickens from a few metres to 700 m (Fig. 11.7). The orebodies generally appear to be associated with areas of considerable serpentinization during which enrichment of the ores seems to have occurred. The ores are dominantly of disseminated type though some massive sulphides occur as at Perseverance. This deposit is now considered by some workers to belong to subclass 1(ii)(a), in which case its large size contrasts strongly with the normal 1–5 Mt orebody size of this subclass. The work of Donaldson *et al.* (1986) suggests that this is indeed the case and that there is a complete continuum between the dunite-hosted intrusive ores and the komatiitic lava-hosted deposits, i.e. that the Groves and Lesher separation into volcanic and intrusive deposit types should be dropped. This suggestion is strongly supported by the work of Barnes & Barnes (1990) who contend that actually these dunite 'intrusives' are coarse-grained olivine adcumulates, which formed in long-lived komatiitic lava rivers that overflowed periodically to form the flanking sequences of orthocumulates and spinifex-textured flows.

Examples of deposits associated with tholeiites [1(i)] occur at Pechenga, USSR, in or related to

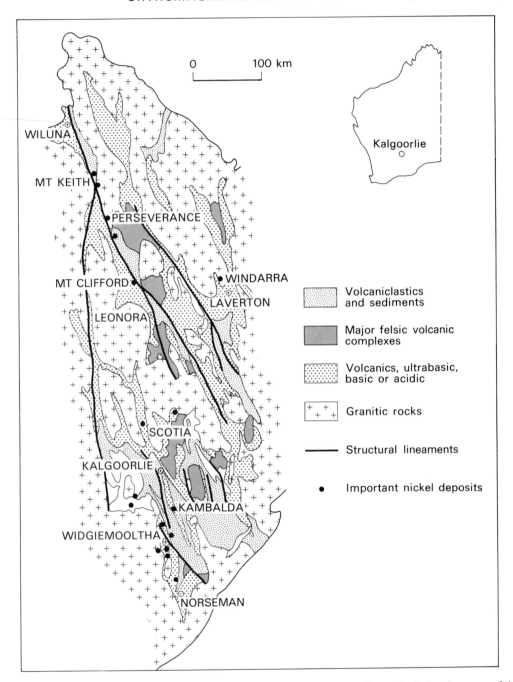

Fig. 11.4 Generalized geological map of the Eastern Goldfields Province of the Yilgarn Block showing some of the important nickel deposits. (Modified from Gee 1975.)

either large intrusions showing an upward change from peridotite to gabbro or smaller relatively undifferentiated masses. Massive ores with nickel as high as 10–12% and copper varying from very low values to 13% occur together with lower grade disseminated ores. The deposits at Lynn Lake,

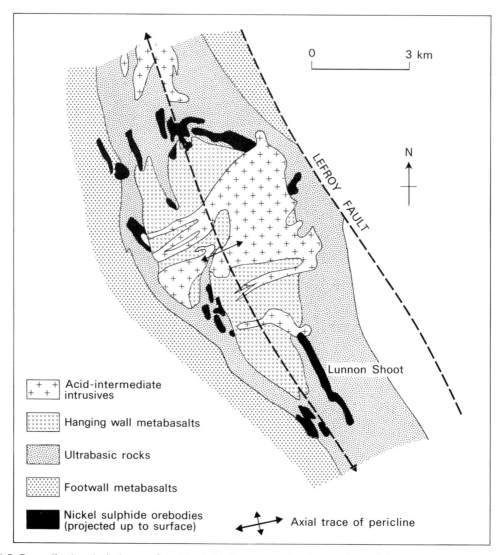

Fig. 11.5 Generalized geological map of the Kambalda Dome showing the positions of the sulphide orebodies.

Manitoba occur in sill-like bodies ranging from peridotite through norite to diorite. Massive and disseminated ores are present.

Plutonic association

Sudbury, Ontario

The Sudbury Structure is a unique crustal feature lying just north of Lake Huron near the boundary of the Superior and Grenville Provinces of the Canadian Shield (Card *et al.* 1984). The structure is about 60 × 27 km (Fig. 11.8) and its most obvious feature is the Sudbury Igneous Complex (1849 Ma), which consists of a Lower Zone of augite–norite, a thin Middle Zone of quartz–gabbro and an Upper Zone of granophyre (Naldrett & Hewins 1984). The Complex is believed to have the shape of a deformed funnel. At the base of the norite there is a discontinuous zone of inclusion and sulphide-rich norite and gabbro known as the sublayer. In the so-called offsets (steep to vertical, radial and concentric dykes that appear to penetrate downwards into the footwall from the base of the Complex) the inclusion-

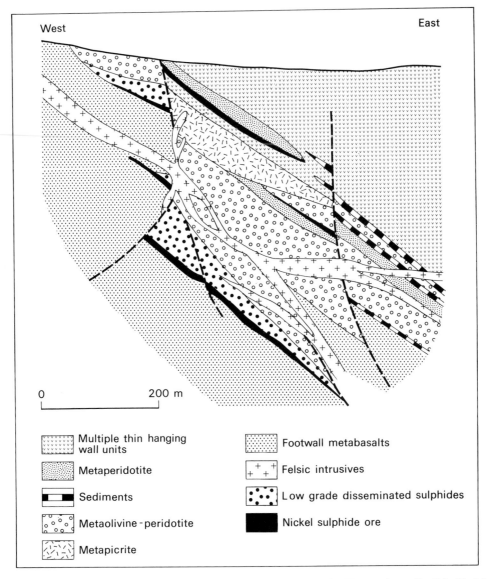

Fig. 11.6 Section through the Lunnon and neighbouring ore shoots, Kambalda, Western Australia. (Modified from Ross & Hopkins 1975.)

rich sulphide-bearing rock is a quartz diorite. The sublayer and offsets are at present the world's largest source of nickel as well as an important source of copper, cobalt, iron, platinum and 11 other elements.

Inside the Complex is the Whitewater Group consisting of a volcaniclastic-like sequence (Onaping Formation), a manganese-rich slate sequence (Onwatin Formation) and the Chelmsford Forma-tion—a carbonaceous and arenaceous proximal turbidite. These formations have suffered only slate grade regional metamorphism. No rocks correlat-able with the Whitewater Group have been posi-tively identified outside the basin.

To the south-east of the Complex there are metasedimentary and metavolcanic rocks of the Huronian Supergroup that were deposited uncon-formably upon the migmatitic Archaean basement,

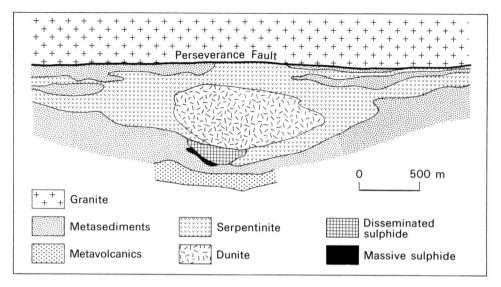

Fig. 11.7 Generalized geology of the Perseverance deposit. Many faults have been omitted. (Modified from Martin & Allchurch 1975.)

which is exposed along the north-western side of the Complex. The Huronian has suffered a much higher grade of metamorphism than the Whitewater Group and it contains a number of basic and acid plutonic intrusions. It forms a southward facing homocline and carries a penetrative foliation caused by flattening. The rocks of the Basin are also considerably deformed (Brocoum & Dalziel 1974) and finite strain analysis suggests that the Chelmsford Formation was shortened by 30% in a north-westerly direction; this means that the Chelmsford Formation and probably the Sudbury Structure were less elliptically shaped prior to deformation.

The origin of the Sudbury Structure and the structure along which the Complex was intruded have been debated since the area was first mapped about the turn of the century. There is now an important school of thought initiated by Dietz (1964) that regards the Structure as having resulted from a meteoritic impact. Shock metamorphic features, including shatter cones, are common in the rocks around the Complex for as much as 10 km from its footwall, and in the breccias of the Onaping Formation, which is regarded by this school as being a fall-back breccia resulting from the impact. Geologists arguing against the meteorite impact hypothesis interpret the Onaping Formation as an ignimbrite deposit and have mapped a quartzite unit, believed to be a basal quartzite, lying beneath the Onaping Formation. This school generally re-

gards the Complex as having been intruded along an unconformity at the base of the Whitewater Group. Another controversial rock type is the 'Sudbury Breccia'. This consists of zones, a few centimetres to several kilometres across, of brecciated country rocks which are sometimes hosts for the orebodies.

The discovery of mineralization during the construction of the Canadian Pacific Railway in 1883 has led to the development of over 40 mines and the total declared ore reserves of the district from the time of the original discovery to the present are of the order of 930 Mt. Of these, about 500 Mt have been exploited but new reserves are constantly being blocked out, just about keeping pace with production. According to Lang (1970), the grade of ores worked in the past was about 3.5% nickel and 2% copper. Today, with large-scale mining methods, the grade worked is around 1% for both metals.

Most of the orebodies occur in the sublayer whose magma was rich in sulphides and inclusions of peridotite, pyroxenite and gabbro. In some places the sublayer appears to be older than the main mass of the Complex and in other places younger and presumably intruded along the footwall of the main mass. It appears to have acquired its high content of inclusions during its passage through an underlying hidden basic igneous complex whose existence is suggested by geophysical data. The sulphides tended to sink into synclinal embayments in the footwall giving a structural control of the mineralization. The

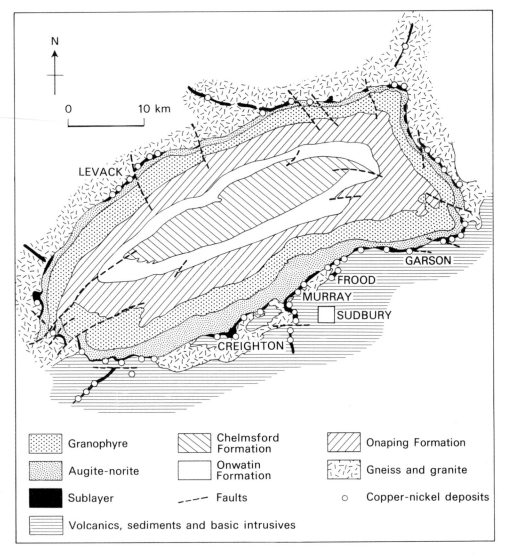

N

0   10 km

LEVACK

GARSON

FROOD

MURRAY

SUDBURY

CREIGHTON

Granophyre

Augite-norite

Sublayer

Chelmsford Formation

Onwatin Formation

Faults

Onaping Formation

Gneiss and granite

o   Copper-nickel deposits

Volcanics, sediments and basic intrusives

Fig. 11.8 Geological map of the Sudbury district. (After Souch *et al.* 1969 and Brocoum & Dalziel 1974.)

Creighton ore zone has the greatest number of ore varieties (Souch *et al.* 1969). It plunges north-westward down a trough at the base of the Complex for at least 3 km (Fig. 11.9) and consists of a series of ore types. The hanging wall quartz–norite above the sublayer occasionally contains enough interstitial sulphide to form low grade ore. In the upper part of the sublayer, ragged disseminated sulphide ore occurs, consisting of closely packed inclusions (several millimetres to 10 cm in size) in a matrix of sulphides and subordinate norite. The sulphide content increases downwards as does the ratio of

matrix to inclusions, with a concomitant increase in inclusion size up to 1 m, resulting in an ore called gabbro–peridotite inclusion sulphide. This ore changes towards the footwall into massive sulphide containing fragments of footwall rocks. It is called inclusion massive sulphide and it is discontinuous, and also forms stringers and pods in the footwall.

The Frood–Stobie orebody is an example of an orebody in an offset dyke. This parallels the footwall of the Complex and it has been suggested that it might once have been continuous with the Complex at a higher level. It is a huge orebody, 1.3 km long,

Surface

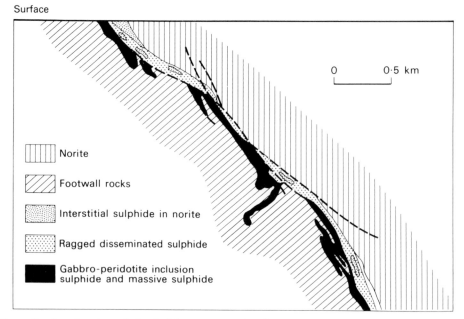

Norite

Footwall rocks

Interstitial sulphide in norite

Ragged disseminated sulphide

Gabbro-peridotite inclusion sulphide and massive sulphide

Fig. 11.9 Generalized section through the Creighton ore zone looking west. (After Souch *et al.* 1969.)

1 km deep and nearly 300 m across at its widest point. It consists of a wedge-shaped body of inclusion-bearing quartz–diorite (Fig. 11.10) with disseminated sulphide partially sheathed by inclusion massive sulphide. In the lower half, the ore described by Hawley (1962) as immiscible silicate–sulphide ore occurs (Fig. 4.1). This ore type grades into massive ore outwards and downwards. Inclusions in the quartz–diorite vary from a few centimetres to many metres in length. The largest found was one of peridotite 45 m long!

For further information on the Sudbury deposits the reader is referred to Pye *et al.* (1984), a veritable storehouse of data and ideas on the deposits and their host rocks.

Deposits in intrusions related to flood basalts

Rifting of cratons has resulted in the development of flood basalts at various times during the earth's history, and some of the associated basic intrusions carry important nickel sulphide mineralization. These include the Jurassic Insizwa Complex, RSA, which is part of the Karoo Magmatic Event; the Lower Triassic ores of the Noril'sk-Talnakh Region in Siberia, and the 1100 Ma mineralization of the Duluth Complex, Minnesota plus the Great Lakes deposit of Ontario, which are both related to Keweenawan magmatism.

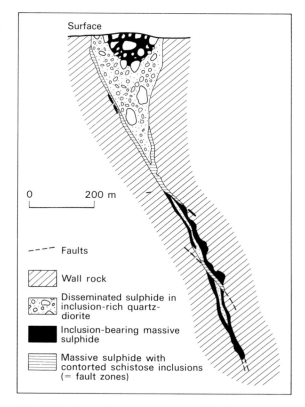

Surface

0   200 m

Faults

Wall rock

Disseminated sulphide in inclusion-rich quartz-diorite

Inclusion-bearing massive sulphide

Massive sulphide with contorted schistose inclusions (= fault zones)

Fig. 11.10 Generalized section through the Frood orebody looking south-west. (After Souch *et al.* 1969.)

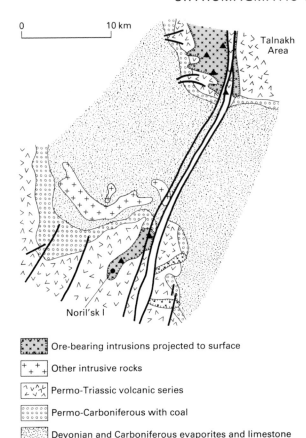

0   10 km

Talnakh Area

Noril'sk I

▨▨ Ore-bearing intrusions projected to surface

+ + + Other intrusive rocks

�∨∧ Permo-Triassic volcanic series

○○○○ Permo-Carboniferous with coal

▦ Devonian and Carboniferous evaporites and limestone

—— Faults

Fig. 11.11 Geology of the Noril'sk-Talnakh region. For location see Fig. 7.7. (After Naldrett 1981.)

The Noril'sk-Talnakh Region (Fig. 11.11) has the largest reserves of copper–nickel ore in the USSR. It lies deep in northern Siberia near the mouth of the Yenisei River (Fig. 7.7). The country rocks are carbonates and argillaceous sediments of the early and middle Palaeozoic overlain by Carboniferous rocks with coals, Permian and a Triassic basic volcanic sequence. The associated gabbroic intrusions form sheets, irregular masses and trough-shaped intrusions depending on their location in the gentle folds of the sediments lying beneath the volcanic sequence (Glazkovsky *et al.* 1977).

The Noril'sk I deposit occurs in a differentiated layered dominantly gabbroic intrusion which extends northwards for 12 km and is 30–350 m thick. In cross section it is lensoid with steep sides (Fig. 11.12). The copper–nickel sulphides form breccia and disseminated and massive ores at the base of the intrusion, and vein orebodies developed in the footwall rocks and the basal portion of the intrusion. Like Sudbury there is a high Cu : Ni ratio and $(Pt + Pd) : Ni = 1 : 500$.

West   East

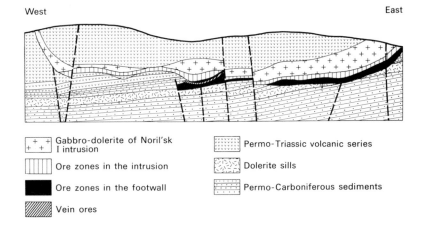

+ + Gabbro-dolerite of Noril'sk I intrusion

▥ Ore zones in the intrusion

■ Ore zones in the footwall

▨ Vein ores

⋮⋮⋮ Permo-Triassic volcanic series

⟋⟍ Dolerite sills

⋯⋯ Permo-Carboniferous sediments

Fig. 11.12 Cross section through the Noril'sk I deposit. (After Glazkovsky *et al.* 1977.)

12 / Greisen deposits

Introduction

Premoli (1985), with prophetic insight, discussed the effect on tin mining of a collapse of the International Tin Council's cartel. His prophecy was that, in the event of a severe fall in the price of tin, the small and medium-sized producers operating high cost, low tonnage vein deposits, e.g. in Cornwall, would be forced to close, and that output from the old established placer deposit producers would also fall drastically as these are often under-capitalized, work deposits of declining grade and are faced with rising production costs. Premoli put forward, however, the caveat that the increasing flood of tin from Brazil, although mainly from placers, 'seems all but unstoppable'. This being the case, he forecast that the well capitalized, technically efficient miners will develop a new generation of tin mines based on comprehensive exploration and evaluation of primary tin deposits, economy of scale, and modern, cost-effective, bulk mining methods. Tin explorationists will now have to think big and their targets will be something like this: 40–60 Mt orebodies with a minimum mill grade of 0.3–0.4% Sn and a recoverability of at least 60%. By-product tungsten or other elements or minerals will of course improve the economic prospects of such deposits, whose economic viability will be somewhat variable depending on mining and mineral processing costs and the available infrastructure. Rio Algom's East Kemptville Mine in Nova Scotia, which came on stream late in 1985, has tested Premoli's thesis admirably. Mineable ore reserves in 1985 were estimated at 58 Mt grading 0.165% Sn and about 0.1% Cu and 0.15–0.2% Zn as by-products. Production was planned to be 9000 t of ore p.d. East Kemptville has of course suffered considerably from the collapse of the tin cartel but has managed to stay in production by mining the higher grade sections of the orebody, run-of-the-mill ore in 1990 being 0.179% Sn, and by improving recoveries, which rose from 63.1% in 1989 to 71.5% in 1990. A new mining plan has been drawn up which reduces proved and probable reserves to 15.2 Mt with an average grade of 0.206% Sn and

which will result in a much shorter mine life. The mine's further progress will be watched with great interest by the whole industry.

The type of primary tin deposits that can come into the grade-tonnage limitations given above are:
1 subvolcanic stockwork deposits (porphyry tins), referred to on pp. 188–9 and 219;
2 greisen deposits;
3 large pegmatite deposits, see Chapter 9;
4 large limestone-replacement or exhalative deposits in Dachang, China (Tanelli & Lattanzi 1985).

In this chapter we will look at greisen deposits.

Tin greisen deposits

Greisen may be defined as a granoblastic aggregate of quartz and muscovite (or lepidolite) with accessory amounts of topaz, tourmaline and fluorite formed by the post-magmatic metasomatic alteration of granite (Best 1982, Štemprok 1987). Greisens are important mainly for their production of tin and tungsten. Usually one element is predominant but there may be by-product output of the other. They are usually developed at the upper contacts of granite intrusions and are sometimes accompanied by stockwork development. The mineralization occurs as large irregular, or sheet-like bodies immediately beneath the upper contact of late stage, geochemically specialized granites and may extend downwards for some 10–100 m before grading through a zone of feldspathic alteration (albitization, microclinization) into fresh granite below (Pollard *et al.* 1988). Pegmatitic fluids often migrate to these apical portions of granite intrusions and may form intrusions along the margins of greisen bodies. In the Erzgebirge these are known by the traditional name stockscheider (Fig. 12.1b).

Deposits of this type have been worked in the Erzgebirge on either side of the Czech–German border for many years, and cross sections of some typical deposits are given in Fig. 12.1 together with a plan and section of the East Kemptville orebody. The Erzgebirge deposits occur as massive greisens and greisen-bordered veins in the uppermost endocontacts and exocontacts of small lithium

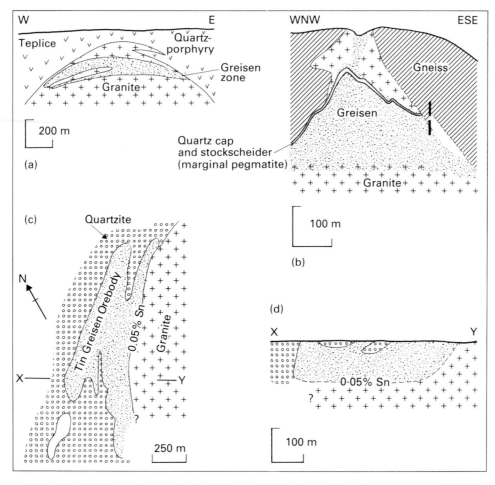

Fig. 12.1 Some tin greisen orebodies. (a) Section through Cínovec, Czechoslovakia. (b) Sadisdorf, Germany. (c,d) Map and section of East Kemptville, Nova Scotia, Canada. ((a,b) are after Baumann (1970) and (c,d) after Richardson *et al.* (1982).)

mica–albite granite cupolas. These coalesce at depth to form a large batholith (Štemprok 1985). Bulk mining of these ores in the late 1980s, and before market economics were introduced, took place at Altenberg (Germany) and Cínovec (Czechoslovakia). The grades of ore then being worked were: Altenberg—0.2–0.3% Sn, 0.01% W, 0.01% Mo, 1% F; Cínovec—0.2% Sn, 0.35% W, 0.35% Li; as reported by Lehmann (1990) who gives a useful up-to-date summary of tin mineralization in the Erzgebirge. In 1985 the world's largest cassiterite mill was set up at one of the world's longest worked tin deposits—Altenberg—which has been worked since the Middle Ages (Mosch & Becker 1985). Similar deposits have been worked in Namibia

(Pirajno & Jacob 1987), Indonesia (Schwarz & Surjono 1990) and other tin provinces.

The mineralization at East Kemptville is developed beneath an inflection in the granite–metasedimentary contact of a discrete pluton (Davis Lake Pluton). The western margin of the deposit is defined by the steeply dipping contact but the eastern boundary is an assay boundary as is the lower margin, which is at about 80–100 m below surface. The greisen is rich in fluorine, which rises to 6.5%. During crystallization of the host granite a fluorine- and tin-rich hydrothermal fluid evolved and scavenged Cl, P, Sn and other metals from the magma concentrating them beneath the contact and thus giving rise to topaz-rich greisens. The Davis

Lake Pluton is part of the South Mountain Batholith (Richardson *et al.* 1982, Chatterjee & Strong 1984, Richardson *et al.* 1990). This batholith was intruded into a deformed Lower Palaeozoic metasedimentary sequence during the Acadian Orogeny and it contains a number of zoned, coalescent plutons that may average more than 10 ppm tin (Smith *et al.* 1982). They are, like the Erzgebirge granites, tin-specialized, S-type granites showing such features as high Rb/K, a marked decrease in Ba and Sr and an increase in Rb from marginal granodiorite to late stage alaskite; a pattern interpreted by Groves & McCarthy (1978) as indicative of *in situ* crystallization. These authors suggest that tin greisens and other tin deposits may develop in the upper parts of granite plutons when an impermeable roof of early crystallized cumulates has formed. Beneath this roof, water-saturated melt accumulates and eventually crystallizes. Incompatible elements, tin and other elements are concentrated in the late intercumulus liquid, which eventually loses equilibrium with the cumulus minerals, and greisenization proceeds. Liquids of this nature appear to have been trapped towards the top of the Bobbejaankop Granite, RSA, where they have formed Sn–W mineralization in a sheet-like horizon within the granite (Taylor & Pollard 1988).

This model suggests some useful exploration indicators. Tin is likely to have been concentrated in zones with the lowest Ba and Sr values and confirmation of such an area, as being one where late stage crystallization occurred, can be obtained by examining the variation of concentration changes of incompatible elements such as Li and Rb with height within the granite. In addition the granites should be tin-specialized, S-type (though not necessarily in a collision tectonic setting, e.g. the tin granites of the Bushveld, RSA) with cupolas and ridge zones largely preserved from erosion, or with other traps for water-saturated melts, such as the contact inflexion at East Kemptville. Further discussions of the nature of tin granites and their recognition can be found in Evans (1982) and Taylor (1979); of granophile mineral deposits in Strong (1981); and of tin and tungsten greisen deposits and other granite-related mineral deposits in Taylor (1979) and Taylor & Strong (1985), whilst Burt (1981) has a useful discussion of the process of greisenization.

A useful discussion of Premoli's ideas, with grades and tonnages of various potential deposits, can be found in Anon. (1985b). In the future greisens may become important sources of Be, Nb–Ta, REE, Y and other elements (Highley *et al.* 1988). A new metallogenic province in the western USA, which may contain the second largest Be resource in the United States as well as many other deposits, was described by Barton (1987) and the Be, Nb–Ta, REE greisens at Thorr Lake, Northwest Territories, Canada are under investigation (Trueman *et al.* 1988).

13 / The skarn environment

Some general points

These deposits have been termed hydrothermal metamorphic, igneous metamorphic and contact metamorphic but the most common term for these deposits (p. 31) used to be Lindgren's (1922) 'pyrometasomatic'; however, at the present time it seems to have been dropped by almost universal consent for the less satisfactory term 'skarn'. Skarn is an old Swedish mining term for silicate gangue, and who might want to mine gangue except perhaps for use as an aggregate?! These deposits were formed at elevated temperatures with the addition and subtraction of material (metasomatism). Their general morphology and nature have been summarized on p. 31. They are developed most often, but not invariably, at the contact of intrusive plutons and carbonate country rocks. The latter are converted to marbles, calc-silicate hornfelses and/or skarns by contact metamorphic effects.

The calc-silicate minerals, such as diopside, andradite and wollastonite, which are often the principal minerals in these ore-bearing skarns, attest to the high temperatures involved, and various lines of evidence suggest a range of 650–400°C for initial skarn formation (Einaudi *et al.* 1981), but in some skarns, particularly Zn–Pb, lower temperatures appear to have obtained (Kwak 1986). The pressures at the time of formation were very variable as the depths of formation were probably from one to several kilometres. Some of the classic pyrometasomatic deposits of the USA are associated with porphyry copper intrusions, indicating a relatively shallow depth of emplacement.

The tyro must bear in mind that the majority of skarns are devoid of economic mineralization. The first geologists to take over this term were metamorphic petrologists who have long used it to describe contact or regionally metamorphosed rocks composed of calcium, magnesium and iron silicates that have been derived from nearly pure limestones or dolomites, into which large amounts of Si, Al, Fe and Mg were *metasomatically* introduced, frequently but not always, at or near the contact with a plutonic igneous intrusion (Best 1982). Identical rocks formed by isochemical metamorphism of impure limestones and dolomites are termed calc-silicate hornfelses. If they contained pre-existing mineralization then this too will be recrystallized and the resultant hornfels may be mistaken for a skarn, e.g. the magnetite orebody of Dielette in Normandy, France (Cayeux 1906). Also it should be noted that some isochemically metamorphosed submarine exhalative deposits contain sulphides in association with CaFeAlMgMn silicates and look remarkably like true skarns. These have been termed reaction skarns and their existence has led to controversy as to whether certain previously recognized skarns are actually metamorphosed exhalites in which no significant metasomatism occurred during their formation (Ashley & Plimer 1989).

Calc-silicate hornfels can usually be distinguished from skarn by examining its field relationships, except in the difficult case of skarn development by reaction between interlayered silicate and carbonate sediments during metamorphism—such rocks are termed 'reaction skarns'. Skarn-like rocks of uncertain origin are often referred to as skarnoids.

Skarns can be classified according to the rocks they replace, and the terms exoskarn and endoskarn were applied originally to replacements of carbonate metasediments (usually marbles) and intrusive rocks, respectively, in contact zones, but some authors have extended the use of the term endoskarn to skarn formed in any aluminous rock, shales, volcanics, etc. (Einaudi & Burt 1982). Both endoskarns and exoskarns may contain ore, but in an endoskarn–exoskarn couplet where the exoskarn is a converted marble then it usually contains most or all the economic mineralization. Exoskarns may be classified according to the dominant mineralogy: as magnesian if they contain an important component of Mg silicates such as forsterite, or as calcic when calcium silicates, e.g. andradite, diopside, are predominant. The majority of the world's economic skarn deposits occur in calcic exoskarns (Einaudi *et al.* 1981). This paper is a classic containing a wealth of information on metalliferous skarns. The general nature of major metalliferous skarn types is shown in Table 13.1.

Table 13.1 General nature of major metalliferous skarn types. (Modified from Einaudi et al. 1981 and Einaudi & Burt 1982)

Type	Calcic copper	Calcic iron	Magnesian iron	Calcic tungsten	Calcic zinc-lead	Calcic molybdenum	Calcic tin	Magnesian tin
Metal association (minor metals)	Cu, Mo(W, Zn)	Fe(Cu, Co, Au)	Fe(Cu, Zn)	W, Mo, Cu(Zn, Bi)	Zn, Pb, Ag(Cu, W)	Mo, W, (Cu, Bi, Zn)	Sn(Be, W)	Sn(Be)
Principal ore and opaque minerals	Chalcopyrite, bornite, pyrite, hematite, magnetite	Magnetite (chalcopyrite, cobaltite, pyrrhotite)	Magnetite (pyrite, chalcopyrite, sphalerite, pyrrhotite)	Scheelite, molybdenite, chalcopyrite, pyrrhotite, pyrite	Sphalerite, galena, chalcopyrite, arsenopyrite	Molybdenite, scheelite, bismuthinite, pyrite, chalcopyrite	Cassiterite, arsenopyrite, stannite, pyrrhotite	Cassiterite, minor arsenopyrite, pyrrhotite, stannite, sphalerite
Typical size (Mt)	1–100	5–200	5–100	0.1–2	0.2–3	0.1–2	0.1–3	1
Typical grade[a]	2% Cu u.g. 1% Cu o.p.	40% Fe	40% Fe	0.7% WO_3	9% Zn, 6% Pb, 171 ppm Ag	0.15–1% MoS_2	0.1–0.7% Sn	Little data
Largest known deposit	Twin Buttes, Arizona, 500 Mt, 0.8% Cu	Sarbai, USSR, 725 Mt, 46% Fe	Sherogesh, USSR, 234 Mt, 35% Fe	MacMillan Pass, Canada, 63 Mt, 0.95% WO_3	Naica, Mexico, 21 Mt, 3.8% Zn, 4.5% Pb, 0.4% Cu, 150 ppm Ag, 0.3 ppm Au[b]	Little Boulder Creek, Idaho, >100 Mt, 0.15% MoS_2	Moina, Tasmania, 30 Mt, 0.15% Sn	?
Exoskarn gangue mineralogy:								
Early minerals (stages 1 & 2)	Andradite, diopside, wollastonite	Ferrosalite, grandite, epidote, magnetite	Forsterite, calcite, spinel diopside, magnetite	Grandite, hedenbergite, idocrase, wollastonite	Mn-hedenbergite, andradite, bustamite, rhodonite	Hedenbergite, grandite, quartz	Malayaite, danburite, dactolite, grandite, idocrase	Spinel, fassaite, forsterite, phlogopite, magnetite, humite
Late minerals (stage 3)	Actinolite, chlorite, montmorillonoids	Amphibole, chlorite, ilvaite	Amphibole humite, serpentine, phlogopite	Hornblende, biotite, plagioclase, epidote, MnFe garnet	Mn-actinolite, ilvaite, epidote, chlorite, dannemorite, rhodochrosite	Amphibole, chlorite	Amphibole, mica, chlorite, tourmaline, fluorite	Cassiterite, fluoborite, magnetite, micas, fluorite

Associated intrusives	Granodiorite to monzogranite	Gabbro to syenite, mostly diorite	Granodiorite to granite	Quartz diorite to monzogranite	Plutons commonly absent, otherwise granodiorite to granite, diorite to syenite	Monzogranite and syenogranite	Granite	Granite
Intrusive morphology	Stocks, dykes, breccia pipes	Large to small stocks, dykes	Small stocks, dykes, sills	Large plutons, batholiths	If present, stocks and dykes	Stocks	Stocks, batholiths	Stocks, batholiths
Tectonic setting	Continental margins and island arcs, syn- to late orogenic	Oceanic island arcs, rifted continental margins	Continental margin arcs, synorogenic	Continental margin arcs, syn- to late orogenic	Continental margin arcs, syn- to late orogenic	Continental margin arcs, late orogenic	Continental margin arcs, late to post orogenic or anorogenic	Continental margin arcs, late to post-orogenic

[a] u.g. = underground, o.p. = open pit.
[b] Data from Megaw et al. (1988).
The igneous rock names have been modified to conform with IUGS recommendations.

Skarn deposits are usually described according to the dominant economic metal or mineral present, e.g. copper, iron, tungsten, zinc–lead, molybdenum, tin, talc, etc. These deposits are generally smaller than many other deposit types, such as porphyry coppers, porphyry molybdenums and lead–zinc, sediment-hosted sulphide deposits, but they are very important sources of tungsten. In some countries, e.g. the USSR, they are of considerable importance for their iron production. Some particularly rewarding copper skarn deposits, especially those with by-product Au and/or Ag, have been worked in various parts of the world and notably large deposits may occur associated with porphyry copper deposits, e.g. Twin Buttes, Arizona and Bingham Canyon, Utah. Zinc–lead skarn deposits occur throughout the world but are seldom of large tonnage. Skarn deposits of molybdenum and tin are of little importance compared with other deposits of these metals, apart from the San Antonio Mine, Santa Eulalia District, Mexico which contained ~ 3 Mt of 1.5% Sn and is still important for its Zn, Pb and Ag. To these metalliferous skarn types Kwak (1986) added the uranium-rare earth type as exemplified by the Mary Kathleen Mine, Queensland. This is the only significant deposit of this type so far found. Other rare types are calcic molybdenum,

calcic tin and magnesium tin (Einaudi *et al.* 1981).

The principal ore and associated opaque minerals for each of the metalliferous skarn types is given in Table 13.1 together with the major gangue mineralogy. Exoskarns usually show a zoning of both the silicate and ore minerals. This has been well documented for the magnetite deposits of Cornwall, Pennsylvania, by Lapham (1968), and Theodore (1977) has discussed the zoning of skarn deposits associated with porphyry copper deposits. One of his examples is reproduced in Fig. 13.1. The mineral zones in the Ely, Nevada, cupriferous skarns generally parallel the igneous contact and the bulk of the copper in the skarn was deposited in veinlets which cut the andradite–diopside rocks. The alteration envelopes along these veins contain actinolite-calcite–quartz–nontronite assemblages. These relationships indicate that a clay–sulphide stage was superposed on the earlier calc-silicate rocks.

Endoskarns also display mineral zonation resulting from a progressive addition of calcium to the igneous protolith (usually anything from granite to gabbro). A common zonation outwards towards the marble host is biotite →amphibole→pyroxene→ garnet. K-feldspar disappears but plagioclase may survive and there may be such a metamorphic convergence between the mineral assemblages of an

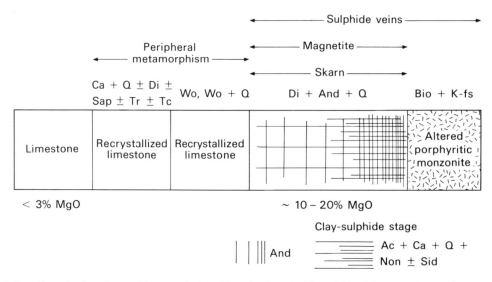

Fig. 13.1 Zonation of mineral assemblages at the Ely, Nevada, deposit. The width of the zones is schematic and the density of the patterns indicates the relative amounts of andradite (And) and the clay–sulphide stage. Abbreviations: Ac = actinolite; And = andradite; Bio = secondary biotite; Ca = calcite; Di = diopside; K-fs = secondary potash feldspar; Non = nontronite; Q = quartz; Sap = saponite; Sid = siderite; Tc = talc; Tr = tremolite; Wo = wollastonite. (Modified from Theodore 1977.)

endoskarn and its adjacent exoskarn that the position of the original intrusive igneous contact is unrecognizable. According to Einaudi & Burt (1982), where skarn occurs near or over the tops of plutonic cupolas, as is the case for most skarns related to porphyry copper plutons and tin skarns, endoskarn is usually absent. It seems therefore that endoskarn development occurred in those areas where fluid flow was dominantly into the pluton or upward along its contacts with marbles, rather than where the metasomatizing fluids flowed up and out of the pluton as in the formation of porphyry copper deposits.

Some examples of skarn deposits

Copper skarn deposits

Most copper skarn deposits are associated with calc-alkaline granodiorite to monzogranite stocks emplaced in continental margin arcs. These are intrusions, themselves often important copper orebodies in the form of porphyry coppers, which are well developed in the Mesozoic and Tertiary continental margin arcs of the western Americas and the similar but Carboniferous arcs of the USSR. A small

number of copper skarns occur in oceanic island arcs associated with quartz diorite to monzogranite plutons, such as the Memé Mine, Haiti. The porphyry copper-associated skarns can be very large, up to 500 Mt of open pit ore, whereas those associated with barren stocks are generally in the 1–50 Mt class. The porphyry copper-associated skarns of the southwestern USA have been comprehensively described by Einaudi (1982).

Copper Canyon, Nevada

This is a porphyry copper-associated deposit where skarn has replaced calcareous shale or argillite beds just above the Golconda Thrust (Fig. 13.2) producing a flat-lying tabular zone of andradite-rich rock in which most of the mineralization occurred. Although the andradite rock stretches at least 400 m from the granite porphyry stock, only that part within 180 m of the contact contains ore. Silicate zones in this deposit are symmetrical about the andradite rock, thus forming zones at right angles to the present igneous contact. For this and other reasons, Theodore (1977) suggests that the porphyry or the present site of the porphyry was not the locus from which the skarn-forming fluids emanated.

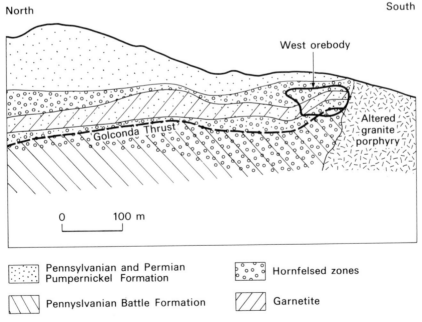

Fig. 13.2 Cross section through the copper–gold–silver skarn deposit at Copper Canyon, Nevada. Rocks peripheral to the hornfelsed zones carry secondary mica. See also Fig. 13.10. (Modified from Theodore 1977.)

The Memé Mine, Northern Haiti

Frequently at the contacts of skarns and intrusions there is a completely gradational contact and this is the case at the Memé copper mine where a large block of Cretaceous limestone has been surrounded by monzogranite. Mineralization was preceded by extensive magmatic assimilation that formed zones of syenodiorite and granodiorite around the limestone. Following the crystallization of the magma, the limestone and neighbouring parts of the intrusion were replaced by skarn. The intrusion-derived endoskarn contains large quantities of diopside, which distinguishes it from the marble-derived exoskarn. There is a completely gradational contact between these skarns (Kesler 1968).

Mineralization followed skarn formation and consisted of the introduction of hematite, magnetite, pyrite, molybdenite, chalcopyrite, bornite, chalcocite and digenite, in that paragenetic order. These occur as replacement zones. The main skarn and ore development is along the lower contact with the limestone block (Fig. 13.3). Skarn formation took place at between 480 and 640°C and exsolution textures suggest that the minimum temperature of copper–iron sulphide deposition exceeded 350°C and the youngest ore minerals crystallized about 250°C. The grade is about 2.5% Cu.

Iron skarn deposits

Skarns have long been important sources of iron ore and the magnetite mine at Cornwall, Pennsylvania, supplied much of the iron used during the industrial revolution in the USA. It is the oldest, continuously operated mine in North America. Mining commenced in 1737 and by 1964, 93 Mt of ore had been produced with an average mill feed grade in 1964 of 39.4% Fe and 0.29% Cu, from which minor by-product cobalt, gold and silver were obtained (Lapham 1968). A pyrite concentrate was used to produce sulphuric acid, and up to 1953, when open pit operations ceased, limestone overburden was crushed and sold as aggregate. A good example of 'waste not want not!' Cornwall is a calcic iron skarn and such skarns are associated with intrusives ranging from gabbro through diorite to syenite, whilst magnesian iron skarns normally are associated with granites or granodiorites (Einaudi *et al.* 1981). Between them these two classes have economic magnetite concentrations in all the major tectonic settings in which skarns are found (Table

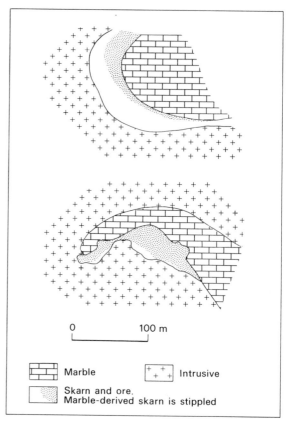

Fig. 13.3 Geological map of the 1500 ft (457.2 m) level (above) and an east–west section (below) of the Memé Mine, Haiti. Note the concentration of skarn and ore beneath the marble. (Modified from Kesler 1968.)

13.1). The largest known deposits in either class occur in the USSR. With magnetite as the major ore mineral, these deposits are often marked by pronounced magnetic anomalies, and the detection of these by aeromagnetic surveys led to the discovery *inter alia* of Sarbai in the USSR and of Marmoraton, Ontario. These deposits typically run 5–200 Mt with a grade of about 40% Fe.

Sarbai is the giant of skarn deposits with 725 Mt grading 46% Fe. It lies in the Turgai Iron Ore Province of the Kazak SSR, i.e. in the south-western part of the Siberian Platform to the east of the southern end of the Urals and about 500 km south-east of the famous and long worked iron skarns of Magnitogorsk. The orebodies (Fig. 13.4) occur in a succession of metamorphosed pyroclastics, marbles and skarns developed from a Carboniferous volcaniclastic-sedimentary succes-

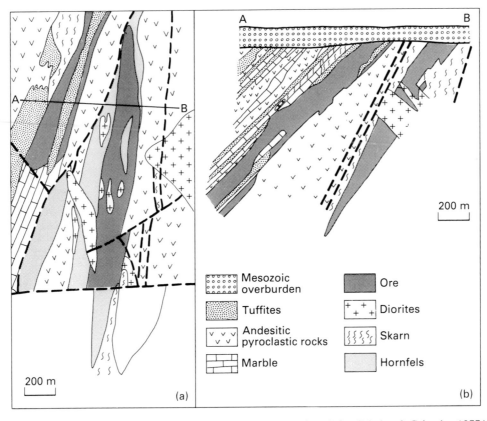

Fig. 13.4 Sarbai iron skarns, USSR: (a) plan of 80 m level; (b) cross section. (After Sokolov & Grigor'ev 1977.)

sion in the western limb of an anticline (Sokolov & Grigor'ev 1977). At Sarbai, as elsewhere in the Turgai Province, there is a marked development of chlorine-bearing scapolite associated with the iron skarn orebodies, indicating the presence of important amounts of brine solution during the metasomatism. The orebody dimensions are impressive: the eastern ore has a strike length of 1.7 km and a thickness of up to 185 m; it has been traced down dip for more than 1 km and the western orebody for 1.8 km. There is a further possible reserve of 775 Mt!

Tungsten skarns

Tungsten skarns, veins and stratiform deposits provide most of the world's annual production of tungsten, with skarns being predominant. In the market economy countries (MEC) most skarn tungsten comes from a few relatively large deposits: King Island, Tasmania; Sangdong, Korea; Canada Tungsten (NWT) and MacMillan Pass (Yukon), Canada; and Pine Creek, California, USA. To put 'relatively large' in perspective it must be remarked that annual MEC production of tungsten is only about 30 000 t, i.e. the equivalent of the production of just one modestly sized copper mine, which in turn is small compared with an iron mine! China is the world's main producer and in 1989 produced about 18 000 t. The USSR ranked next with 7000 t. The leading MEC countries in 1989 were the Republic of Korea with 1560 t, Austria with 1543 t and Australia with 1400 t. (The most important tungsten deposit types are volcanic-associated, strata-bound, deposits, see pp. 210–211.) In this section we will look at one large and one small tungsten mine.

King Island Scheelite

King Island, which lies between Australia and Tasmania at the western approach to Bass Strait, contains a number of important tungsten deposits. These are scheelite-bearing skarns developed in Upper Proterozoic to Lower Cambrian sediments that have been intruded by a granodiorite and a monzogranite of Devonian age. These granites are thought to be related to the tin- and tungsten-bearing granites of the Aberfoyle and Storey's Creek district of Tasmania (Danielson 1975). Mined ore and reserves at 1980 were put at 14 Mt averaging 0.8% WO_3 (Einaudi et al. 1981).

Scheelite-bearing andradite skarns were formed by the selective replacement of limestone beds and as a result they form stratiform orebodies 5–40 m thick. There is a great deal of mineralized skarn below ore grade. The orebodies lie either in the exocontact of the granodiorite or the monzogranite (Fig 13.5). Irregular relicts of marble occur in the skarn demonstrating its metasomatic origin, and Edwards et al. (1956) showed that the replacement was a volume-for-volume process with massive addition of silica, iron and aluminium and subtraction of calcium and CO_2. A most detailed and exhaustive geochemical and fluid inclusion investigation was performed by Kwak (1978a,b), who concluded that the Edwards et al. mass balance calculations are in error owing to poor selection of sampling sites. Kwak's work indicated that the most favourable site for ore formation was where the marbles contained numerous interlayers of hornfels or hornfelsic fragments within 400 m of the granite contact. The stable isotopic, paragenetic and petrological investigations recorded in Wesolowski et al. (1988) indicate the following stages in the development of these orebodies. Intrusion of the granitoids into a sequence of interbedded impure carbonates, shales and volcanic rocks gave rise to contact metamorphism during which the shales became highly impermeable metapelites, but the interbedded dolomitic carbonates probably became more permeable when they were changed into marbles. Pre-existing faults and faults and fractures created by the intrusions channelled the flow of high temperature fluids (up to 800°C; Kwak 1986) into the permeable marbles creating massive calc-silicate replacement skarns and a chemical environment favouring the precipitation of scheelite.

Salau Tungsten Mine, Ariège (central Pyrenees), France

Over 40 significant tungsten occurrences are known

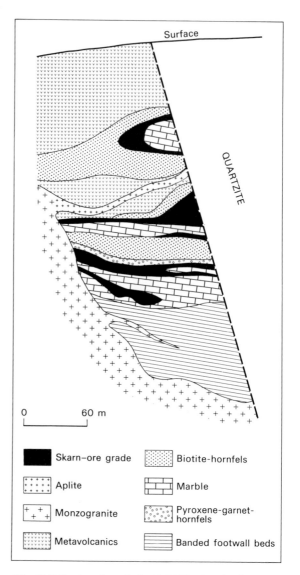

Fig. 13.5 Section through the Bold Head orebodies looking north, Tasmania. (Modified from Danielson 1975.)

Legend:
- Skarn–ore grade
- Aplite
- Monzogranite
- Metavolcanics
- Biotite-hornfels
- Marble
- Pyroxene-garnet-hornfels
- Banded footwall beds

along the length of the Pyrenees with two important deposits—one of which, Salau, was mined (Prouhet 1983, Fonteilles et al. 1989). Salau lies in the axial zone of the Pyrenees near the Spanish frontier where the late Hercynian Salau Granodiorite invaded Devonian limestones and converted them at the contact into grey graphitic marbles. The orebodies occur in exoskarns in a roof pendant in the granite (Fig. 13.6) and large volumes of the ore are rich in

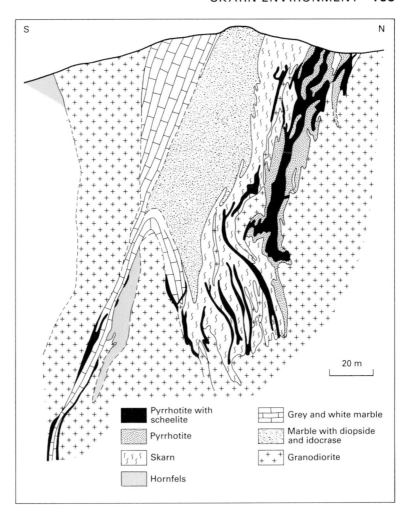

Fig. 13.6 Cross section through the Salau Tungsten Mine, France. (After Prouhet 1983.)

Legend:
- Pyrrhotite with scheelite
- Pyrrhotite
- Skarn
- Hornfels
- Grey and white marble
- Marble with diopside and idocrase
- Granodiorite

20 m

pyrrhotite with minor amounts of chalcopyrite, molybdenite, arsenopyrite and other sulphides. Grades vary from 0.1–20% WO_3 and the probable reserves before mining commenced were 1.3 Mt running 1.5% WO_3. Mining ceased in November 1986 owing to the low tungsten price, even though workable reserves grading nearly 1.75% WO_3 were still available. The scheelite is richly disseminated through the pyrrhotite-bearing skarn (Fig. 13.7) and it generally runs 1–2% WO_3 but in the pyrrhotite-poor hedenbergite–garnet skarns it is only around 0.2%. Geochemical work has failed to discover any source bed for the tungsten and has also shown that the granodiorite is very poor in this element (Barbier 1982). Its ultimate origin is for the moment unknown.

Talc and graphite skarns

Talc

Talc-bearing skarns and similarly altered carbonate and other metasedimentary rocks provide about 70% of world talc production. Good and important examples of these deposits occur in France and Austria (Moine *et al.* 1989).

The Trimouns Mine, an open pit operation, lies towards the eastern end of the French Pyrenees and at an altitude of 1800 m. Talc production is over 300 000 t p.a. and reserves of at least 20 Mt are present. The ores occur along the boundary between a basement of high grade metamorphic rocks and migmatites of the St Barthelemy Massif and an

(a)

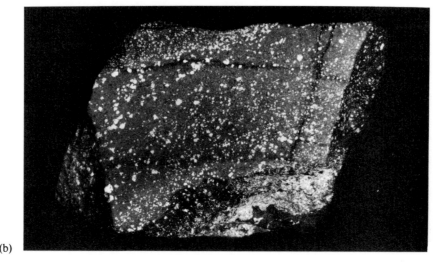

(b)

Fig. 13.7 (a) Pyrrhotite-rich scheelite skarn, Salau Mine, France. Dark coloured grains are calc-silicates, hedenbergite, andradite, etc. (b) The same in ultra-violet light showing the abundant dissemination of scheelite. Scale: actual size.

overthrust cover of lower grade metamorphic rocks of upper Ordovician to Devonian age (Fig. 13.8). The lower part of these hanging wall rocks contains discontinuous dolomitic lenses 5–80 m thick, overlying but also intercalated with mica schists in which small bodies of cross cutting leucogranites, aplites, pegmatites and quartz veins occur.

During the overthrusting intense shearing of the dolomites, schists and other rocks permitted extensive hydrothermal circulation which produced talc-rich ore (80–97% talc) in the dolomites and chlorite-rich ore (10–30% talc) in the silicate rocks. Many residual blocks of the primary rocks are present within the ore. The main orebody is 10–80 m thick and dips eastwards at 40–80°. Rock volumes appear to have remained roughly constant during the metasomatism. From a study of mineral assemblages and compositions Moine *et al.* (1989) showed that the metasomatism took place at about 400°C under a pressure of about 0.1 GPa. Highly saline,

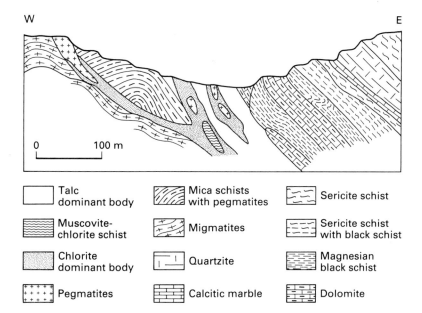

W   E

0   100 m

- Talc dominant body
- Muscovite-chlorite schist
- Chlorite dominant body
- Pegmatites
- Mica schists with pegmatites
- Migmatites
- Quartzite
- Calcitic marble
- Sericite schist
- Sericite schist with black schist
- Magnesian black schist
- Dolomite

Fig. 13.8 Sketch section through the Trimouns Talc Mine, Pyrenees, France. (After Moine *et al.* 1989.)

CO_2-poor solutions with high Ca and Mg concentrations played an important role in this metasomatism but their source is still conjectural.

Graphite

A minor amount of graphite production comes from skarns, e.g. the Norwegian Skaland Mine deep within the Arctic Circle just south of Tromso, where lenses of skarn up to 200 m long and usually 5–6 m thick (maximum 24 m), carry 20–30% graphite and occur within mica schists surrounded by metagabbros and granites. Gangue minerals include diopside, hornblende, labradorite, sphene, garnet, scapolite and wollastonite. The deposits are thought to have resulted from the concentration of pre-existing carbon in the sediments (Bugge 1978) and these could be calc-silicate hornfelses or reaction skarns.

Genesis of skarn deposits

A common pattern in the evolution of proximal skarns (skarns near or at an igneous contact) has been recognized which takes the form of (1) initial isochemical metamorphism, (2) multiple stages of metasomatism, and (3) retrograde alteration (Fig. 13.9)

Stage 1

This involves the recrystallization of the country rocks around the causative intrusion, producing marble from limestone, hornfels from shales, quartzites from sandstones, etc. Reaction skarns may form along lithological contacts. If the marbles are impure then various calcium and magnesian silicates may form and we have a calc-silicate hornfels that might contain minerals of economic interest, such as talc and wollastonite. The principal process involved in this isochemical metamorphism is diffusion of elements in what can be an essentially stationary fluid, apart from the driving out of some metamorphic water. The rocks as a whole may become more brittle and more susceptible to the infiltration of fluids in stage 2.

Stage 2

The infiltration of the contact rocks by hydrothermal-magmatic fluids leads to the conversion of pure and impure marbles, and other rock types, into skarns and the modification of calc-silicate hornfels of stage 1. This is a prograde metamorphic and metasomatic process operating at temperatures of about 800–400°C (Kwak 1986) during which an ore fluid evolves, initial ore deposition takes place and

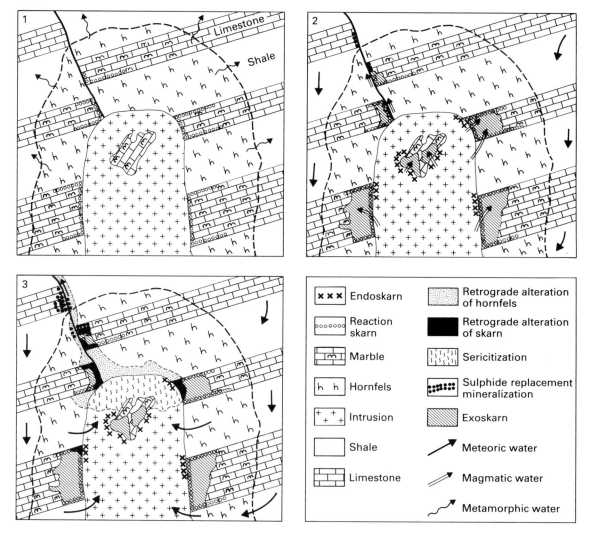

Fig. 13.9 Stages in the development of skarn deposits. (1) The intrusion of hot magma drives out ground, formation and metamorphic waters, produces a metamorphic aureole and reaction skarns. (2) Hydrothermal solutions exsolve from the crystallizing magma to form endo- and exoskarns, some sulphide replacement mineralization may occur. (3) The system cools permitting meteoric water to enter the intrusion and the skarns, producing sericitization in the intrusion and retrograde reactions in the skarns and hornfelses. Sulphide replacement bodies may develop during this stage.

the pluton begins to cool. The new minerals developed (Table 13.1) are dominantly anhydrous. Deposition of oxides (magnetite, cassiterite) and sulphides commences late in this stage but generally peaks during stage 3.

Stage 3

This is a retrograde (destructive) stage accompany-

ing cooling of the associated pluton and involving the hydrous alteration of early skarn minerals and parts of the intrusion by circulating meteoric water. Calcium tends to be leached and volatiles introduced with the development of minerals such as low-iron epidote, chlorite, actinolite, etc. (see Table 13.1). Declining temperatures lead to the precipitation of sulphides. The alteration is usually structurally controlled and cuts across earlier skarn patterns with the sulphide deposition often extending beyond the skarn boundaries into marble or hornfels. Here reactions at the marble contact may lead to neutralization of the hydrothermal solutions and the development of high grade sulphide ores.

In distal skarns, stage 1, or even stage 2, may not be developed and fluid inclusion work suggests formational temperatures of 350–210°C (Kwak 1986). As Meinert (1983) pointed out the degree to which a particular stage is developed in a particular skarn will depend upon its geological environment. The metamorphism during stages 1 and 2 is likely to be more extensive and higher grade around a skarn developed at deep crustal levels than one formed at shallow depths. Conversely retrograde alteration during cooling and the possible influx of meteoric water (stage 3) will probably be more intense at shallow rather than at deeper levels.

The origin of all the introduced material in certain skarns, e.g. vast tonnages of iron, has been much debated. The great majority of workers who have investigated these deposits consider that in most cases the pluton responsible for the contact metamorphism was also the source of the metasomatizing solutions. Whilst it is conceivable that a granitic pluton might supply much silica, it might be thought unlikely that it could have supplied the amount of iron that is present in some deposits. However, Whitney *et al.* (1985) have shown that it is probable that in natural magmatic systems, the concentration of iron in chloride solutions coexisting with magnetite or biotite is very high. This high solubility may explain the large quantities of iron in some skarns associated with granitic intrusions. On the other hand, where the pluton concerned is basic, the supply of iron does not present such great problems. These difficulties do become insurmountable, however, for the small class of pyrometasomatic

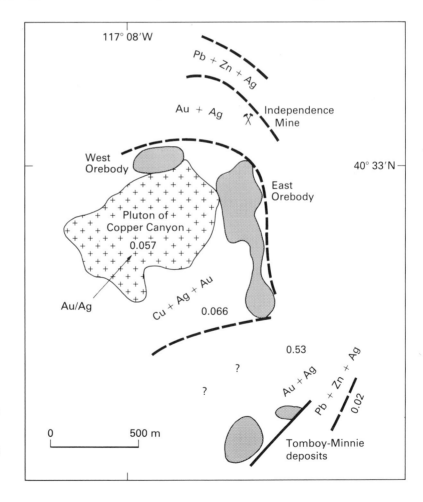

Fig. 13.10 Zonal distribution of metals and Au/Ag ratios in skarn deposits of the Copper Canyon area, Nevada. The metal zonation and ratios are uncertain in the area with question marks. (After Theodore *et al.* 1986.)

deposits, such as the Ausable Magnetite District, New York State, which have no associated intrusions. Perhaps the main function of the intrusion in some examples is that of a heat engine. For Ausable, Hagner & Collins (1967) suggested migration of iron from accessory magnetite in granite–gneiss, with concentration in shear zones to form magnetite-rich bodies, together with the release of iron during the recrystallization of clinopyroxene- and hornblende-gneisses.

The experimental replacement of marble by sulphides has been achieved by Howd & Barnes (1975) who found that ore-bearing bisulphide solutions at 400–450°C and 500 MPa when oxidized, produced acid solutions that dissolved the marble and provided sites for sulphide deposition. These experiments and proximity of many skarn deposits to porphyry copper intrusions suggest that circulating magmatic-hydrothermal solutions have played an important part in the ore genesis. This supposition is supported by the results of a number of studies, a few of which can be touched on here. In investigating the development of the proximal skarns at Elkhorn, Montana, Bowman *et al.* (1985) found that the isotopic values of hydrogen and oxygen indicated that no meteoric water was involved in stages 1 and 2 but that it became abundant in stage 3. The skarn-forming fluids of stage 2 were probably of magmatic origin. Kalogeropoulos *et al.* (1989) found that the lead of galena and sulphur isotopic compositions determined on all the sulphides of the Pb–Zn–Au–Ag skarns of Olympias, Greece, indicated that both elements were of igneous origin. When reviewing the work on fluid inclusions from 24 different skarn deposits from various continents, Kwak (1986) discovered that the proportions of elements present in fluid inclusions that do not normally take part in skarn formation (Na, K and Cl) reflect the nature of the associated pluton. High KCl contents occur in skarns adjacent to high-K granitoids, whereas high NaCl contents are present in skarns adjacent to calcic granitoids. In one deposit, where a drill hole had penetrated the granite underlying the skarn, the fluids in the inclusions were rich in the skarn-forming elements Fe and F and the homogenization temperatures and salinities were compatible with those in the overlying skarn.

Kwak's work also demonstrated very convincingly that falling temperature gradients within skarns from the igneous contact outwards, e.g. the same zones of garnet grains from King Island Scheelite, show a temperature range of 800–400°C over 500 m. Kwak's review revealed that the formation temperatures of Zn–Pb skarns are commonly in the range of 400–150°C in keeping with their usual distal development and the zoning of mineralization that is found in some areas (Fig. 13.10).

Other skarn deposits

Useful short summaries of skarn deposits can be found in Sawkins (1984) and Edwards & Atkinson (1986). The best exhaustive summary is Einaudi *et al.* (1981), and part 4 of *Economic Geology,* 77 for 1982 is devoted to skarn deposits. A comprehensive discussion of W–Sn skarns can be found in Kwak (1987).

14 / Disseminated and stockwork deposits associated with plutonic intrusives

We are concerned in this chapter with low grade, large tonnage deposits which are mined principally for copper, molybdenum and tin. These deposits are normally intimately associated with intermediate to acid plutonic intrusives and all are characterized by intense and extensive hydrothermal alteration of the host rocks. The ore minerals in these deposits are scattered through the host rock either as what is called disseminated mineralization, which can be likened to the distribution of seeds through raspberry jam, or they are largely or wholly restricted to quartz veinlets that form a ramifying complex called a stockwork (Fig. 14.1). In many deposits or parts of deposits both forms of mineralization occur (Fig. 14.2).

The first copper deposits of this type to be mined on any scale are in some of the south-western states of the USA and it was here that the cost effectiveness of bulk mining methods was first demonstrated in the 1920s and the mining of much lower grade copper ores than had hitherto been exploited be-

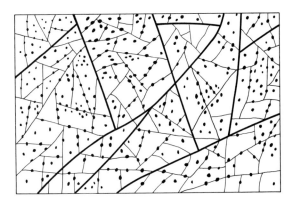

Fig. 14.2 Schematic drawing of a stockwork in a porphyry copper deposit. Sulphides occur in veinlets and disseminated through the highly altered host rock. (After Routhier 1963.)

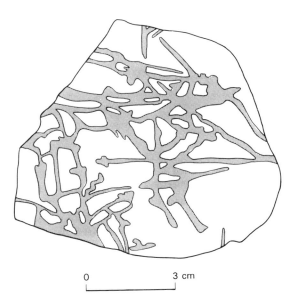

Fig. 14.1 Stockwork of molybdenite-bearing quartz veinlets in granite that has undergone phyllic alteration. Run of the mill ore, Climax, Colorado.

came possible. It should, however, be stressed that our civilization would not have been able to produce metals so abundantly and so cheaply by these methods without the important invention, by Francis and Stanley Elmore in 1898, of that most efficient method of mineral processing, flotation. The fame and unique importance of this invention has long gone largely unsung and unrecognized (Wolfe 1984).

These American copper deposits are associated with porphyritic intrusives, often mapped as porphyries. The deposits soon came to be called copper porphyries, the name by which they are still generally known. On the other hand, more or less identical molybdenum deposits have been known as disseminated molybdenums although they are also called molybdenum stockworks or porphyry molybdenums. Similar tin deposits are more usually called tin stockworks, though the term porphyry tins has been used. The student will find all these names in present use. Whereas economic porphyry coppers and molybdenums are usually extremely large orebodies (50–500 Mt is the common size range), tin stockworks are much smaller, 2–20 Mt being the common size range. All three types of metal deposit may yield important by-products. Amongst these are molybdenum and gold from porphyry coppers; tin, tungsten and pyrite from the Climax molybdenum deposit (other porphyry molybdenums tend to be without useful by-products); and tungsten, molybdenum, bismuth and fluorite from tin stockworks.

Porphyry coppers annually provide over 50% of the world's copper and there are many deposits in production. They are situated in orogenic belts in many parts of the world. Porphyry molybdenums in production are far fewer, about 10, and account for over 70% of world production. Tin stockworks are much less important, most tin production coming from placer and vein deposits.

Porphyry copper and molybdenum deposits are closely related and, although the next section is mainly devoted to porphyry coppers, mention will be made of some salient points concerning porphyry molybdenum.

Porphyry copper deposits

General description

As has been indicated above, these are large low grade stockwork to disseminated deposits of copper that may also carry minor recoverable amounts of molybdenum, gold and silver. Usually they are copper–molybdenum or copper-gold deposits. They must be amenable to bulk mining methods, that is open pit or, if underground, block caving. Most deposits have grades of 0.4–1% copper and total tonnages range up to 1000 million with a few giants being even larger than this (Fig. 14.3 and Table 14.1).

Grade and tonnage define the total amount of metal in the ground but with the continual drop in the price of copper in real terms, the emphasis in recent years has shifted to grade. At the low grade, high tonnage end of the scale, development costs can turn an orebody into a wastebody, and few of the porphyry copper deposits developed in the 1970s would have covered their capital costs had today's prices prevailed when they started production. Selective mining is of course impossible and host rock, stockwork and disseminated mineralization have to be extracted *in toto*. In this way, some of the largest man-made holes in the crust have come into being.

The typical porphyry copper deposit occurs in a cylindrical stock-like, composite intrusion having an elongate or irregular outcrop about 1.5×2 km, often with an outer shell of equigranular medium-grained rock. The central part is porphyritic—implying a period of rapid cooling to produce the finer grained groundmass—the porphyry part of the intrusion. This raises the problem of how a late phase of rapid cooling could occur in the hot centre of the intrusion, insulated as it would be by the only just solidified and still hot outer portion. This problem will be considered in a later section.

Comprehensive reviews of porphyry copper deposits can be found in Titley & Beane (1981) and Titley (1982a) and there is a useful summary in McMillan & Panteleyev (1989). One of the most comprehensive descriptions of a single deposit is that of the El Salvador deposit, in Gustafson & Hunt (1975), which is extensively utilized in the Open University case study (Open University S333 Course Team 1976) and tyros will find this to be a valuable aid to study. Another wide ranging

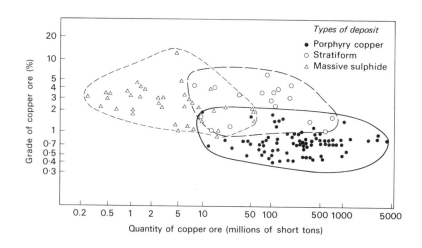

Fig. 14.3 Tonnage–grade relationships in porphyry copper, stratiform and massive sulphide deposits. (After COMRATE 1975.)

Table 14.1 Some giants and dwarfs of the porphyry copper world. (Data from Gilmour 1982 and other sources)

Country	Deposit	Tonnage[a] (Mt)	Grade[b]			
			Cu%	Mo%	Au g t^{-1}	Ag g t^{-1}
USA	Bingham (Utah)	2733	0.71	0.053	—[c]	—
	Morenci (Arizona)	848	0.88	0.007	—	—
	Ray (Arizona)	172	0.85	—	—	—
	San Manuel–Kalamazoo (Arizona)	980	0.74	0.015	—	—
	Santa Rita (New Mexico)	662	0.97	—	—	—
Canada	Lornex (British Columbia)	544	0.374	0.013	—	Tr
	Valley Copper	853	0.48	—	—	—
Mexico	Cananea	1542	0.79	—	—	—
Panama	Cerro Colorado	2000+	0.6	0.015	0.062	4.35
Chile	Chuquicamata	9423	0.56	0.06	—	—
	El Salvador	283	1.17	0.033	—	—
Papua New Guinea	Ok Tedi	543	0.6	—	0.5	—
	Panguna	1085	0.48	—	0.55	1.5
Iran	Sar Cheshmeh					
	Supergene	92	1.996	Tr	Tr	Tr
	Hypogene	334	0.896	Tr	Tr	Tr
UK	Coed-y-Brenin	200	0.3	0.003+	0.082	
Yugoslavia	Bor	383	0.428	~0.1	0.065	~4.26

[a] Past production plus reserves.
[b] For present reserves, often higher grade worked in the past.
[c] Gold by-product is important, grade figures not available.

investigation is that of the Yerington area by the Stanford School under Einaudi—see Dilles (1987) and references therein.

Petrography and nature of the host intrusions

The most common hosts are acid plutonic rocks of the granite clan ranging from granite through grano-diorite to tonalite, quartz monzodiorite and diorite. However, diorite through monzonite (especially quartz-monzonite) to syenite (sometimes alkalic) are also important host rock types. Suggestions made in the past that diorite hosts occur only in island arcs have proved to be incorrect. Silica-poor hosts occur in both British Columbia and the central Andes.

Many authors agree that porphyry copper deposits are normally hosted by I-type granitoids, within which category it is important to distinguish the I(Cordilleran) from the I(Caledonian) intrusives, as the latter rarely carry economic mineralization

(Pitcher 1983). Host intrusions in island arc settings have primitive initial strontium isotope ratios of 0.705–0.702 and are presumably derived from the upper mantle or recycled oceanic crust. The same ratios from mineralized intrusions in continental settings are generally higher indicating derivation from, or more probably, contamination by, crustal material (McMillan & Panteleyev 1989). Multiple intrusive events are common in areas with porphyry copper mineralization, with the host intrusions normally being the most differentiated and youngest of those present.

Host intrusions with associated volcanism generally formed late in the volcanic cycle and mineralization usually followed one or more pulses of magma intrusion. Thus at Ray, Arizona, early quartz–diorite was emplaced at 70 Ma, a porphyritic phase at 63 Ma and the mineralized porphyry at 61 Ma (Cornwall & Banks cited in McMillan & Panteleyev 1989). All the above points have important implications for exploration.

Attempts have been made to discriminate between barren and mineralized intrusions but without anything like universal success. Baldwin & Pearce (1982) recommended the use of a Y–MnO plot to discriminate between them and Hendry *et al.* (1985) not only showed that mineralized intrusions and their deeper level equivalents are depleted in copper compared with neighbouring barren intrusions, but also found that the latter characterized by amphiboles of constant composition whilst fertile intrusions have amphiboles of variable composition, even within single grains.

Intrusion geometry

The host intrusions usually appear to be passively rather than forcefully emplaced, stoping and assimilation being the principal mechanisms. They can be divided into three classes.

1 A class in which the ore-related intrusive is simply an isolated stock. A variation on this theme could be a sill or a series of dykes or irregular bodies.

2 In this class we no longer have a discrete isolated stock. The host is now a late stage unit of a composite, co-magmatic intrusion, often batholithic in dimensions. Examples belonging to this class occur in both continental and island arc settings.

3 This class is not yet known to carry economic mineralization but it is clearly related to porphyry copper deposits. Occurrences belonging to this class take the form of extensive alteration zones carrying weak mineralization and occurring in the upper parts of equigranular intrusions.

Hydrothermal alteration

In 1970, Lowell & Guilbert described the San Manuel–Kalamazoo orebody (Arizona) and compared their findings with 27 other porphyry copper deposits. From this study they drew up what is now known as the Lowell–Guilbert model. In this invaluable and fundamental work they demonstrated that the best reference framework to which we can relate all the other features of these deposits is the nature and distribution of the zones of hydrothermal wall rock alteration. They claimed that generally four alteration zones are present as shown in Fig. 14.4. These are normally centred on the porphyry stock in coaxial zones that form concentric but often incomplete shells and they are frequently used as a guide to ore in exploring porphyry copper deposits. In the Lowell–Guilbert model they are as follows.

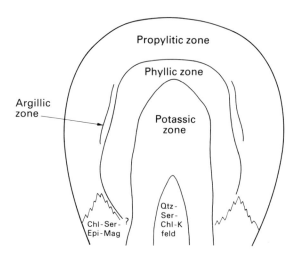

Fig. 14.4 Hydrothermal alteration zoning pattern in the Lowell–Guilbert model of porphyry copper deposits. (After Lowell & Guilbert 1970.)

Potassic zone. This zone is not always present. When present it is characterized by the development of secondary orthoclase and biotite or by orthoclase–chlorite and sometimes orthoclase–biotite–chlorite. Sericite also may be present. These secondary minerals replace the primary orthoclase, plagioclase and mafic minerals of the intrusion. Anhydrite may be prominent in this zone. The secondary potash feldspar is generally more sodic than the primary potash feldspar. It may also be present in the quartz veinlets forming the stockwork. There is often a low grade core to this zone in which chlorite and sericite are prominent.

Phyllic zone. This is alteration of the type known in other deposits as sericitization and advanced argillic alteration. It is characterized by the assemblage quartz–sericite–pyrite and usually carries minor chlorite, illite and rutile. Pyrophyllite may also be present. Carbonates and anhydrite are rare. The inner part of the zone is dominated by sericite while further out, clay minerals become more important. The sericitization affects the feldspars and primary biotite, alteration of the latter mineral producing the minor rutile. These are silica-generating reactions, so much secondary quartz is produced (silicification). The contact with the potassic zone is gradational over tens of metres. When the phyllic zone is present it possesses the greatest development of disseminated and veinlet pyrite.

Argillic zone. This zone is not always present. It is the equivalent of what is called intermediate argillic alteration in other deposits. Clay minerals are prominent with kaolin being dominant nearer the orebody, and montmorillonite further away. Pyrite is common, but less abundant than in the phyllic zone. It usually occurs in veinlets rather than as disseminations. Primary biotite may be unaffected or converted to chlorite. Potash feldspar is generally not extensively affected.

Propylitic zone. This outermost zone is never absent. Chlorite is the most common mineral. Pyrite, calcite and epidote are associated with it. Primary mafic minerals (biotite and hornblende) are altered partially or wholly to chlorite and carbonate. Plagioclase may be unaffected. This zone fades into the surrounding rocks over several hundreds of metres.

Obviously, in many deposits the behaviour of these zones in depth is poorly known and for some deposits there are no data at all. What evidence there is suggests that the zones narrow in depth and quartz–potash-feldspar–sericite assemblages become more frequent, with chlorite replacing biotite.

Hypogene mineralization

The ore may be found in three different situations. It may be (a) totally within the host stock, (b) partially in the stock and partially within the country rocks (Fig. 2.10), or (c) in the country rocks only. The most common shape for the orebody in the examples analysed by Lowell & Guilbert (1970) is that of a steep-walled cylinder (Fig. 14.5). Stubby cylindrical to flat conical forms and gently dipping tabular shapes are also known. The orebodies are usually surrounded by a pyrite-rich shell.

Like the alteration, the mineralization also tends to occur in concentric zones (Fig. 14.5). There is a central barren or low grade zone with minor chalcopyrite and molybdenite, pyrite usually forming only a few per cent of the rock, but occasionally ranging up to 10%. The mineralization appears to be disseminated rather than fracture-controlled, but Titley & Beane (1981) contend that this is simply a matter of scale and the 'disseminated' mineralization was located by microfractures and hair-line cracks not noticed by some observers. Passing outwards, there is an increase first in molybdenite and then in chalcopyrite as the main ore shell is encountered. Veinlet mineralization is now more

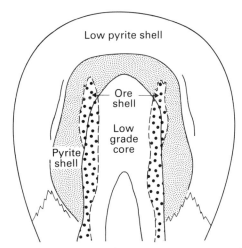

Fig. 14.5 Schematic diagram of the principal areas of sulphide mineralization in the Lowell–Guilbert model of porphyry copper deposits. Solid lines represent the boundaries of the alteration zones shown in Fig. 14.4. (After Lowell & Guilbert 1970.)

important. Pyrite mineralization also increases in intensity outwards to form a peripheral pyrite-rich halo with 10–15% pyrite but only minor chalcopyrite and molybdenite. The shells show a spatial relationship to the wall rock alteration zones (see figures) with the highest copper values often being developed at and near the boundary between the potassic and phyllic zones. Mineralization is not commonly found in the argillic zone, which usually appears to have been formed later and to be superimposed on the other alteration zones (Beaufort & Meunier 1983). Weak, non-economic mineralization continues outwards into the propylitic zone.

Breccia zones and pipes

Breccia zones and pipes are common in a number of deposits and are often mineralized, while some deposits consist mainly of mineralized breccia pipes. These breccias may occur within the porphyry body or in its wall rocks. Some appear to be the result of hydrothermal activity—fluidized breccias with rounded clasts and rock flour cements—whilst others appear to be angular collapse breccias. The former are frequently referred to as pebble dykes.

Vertical extent of porphyry bodies

Sillitoe (1973) suggested that porphyry copper

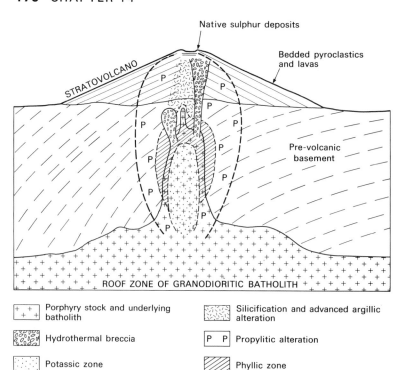

Fig. 14.6 Diagrammatic representation of a simple porphyry copper system on the boundary between the volcanic and plutonic environments. (After Sillitoe 1973.)

Legend:

- [+ +] Porphyry stock and underlying batholith
- Hydrothermal breccia
- Potassic zone
- Silicification and advanced argillic alteration
- [P P] Propylitic alteration
- Phyllic zone

deposits occur in a subvolcanic environment associated with small high level stocks and he emphasized their close association with subaerial calc-alkaline volcanism. He envisaged the host pluton as being overlain by a stratovolcano (Fig. 14.6), and contended that evidence from Chilean and Argentinian deposits shows that the propylitic alteration extends upwards into the stratovolcano, which is itself likely to carry native sulphur deposits. The zones of alteration close upwards. The potassic zone dies out, sericitic and argillic alteration become important and the upper limit of economic mineralization is reached. At the same time, the porphyry stock becomes smaller in size and hydrothermal breccias appear over large areas.

Francis *et al.* (1983) proposed a modification of this picture. Whereas Sillitoe appeared to infer that the overlying stratovolcanoes were all large structures, these authors contend that the volcanic tops of porphyry systems may vary a lot in size and that many may be capped by small dacitic domes. Christie & Braithwaite (1986) and other workers have pointed out that these upper portions of porphyry systems may carry epithermal gold mineralization.

In the lower parts of porphyry copper deposits the available evidence suggests that there is often a downward transition from porphyry into an equigranular plutonic rock of similar composition which forms part of a pluton of much larger dimensions. With this textural change the mineralization dies out in depth. Sodic-calc alteration (development of oligoclase and actinolite) has been recorded in the root zones of a number of deposits and may be a fairly common alteration type at this level (Carten 1986).

The diorite model

Subsequent to Lowell & Guilbert's classic work it has been recognized that some porphyry copper deposits are associated with intrusives having low silica : alkali ratios. Various names have been suggested for this type. The one that has won general recognition is 'diorite model', although the host pluton may be a syenite, monzonite, diorite or alkalic intrusion (Hollister 1975). Diorite model deposits differ in a number of ways from the Lowell–Guilbert model; one of the main reasons appears to be that sulphur concentrations were relatively low in the mineralizing fluids. As a result, not all the iron oxides in the host rocks are converted to pyrite and much iron remains in the chlorites and biotites while excess iron tends to occur as magnetite, which may be present in all alteration zones.

Alteration zoning

The phyllic and argillic alteration zones are usually absent so that the potassic zone is surrounded by the propylitic zone. This zonal pattern is present in both island arc and continental porphyry copper deposits. In the potassic zone, biotite may be the most prominent potassium mineral and when orthoclase is not well developed, plagioclase may be the principal feldspar.

Mineralization

The main difference from the Lowell–Guilbert model is that significant amounts of gold may now occur and molybdenum/copper is usually low. The fractures containing gangue silicate minerals and copper sulphides may be devoid of quartz. On the other hand, chlorite, epidote and albite are fairly common.

Comparison of the Lowell–Guilbert and diorite models

Table 14.2 (based on Hollister 1975) contrasts the principal features of each model and lists further features of the diorite model.

Metal abundances in porphyry copper deposits

The vast majority of deposits can be divided into two classes according to whether the principal accessory metal is molybdenum or gold (Kesler 1973). Generally speaking, copper–gold deposits appear to be concentrated in island arc settings and copper–molybdenum where continental crust is present. There are, however, some notable exceptions, e.g. Cerro Colorado, one of the largest island arc deposits, has a high molybdenum : gold ratio and Sillitoe (1980a) has reported the presence of porphyry molybdenum mineralization in the Philippines. A plot of these metals for porphyry deposits in western Canada shows a gradation from Cu–Mo to Cu–Au types rather than a clear-cut difference (Sinclair *et al.* 1982) and other exceptions were

Table 14.2 Comparison of Lowell–Guilbert and diorite models of porphyry copper deposits

Feature	Lowell–Guilbert model	Diorite model
Host pluton		
Common rock types	Monzogranite granodiorite, tonalite	Syenite, monzonite
Rarer rock types	Quartz–diorite	Diorite
Alteration		
Central core area	Potassic	Potassic
Peripheral to core	Phyllic	Propylitic
	Argillic	
	Propylitic	
Mineralization		
Quartz in fractures	Common	Erratic
Orthoclase in fractures	Common	Erratic
Albite in fractures	Trace	Common
Magnetite	Minor	Common
Pyrite in fractures	Common	Common
Molybdenite	Common	Rare
Chalcopyrite/bornite	3 or greater	3 or less
Dissemination of chalcopyrite	Present	Important
Gold	Rare	Important
Structure		
Breccia	May occur	Rare
Stockwork	Important	Important

noted previously by Titley (1978). Although Kesler's contention is still generally true as viewed on a world-wide basis, we must keep an open mind on this subject.

Regional characteristics of porphyry deposits

The distribution of porphyry copper and molybdenum deposits is shown in Fig. 14.7. From this map it can be seen that the majority of porphyry deposits are associated with Mesozoic and Cenozoic orogenic belts in two main settings—island arcs and continental margins. The major exceptions are the majority of the USSR deposits and the Appalachian occurrences of the USA. These exceptions belong to the Palaeozoic. Only a few porphyry deposits have so far been found in the Precambrian. These facts are of great importance from the exploration point of view.

South-western USA and Mexico

The deposits of this region form an oval cluster in the USA which contrasts with the the linear belts of deposits in the Andes and elsewhere (Lowell 1974), but which continues south-eastwards as a linear belt into Mexico for at least 1900 km (Damon *et al.* 1983).

There are over 100 deposits, with about 30 in production. The ages range from 20 to 195 Ma with a peak around 60 Ma (Laramide Orogeny). Some deposits lie along marked lineaments with a tendency to be developed at their intersections, but this is not true of all the deposits. They are spread over a considerable area and some are so far inland that, if the genesis of the host intrusives is to be linked to a subduction zone, then it is necessary to postulate very low dipping or multiple subduction zones.

Northern (Canadian) Cordillera of America

In this complex region there is a range of copper, copper–molybdenum and molybdenum porphyries. From their study of a large area in central British Columbia, Griffiths & Godwin (1983) have related these to magma generation connected with Benioff Zones active during late Triassic–Jurassic and Cretaceous to early Tertiary times. The host granitoids are I-type and the Cu : Mo ratios of the deposits correlate with the composition of the host plutons, decreasing from diorite to granite and implying a genetic link between the major element chemistry and the metal content of the magmas.

Appalachian orogen

According to Hollister *et al.* (1974) and Hollister (1978), porphyry deposits were first developed in this region during the Cambrian and Ordovician and later in the Devonian and Carboniferous. Considering all the deposits together, copper porphyries formed first, followed by the coeval development of separate copper and molybdenum porphyries, with a final period when only molybdenum porphyries were formed. Accompanying this change was a variation in the composition of the magmas associated with the mineralization from quartz-monzonitic to granitic. The copper and molybdenum porphyries either lack or have a very small development of the phyllic zone. This may be an effect of deep erosion. On the other hand, this fact and the low content of pyrite may indicate a deposit type transitional to the diorite model.

Prior to continental drift, this orogen was continuous with the Caledonian Province of the British Isles where porphyry copper mineralization has been found in North Wales (Cambro-Ordovician) and Scotland (Devonian).

The Andean province

As Fig. 14.8 shows, this has a most marked linear distribution of deposits. These occur in a region where erosion has not cut down as deeply as it has in the coastal range batholiths to the west, but has cut down more deeply than in the volcanic belt to the east. The age range of most of the deposits is 59–4.3 Ma, nearly all have a significant molybdenum content and the underlying crust appears to be entirely continental.

Most deposits are related to stocks of intermediate composition and calc-alkaline affinity, although a few are associated with stocks of the shoshonitic suite. There are 15 major porphyry copper deposits and at least 50 occurrences in Chile and Argentina. The belt continues northwards into Colombia and Central America. The exploitation of the Chilean deposits has made that country the western world's leading copper producer. A good summary of the Chilean deposits can be found in Sillitoe (1981a).

The South-western Pacific island arcs

The island arcs of this province generally lack continental crust; gold is usually an important by-product and molybdenum uncommon. Lowell–

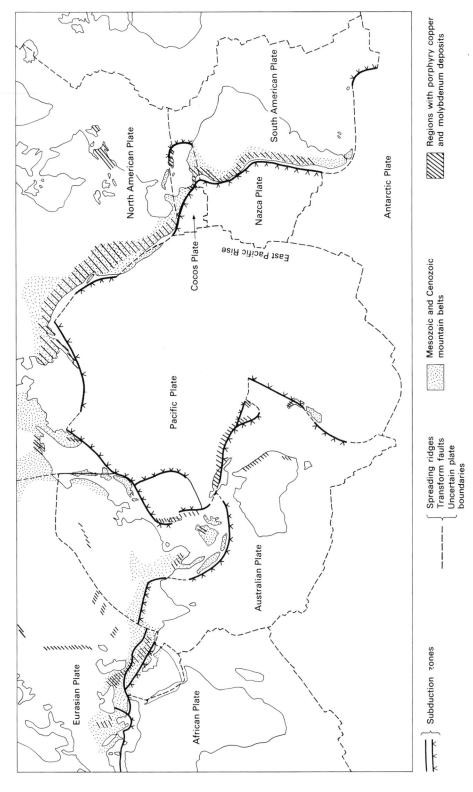

Fig. 14.7 The principal porphyry copper and molybdenum regions of the world. Also shown are present plate boundaries and Mesozoic–Cenozoic mountain belts.

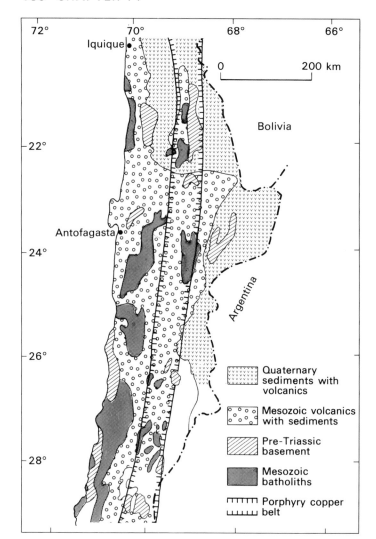

Fig. 14.8 Map of part of the Chilean porphyry copper province.

Guilbert and diorite model deposits occur together. Most of the deposits occur in arcs containing thick sequences of pre-ore rocks. The host porphyries have penetrated to high structural levels and are all very young (less than 16 Ma). The deposits formed at a late stage of arc evolution just before volcanic activity ceased.

Useful descriptions of the deposits in Papua New Guinea with invaluable summaries of economic data can be found in Amade (1983). Readers interested in the controversy surrounding the development of one of these, Ok Tedi, should read Jackson (1982).

Porphyry copper and molybdenum deposits in the USSR

The majority of USSR deposits are Palaeozoic with a peak in the Carboniferous. The distribution of the major fields is shown in Fig. 14.7. The most important are in Kazakstan, the Caucasus, Uzbekistan and the Batenevski Range in Siberia. Most, if not all, deposits are related to present or former subduction zones and about a quarter are molybdenum-poor copper porphyries that have formed in island arc settings. Copper–molybdenum deposits lie along continental or microcontinental margins, like those of the south-western USA and the Andes. Porphyry molybdenum deposits in any given area are always younger than associated porphyry copper deposits (Laznicka 1976). The north-eastern USSR is a good example of an area where it is important from the prospecting point of view to distinguish between I(Cordilleran) and I(Caledonian) type granites for, whilst the former type in Kamchatka carries porphyry copper-type mineralization, the I_{CAL} granitoids of the Verkhoyansk Belt to the west and north of the Kolyma-Omolon Plate, like the type I_{CAL} granitoids of Scotland, do not carry any economic porphyry style mineralization (Shilo *et al.* 1983, Pitcher, pers. comm. 1983). The I_{CAL} granitoids of Scotland appear to belong to the third type of host intrusion described on p. 174, and a good example is the Ballachulish Granite which carries areas of weak Cu–Mo mineralization in extensive phyllic alteration (Evans *et al.* 1979).

Genesis of porphyry copper deposits

The principal arguments over recent years have been concerned with a magmatic versus a meteoric derivation for the mineralizing fluids and the origin of the metals and sulphur. In considering the formation of these deposits we must remember that the most striking characteristic of porphyry copper deposits when compared with other hydrothermal orebodies is their enormous dimensions. The size and shape of these deposits imply that the hydrothermal solutions permeated very large volumes of rock, including country rocks, as well as the parent pluton. That at least some of these solutions originated in the host pluton is suggested by the existence of crackle brecciation.

Crackle brecciation and its origin

Crackle brecciation is the name given to the fractures that are usually healed with veinlets to form the stockwork mineralization. The zone of crackle brecciation is usually circular in outline, always larger than the orebodies and it fades out in the propylitic zone. It is often less well developed near the centre of the deposit, particularly if potassic alteration is present. This brecciation is thought to be due to the expansion resulting from the release of volatiles from the magma (Phillips 1973).

The host magmas of porphyry copper deposits appear to have reached to within 0.5–2 km of the surface before equigranular crystallization commenced in their outer portions. The intrusions would then be stationary and the confining pressure would not fluctuate. With the steady development of crystallization, however, anhydrous minerals form and the liquid magma becomes richer in volatiles, leading to an increase in the vapour pressure. If the vapour pressure rises above the confining pressure, then what is called retrograde boiling will occur and a rapidly boiling liquid will separate. If retrograde boiling occurs in a largely consolidated rock, the vapour pressure has to overcome the tensile strength of the rock as well as rising above the confining pressure. This will result in expansion and extensive and rapid brecciation (Fig. 14.9). The reason for this is that water released at a depth of about 2 km at 500°C would have a specific volume of 4 and, if 1% by weight formed a separate phase, it would produce an increase in volume of about 10%. At shallower depths the increase would be even greater and the degree of fracture intensity higher (Burnham & Ohmoto 1980). Evidence for the development of retrograde boiling in porphyry copper deposits is common in the form of the widespread occurrence of liquid-rich and gas-rich fluid inclusions in the same thin section (Chapter 3).

It is important to notice that whilst the crystallization of solid phases is an exothermic one, bubble formation is an endothermic process. Rapid nucleation and the adiabatic expansion of the vapour would absorb a great deal of heat, taking up the latent heat of crystallization and significantly lowering the temperature of the system. This would result in a second phase of rapid cooling in the central part of the intrusion which would considerably increase the number of nucleation sites producing a period of rapid crystallization, which, in turn, would be responsible for the fine-grained groundmass and hence the porphyritic nature of the intrusion.

Some chemical processes in the formation of porphyry copper deposits

Retrograde boiling produces an aqueous phase (hydrothermal solution) in a porphyry system and chloride ion is partitioned strongly into it as is bisulphide ion, provided a sulphide mineral, such as pyrrhotite, is not stable (Burnham 1979). The presence of chloride ion supplies a transporting mechanism for the base metals that also fractionate strongly into the aqueous phase, and the bisulphide provides the sulphur for the eventual precipitation of sulphides. However, there are important controls

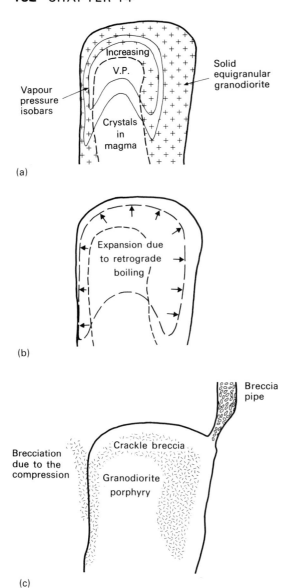

(a)

(b)

(c)

Fig. 14.9 Three stages in the development of crackle brecciation. (a) Vapour pressure building up in and around upper portion of magmatic fraction. (b) Retrograde boiling occurs causing expansion. (c) Distribution of resulting brecciation. (After Phillips 1973.)

on the fractionation of sulphur into the aqueous phase, as Burnham & Ohmoto (1980) have pointed out. One of these is the fugacity of oxygen (f_{O_2}) in the magma. Although sulphur is in solution in hydrous silicate melts as HS^-, it exists in the aqueous phase as both H_2S and SO_2, and the partition coefficient for total sulphur increases with an increase in the $SO_2 : H_2S$ ratio, which itself goes up with an increase in f_{O_2}. Now f_{O_2} in a magma prior to retrograde boiling is largely determined by the $Fe^{3+} : Fe^{2+}$ ratio of the magma, which in turn is largely dependent on the source rock from which the magma was generated. Thus aqueous fluids that separate from I-type magmas with relatively high f_{O_2} tend to produce sulphur-rich porphyry copper mineralization, whereas fluids from S-type magmas may deposit the more sulphur-poor tin oxide mineralization. Sulphur isotopic studies indicate that sulphur in porphyry copper deposits is largely of upper mantle or remelted oceanic crust origin (McMillan & Panteleyev 1989), which emphasizes the importance of the above control.

Evidence from isotopic and geochemical investigations

Further evidence of a magmatic derivation of at least some of the hydrothermal solutions comes from stable isotope investigations (Sheppard 1977). Waters in equilibrium with potassium silicate alteration assemblages and formed at 550–700°C are isotopically indistinguishable from primary magmatic waters (Fig. 14.10). On the other hand, waters associated with sericites from the phyllic zone of alteration are depleted in ^{18}O relative to the biotites of the potassic zone. Comparison with Fig. 4.15 suggests that formation waters from the country rocks were involved in the sericitization; in other words, meteoric water played a significant role in the hydrothermal fluids responsible for the phyllic alteration. The isotopic data for advanced and intermediate argillic alteration show an identical pattern to that for the phyllic alteration data. Field and microscopic evidence suggest that the phyllic and argillic alterations were later than the potassium silicate and propylitic alterations and were superimposed to varying degrees upon them. These two stages of development are depicted in Fig. 14.11.

It appears that after intrusion of the porphyry body, solidification occurs and a magmatic-hydrothermal solution evolves. This solution reacts with the porphyry and to a varying extent with the surrounding country rocks, giving rise to the development of a central zone of potassium silicate alteration. The introduction of much of the metals and sulphur probably accompanies this stage. Further out from the intrusion, thermal gradients set up

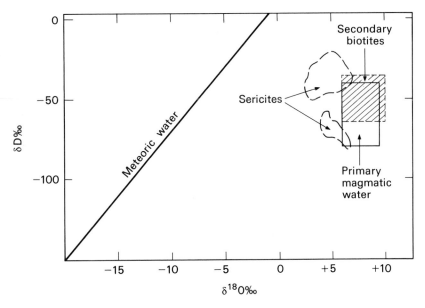

Fig. 14.10 Isotopic compositions of hydrothermal waters associated with secondary biotites from the potassic zones and sericites from the phyllic zones of five porphyry copper deposits. (Modified from Sheppard 1977.)

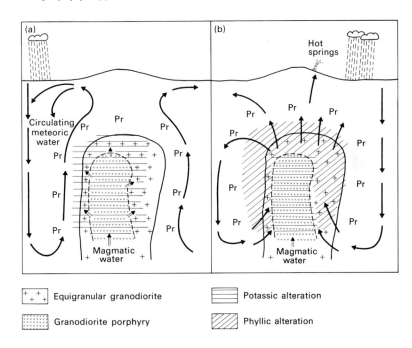

Fig. 14.11 Diagrammatic sections through a porphyry copper deposit showing two stages in the development of the hydrothermal fluids leading to the formation of a Lowell–Guilbert model deposit (b). Pr = propylitic alteration. (Modified from Sheppard 1977.)

a convective circulation of water in the country rocks, and this is responsible for the propylitic alteration (diorite model conditions), as Taylor (1981) has shown for the diorite model deposit of Bakırçay, Turkey. In this study he used Sr isotopes to demonstrate that fluids that formed the potassic alteration were magmatic-hydrothermal and those that caused the propylitic alteration were meteoric-hydrothermal.

When the intrusion cools, this meteoric–

formation hydrothermal system may encroach upon and mix with the waning magmatic system leading to the development of lower temperature minerals: sericite, pyrophyllite and clay minerals. These would replace in particular the feldspar and biotite of the outer part of the original potassium silicate zone. The relatively rapid gradients in pH, temperature, salinity, etc., across the interface between these two hydrothermal systems probably account for the concentration of copper around the boundary zone between the potassium silicate and phyllic zones. With this second stage of alteration, a Lowell–Guilbert model deposit comes into being.

Supporting evidence for these two stages in the formation of Lowell–Guilbert model deposits comes from the work of Taylor & Fryer (1982, 1983), who have used REE geochemistry to demonstrate that the meteoric-hydrothermal fluids had the potential to remobilize and reconcentrate copper and molybdenum, giving rise to the second stage of hypogene leaching and enrichment at both Bakırçay and Santa Rita, New Mexico. Although this sequence of events probably happened during the formation of many porphyry copper deposits, the very detailed work of Bowman *et al.* (1987) on the Bingham Canyon, Utah deposit showed that the waters that produced the potassic and propylitic alteration there were magmatic waters whose temperature had been lowered and isotopic character changed by mixing with

formation water (Fig. 14.12). A later influx of meteoric water produced a second isotopic–temperature trend (Fig. 14.12) and gave rise to the later phyllic alteration.

Thus, in the light of our present knowledge, the isotopic evidence indicates that magmatic water exsolved from consolidating plutons is responsible for the potassic alteration. However, Bowman *et al.* (1987) pointed out that the original 'magmatic' water at Bingham (and other localities) could be meteoric water that had acquired magmatic-like isotopic values by fluid–rock interactions at very low water : igneous rock ratios—less than 0.07 to reproduce the hydrothermal fluids responsible for the outer propylitic zone and less than 0.001 to produce the fluids in the inner potassic zone. If this were the mode of formation of the mineralizing fluid, then insufficient fluid to account for the enormous volume of hydrothermal alteration and development of quartz veins could have been generated within the Bingham Stock itself. Enough fluid with the required isotopic values could only have been formed by meteoric water–igneous rock interaction involving a vastly greater volume of igneous rock at a deeper level. The fluids would then have had to be focused upwards into what became the potassic alteration zone within the Bingham Stock. This could conceivably have been the case at Bingham but it seems most unlikely that such fluids

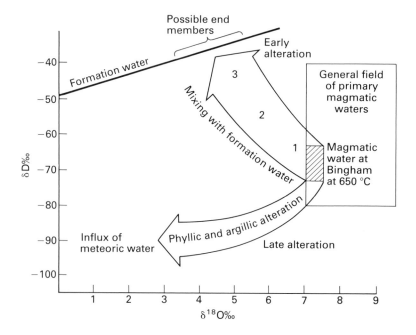

Fig. 14.12 Changes in the calculated values of δD and δ^{18}O of water during the development of early and late hydrothermal alteration in the Bingham Canyon porphyry system. (1) potassic zone, (2) potassic–propylitic border, (3) propylitic zone. (After Bowman *et al.* 1987.)

would be almost unerringly channelled into the suitable host stocks at higher structural levels during the genesis of this relatively common type of deposit.

Fluid inclusion evidence

This has been comprehensively reviewed by Roedder (1984) who emphasized the remarkable uniformity of the three characteristics of high maximum temperatures (as high as 725°C), high salinities (up to 60 wt % alkali chlorides) and evidence of boiling for all deposits. He considered that the extensive fluid inclusion data favours a magmatic hypothesis of origin for the mineralizing fluids. The highest temperature inclusions, which are rich in copper and other metals, characterize the central portions of porphyry systems (e.g. Chivas & Wilkins 1977, Easthoe 1978), and their distribution patterns mimic the zonal alteration–mineralization patterns. Fluid inclusion temperatures and salinities decrease both away from the central portions and with time of formation. All this evidence agrees well with the concept depicted in Fig. 14.11.

Source of the copper—is it leached from the magma or the wall rocks of the intrusion?

Economic porphyry copper deposits are more likely to develop if the parent magma is enriched in copper; can this happen during calc-alkaline differentiation? Feiss (1978) and Mason & Feiss (1979) suggested that, as copper prefers octahedral sites in crystallizing magmas, be they in melt or crystalline phases, then magmas with more octahedral sites will contain more copper in the melt at the retrograde boiling stage and thus be more likely to give rise to porphyry copper-type mineralization. Now the number of octahedral sites is proportional to the alumina : alkalies ratio and a study of mineralized versus barren intrusions in porphyry copper belts showed that the former are characterized by a higher ratio. This is one mechanism that can lead to the concentration of copper into certain magmatic differentiates.

Another is the enrichment that may occur when copper behaves as an incompatible element early in the development of magmatic systems as a result of the relatively poor development of magnetite and augite during fractional crystallization, because these minerals abstract copper from melts when they crystallize. The discovery, in a volcano, of mafic to intermediate lavas that are relatively poor in magnetite and augite, may therefore indicate the presence of a porphyry copper deposit in its roots, and such lavas, which are enriched in copper, have been recognized in Central America (Eilenberg & Carr 1981). It therefore seems very likely that some calc-alkaline magmas may be enriched in copper during differentiation, and this could help explain why only some of a series of similar intrusions become porphyry copper deposits.

If we pursue the possibility that the copper (and other wanted metals) are derived from the host intrusion, then what evidence supports this hypothesis? Only indirect evidence we must admit, because although we do not have any radioactive tracers to monitor the movement of copper, molybdenum, gold and silver, we do none the less have fairly compelling evidence. For example, Anderson (1980) analysed the metal distribution of porphyry deposits in the Canadian Cordillera and showed that it correlates strongly with the regional time–space distribution of the host intrusions and not with the rock types hosting those intrusions. This conclusion was confirmed in a comprehensive statistical study by Griffiths & Godwin (1983), indicating that the fundamental processes governing the metallization of porphyry copper deposits are probably subduction-controlled rather than upper crustal, a conclusion supported by the work of Sillitoe & Hart (1984), which showed that the lead isotopes in Colombian porphyry copper deposits probably came from subducted pelagic sediment. The final concentration of the copper into a stockwork is of course an upper crustal process, but further evidence that it is derived from the host intrusion and not the country rocks comes from the work of Hendry et al. (1985), who have shown that the deep level and temporal equivalents of North American porphyry copper intrusions are depleted in copper compared with similar barren intrusions. This work suggests that copper is abstracted not only from apical portions of the porphyry host but also from deeper parts of the parent intrusions. Evidence is accumulating through the work of Ilton & Veblen (1988) and others that the copper is not leached out of already precipitated silicates, such as biotite, since it was present in the fluid that produced the potassic alteration. Using transmission electron microscopy, Ilton & Veblen showed that in the two deposits they studied trace copper is concentrated in the secondary hydrothermal biotites (up to 10 wt %) and not in the primary magmatic biotites (< 0.3 wt %). In other words the sheet silicates in the host rocks acted as traps for copper and not sources of that metal.

Porphyry molybdenum deposits

General description

These have many features in common with porphyry copper deposits; some of these features have been touched on above and useful summary accounts have been given by White *et al.* (1981) and Wallace (1991). Average grades are 0.1–0.45 MoS$_2$ (molybdenum grades are *more usually* given as MoS$_2$) and one deposit produces by-product tin and tungsten. Host intrusions vary from quartz monzodiorite through granodiorite to granite. Stockwork mineralization is more important than disseminated mineralization and the orebodies are associated with simple, multiple or composite intrusions or with dykes or breccia pipes. There are three general orebody morphologies (Ranta *et al.* 1984, see Fig. 14.13), and tonnages range from 50 to 1500 Mt. Sections through the two biggest deposits are given in Figs 14.14 and 14.15. The molybdenite occurs in (a) quartz veinlets carrying minor amounts of other sulphides, oxides and gangue, (b) fissure veins, (c) fine fractures containing molybdenite paint, (d) breccia matrices and, more rarely, (e) disseminated grains. Supergene enrichment, which can be very important in porphyry coppers, is generally absent or minor.

White *et al.* (1981) have divided the porphyry molybdenums of North America into Climax and Quartz Monzonite Types and Laznicka (1985) has termed the second type the granodiorite–quartz monzonite class. Confusion is building up here, which can be resolved only by strict adherence *and reference* to one igneous rock classification scheme. White *et al.* appear to have used the term quartz monzonite as defined in time honoured, North American usage, the equivalent rock name in Europe being adamellite. Following the IUGS classification (Streckeisen 1976), now in all recent textbooks on igneous rocks, it must be stated that the majority of porphyry molybdenum deposits are associated with granites, including all Climax types and many 'quartz monzonite types'. Some 'quartz monzonite types' are associated with granodiorite and a few with quartz monzodiorite. Probably none is associated with quartz monzonites as defined in the IUGS classification!

This is all very confusing for the student *and* the mineral explorationist selecting targets in a new concession area. It is unfortunately a difficulty that will diminish only slowly over the years as those in the applied field begin to discipline themselves more strongly in their use of up-to-date nomenclature. The reader should remember that this is a problem that permeates the whole of the science and not just the subject of porphyry molybdenums! An interesting little exercise in this connection is to superimpose the IUGS classification boundaries and approved rock names on Fig. 2 of Griffiths & Godwin (1983).

The principal differences between these two types are that the Climax type generally has high trace or accessory contents of tin and tungsten, multiple intrusion of highly evolved magmas enriched in F, Rb and incompatible elements and several mineralizing events, intense silicification associated with its

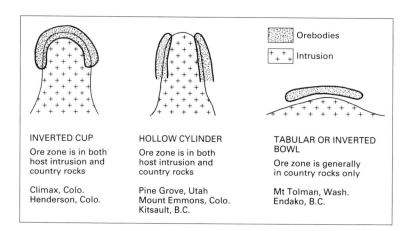

INVERTED CUP

Ore zone is in both host intrusion and country rocks

Climax, Colo.
Henderson, Colo.

HOLLOW CYLINDER

Ore zone is in both host intrusion and country rocks

Pine Grove, Utah
Mount Emmons, Colo.
Kitsault, B.C.

TABULAR OR INVERTED BOWL

Ore zone is generally in country rocks only

Mt Tolman, Wash.
Endako, B.C.

Fig. 14.13 Porphyry molybdenum orebody morphologies.

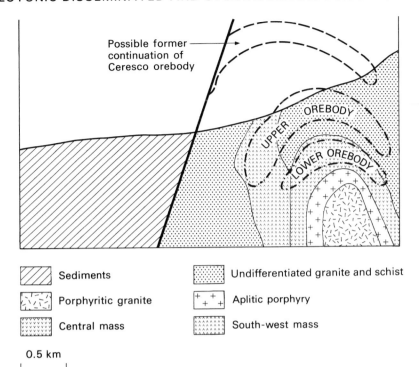

Fig. 14.14 Generalized geological section through the Climax molybdenum mine, Colorado. (Modified from Hall *et al.* 1974.)

Possible former continuation of Ceresco orebody

Sediments

Porphyritic granite

Central mass

Undifferentiated granite and schist

Aplitic porphyry

South-west mass

0.5 km

wall rock alteration, high average ore grades and is low in copper compared with the other type, which lies in the transition zone to molybdenum-bearing porphyry coppers. Perhaps it would be better if the second type was named after a well-known deposit of this type, such as Endako—*if* there is merit in this division for, as Laznicka (1985) has remarked, there appear to be many deposits outside North America that are transitional between these two types. These are referred to as sub-Climax type. The quartz monzonite type is more broadly distributed throughout the Western Cordillera of North America and similar deposits occur in China, Peru, Sulawesi and Yugoslavia. Climax-type deposits are fewer in number and restricted to the Colorado Mineral Belt.

The multiple intrusions present in some deposits have produced very complicated effects. At the Henderson deposit recent work (Carten *et al.* 1988, Seedorff 1988) has revealed the existence of 12 stocks related to three intrusive centres. High temperature, hydrothermal alteration assemblages, with which most of the molybdenite is associated, were developed in numerous cycles, each corresponding to the emplacement of an ore shell about the stock. The superposition of ore shells gave rise to broad ore zones about each of the three intrusive centres.

Hydrothermal alteration

The alteration patterns are very similar to those found in porphyry copper deposits, with potassic alteration and silicification being predominant. The most detailed study is on the Urad and Henderson deposits (Wallace *et al.* 1978, Carton *et al.* 1988, Seedorff 1988) where, associated with the Henderson orebody, there are potassic (secondary K-feldspar ± fluorite ± quartz ± molybdenite) and silicic (quartz + fluorite ± molybdenite) assemblages spatially and genetically related to the crystallization of individual stocks. The lower temperature zones of alteration envelop intrusive centres rather than single stocks. They have been summarized as quartz–topaz, phyllic, argillic and propylitic zones, but Seedorff (1988) has shown that the detailed picture is much more complex. There is a silicified zone which lies largely within the potassic zone. The Henderson orebody is roughly coincident with the potassic and silicified zones. A prominent pyrite zone, carrying 6–10% pyrite, is developed around the Henderson orebody and a similar less

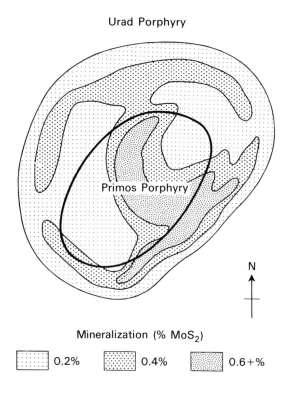

Urad Porphyry

Primos Porphyry

N

Mineralization (% MoS$_2$)

0.2%   0.4%   0.6+%

Fig. 14.15 Locality map (top right), plan (8000 ft or 2438.4 m level, above) and section (lower right) of the Henderson molybdenum orebody, Colorado. (Modified from Wallace *et al.* 1978.)

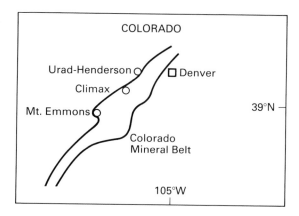

COLORADO

Urad-Henderson ○   □ Denver
Climax ○
Mt. Emmons ○   39°N

Colorado
Mineral Belt

105°W

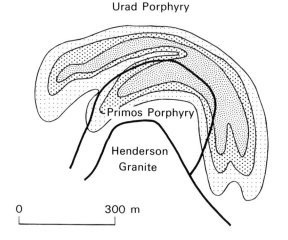

Urad Porphyry

Primos Porphyry

Henderson
Granite

0   300 m

distinct zone is present at Climax. Peripheral pyrite zones, which appear to coincide mainly with the phyllic zone, have been reported from a number of other porphyry molybdenums.

Genesis of porphyry molybdenum deposits

The close spatial association of the orebodies of most deposits with the potassic zone of alteration in a number of small stocks suggests a magmatic source and Wallace *et al.* (1978) argue that, for the Henderson and Climax deposits, this must have been unexposed master reservoirs which fed the columns of exposed host intrusions. They consider that the volumes of the host intrusions are far too small to have supplied the large tonnage of molybdenum that is present in the deposits, a hypothesis that is further developed by Carten *et al.* (1988). These ideas are supported by the extensive isotopic researches of Stein (1988), which indicated *inter*

alia: (a) that the Tertiary granite–molybdenum systems were derived by partial melting of lower crust, probably 1400 Ma old granulite, during the development of the Rio Grande rift system; (b) that meteoric water was not involved in the ore genesis. Previously many other authors have suggested that the source of the molybdenum and its host magma may have been continental crust (e.g. Hollister 1975, Stein 1985), but Westra & Keith (1981) and Knittel & Burton (1985) have advanced evidence indicative of a mantle origin, which is surely a very likely explanation for the deposits in the Philippines.

Porphyry tin and porphyry tungsten deposits

Much primary tin has been won in the past from stockworks at Altenberg in Germany and Cínovec in Czechoslovakia, from many deposits in Cornwall, England, and from deposits in New South Wales,

Tasmania and South Africa. Eroded stockworks in Indonesia and Malaysia have provided much of the alluvial cassiterite in those countries. Exploitation of such deposits in recent times has been mainly in Germany and New South Wales. At Ardlethan in New South Wales a quartz–tourmaline stockwork occurs in altered granodiorite carrying secondary biotite, sericite and siderite (Ren *et al.* 1988). With a grade of 0.45% tin this deposit is economic. The majority of tin stockworks in the world, however, only run about 0.1% tin and they are not at present mineable at a profit.

Tin stockworks in the countries mentioned above belong to the plutonic environment. Recently, Sillitoe *et al.* (1975) and Grant *et al.* (1980) have described porphyry tin deposits from the subvolcanic section of the Bolivian tin province south of Oruro and have shown that these deposits have much in common with porphyry copper deposits— large volumes of rock grading 0.2–0.3% tin. There is pervasive sericitic alteration that grades outwards into propylitic alteration and pyrite halos are present in two deposits. Major differences include the absence of a potassic zone of alteration, the association with stocks having the form of inverted cones rather than upright cylinders and the presence of swarms of later vein deposits.

Porphyry tungsten (usually W–Mo) deposits have now been described from various parts of the world, especially North America (e.g. Noble *et al.* 1984, Davis & Williams-Jones 1985) and China (Clarke 1983). Hydrothermal alteration is not always present as pervasive zones and when present may not show a regular concentricity. Grades are low: at Mount Pleasant, New Brunswick, 9.4 Mt running 0.39% WO_3 and 0.2% MoS_2 (Kooiman *et al.* 1986), and at Logtung, Yukon, 162 Mt running 0.13% WO_3 and 0.052% MoS_2. Geological settings are very variable: subvolcanic quartz–feldspar porphyry at Mount Pleasant, diorite to biotite granite in China, and granite, but principally its contact aureole, at Logtung.

15 / Stratiform sulphide, oxide and sulphate deposits of sedimentary and volcanic environments

In this chapter we are concerned with a class or classes of deposit whose origin is at present hotly debated. As was pointed out in Chapter 2, there are many types of stratiform deposit. This chapter is mainly devoted to sulphide deposits and the related oxide deposits can only be mentioned *en passant.* The latter do not include bedded iron and manganese deposits, placer deposits and other ores of undoubted sedimentary origin, which are dealt with in Chapter 18. The related oxide deposits are in tin and uranium (and possible iron) with which may be grouped certain tungsten deposits.

Concordant deposits referred to in Chapter 2 which belong to this class include the Kupferschiefer of Germany and Poland; Sullivan, British Columbia; the Zambian Copperbelt, and the large group of volcanic-associated massive sulphides. There appears to be a possible gradation in type and environment from deposits such as the Kupferschiefer, which are composed dominantly of normal sedimentary material and which occur in a non-volcanic sedimentary environment, through deposits such as Sullivan, which are richer in sulphur and may have some minor volcanic formations in their host succession, to the volcanic-associated massive sulphide deposits which are composed mainly of sulphides and occur in host rocks dominated by volcanics. Stanton (1972) considered that these and other deposits formed a 'spectrum of occurrence' and treated them as one class. Other workers, e.g. Barnes (1975), Solomon (1976), Gustafson & Williams (1981), Franklin *et al.* (1981), Sangster (1983a) and Eckstrand (1984) have felt it better to divide these deposits into two classes, the first class being those developed in a sedimentary environment where sedimentary controls are important, and the second being the volcanic-associated massive sulphide deposits in which exhalative processes were important during genesis. Such a division introduces difficulties when dealing with deposits showing only a weak link with volcanism but where exhalative processes may have been important, e.g. Sullivan. However, this division will be followed in this chapter as the author feels that deposits such as the Kupferschiefer and the Zambian Copperbelt are

sufficiently different from the volcanic-associated massive sulphide deposits to warrant some differentiation. Among the sediment-hosted, stratiform, base metal deposits two major groups are recognized by the majority of workers: sediment-hosted copper and sediment-hosted lead–zinc deposits. The latter are often termed sedex (i.e. sedimentary exhalative) deposits.

Stratiform sulphide deposits of sedimentary affiliation

General characteristics

The majority of these deposits occur in non-volcanic marine or deltaic environments. They are widely distributed in space and time, i.e. from the Proterozoic to the Tertiary, and can vary in tonnage from several hundred millions down to subeconomic sizes. In shape, they are broadly lensoid to stratiform with the length at least ten times the breadth. There is often more than one ore layer present. Feeder zones (cf. Fig. 2.14) have been identified below some deposits and may be present below many more, but as mining operations rarely penetrate into footwalls on a large scale, they will probably never be seen. The degree of deformation and metamorphism varies with that of the host rocks, suggesting a pre-metamorphic formation. They are frequently organic-rich, particularly those in shales, and usually contain a less complex and variable suite of minerals and recoverable metals than volcanic-associated massive sulphide deposits. The sulphides have a small grain size so that fine and often costly grinding is necessary to liberate them from the gangue. They may show a shore to basinward zoning of $Cu + Ag \rightarrow Pb \rightarrow Zn$ (Barnes 1975). It is possible and probably valid to erect various subclasses (cf. Eckstrand 1984), by emphasizing differences in metal ratios, geological environments and so on. Space does not permit this treatment in detail here, but attention must be drawn to the tendency for copper and lead–zinc deposits to be separate from each other and to have markedly different metal ratios.

The geological settings of these deposits are mostly intracratonic and the majority do not appear to be related directly to orogenic events or plate margin activity. Regional settings include (a) first marine transgressions over continental deposits (Kupferschiefer, Zambia, White Pine), (b) carbonate shelf sequences (Ireland), (c) fault controlled, sedimentary basins (Selwyn Basin, Yukon; Belt-Purcell Basin, British Columbia). Some of these environments appear to be aulacogens (cf. Elmore 1984).

Economically both the copper and lead–zinc deposits are of great importance on a worldwide scale; indeed sediment-hosted, stratiform copper deposits are second only to porphyry coppers as producers of the metal.

Copper deposits

These are discussed in great detail in Boyle *et al.* (1989), which is veritable mine of information on this deposit type, and ideas on their genesis are discussed in Jowett (1991). In addition to the general points listed above, it should be noted that this deposit type is the world's most important source of cobalt (from the Central African Copperbelt) and is becoming an important producer of by-product silver (Poland and USA). Most grades of recently worked and working deposits vary from 1.18 to 5% Cu, but the lower grade deposits have sweeteners. Tonnages can be enormous, e.g. Lubin, Poland 2600 Mt running about 2% Cu, 30–80 g t^{-1} Ag and 0.1 g t^{-1} Au.

Most major deposits occur in reduced, pyritic, organic-rich, calcareous shales, or their metamorphic equivalents but the remaining approximately one third occur in sandstones. These immediate host rocks occur in anoxic, paralic marine (or large scale, saline lacustrine) sediments immediately above typically red, oxidized, continental clastic sediments, and deposits of this type are found in rock sequences post-dating the first appearance of red beds (*c.* 2400 Ma ago) and range in age to Recent. The most important and abundant deposits are in Upper Proterozoic and Upper Palaeozoic rocks that were deposited in arid and semi-arid areas within continental rift environments not further than 20–30° from the palaeoequator. In many areas those rocks are interbedded with evaporites. At the oxidation–reduction boundary the ascending sequence of minerals in the mineralized ground contains all or some of the following: hematite, native copper, chalcocite, bornite, chalcopyrite, galena, sphalerite and pyrite. These occur in mineral zones that overlap upward and outward (Jowett 1989, Kirkham 1989).

Some examples of copper deposits

The European Kupferschiefer

This is probably the world's best known copper-rich shale. It is of late Permian age and has been mined at Mansfeld, Germany for almost 1000 years. The Kupferschiefer underlies about 600 000 km^2 in Germany, Poland, Holland and England (Fig. 15.1). Copper concentrations greater than 0.3% occur in about 1% and zinc concentrations greater than 0.3% in about 5% of this area. Thus, although all the Kupferschiefer is anomalously high in base metals, ore grades are only encountered in a few areas. The most notable recent discoveries have been in southern Poland where deposits lying at a depth of 600–1500 m have been found during the last two decades. Here, the Kupferschiefer varies from 0.4 to 5.5 m in thickness. Average copper content is around 1.5% and reserves at 1% Cu amount to nearly 3000 Mt making Poland the leading copper producer in Europe. The area underlain by these deposits is approximately 30 × 60 km.

The Kupferschiefer consists of thin alternating layers of carbonate, clay and organic matter with fish remains which give it a characteristic dark grey to black colour. The Kupferschiefer is the first marine transgressive unit overlying the non-marine Lower Permian Rotliegendes, a red sandstone sequence, and it is succeeded by the Zechstein Limestone which in turn is overlain by a thick sequence of evaporites. The Kupferschiefer and the Zechstein evaporites may represent a tidal marsh (sabkha) environment which developed as the sea trangressed desert sands.

Sulphide mineralization occurs across the contact between the Upper Permian Zechstein marine sequence and the Lower Permian Rotliegendes red beds. Ore is contained in the Kupferschiefer, the overlying limestone and the underlying sandstone. The copper and other metals are disseminated throughout the matrix of the rock as fine-grained sulphides (principally bornite, chalcocite, chalcopyrite, galena, sphalerite), commonly replacing earlier calcite cement, lithic fragments and quartz grains as well as other sulphides. Typical features of the mineralization are shown in Figs 15.2 and 15.3. Horizontal and vertical veinlets of gypsum, calcite

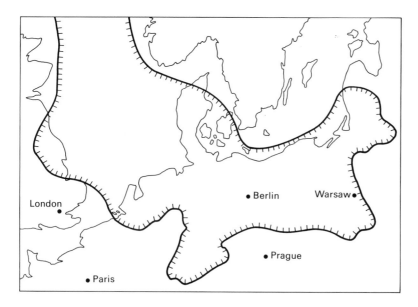

Fig. 15.1 Extent of the Zechstein Sea in Central Europe. The Kupferschiefer occurs at the base of the Upper Permian (Zechstein).

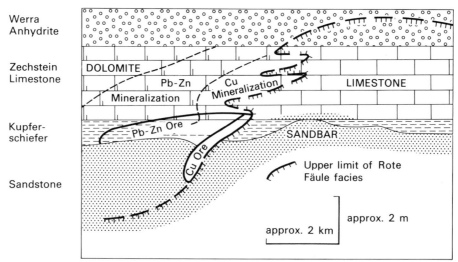

Fig. 15.2 Diagrammatic section through orebodies in the basal Zechstein with the Rote Fäule facies alteration gently transgressing the bedding above an area of sandbars formed by marine reworking of the Rotliegendes. Sulphide mineralization occurs in the unoxidized zone adjacent to the Rote Fäule with copper nearest to it and lead–zinc further away. (After Brown 1978.)

and base metal sulphides are common in the Lubin district and increase the grade of the ore significantly. They are interpreted by Jowett (1987) as resulting from hydraulic facturing. A zone of superposed diagenetic reddening known as the Rote Fäule facies transgresses the stratigraphical horizons. Copper mineralization lies directly above the Rote Fäule and the copper zone is overlain, in turn, by lead–zinc mineralization. This relationship to the Rote Fäule has meant that the delineation of this facies is the most important feature of the search for new orebodies. The Rote Fäule copper zones are coincident with underlying highs in the buried basement, and the metal zoning dips away from the highs toward the basin centres. A recently discovered but as yet unexploited aspect of the Polish

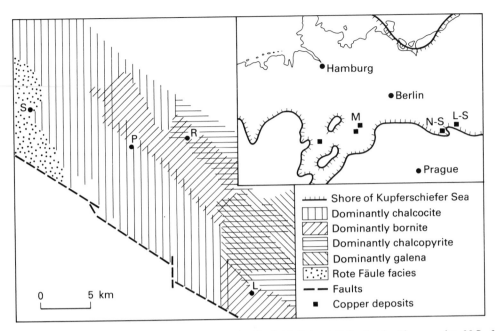

Fig. 15.3 Suphide mineral zones in the Lubin–Sieroszowice Orefield, Poland. L-S = Lubin–Sieroszowice, N-S = North Sudetic Syncline, M = Mansfeld, S = Sieroszowice, P = Polkowice, R = Rudna, L = Lubin. (Modified from Sawlowicz 1990.)

Kupferschiefer is the occurrence of Pt-rich (> 10 ppm) shales along strike lengths in excess of 1.5 km and values of > 200 ppm Pt have been found over 50 m strike lengths (Kucha 1982). (PGM-rich shales have recently been recognized in Canada, China, Czechoslovakia and the USA and may represent a possible source of these metals (Coveney & Nansheng 1991, Pašava 1991).)

The Zambian Copperbelt

This is part of the larger Central African Copperbelt of Zambia and Shaba (Zaïre) which produced about 17% of the western world's copper in the early 1980s. In 1984 Zambia produced 531 000 t of copper and the mill grade at the principal mines varied from 1.49 to 2.81%, with appreciable by-product cobalt from some. The industry in Zambia is contracting rapidly. The harsh realities are that the mines are old, with declining grades and an average of 15 years life ahead of them, and the country simply does not have the cash for investment to increase their efficiency and operate them profitably. Shaba produced about 500 000 t copper in 1984 from ore averaging around 4% Cu + Co; but here too the industry is suffering and production in 1989 was

down to 441 000 t. With 1989 copper production in Zambia down to 460 000 t these two countries produced only 7% of the western world's copper output.

Almost all the copper mined in 1984 came from restricted horizons within the late Proterozoic Katangan sediments of the Lufilian Arc (Fig. 15.4). The Katangan rests unconformably on a granite–schist–quartzite basement and the lowermost Katangan sediments fill in the valleys of the pre-Katangan land surface. Most mineralization in Zambia and south-eastern Shaba occurs in the Ore Formation which lies a few metres above the level at which the pre-Katangan topography became filled in. Shale or dolomitic shale forms the host rock for about 60% of the mineralized ground, and the shale orebodies form a linear group to the south-west of the Kafue Anticline (Fig. 15.5). Arkose–arenite hosted ores occur mainly to the north-east of the anticline, e.g. Mufulira (see pp. 35–36). The footwall succession consists of quartzites, feldspathic sandstones and conglometrates of both aquatic and aeolian origin. The Ore Formation, generally 15–20 m thick, is succeeded by an alternating series of arenites and argillites which, with the rocks below them, make up the Lower Roan Group. All the rocks and their

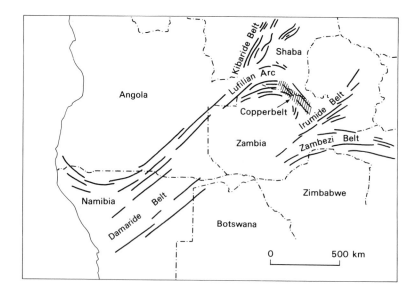

Fig. 15.4 Location of the Copperbelt in relation to the main tectonic trends of Central Africa. (After Raybould 1978.)

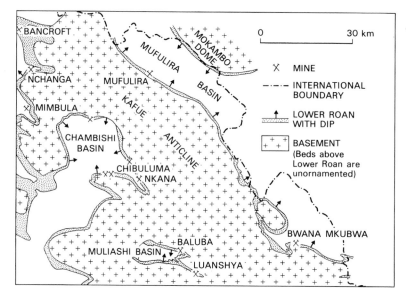

Fig. 15.5 Location map for the Zambian Copperbelt showing the regional geology. (Modified from Fleischer *et al.* 1976.)

contained copper minerals have suffered low to high grade greenschist facies metamorphism and many of the so-called shales are biotite-schists. In places they are tightly folded (Fig. 15.6).

Copper, together with minor amounts of iron and cobalt, occurs mainly in the lower part of the Ore Formation as disseminated bornite, chalcopyrite and chalcocite. In places, mineralization passes for short distances into the underlying beds. Both the upper and lower limits of mineralization are usually sharply defined. The sulphide minerals show a consistent zonal pattern with respect to the strand-line from barren near-shore sediments to chalcocite in shallow water, to bornite with carrollite and chalcopyrite, then chalcopyrite and finally pyrite (in places with sphalerite) in the deeper parts of marine lagoons and basins (cf. Fig. 15.3).

The White Pine copper deposit, northern Michigan

A different type of copper deposit is found in Precambrian strata at White Pine (Fig. 15.7). This

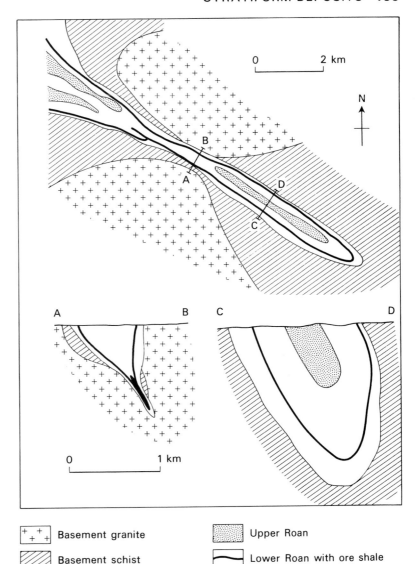

Fig. 15.6 Sketch map and sections of the Luanshya Deposit, Zambia. (Modified from Dixon 1979.)

Basement granite

Basement schist

Upper Roan

Lower Roan with ore shale

district supplies about 5% of the copper mined in the USA. When large scale production commenced in 1953 reserves were about 230 Mt grading 1.2% copper as chalcocite and native copper, which occur in the Nonesuch Shale. Despite extensive exploitation since then reserves still stand at about 200 Mt grading 1.1% Cu and 7.46 g t^{-1} Ag. The Nonesuch Shale, deposited about 1000 Ma ago, is about 150–200 m thick but only the lower 8–15 m contain significant copper mineralization (Burnie *et al*, 1972). The name 'Nonesuch Shale' is misleading because much of the formation is siltstone or

sandstone. Copper mineralization is confined almost invariably to individual lithological units and the content changes with sedimentary facies variations. The Nonesuch Shale conformably overlies the Copper Harbor Conglomerate, the upper beds of which are also locally cupriferous.

The Nonesuch Shale has been divided into three subzones. A basal cupriferous zone contains chalcocite and native copper, but only traces of bornite, chalcopyrite and pyrite. Next comes a transition zone with a gradation in the mineralogy from chalcocite through bornite and chalcopyrite to

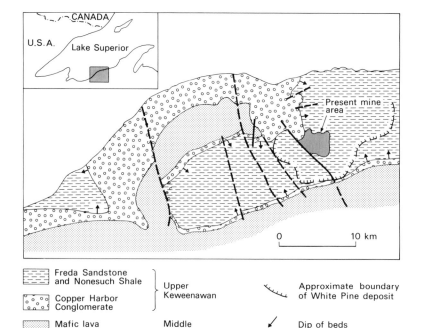

Freda Sandstone
and Nonesuch Shale } Upper
Keweenawan

Copper Harbor
Conglomerate

Mafic lava
flows — Middle
Keweenawan

Approximate boundary
of White Pine deposit

Dip of beds

Fig. 15.7 Geological sketch map of the area containing the White Pine copper deposit. (Modified from Burnie *et al.* 1972 and White 1971.)

pyrite; lead–zinc minerals also occur. This is overlain by copper-poor pyritous shale and siltstone petrologically similar to the copper-bearing rocks. Within the cupriferous zone, maximum copper concentrations are found in siltstone and shale units immediately overlying sandstones, which are usually copper-poor.

Genesis

Ideas concerning the origins of these deposits show, in general, a similar evolution from epigenetic, to syngenetic, to what might be termed 'just epigenetic'. For example the original White Pine Mine worked ore from both the Nonesuch Shale and the Copper Harbor Conglomerate along the White Pine Fault and, because the mine was on a structural feature, it was assumed that the ore was epigenetic and deposited by hydrothermal solutions that rose up the fault. The discovery that the ore persisted along the base of the Nonesuch Shale over many tens of kilometres squared led to a concept of syngenetic origin.

Further study showed that ore cuts progressively but gently across bedding, leading to a more sophisticated concept of genesis—that copper-bearing solutions circulated through beds which contained syngenetic pyrite, and that the copper replaced the

iron, for which there is considerable ore microscopic evidence. Seasor & Brown (1989) suggested that the source of the copper was a reservoir of oxidized, chlorine-rich solutions in the underlying hematitic red bed, the Copper Harbour Conglomerate, from which the copper was leached.

For the Zambian deposits many workers have proposed and still propose a syngenetic origin, e.g. Fleischer *et al.* (1976) and Garlick (1989) who cite many lines of evidence, particularly sulphidic bedding planes eroded by scour channels and slump folded ore horizons, for mineralization having occurred during sedimentation. These authors postulated that the supply of copper came from springs or streams draining a hinterland of aeolian sands and terrestrial red beds, and that it was precipitated selectively by hydrogen sulphide in bodies of standing water to form the zones of mineralization described above. Binda (1975) described detrital grains of bornite and bornite-bearing rock fragments from Mufulira and suggested that clastic sedimentation played a role (of yet unknown importance) as an ore-forming process in the Copperbelt. From an epigenetic viewpoint, Raybould (1978) has pointed out that the Zambian Copperbelt and other major Proterozoic stratiform copper and lead–zinc deposits, such as McArthur River, Northern Territory; Mount Isa, Queensland; Sullivan, British Columbia;

White Pine, Michigan, etc., appear to have developed in intracratonic and cratonic-margin rift systems of mid to late Proterozoic age. He suggested that the mineralization resulted directly from deep-seated processes coeval with the rifting and not from surface weathering. Annels (1979) suggested that diagenetic metal chloride-enriched brines were responsible for the mineralization at Mufulira.

Since then, work on deposits of this type all over the world has led to the conclusion by most observers that copper and associated metals have been added to their host rocks after sedimentation and after at least very early syndiagenetic accumulations of sulphate and sulphide were formed, some of which, particularly pyrite, have been replaced by later copper and cobalt minerals. Features such as this and the slightly transgressive nature of the mineralization have now been reported from many deposits. Brown (1978) and Chartrand & Brown (1985) illustrate the former observations well and the latter feature has been described above. In the Kupferschiefer the clear spatial relationship between sulphide and hematite mineralization suggests that the ore genesis was closely related to processes responsible for the Rote Fäule formation. Metal zoning transgressing the strata, localization of copper deposits around the Rote Fäule facies and the coexistence of hematite grains, hematite pseudomorphs after pyrite, metal sulphides and copper sulphide replacements of pre-existing pyrite within the outer part of the Rote Fäule facies suggest a post-sedimentary origin of the Rote Fäule ore system. Oszczepalski (1989) postulated that the ore solutions were metalliferous formational waters expelled from the Rotliegendes after leaching much of its copper content.

The proposition that the Central African Copperbelt sedimentation and mineralization took place in a rift zone has been elaborated by, among others, Annels (1984, 1989), who suggested that the mineralization was due to hydrothermal leakage from the bounding fractures of saline formation waters that had leached Fe, Co and Cu, plus a wide range of minor elements, from basement rocks particularly basalts. On reaching the Lower Roan cover rocks via basement fractures and faults these solutions, depending on the degree of cementation and relative permeabilities, are considered by Annels to have followed the basement unconformity upwards towards the base of the Ore Formation or penetrated the footwall arenites. On reaching the footwall of the ore shales they migrated laterally through the permeable footwall conglomerates and arenites at the same time infusing upwards into the shales to produce rich, widespread mineralization following the early diagenetic events. Isotopic research indicates that the sulphur in the sulphides was derived from sea water sulphate (Sweeney & Binda 1989) already present in the arenites in the form of anhydrite and released from this mineral by high-temperature (> 250°C) inorganic reduction. In the shales the metals were probably precipitated under the influence of bacteriogenically reduced sulphur at temperatures of 140–215°C (Annels 1989).

Gustafson & Williams (1981) suggested that the unifying feature for all sediment-hosted, stratiform, base metal deposits was that each was developed in a structural situation that permitted heated basinal brines to be moved to a shallow site of sulphide deposition. The relative timing of this movement and hence of the mineralization would of course be variable, which accounts for the evidence of three stages of mineralization in the Kupferschiefer and its European equivalents ranging from syngenetic to late diagenetic (Vaughan *et al.* 1989) and the evidence of early and/or late diagenetic mineralization from a number of different deposits (Huyck & Chorey 1991).

Sediment-hosted lead–zinc deposits (sedex deposits)

These have a worldwide distribution and are very important metal producers. The grades and tonnages of a number of deposits are given in Fig. 15.8 and Table 15.1. For multi-lensoid deposits these represent the average grade and the total tonnage for all component lenses. Figure 15.8 demonstrates the impressive size (average of 70 Mt) and grade (average Pb + Zn = 12%) of these deposits, some of which have important by-products (Table 15.1). Thus, like the sediment-hosted copper deposits, these too can form giant orebodies, or groups of stacked orebodies. They appear, as Russell *et al.* (1981) and Large (1983) *inter alia* have suggested, to form a distinctive group of ores formed in local basins on the sea floor as a result of protracted hydrothermal activity accompanying continental rifting and they constitute the important ore deposit type known as sedex deposits. The host rocks for these are generally shales, siltstones and carbonates and not just black shales as was formerly thought. The host sediments represent both low and high energy depositional environments with locally derived breccias becom-

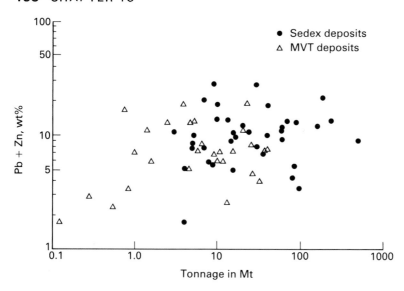

Fig. 15.8 Grade–tonnage plot for sedex and Mississippi Valley-type deposits. (Modified from Sangster 1990.)

Table 15.1 Size and grade of some sediment-hosted, stratiform, lead–zinc deposits (Gustafson & Williams 1981 and other sources)

Country	Deposit	Tonnage of ore in Mt (reserves and past production)	Grade Cu%	Pb%	Zn%	By-products	Age
Australia	Broken Hill	180	0.2	11.3	9.8	Ag 175 g t^{-1}	Lower to Middle Proterozoic
	McArthur River	237	0.2	4.1	9.2	Ag 41 g t^{-1}	Middle Proterozoic
	Mount Isa	88.6	0.06	7.1	6.1	Ag 160 g t^{-1}	Middle Proterozoic
Canada	Howard's Pass	100	—	1.5	6.0	—	Silurian
	Sullivan	160	—	6.6	5.9	Ag 68 g t^{-1} Sn, Cd, Cu, Au	Middle Proterozoic
Germany	Meggen	60	0.2	1.3	10.0	Baryte	Devonian
	Rammelsberg	30	1.0	9.0	19.0	Ag 103 g t^{-1} Baryte	Devonian
Ireland	Navan	70	—	2.6	10.1	—	Carboniferous
	Silvermines	18.4	—	2.8	7.4	Ag 21 g t^{-1} Baryte	Carboniferous
	Tynagh	12.3	0.4	4.9	4.5	Ag 58 g t^{-1}	Carboniferous
RSA	Gamsberg	93.5	—	0.6	7.4	—	Middle Proterozoic

ing important in the latter (Large 1983). The ores are finely layered and often interbanded with rock material (Figs 2.15, 15.9), with both being affected by soft sediment deformation suggestive of a syn-sedimentary origin of the sulphides. At Silvermines and Tynagh, fossil hydrothermal chimneys have been found which are similar to those known from present day hydrothermal vents on the East Pacific Rise and other constructive plate margins (Banks

1985). These discoveries, together with the evidence of feeder zones and other observations described above, suggest that some (probably all) of these deposits have been formed by hydrothermal solutions venting into restricted basins on the sea floor. The environment was not, however, that of the deep ocean, but more like that of the Gulf of California, where sulphide and baryte deposits are forming today in the axial rifts of a region of active

(b)

Fig. 15.9 Photographs of ore specimens from Mount Isa, Queensland. The light coloured material is sulphide and the dark is silicate. Note the slump folds, associated fracturing and synsedimentary faults affecting both sulphide-rich and silicate-rich layers.

sedimentation (Lonsdale & Becker 1985). However, the sea depths are still much greater than those postulated for the formation of these giant lead–zinc deposits—about 50–800 m (Sangster 1990). Zoning in these Pb–Zn deposits is discussed on p. 92.

Large (1983) categorized the settings of sedex deposits in terms of a hierarchy of sedimentary basins of decreasing size. His first order basins with lateral extents of hundreds of kilometres may be either epicratonic embayments into continental margins or, less importantly as hosts, intracratonic basins. Some first order basins appear to have been aulacogens (Chapter 23). Second order basins are tens of kilometres in size, occur within first order basins and contain third order basins, less than 10 km in diameter, within which the stratiform sulphides accumulated. Both second and third order basins are bounded by synsedimentary faults leading to rapid changes in lithological thicknesses and the development of debris-flow breccias. The environments within these third order basins appear to have varied from deep water, sediment-starved and an-

oxic, like that of the Black Sea (Howard's Pass deposit), through those with a high sediment input of turbidites and shales (Sullivan and Broken Hill), to shallow water, oxygenated, carbonate-filled basins (Silvermines, Tynagh), which may even contain anhydrite beds indicating evaporitic environments (Balmat, USA), or even to lacustrine environments (Sangster 1990). These third order basins acted as morphological traps into which dense, metal-bearing brines appear to have flowed from distant vents to form brine pools, which then lost heat and salt to the overlying sea water, leading to the precipitation of fine-grained, layered sulphides. The majority of sedex deposits do not appear to be underlain by stockwork feeder zones and are therefore regarded as distal exhalative, unlike the volcanic-associated massive sulphide deposits which are dominantly proximal exhalative.

How were these metal-bearing brines generated? Sangster (1990) drew attention to the similarity of temperature, salinity and pH of the solutions and the sulphur isotopic compositions of the sulphide of both sedex and Mississippi Valley-type deposits and suggested that the solutions responsible for the formation of both deposit types originated in, and were expelled from clastic sedimentary basins—a mode of genesis discussed at length in Chapter 4.

On the other hand Russell *et al.* (1981) contended that the exhalations were formed by sea water convection cells similar to those shown in Fig. 15.10, which dissolved base metals from the rocks they traversed (see Chapter 4). Such cells would only have to penetrate a few kilometres into the oceanic crust in order to be heated to and sustained at the necessary temperatures, *if* supplied with the latent heat of crystallization from a magma chamber (Cann & Strens 1982).

For an intracratonic situation, like that of Ireland, it is of course necesary to postulate much deeper penetration of sea water. Russell *et al.* suggested a penetration of about 10 km if the deposit was mainly Pb–Zn, with a depth of about 15 km being necesary to permit the leaching of copper. In a terrane having a high geothermal gradient under rifting conditions, continued cooling of the rocks would allow convection to reach deeper and deeper into the crust as shown in Fig. 15.10. In the early stages of convection the shallow penetration would lead in the main to leaching of iron, manganese and silica, which would account for the manganese enrichment in the footwall rocks of many deposits. With deeper

circulation the temperature and the time for water–rock interaction would increase and lead and zinc also would be leached. For many deposits this might be the limit of the convective system, leading to the formation of essentially copper free deposits. With still deeper penetration copper too might be leached to make that element an important by-product in the deposit (see Table 15.1). A possible evolutionary path of the mineralizing fluid is shown in Fig. 15.11. It can be seen that passing through stage 2 the fluid is capable of leaching out zinc (lead is assumed to have similar solubility) and that moving into stage 3 (at higher temperatures) solutions capable of carrying 1 ppm or so of copper are generated.

Much of the exhaled iron is probably oxidized and may be precipitated to give rise to iron formation adjoining the sulphide deposit. Oxidation of manganese is very sluggish and so this element may spread out in the sea water surrounding the hydrothermal plume to be incorporated in the sea bottom sediments. In this manner a detectable manganese halo around the deposit may be formed such as that around the Tynagh Mine in Ireland, where the halo at the 100 ppm Mn level is over 15 km in diameter (Russell 1983).

This is a promising model, as it accounts not only for the formation of these deposits but also for their distribution, and hence the prediction of the approximate locations of other possible deposits. The

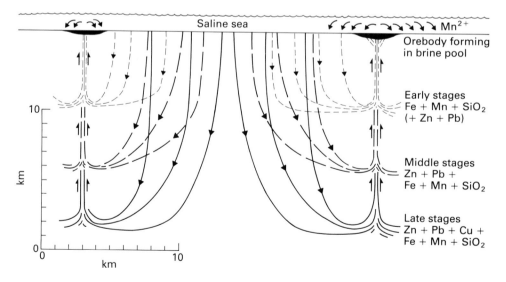

Fig. 15.10 Diagram showing how sea water circulation through the crust might give rise to the formation of an exhalative, sediment-hosted, stratiform ore deposit. (After Russell *et al.* 1981.)

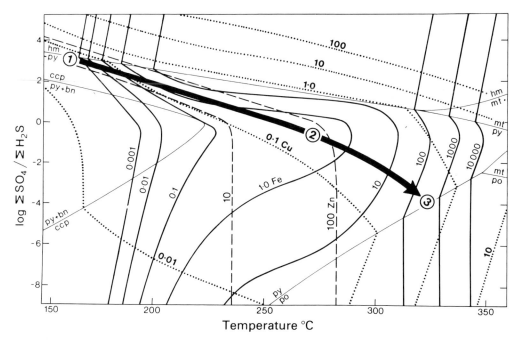

Fig. 15.11 Diagram illustrating solubility of Fe (——), Cu (· · ·) and Zn (– – –) in parts per million assuming equilibrium with phases in the Cu + Fe + Zn + S + O system and probable solution conditions. A possible evolutionary path for the composition of the convecting fluids is shown. Points 1, 2 and 3 correlate approximately with the early, middle and late stages of Fig. 15.10. (Reproduced with permission from Russell *et al.* 1981, *Mineralium Deposita*, **16**.)

reader is strongly recommended to study Russell *et al.* (1981) or further developments of this hypothesis in Russell (1983). Russell's model has been tested quantitatively by Strens *et al.* (1987) who showed that it is capable of producing the very largest Pb–Zn deposits during a single episode of circulation provided favourable conditions obtain. These conditions include a large recharge area (300 km²), a permeability > 1 mD, a circulation depth of 10 km, a time span of at least 50 ka and high depositional efficiency. The infrequency with which such a set of conditions coincide may account for the rarity of large deposits.

In concluding this section a little space must be devoted to Broken Hill and similar deposits, which a number of workers believe should be considered as a separate subclass within the sedex group. Their reasons were summarized by Beeson (1990). Many of the distinguishing characteristics of the subclass are given in Table 15.2. The high metamorphic grade of these deposits increases the grain size and improves their mineral processing properties; this and their large size and tendency to carry high grades

of Pb–Zn–Ag make them attractive exploration targets. The occurrence of volcanics in the footwall sequence that are Na-rich in their lower portion but grade upwards into K-rich facies is important; however, a very marked feature is the development of chemical sediments, which include such diverse lithologies as iron formation, metachert, garnet-, gahnite-, and lead-bearing feldspar, tourmaline- and apatite-bearing rocks. These rocks resulted from the regional metamorphism of chemical precipitates of unusual composition (Stanton 1986). Similar exhalites occur at Bergslagen, Sweden and other Broken Hill-type deposits. The development of tourmaline-bearing rocks appears to be a particularly valuable exploration guide and these Pb–Zn deposits have a consistent position within an overall stratigraphy of clastic and chemical sediments.

The source of the ore fluids is much debated. Plimer (1985) maintained that the lead and sulphur isotopic compositions indicated a mantle source, but sulphur values for other deposits indicate a sea water origin for this element. The proximal (vent-related) position of the Broken Hill-type Zinkgruvan

Table 15.2 Comparison of Broken Hill-type and other sediment-hosted lead–zinc deposits. (From Beeson 1990)

	Broken Hill	Sediment-hosted stratiform
Metamorphic grade	High	Low
Intracratonic setting proposed	Yes	Yes
Redox characteristics of host rocks	Oxidizing	Reducing
Bimodal volcanics in underlying rocks	Yes	Sometimes
Change from volcanic-dominated to sediment-dominated sequence	Common	No
Na-rich volcanics:		
Below base metals	Yes	No
Lateral to base metal	Yes	Rare
K-rich rocks around base metal ore	Yes	Common
Iron formation facies change from oxide to silicate to sulphide:		
Below base metals	Yes	No
Lateral to base metals	Yes	No
Quartzite/chert associated with ore	Yes	Sometimes
Widespread Mn-rich chemical sediments	Yes	Yes
Tourmaline rocks lateral to base metals	Yes	Not known
Pyrite associated with Na-rich rocks below base metal ores	Yes	Rare
Mineralization extensive along strike at same stratigraphical level	Yes	Possibly

deposit, Sweden, in a distal volcanic facies and other features led Hedström *et al.* (1989) to propose that it represents a variety transitional to volcanic-associated massive sulphide deposits in its setting and mode of genesis.

Volcanic-associated massive sulphide (VMS) deposits

Some attention has already been paid to these deposits in Chapters 2 and 4. This section will therefore be used to amplify and add to what has already been written. The reader is consequently recommended to read pp. 38–39 and 71–79 before reading this section.

Size, grade, mineralogy and textures

Data from five metallogenic provinces are given in Table 15.3. The majority of world deposits are small and about 80% of all known deposits fall in the size range 0.1–10 Mt. Of these about a half contain less than 1 Mt (Sangster 1976). Average figures of this type tend to hide the fact that this is a deposit type than can be very big or rich or both and then very profitable to exploit. Examples of such deposits are given in Table 15.4.

The mineralogy of these deposits is fairly simple and often consists of over 90% iron sulphide, usually as pyrite, although pyrrhotite is well developed in some. Chalcopyrite, sphalerite and galena may be major constituents, depending on the deposit class, bornite and chalcopyrite are occasionally important and arsenopyrite, magnetite and tetrahedrite-tennantite may be present in minor amounts. With increasing magnetite content these ores grade to massive oxide ores (Solomon 1976). The gangue is principally quartz, but occasionally carbonate is developed and chlorite and sericite may be important locally. Their mineralogy results in these deposits having a high density and some, e.g. Aljustrel and Neves-Corvo in Portugal, give marked gravity anomalies, a point of great exploration significance.

The vast majority of massive sulphide deposits are zoned. Galena and sphalerite are more abundant in the upper half of the orebodies whereas chalcopyrite increases towards the footwall and grades downward into chalcopyrite stockwork ore (Figs 2.20, 4.13). This zoning pattern is only well developed in the polymetallic deposits. As the number

Table 15.3 Average grade and tonnage data for VMS deposits of selected regions. (Data from Large *et al.* 1987 and Lydon 1989)

Region	Dominant deposit type	Number of deposits	Average grade[a]						Tonnage (Mt)
			Cu (%)	Zn (%)	Pb (%)	N –	Ag (g t^{-1})	Au (g t^{-1})	
Abitibi Belt, Canada (Archaean)	Cu–Zn	52	1.47	3.43	0.07	47	31.9	0.8	9.2
Norwegian Caledonides (Palaeozoic)	Cu–Zn	38	1.41	1.53	0.05	0	na	na	3.5
Bathurst, New Brunswick, Canada (Palaeozoic)	Zn–Pb–Cu	29	0.56	5.43	2.17	28	62.0	0.5	8.7
Green Tuff Belt, Japan (Tertiary)	Zn–Pb–Cu	25	1.63	3.86	0.92	7	95.1	0.9	5.8
Tasman Geosyncline, Eastern Australia (Palaeozoic)	Cu–Au	} 42	1.3			5	17.2	2.1	} 6.4
	Zn–Pb–Cu[b]		0.48	15.15	6.28	4	160.0	3.0	

[a] N = Number of deposits for which data available to calculate average Ag and Au grades.
[b] Data in this line is for some Tasmanian deposits only.

Table 15.4 Tonnages and grades for some large or high grade VMS deposits

Mined ore + reserves	(Mt)	Cu%	Zn%	Pb%	Sn%	Cd%	Ag g t^{-1}	Au g t^{-1}
Kidd Creek, Ontario	155.4	2.46	6.0	0.2	r[c]	r	63	—
Horne, Quebec	61.3	2.18	—	—	—	—	—	4.6
Rosebery, Tasmania	19.0	0.8	15.7	4.9	—	—	132	3.0
Hercules, Tasmania	2.3	0.4	17.8	5.7	—	—	179	2.9
Rio Tinto, Spain[a]	500	1.6	2.0	1.0	—	—	r	r
Aznalcollar, Spain	45	0.44	3.33	1.77	—	—	67	1.0
Neves-Corvo	30.3	7.81	1.33	—	—	—	—	—
Portugal[b]	2.8	13.42	1.35	—	2.57	—	—	—
	32.6	0.46	5.72	1.13	—	—	—	—

[a] Rio Tinto, originally a single stratiform sheet, was folded into an anticline whose crest cropped out and vast volumes were gossanized. The base metal values are for the 12 Mt San Antonio section.
[b] The three lines of data represent three ore types and not particular orebodies of which there are four. Ag is present in some sections.
[c] r indicates that this metal is recovered, but the average grade is not available.

of mineral phases decreases, so the zonation tends to become obscure and may not be in evidence at all in deposits dominated by pyrite or pyrite–chalcopyrite. From a study of 19 deposits, Huston & Large (1989) showed that gold tends to have two distinct occurrences: (a) an Au–Zn–Pb–Ag association found in the upper part of zinc-rich massive sulphide lenses, and (b) an Au–Cu association commonly present in the stockwork and lower portions of copper-rich deposits. In each of the 19 deposits one of these occurrences dominates, almost to the exclusion of the other. From their study of gold distribution in some orebodies of the Iberian Pyrite Belt, Strauss & Beck (1990) arrived at very similar conclusions.

Textures vary with the degree of recrystallization. The dominant original textures appear to be colloform banding of the sulphides with much development of framboidal pyrite, perhaps reflecting colloidal deposition. Commonly, however, recrystallization, often due to some degree of metamorphism, has destroyed the colloform banding and produced a granular ore. This may show banding in the zinc-rich section, whereas the chalcopyrite ores are rarely banded. Angular inclusions of volcanic host rocks are occasionally present and soft sediment structures (slumps, load casts) are sometimes seen. Graded bedding has also been reported from some deposits (Sangster & Scott 1976, Ohmoto & Skinner 1983).

Wall rock alteration

Wall rock alteration is usually confined to the footwall rocks. Chloritization and sericitization are the two commonest forms. The alteration zone is pipe-shaped and contains within it and towards the centre the chalcopyrite-bearing stockwork. The diameter of the alteration pipe increases upward until it is often coincident with that of the massive ore. Metamorphosed deposits commonly show alteration effects in the hanging wall. This may be due to the introduction of sulphur released by the breakdown of pyrite in the orebody. The sulphur, by reacting with pore solutions, could give rise to extensive hydrogen ion production. On the other hand the thermal convection cells extant at the time of deposit formation may have continued for some time afterwards and produced alteration effects in the hanging wall rocks (Solomon *et al.* 1987).

A study of the distribution patterns of economically important elements in wall rock alteration envelopes can be valuable in planning the exploration strategy for a deposit. Using this approach Elliott-Meadows & Appleyard (1991) in their work on the Lar deposit, Manitoba located a zone of dispersed mineralization about 120 m beneath the main orebody and also indicated the possibility that mineralization extends to deeper levels than have so far been explored.

Classifications

The geochemical division into iron, iron–copper, iron–copper–zinc and iron–copper–zinc–lead deposits has been touched on already (pp. 39, 71), but it must be emphasized that while we may find pyrite deposits without any appreciable copper, copper is never found on its own. Similarly, if we find lead, we will have zinc and at least accessory copper too. With zinc will come copper and perhaps lead. Using a different approach from the simple chemical one,

Lydon (1989) has shown that, if each point on a ternary diagram of the grades of the deposits in the first four regions of Table 15.3 is weighted in terms of tonnes of ore metal within the deposit, then clearly there are just two major groups, Cu–Zn and Zn–Pb–Cu (Fig. 15.12). Indeed, as Lydon stresses, there are few so-called copper deposits without some zinc.

Some of the names commonly given to these different types have been mentioned on p. 71. Although for a number of years the dominantly iron–copper–zinc deposits of the Canadian Archaean were considered by many workers to be a variant of the Kuroko type, it is now generally agreed that they are best considered as a separate type, which Hutchinson (1980) termed Primitive. Morton and Franklin (1987) suggested that the Primitive type can be divided into two subtypes—the Noranda and Mattabi on the grounds of differences in alteration characteristics, associated volcanic rocks and environments. The Noranda subtype has a well defined, zoned alteration pipe beneath it, and if

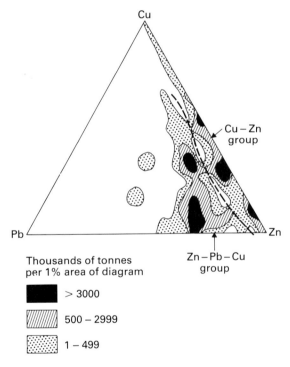

Fig. 15.12 Data from the first four regions of Table 15.3 plotted on a Cu–Pb–Zn ternary diagram weighted in tonnes of ore metal contained in the deposit and then contoured. (Modified from Lydon 1989.)

there is a subjacent semi-conformable widespread alteration zone, then this is cut by the sharp boundaries of the pipe. The volcanics are dominantly mafic and felsic flows and hyaloclastites believed to have formed at water depths greater than 500 m. By contrast the Mattabi subtype has a broad alteration pipe without sharp boundaries that is gradational into the semi-conformable alteration below. The footwall succession has a much higher proportion of felsic volcanics which are dominantly fragmental and thought to have been emplaced at depths of less than 500 m. Both subtypes include large and valuable deposits and the recognition of the existence of these two subtypes will be important in any exploration programme. A summary of the nature of the different types of VMS deposits is given in Table 15.5.

Stanton (1978) and Vokes (1987) have objected to the division of these deposits into various types. Stanton considered these ores to be part of one continuous spectrum, showing a progressive geochemical evolution which accompanies that of the associated calc-alkaline rocks in island arcs. It is certainly difficult to assign some deposits to particular types except on a purely geochemical basis. For example, the deposits of West Shasta, California, which are believed to have formed in a Devonian bimodal suite of island arc volcanics (Lindberg 1985, South & Taylor 1985), are, like Kuroko deposits, intimately associated with rhyolite lavas and domes, but they are Cu–Zn(+ Au–Ag) type and lack significant lead. Thus although they have many of the properties of Kuroko-type deposits, a case could be made for correlating them with the Primitive type, of which it must be remarked that some deposits, e.g. Kidd Creek, Ontario (0.2%) and Mons Cupri, Western Australia (2.5%) do carry recoverable lead. Yes, reader, the situation is confusing!—we may indeed be dealing with a continuous spectrum that we have not yet fully recognized, as Lydon's work suggests (Fig. 15.12), but meanwhile mining geologists will continue to use the terms listed in Table 15.5 because in the present state of our knowledge they imply useful summary descriptions of the deposits, helpful in formulating exploration models.

It is important to note that precious metals are also produced from some of these deposits, indeed in some Canadian examples of the Primitive type they are the prime products. Both Besshi and Kuroko types may also produce silver and gold whilst the Cyprus type may have by-product gold.

Some important field occurrence features

Association with volcanic domes

This frequent association is stressed in the literature and the Kuroko deposits of the Kosaka district, Japan, are a good example (Fig. 15.13). Many examples are cited from elsewhere, e.g. the Noranda area of Quebec and some authors infer a genetic

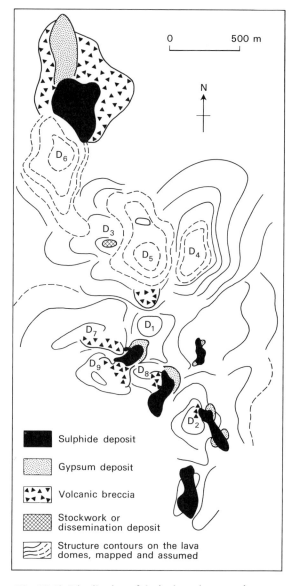

Fig. 15.13 Distribution of dacite lava domes and Kuroko deposits, Kosaka District, Japan. (Modified from Horikoshi & Sato 1970.)

Table 15.5 Volcanic-associated massive sulphide deposit types. (Modified from Hutchinson 1980)

Type	Volcanic rocks	Clastic sedimentary rocks	Depositional environment	General conditions	Plate tectonic setting	Known age range
Besshi (= Kieslager) Cu–Zn±Au±Ag	Within plate (intraplate) basalts	Continent-derived greywackes and other turbidites	Deep marine sedimentation with basaltic volcanism	Rifting	Epicontinental or back-arc	Early Proterozoic Palaeozoic
Cyprus Cu(±Zn)±Au	Ophiolitic suites, tholeiitic basalts	Minor or absent	Deep marine with tholeiitic volcanism	Tensional, minor subsidence	Oceanic rifting at accreting margin	Phanerozoic
Kuroko Cu–Zn–Pb±Au±Ag	Bimodal suites, tholeiitic basalts, calc-alkaline lavas and pyroclastics	Shallow to medium depth clastics, few carbonates	Explosive volcanism, shallow marine to continental sedimentation	Rifting and regional subsidence, caldera formation	Back-arc rifting	Early Proterozoic Phanerozoic
Primitive Cu±Au±Ag	Fully differentiated suites, basaltic to rhyolitic lavas and pyroclastics	Immature greywackes, shales, mudstones	Marine, < 1 km depth. Mainly developed in greenstone belts	Major subsidence	Much debated: fault-bounded troughs, back-arc basin?	Archaean–early Proterozoic

connexion. Certain Japanese workers, however, consider that the Kuroko deposits all formed in depressions and that the domes are late and have uplifted many of the massive sulphide deposits. In other areas, e.g. the Ambler District, North Alaska no close association with rhyolite domes has been found (Hitzman *et al*. 1986).

Cluster development

Although the Japanese Kuroko deposits occur over a strike length of 800 km with more than 100 known occurrences, these are clustered into eight or nine districts. Between these districts lithologically similar rocks contain only a few isolated deposits and this tends to be the case, with a few notable exceptions, for massive sulphide occurrences of all ages.

Favourable horizons

The deposits of each cluster often occur within a limited stratigraphical interval. For Primitive and Kuroko types, this is usually at the top of the felsic stage of cyclical, bimodal, calc-alkaline volcanism related to high level magma chambers rather than to a particular felsic rock type (Leat *et al*. 1986, Rickard 1987). Sometimes this favourable horizon is hosted by relatively quite thin developments of volcanic rocks, as in the Iberian Pyrite Belt, where the volcanic–sedimentary complex underlying the many enormous orebodies of this region is only 50–800 m thick. The Japanese Kuroko mineraliza-

tion and associated volcanism occurred during a limited period of the Middle Miocene in the Green Tuff volcanic region. By contrast the deposits of the Noranda area, Quebec (Fig. 15.14) lie in a narrow stratigraphical interval within a vast volcanic edifice at least 6000 m thick and in a similar manner the Scuddle Prospect and nearby VMS deposits of the Yilgarn Block, Western Australia occur in a thin regionally extensive horizon (Ashley *et al*. 1988).

Deposit stratigraphy

As has already been indicated massive sulphide deposits tend to have a well developed zoning or layering. Kuroko deposits have the best and most consistent stratigraphical succession of ore and rock types, and an idealized deposit (Fig. 15.15) contains the following units:
1 hanging wall—upper volcanics and/or sedimentary formation;
2 ferruginous quartz zone—chiefly hematite and quartz (chert);
3 baryte ore zone;
4 Kuroko or black ore zone—sphalerite–galena–baryte;
5 Oko or yellow ore zone—cupriferous pyrite ores; about this level, but often towards the periphery of the deposit, there may be the Sekkoko zone of anhydrite–gypsum–pyrite;
6 Keiko or siliceous ore zone—copper-bearing, siliceous, disseminated and/or stockwork ore;
7 footwall—silicified rhyolite and pyroclastic rocks.

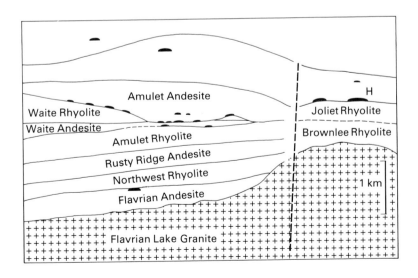

Fig. 15.14 Stratigraphical distribution of volcanic-associated massive sulphide deposits in the Noranda area, Quebec. (After Lydon 1989.)

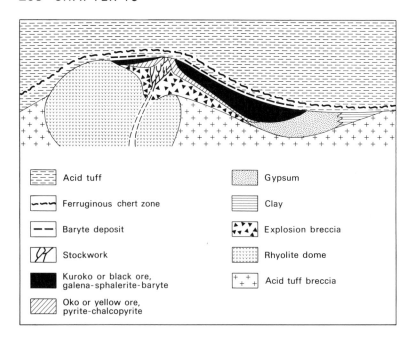

Acid tuff

Ferruginous chert zone

Baryte deposit

Stockwork

Kuroko or black ore, galena-sphalerite-baryte

Oko or yellow ore, pyrite-chalcopyrite

Gypsum

Clay

Explosion breccia

Rhyolite dome

Acid tuff breccia

Fig. 15.15 Schematic section through a Kuroko deposit. (Modified from Sato 1977.)

Genesis

The genesis of these deposits is discussed on pp. 71–79.

Volcanic-associated oxide deposits

Iron deposits

Solomon (1976) suggested that stratiform oxide ores such as the magnetite–hematite–apatite ores of Kiruna and Gällivare, northern Sweden, are oxide end-members of the massive sulphide group—a view echoed by Parák (1985). These ores sometimes contain as much as 15% P in the form of apatite, and magnetite (±hematite)–apatite ores are often referred to as Kiruna-type ores, but some in the type area are virtually free from apatite. There does indeed appear to be a range of ore types from massive sulphide deposits containing minor magnetite, through magnetite–pyrite ores with minor chalcopyrite and trace sphalerite such as Savage River, Tasmania (Coleman 1975), to stratiform iron oxide deposits, which may, or may not, carry appreciable phosphorus. Some possibly related deposits, e.g. in Iran, South Australia and the USSR, appear to have the form of mineralized pipes. These oxide deposits can be immense. The Kiruna orebody crops out over a strike length of 4 km, extends down dip for at least 1 km and is 80–90 m thick.

Phosphorus-poor ore runs 67% Fe and phosphorus-rich 2–5% P and about 60% Fe. Production is now entirely from underground workings and at 18 Mt p.a. makes Kiruna the world's largest underground mine. At Savage River, ore reserves in 1975 were 93 Mt in one of many deposits in a zone up to 23 km long. Among the pipe deposits there are 14 in the Bafq district of Iran together totalling > 1000 Mt (Förster & Jafarzadeh 1984), the single deposit of Korshunovsk in Irkutsk, USSR carries 428 Mt of ore (Sokolov & Grigor'ev 1977) and the enormous Cu–U–Au–Ag Olympic Dam deposit in South Australia contains over 2000 Mt of mineralized hematite breccias having an average grade of 1.6% Cu, 0.06% U_3O_8, 3.5 g t^{-1} Ag and 0.6 g t^{-1} Au. The proved and probable ore totals 450 Mt at 2.5% Cu, 0.08% U_3O_8, 6 g t^{-1} Ag and 0.6 g t^{-1} Au (Oreskes & Einaudi 1990). The breccias are also highly enriched in F and Ba and contain ~ 5000 ppm REE. The amount of iron in the hematite and copper–iron sulphides is huge. Here we have space only for a few remarks on the geology and genesis of these deposits. They all occur in volcanic or volcanic–sedimentary terrane, but there is no suggestion of some correlation of deposit type with volcanic rock type, as may be the case with the massive sulphides. This is because Savage River is associated with metamorphosed tholeiitic basalts, Kiruna with keratophyres, Iron Mountain, Missouri with andesites, the Bafq district and Cerro de Mercado, Mexico with rhyolites and Olympic Dam apparently with intermediate or silicic volcanics.

A sketch map of the Kiruna area is given in Fig. 15.16. The stratiform and concordant nature of the orebodies and their volcanic setting are apparent. The ores are both massive in part and well banded elsewhere—a banding that in places looks very much like bedding, sometimes with cross stratification. At Luossavaara, what appears to be a stockwork underlies the orebody (Fig. 15.17) and explosive volcanic activity has formed a hanging wall breccia containing fragments of the ore. All this evidence suggests that these are exhalative deposits (Parák 1975, 1985, 1991).

At Olympic Dam, which appears to be an explosion pipe (Selby 1991), barytic and hematitic sedimentary rocks and volcaniclastics occur as fragments in the upper portions of the breccia complex. These appear to be from exhalites and sediments formed on the Proterozoic palaeosurface immediately above the breccia body (Oreskes & Einaudi 1990). These authors emphasized that there is no evidence of intrusion of unusual, Fe-rich magmas at the present level of exposure. The hydrothermal activity was probably driven by a magmatic heat source. A pronounced magnetic anomaly over the

deposit suggests either that the mineralization becomes magnetite-rich at depth, or that a mafic intrusive underlies the deposit (Selby 1991). The mineralizing fluids may have been similar to the deuteric (hydrothermal) fluids that evolved from high level quartz-monzonite to diorite intrusions at Great Bear Lake, Canada and formed pods, veins and disseminations of magnetite–apatite–actinolite in and above the intrusions (Hildebrand 1986). Cliff *et al.* (1990), using radiometric dating methods, claimed that the Kiruna ores were probably formed during orogenesis. They cite much evidence of hydraulic fracturing by magnetite-rich fluids in the footwall (like the stockwork illustrated in Fig. 15.17) and quote as yet unpublished evidence from oxygen isotope geothermometry indicating a formation temperature of the deposit of about 600°C—well below that required to keep an iron oxide melt fluid.

Despite the weight of evidence favouring an epigenetic and/or exhalative hydrothermal origin for these deposits, a number of workers have suggested that the conformable deposits are oxide lava flows that have undergone some reworking at the surface. This hypothesis was put forward by Willams (1969) in view of the interesting occurrences of undoubted magnetite–hematite–apatite lava flows in Chile (Park 1961). A variant of this

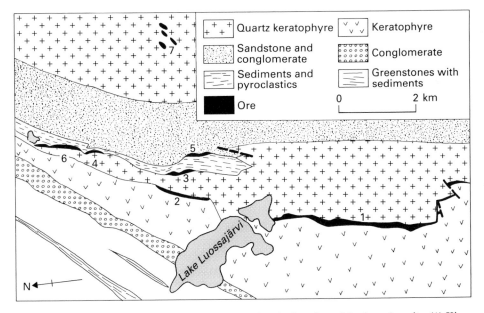

Fig. 15.16 Sketch map of the geology of the Kiruna area showing the location of the iron deposits. (1) Kirunavaara; (2) Luossavaara; (3) Rektorn; (4) Henry; (5) Haukivaara; (6) Nukutusvaara; (7) Tuollovaara. The formations dip and young eastwards. (After Parák 1975, 1985.)

Fig. 15.17 Stockwork of magnetite veins cutting keratophyre in the footwall of the Luossavaara iron orebody.

magmatic origin is the older view that the Kiruna ores are intrusive magmatic segregations. A recent persuasive advoate was Frietsch (1978). Support for this view came from Nyström (1985) and Nyström & Henriquez (1989) who reported the presence of columnar and dendritic magnetite. They contend that these are igneous textures and suggest that the reported cross and graded bedding may be igneous sedimentation features. The columnar magnetite matches that from the El Laco lavas flows very closely. The dendritic magnetite texture is similar to the skeletal platy olivines in the spinifex texture of komatiites.

Lyons (1988) described subaerial hematite–magnetite lavas in Durango, Mexico with associated intrusive, replacement and sedimentary iron oxide deposits. The juxtaposition of these very different deposit types might, he felt, explain the heated debates over the origin of many of these deposits. Are all the debaters right for part of the time?! Certainly many of the features clearly displayed in Durango may be present at Kiruna, but obscured by the later deformation and metamorphism.

Other metalliferous deposits

Other possible exhalative oxide deposits include the Rexspar uranium deposits, British Columbia (Preto 1978) certain uranium deposits in Labrador (Gandhi 1978) and some tin ores. Exhalative tin ores were first described from the Erzgebirge of Germany (Baumann 1970) but this interpretation is now disputed (Lehmann 1990). However the most important deposits so far found are probably those of Changpo in Dachang, China (Tanelli and Lattanzi 1985). Exhalative tungsten ores have also come into prominence during recent years particularly as a result of the pioneering work of Höll & Maucher (1976) and Höll *et al.* (1987) who have traced Sb–W–Hg mineralization along the Eastern Alps through the whole length of Austria. The mineralization is stratiform, occurs in Lower Palaeozoic inliers and is associated with metatholeiites. The scheelite mineralization is mostly present in fine-grained metachert bands that probably represent an original exhalite and which is very finely banded (Fig. 15.18). The scheelite bands appear to be synsedimentary and have suffered the same deformation and metamorphism as the enclosing rocks; one of the world's largest tungsten mines (Felbertal, to the south of Salzburg), exploits this mineralization. Mineralization of this type has now been reported from a number of areas of the world, e.g. the Italian Alps (Brigo & Omenetto 1983), Broken Hill, New South Wales (Barnes 1983), Sante Fe, New Mexico (Fulp & Renshaw 1985), Sangdong, South Korea (Fletcher 1984), La Codosera Spain (Arribas & Gumiel 1984) and Pakistan (Leake *et al.* 1989). The general volcanic association appears to be with basalts or rhyolites. Plimer (1987b) reviewed

Fig. 15.18 Banded scheelite ore, Felbertal, Austria, showing folding of both scheelite-bearing and silicate layers. Photograph taken using ultraviolet light and supplied through the courtesy of Prof. Dr R. Höll (scale in cm).

the stratiform scheelite deposits and their spatial association with rocks containing sufficient tourmaline to be termed tourmalinites. He recognized three deposit types: calc-silicate-hosted scheelite, amphibolite-hosted scheelite and tourmalinite-hosted scheelite.

Modern analogues of stratiform scheelite deposits occur in the Waimangu Geothermal Field, New Zealand (Seward & Sheppard 1986) in an area of basaltic and rhyolitic volcanic activity. Here hot spring waters are producing siliceous precipitates carrying up to 4% W. The ascending fluids of the Waimangu system carry $\sim 200\,\mu g\,kg^{-1}$ of W and could therefore transport 70 t of tungsten to the Frying Pan Lake, into which most of them pass, over a period of 100 years. This amount of tungsten could be obtained from the leaching of $\sim 0.03\,km^3$ of underlying rhyolitic ignimbrite.

The mercury end-member of the Sb–W–Hg association brings us back to sulphide mineralization in the form of cinnabar—the principal mineral of the famous Almaden mercury deposits of Spain (Saupé 1990).

Exhalative baryte deposits

Stratiform baryte deposits occur in both dominantly sedimentary settings, often as part of, or associated with, sedex deposits, and volcanic environments where they may form part of a volcanic-associated massive sulphide deposit.

Sediment-hosted stratiform baryte deposits

Some of these deposits, e.g. Ballynoe at Silvermines, Ireland, Meggen and Rammelsberg, Germany (see Table 15.1), are associated so intimately with sedex sulphide deposits that they must have been generated by the same period of hydrothermal activity. The Ballynoe deposit is a particularly clear example. At Silvermines the exhalative pyritic Zn–Pb ores of the B and Upper G orebodies (Fig. 2.13) grade up dip, and up the palaeoslope of the basin of deposition, into the Ballynoe baryte orebody (Fig. 15.19). The baryte body has a pyritic footwall and capping and, in material from the footwall, Larter et al. (1981) discovered pyrite tubes comparable with those associated with the black smokers of the deep oceans. Presumably these mark the position of vents through which exhalative solutions passed. This implies that the baryte deposit is not a distal facies of the metallic ores but was deposited from local hydrothermal exhalations in a shallower part of the basin and from its own feeder system (Fig. 15.19). Similar pyrite chimneys have been found in pyritic ore from the Zn–Pb deposits.

Deposits of this type have a worldwide distribution (Clark et al. 1990) and they provide much of the

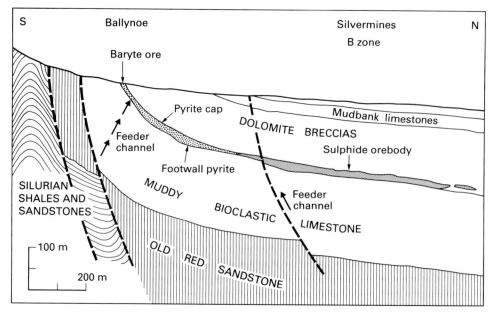

Fig. 15.19 Geological section through part of the Silvermines ore district, Ireland showing the Ballynoe baryte deposit and the B zone sulphide orebodies. The limestones are Carboniferous, the Old Red Sandstone Devonian. (After Larter *et al.* 1981.)

baryte used in drilling fluids in oil and gas drilling rigs. This is because generally they are insufficiently pure to be used for applications requiring highly refined products, e.g. the paint industry. However, the deposits themselves provide large tonnages that can often be mined in open pits to yield a cheap product.

Volcanic-associated stratiform baryte deposits

Baryte is found with some volcanic-associated mas-

sive sulphide deposits, particularly those of Kuroko type, and in a number of these it is mined. Being a low priced product (about £50 t^{-1} for oil fluid grade in September 1991) the deposits must be amenable to cheap mining methods to be economic. This is the case in the Hokuroku and Hokkaido districts of Japan, which have supplied much of that country's baryte, and Peru, where the deposits are conveniently sited near the port of Callao and the Talara Oilfields.

16 / The vein association and some other hydrothermal deposit types

Introduction

Vein, manto and pipe deposits have already received considerable attention in some earlier chapters. Their morphology and nature were described in Chapter 2 when such subjects as pinch-and-swell structure, ribbon ore shoots, mineralization of dilatant zones along faults, vein systems and orebody boundaries were outlined. The reader may find it convenient to review what is written on these subjects on pp. 26–29. Similarly in Chapter 3, in a discussion of precipitaion from aqueous solutions, crustiform banding (that characteristic texture of veins) was described, as were fluid inclusions and wall rock alteration, whose relationship to veins is most marked. It is generally agreed that vein filling minerals were deposited from hydrothermal solutions, and in Chapter 4 some consideration was given to the origin and nature of hydrothermal solutions, including metamorphic processes such as lateral secretion. The relevant sections are on pp. 55–67 and 67–69, and on pp. 79–83 the role of hydraulic fracturing in the formation of veins and other mineral deposits is reviewed. In Chapter 5, paragenetic sequence and zoning were discussed with special reference to vein deposits; see pp. 89–93.

If this chapter had been written 50 or so years ago it would probably have been by far the longest chapter in the book. However, since the 1940s the importance of veins has been steadily diminishing. The reasons for this are twofold. Firstly, a number of deposits formerly thought to belong to this association have been recognized as belonging to other ore classes. For example the Horne orebody of Noranda, Quebec was, for decades, believed to be a hydrothermal replacement pipe but is now considered to be a deformed volcanic massive sulphide deposit. The second and more important reason is economic. Taking copper as an example, only those ores containing more than 3% Cu were economically workable until the nineteenth century and such high metal levels were generally only reached in hydrothermal veins, so exploration and mining concentrated on this class of orebody. Today, rocks containing 0.5% Cu can be economically exploited by large scale mining methods and consequently veins do not hold their pre-eminent position any more. They are, however, still important for their production of gold, tin, uranium and a number of other metals and industrial minerals, such as fluorspar and baryte. Of course thick, high grade, base metal veins are still well worth finding and a good example is the El Indio Mine high in the Chilean Andes, where massive sulphide veins can be over 10 m thick with average grades of 2.4% Cu. As a result of their past importance, studies of vein ores have had a very profound influence on theories of ore genesis almost up to the present day, an influence out of all proportion to their true value.

Vein deposits and the allied, but less frequent, tubular orebodies show a great variation in all their properties. For example, thickness can vary from a few millimetres to more than 100 m and so far as environments are concerned, veins can be found in practically all rock types and situations, though they are often grouped around plutonic intrusions as in Cornwall, England. Mineralogically they can vary from monominerallic to a mineral collector's paradise, such as the native silver–cobalt–nickel–arsenic–uranium association of the Erzgebirge and Great Bear Lake, Northwest Territories.

Kinds of veins

The majority of veins have probably formed from uprising hydrothermal solutions which precipitated metals under environments extending from near magmatic high temperature–high pressure conditions, to near surface low temperature–low pressure conditions. A few veins are pegmatitic and represent end stage magmatic activity; a few are clearly volcanic sublimates.

Mineralogically, gangue minerals can be the dominant constituents as in auriferous quartz veins. Quartz and calcite are the commonest, with quartz being dominant when the host rocks are silicates and calcite with carbonate host rocks. This suggests derivation of gangue material from the wall rocks. Sulphides are most commonly the important metallic minerals but in the case of tin and uranium oxides are predominant. Here and there, native metals are abundant, especially in gold- and silver-bearing veins.

Zoning

This important property has already been considered. It is so commonly developed within single veins or groups of veins (district zoning) that some further mention of it must be made here. The base metal veins of Cornwall show some of the best and earliest studied examples of orebody and district zoning (Fig. 5.6 and Table 5.1), which has had important economic implications. For example, after the recognition of zoning in this tinfield, probably some time before de la Beche wrote about it in 1839, a number of abandoned copper mines were deepened and the delighted owners found tin mineralization beneath the copper zone. This tin had not been discovered earlier because there can be a barren zone of as much as 100 m or so between the tin and copper zones. On the other hand, as at the old Dolcoath Mine, there may be an overlap and a mine which commenced life as a copper producer could in time become a copper–tin producer and lastly a tin producer. Such was the history of this, the richest individual tin vein that has been worked anywhere in the world. In all, Dolcoath produced 80 000 t Sn and 350 000 t Cu and it is an interesting exercise to calculate the value of this metal production at present day prices! As can be seen from Table 5.1, the zoning generally takes the form of a progressive change in composition with depth. Gradual changes in metallic minerals are accompanied by somewhat less pronounced changes in the gangue minerals, with quartz being present at all levels. This zoning has been attributed to a slow fall in temperature with increasing distance from the granite intrusions, though the zonal boundaries, if they are isothermal surfaces, are not parallel to the granite contacts. This discrepancy is, however, simply due to their adopting a compromise attitude between that of the source of heat and the cooling surface (ground level).

Some important vein deposit types

As may be imagined from the variable nature of veins, the definition of types would be a fruitful field for the classification-mad scientist who delights in pigeon-holing nature, even if he can find only one

example for some of his holes! Eckstrand (1984) lists eight vein deposit types in his valuable volume and five of these, plus epithermal gold deposits in volcanic rocks, are briefly discussed here. The first three will be considered together, for economy of space and because they are all gold vein types.

As gold is very much in mineral explorationists' sights at the time of writing, the student is strongly recommended to study and compare Eckstrand's mineral deposit types 8.2, 11 and 15. The first and least important is the turbidite-hosted vein gold found in the Archaean (e.g. Yellowknife, Northwest Territories) and in the Phanerozoic (e.g. Nova Scotia; Ballarat, Bendigo, Castlemaine in Australia). The veins may be concordant with the development of saddle reefs as well as being thoroughly discordant and developed in reverse faults, ladder veins, etc. The importance of hydraulic fracturing in their development has been indicated by Cox *et al*, (1987). On a worldwide basis this has not yet been established as an important deposit type. Volcanic-associated vein (11) and intrusion-associated gold (15) are much more important. These are among the big gold producers of the Archaean greenstone belts and are discussed below. In addition to these three vein gold deposit types, epithermal gold deposits in volcanic terranes and Carlin-type deposits, which are both described later in this chapter, are at the present time important exploration targets.

Vein uranium has a worldwide significance. The classic vein deposits of Joachimsthal in the Erzgebirge (*Jáchymov* in Czech), from which the Curies separated radium, produced large tonnages of uranium concentrates for USSR consumption during the two decades following the explosion of the first atomic bombs (Vesely 1985). These are the famous Ag–Co–Ni–As–U veins described by Georgius Bauer, who wrote under the Latin name of Agricola, in his book *De Re Metallica* (1556). Agricola's works laid the foundation of modern mining science and of the geology of ore deposits. Vein uranium is important throughout the Hercynian Massifs of central and western Europe (see Fig. 5.5), and indigenous deposits make a significant contribution to the French nuclear industry. The veins are frequently emplaced in two-mica granites and the vein material is regarded as having been leached out of these 'fertile' granites by circulating hydrothermal solutions (Cuney 1978, Leroy 1978, 1984).

Vein tin deposits are descibed below and in various other parts of this book, e.g. pp. 61–62, 89–91, 313–316. Apart from placer tin deposits, they have been the world's most important source of tin.

Some examples

These examples have been chosen to illustrate not only different vein types but also the Lindgren classification (Chapter 6) into hypothermal, meso-thermal and epithermal classes.

Hypothermal Archaean vein gold deposits

These big (and small) gold producers are found and sought for in Archaean greenstone belt terrane all over the world. Many famous names belong here, such as the Golden Mile at Kalgoorlie, Western Australia, the Kolar Goldfield of India and the Kirkland Lake and Timmins areas of Ontario. Groves & Foster (1991) have given us an excellent review of this deposit type and much of the following information is culled from their paper.

Size and grade

Those greenstone belts that are well mineralized and lie within major Archaean provinces, e.g. the Superior Province of Canada and the Yilgarn Block of Western Australia, contain hundreds to several thousands of individual gold deposits. The majority of these have, or had, < 1 t Au. However, such belts often contain several large deposits often with one or more giant gold districts (or camps), e.g. the Superior Province with 25 orebodies that each contain (or contained) 45 t Au, including the Timmins (Porcupine) Camp which has produced > 1530 t Au.

Grades of mined ore vary widely even within individual camps and show a dramatic decrease with time, e.g. in the Yilgarn Block from 40 ppm in the 1890s to 5 ppm in 1988. This is largely owing to improved mining and mineral processing technology and changes in the gold price. At the present day most underground deposits run 4–8 ppm but some grade 10–15 ppm. Open pit mines are exploiting grades as low as 1–2 ppm.

Host rocks and structural control of mineralization

These deposits are usually present in greenschist facies terranes (Fig. 16.1) in structures of brittle–ductile transition regime. The original rocks are very variable but the dominant hosts in all the world's Archaean cratons are tholeiitic pillowed basalts and komatiites and their pyroclastic equivalents or felsic to mafic intrusives. The gold mineralization is associated with laterally zoned, wall rock alteration haloes whose formation involved the metasomatic addition of SiO_2, K_2O, CO_2, H_2O and Au. As the structural control of mineralization changes from more widely spaced faults or fractures, to closely spaced minor fractures, these deposits shade into one of the types of disseminated gold deposits described below. Good descriptions of vein gold deposits are to be found in Hodder & Petruk (1982), Colvine (1983) and Foster (1984).

The controls on the location of Archaean gold deposits are becoming better known, and many vein sets have been recognized as being commonly strata-bound and often confined to competent less basic basaltic units in metamorphosed tholeiitic–komatiitic sequences. The physical properties of these rocks have favoured hydraulic fracturing and fluid access, and their mineralogy and geochemistry have controlled gold deposition within the veins. These considerations led Phillips *et al.* (1983) to single out rocks of this type in greenschist to transitional amphibolite facies areas as favourable exploration targets. The host structures appear to be related to movement on regional and transcraton shear zones, with different structural styles of mineralization resulting from variations in the orientation of the regional stress field and the strength of the host rocks. Common hosts are cross-cutting or layer-parallel shear zones, strike-extensive laminated quartz veins (more rarely saddle reefs), extensional quartz-vein arrays and/or breccias. This variety of structural styles is illustrated in Fig. 16.1. We find the best mineralized terranes where there is a high density of anastomosing, oblique- to strike-slip, transcraton and regional shear zones that control the geometry of adjacent subsidiary mineralized structures. These high strain zones focused fluid flow in otherwise regionally low strain greenstone belts. Gold mineralization tends to occur where these structures are accompanied by regional carbonation (carbonatization) and the emplacement of swarms of minor intrusions of porphyries and lamprophyres.

Genesis

The gold mineralization resulted from this highly focused flow of low salinity, H_2O–CO_2 fluids along regional shear zones with transient flow into adjacent brittle–ductile structures as a result of the

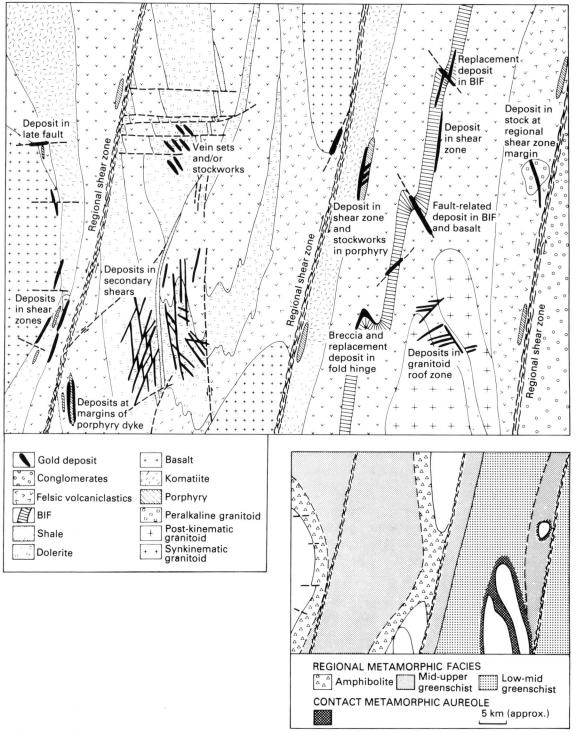

Fig. 16.1 A schematic representation of the nature of epigenetic, Archaean gold mineralization showing many of the variable structural styles, host rocks and, in the lower diagram, the metamorphic setting. (The figure is based largely on Western Australian examples and is taken with permission from Groves *et al.* (1988), fig. 2, *Geology Department and University Extension, University of Western Australia, Publication 12.*)

lowering of pressure during failure. Gold precipitation appears to have normally occurred at $300 \pm 50°C$ and $0.1–0.3$ GPa as a result of fluid–wall rock reactions, particularly the sulphidation of iron-rich host rocks, or phase separation within the fluid. Further discussion of the origin of these deposits may be found on pp. 69–71.

Butte, Montana—a mesothermal orefield

This is one of the world's most famous vein mining districts. From 1880 to 1964 Butte produced 300 Mt of ore yielding 7.3 Mt Cu, 2.2 Mt Zn, 1.7 Mt Mn, 0.3 Mt Pb, 20 000 000 kg Ag, 78 000 kg Au, together with significant amounts of bismuth, cadmium, selenium, tellurium and sulphuric acid. Because of this wealth of mineral production from a very small area (little more than 6×3 km), with more than a score of

mines, Butte has been aptly called the richest hill on earth, for the monetary value of its production has been exceeded only by the much larger Witwatersrand Goldfield of South Africa. Cut-and-fill stoping of the veins was the main mining method up to 1950, then it was joined by the block caving of veined ground, and in 1955 large scale, open pit working of low grade porphyry-type mineralization commenced. Reserves are still extensive, of the order of 10 Mt of high grade vein copper and silver ore and 500 Mt of low grade copper mineralization.

The Butte field is in the south-western corner of the Cretaceous Boulder Batholith. Soon after the emplacement of the intrusion large parts of the area were covered by rhyolitic and dacitic eruptions (Fig. 16.2). The veins occur in a granodiorite which has given a radiometric age of 78 Ma. There were two mineralization stages: pre-main and main. The

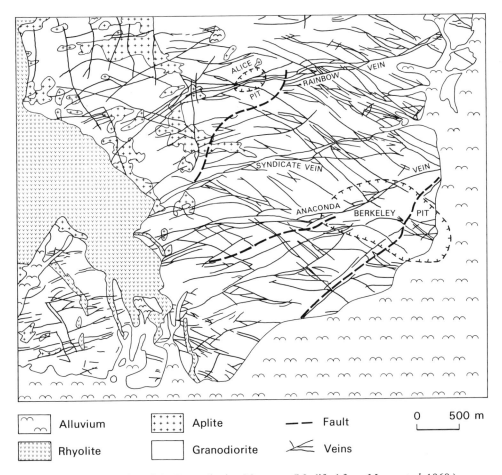

Fig. 16.2 Surface geology and veins of the Butte district, Montana. (Modified from Meyer *et al.* 1968.)

pre-main stage consists of small quartz veins carrying molybdenite and chalcopyrite found in the deeper central parts of the mineralized zone. They are bordered by alteration envelopes carrying potash feldspar, biotite and sericite. The biotite has been dated at 63 Ma.

The main mineralization stage occurs in several vein systems of which the most important are the easterly trending Anaconda and the later north-westerly trending Blue veins. The Anaconda veins were the major producers in the western third of the mineralized zone and also in the eastern third, where they divide into myriads of closely spaced south-easterly trending minor veins. This is called horse-tailing and gives rise to porphyry-type mineralization suitable for mass mining methods. The Anaconda veins were the largest and most productive, averaging 6–10 m in thickness with local ore pods up to 30 m thick. The Blue veins usually offset the Anaconda veins with a sinistral tear movement and sometimes ore has been dragged from the Anaconda into the Blue veins by this fault movement. Individual oreshoots persist along strike and in depth for hundreds of metres.

All the veins contain similar mineralization and this is strongly zoned (Fig. 16.3). There is a Central Zone of copper mineralization, which at depth contains the quartz-molybdenite veins of the pre-main stage and which is particularly rich in chalcocite–enargite ore. This gradually gives way outwards to ore dominated by chalcopyrite and

containing minor amounts of sphalerite—the Intermediate Zone. The Peripheral Zone is principally sphalerite–rhodochrosite mineralization with small quantities of silver. All the veins are bordered by zones of alteration, which usually consists of sericitization next to the vein followed outwards by intermediate argillic alteration and then by propylitization. In deeper parts of Butte, advanced argillic alteration is present next to the veins. Sericite produced by this alteration dates at 58 Ma, distinctly later than the biotite of the quartz-molybdenite mineralization. The zoning suggests that the main mineralization was effected by hydrothermal solutions which passed upwards and outwards during a long period of time through a steadily evolving fracture system.

Brimhall (1979) has demonstrated that the mineralization of the pre-main stage consisted of widespread, zoned, low grade, disseminated copper–molybdenum mineralization of porphyry type formed at 600–700°C. During the later main stage mineralization a geothermal system, involving hydrothermal fluids of meteoric origin and possibly set in motion by a younger intrusive system, achieved temperatures of 200–350°C and leached much of the disseminated copper, redepositing it to form the rich vein orebodies.

Butte is an excellent example of mesothermal mineralization (Chapter 6). A useful review of some important mesothermal gold deposits is given by Kerrich (1989).

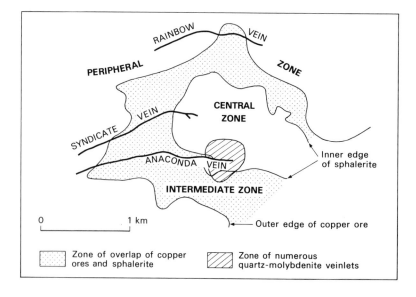

Fig. 16.3 Zoning on the 2800 ft (853.44 m) level at Butte, Montana. (Data from Meyer *et al.* 1968.)

The Llallagua tin deposits, Bolivia—an example of telescoping

Bolivia possesses the greatest known reserves of tin outside the countries of south-east Asia, most of her reserves being in vein and disseminated deposits. There is a long history of mining dating back to the Spanish colonial days of the sixteenth century when most of the important tin deposits were discovered and were mined initially for the fabulously rich silver ores that had formed, partly by supergene enrichment processes, in the upper parts of the vein systems. Some of these deposits are still being mined, mainly for tin ores beneath the silver-rich zones. In common with other tin-producing countries whose output came largely from vein deposits, Bolivia has been hit by the low tin prices of recent years (see pp. 10, 154). Tin production fell from 16 472 t in 1985 to 4039 in 1989. Fox (1991) expressed the view that no Bolivian mine whose main product is tin can be viable economically; nevertheless Comibol (the state mining company) brought production up to 5938 t in 1990 and has a target of 7427 t for 1991. One or two Bolivian tin mines producing substantial amounts of by-product zinc and silver still make a profit. In the future Bolivian tin production, if it is to be economic, must come from such mines, and as a by-product from deposits such as the fabulous silver porphyry of Cerro Rico de Postosi, which is estimated to contain 828 Mt running 150–250 ppm silver and 0.3–0.4% Sn (Suttill 1988).

The Bolivian tinfield extends along the Andean ranges east of the high plateau of Bolivia from the Argentine border northwards past Lake Titicaca and into Peru (Fig. 16.4), where many new tin deposits may be found (Kontak & Clerk 1985). North-west of Oruro the deposits are mainly tin and tungsten veins associated with granodioritic batholiths. The batholiths range in age from Triassic to Miocene, with the Miocene ones being best mineralized.

South of Oruro there is a tin–silver association spatially related to high level subvolcanic intrusions. At some of these volcanic centres both the intrusives and the coeval volcanics are preserved, at others erosion has removed the volcanics completely leaving only the intrusives. This is the case at Llallagua, the world's largest tin mine working primary tin deposits, which is estimated to have produced over 500 000 t of tin since the beginning of this century. The mine occurs in the Salvadora Stock, which occupies a volcanic neck cutting the

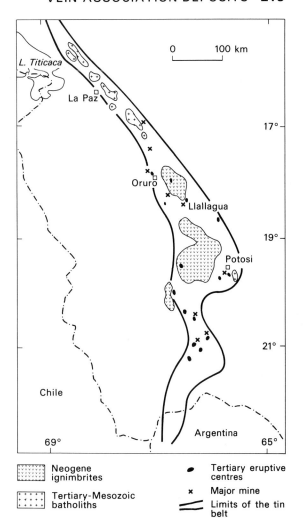

Fig. 16.4 The Bolivian Tin Belt. (After Grant *et al.* 1977.)

core of an anticline in Palaeozoic rocks. The stock, which is made up of xenolithic and highly brecciated porphyry, narrows with depth. The original texture and rock composition (probably quartz-latite) are obscured by pervasive alteration. The ubiquitous brecciation is important because it has produced an increased permeability of the stock, which has given rise to alteration and mineralization independent of the geometry of the later vein systems. This dispersed mineralization may be of high enough grade to permit bulk mining methods to be used, i.e. it may be a porphyry tin deposit (Grant *et al.* 1977). The alteration has produced a host rock consisting of primary quartz, tourmaline, sericite and secondary quartz.

There is a network of veins in and around the stock, some of which are shown in Fig. 16.5. The major veins trend about 030° and appear to form part of a conjugate system of normal faults. They are typified by the San Jose type, which generally has a dip of 45–80°, a good width and strike persistence, and an average width of about 0.6 m, although widths of up to 1.8 m are known. These are the richest veins and can contain up to 1 m of solid cassiterite. Clay gouge is very common. The Serrano vein type is much thinner (average 0.3 m), nearly vertical and impersistent. These veins can be as richly mineralized as the San Jose type, but with such narrow veins, dilution with country rock material occurs during mining. There is little or no clay gouge. These veins may have formed in vertical tension gashes associated with the normal faulting.

The first stage of mineralization consisted of the formation of crusts of quartz followed by bismuthinite and cassiterite, and many high grade veins are composed almost entirely of these three minerals. Wolframite and tourmaline also belong to this early stage (Turneaure 1960). This was followed by a stage of sulphide mineralization, principally pyrrhotite and franckeite, the pyrrhotite being later largely replaced by pyrite, marcasite and siderite. Arsenopyrite, sphalerite, stannite and rare chalcopyrite were also formed during this later stage of sulphide

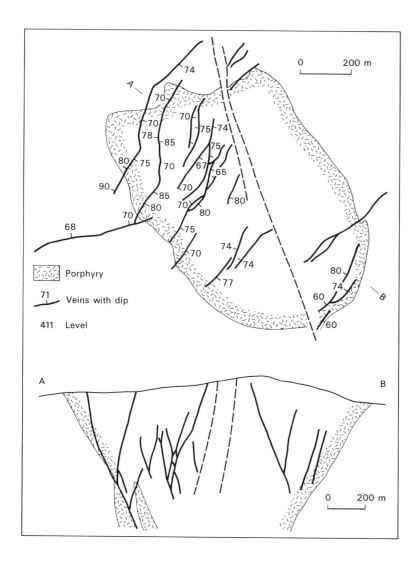

Fig. 16.5 Plan and section of the major veins at Llallagua, Bolivia. (Modified from Turneaure 1960.)

deposition. The discerning reader will have noted that in this vein system there is present, over a very restricted vertical range, a host of minerals which in Cornwall are developed in a zonal sequence spread out vertically over hundreds of metres (Table 5.1). The concentration of low and high temperature, epithermal to hypothermal minerals into the same 'zone' is called telescoping and is thought to be due to the existence of high temperatures in near surface host rocks induced by the volcanic activity and a blanket of hot volcanic deposits. Fluid inclusion work by Grant *et al.* (1977) has shown that the pervasive alteration and early vein growth, including cassiterite deposition, took place at temperatures (uncorrected for pressure) of about 400–350°C. The temperature dropped to 300°C and lower during the sulphide deposition. Data from Llallagua is included in Fig. 3.5, which is based on samples collected along much of the length of this tin belt and which illustrates the great variation in temperatures—from hypothermal down to epithermal—that obtained during mineral deposition in the tin veins.

Epithermal precious metal deposits in volcanic rocks

Some examples of these deposits take the form of discrete veins but others are stockwork or breccia hosted (bulk-tonnage deposits) that could be included among the disseminated gold deposits described later in this chapter. Good general descriptions are to be found in Hayba *et al.* (1986), Heald *et al.* (1987) and Henley (1991).

This precious metal deposit type may carry important quantities of silver as well as gold and it has been important during various historical periods. The quartz–gold veins of Krissites and other areas in northern Greece were the backbone of the Ancient Greek economy and those of Dacia (Romania) helped to sustain the Roman economy. At the present day deposits of this type are the goal of major exploration programmes around the Pacific rim, especially in Japan, the south-western Pacific and western North America. Two recent finds illustrate the range of their grade–tonnage spectrum. Hishikari in Kyushu, Japan, discovered in 1980, consists of a vein system, with veins up to a few metres wide having an average grade of $70 \, g \, t^{-1}$ and high silver values, that forms an ore reserve of 1.4 Mt—a real bonanza find which during 1988 was the world's lowest cost gold mine. By contrast the Ladolam deposit, Lihir Island, PNG, is a breccia

hosted, disseminated deposit with at least 167 Mt of ore grading $3.43 \, g \, t^{-1}$. Bulk tonnage open pit mining, with modern cyanide extraction and heap leach techniques have made even the large, low grade deposits attractive exploration targets. Many of the famous silver–gold deposits of Mexico and Peru that are hosted by terrestrial volcanics belong to this class of deposit, e.g. see Gibson *et al.* (1990) and references therein.

Using their dominant wall rock alteration assemblages Hayba *et al.* (1986) and Heald *et al.* (1987) have divided these deposits into two classes: acid sulphate and adularia–sericite. Henley (1991) prefers and uses the term alunite–kaolinite rather than acid sulphate. The older terminology will be used in this book. Characteristics of these two classes are summarized in Table 16.1. At the time of writing, exploration results suggest that the adularia–sericite class is by far the most common. There is apparently no evidence that they are end members of a spectrum of deposits and all so-called intermediate examples have proved to be the result of one type overprinting the other.

For both classes calderas are important as structural settings, but in the western USA where calderas are well developed few appear to have been economically mineralized in this way. Their general setting, other evidence and data shown in Table 16.1 have led to their being interpreted as forming from near surface geothermal systems. The small amount of isotopic data so far available suggests that the mineralizing fluid was dominantly meteoric, but there is clear evidence in some studies of a magmatic component. However, it must be borne in mind that in such a near surface environment any magmatic fluids would, in most cases, be overwhelmed by the meteoric ground water of the near surface environment.

Heald *et al.* (1987) suggested that these two different classes of deposit reflect in particular their distance from their heat source; the acid sulphate type being spatially and temporally related to a shallow hydrothermal system in the core of a volcanic dome (Fig. 16.6a). These domes often occur along or near the ring fracture faults of calderas and as a result of this location the mineralized zone is usually small, the heat source being the volcanic conduit. Mineralization may be capricious but the El Indio deposit, Chile is a spectacularly rich example with the ore in well-developed veins several metres in width. Grades averaging $225 \, g \, t^{-1}$ Au, $104 \, g \, t^{-1}$ Ag and 2.4% Cu

Table 16.1 Characteristics of acid sulphate and adularia–sericite epithermal deposits. (Based on Hayba *et al.* 1986, Heald *et al.* 1987 and Henley 1991)

	Acid sulphate	Adularia–sericite
Plate tectonic setting	Both found in subduction environments at plate boundaries, particularly in back-arc basins	
Regional structural setting	Calderas, silicic domes	In calderas and other complex volcanic environments
Local structural setting	Complex systems of faults or fractures developed in several generations	
District and deposit dimensions	Smaller than adularia–sericite deposits. Vertical extent usually < 500 m, often equidimensional	12–190 km², vein length: width ~ 3 : 1, strike length several kilometres, vertical extent 100–700 m
Host rocks	Mainly rhyodacite (also rhyolite, trachyandesite) forming domes and ash-flows	Rhyolite to andesite + associated intrusions and some sediments
Time relationships	Ore + host similar ages (< 0.5 Ma)	Distinctly different ages (usually > 1 Ma)
Mineralogy	Enargite, pyrite, covellite, native gold, electrum, base metal sulphides, sulphosalts, tellurides. Adularia and chlorite absent or rare. Sometimes bismuthinite	Adularia, sericite, chlorite, acanthite, tetrahedrite, native silver and gold, base metal sulphides. No hypogene alunite. No bismuthinite
Metal production	Both gold- and silver-rich deposits. Significant copper production	Both gold- and silver-rich deposits. Base metal production variable
Wall rock alteration	Advanced argillic. Much hypogene alunite and kaolinite. No adularia. Outer (upper) zone of intermediate argillic + sericitization	Sericitic[a] to intermediate argillic. Abundant adularia. Only supergene alunite
	Outer zone of propylitic alteration	
Temperature of ore deposition	200–300°C[b]	Ore: 200–300°C, gangue down to 140°C, boiling in some cases only
Salinity	Little data, perhaps 1–6 wt % NaCl equiv.	Usually < 3 wt % NaCl equiv., can be up to 13% if base metals important
Depth of formation	300–600 m, can be > 1200 m	100–1400 m, mostly 300–600 m
Source of mineralizing fluids	Little data	Dominantly meteoric. Magmatic component
Sulphide sulphur source	Little data, possibly magmatic	Deep-seated magmatic or volcanic country rocks
Source of lead	Enclosing volcanics or magmatic fluids indicated for two deposits, subvolcanic Precambrian basement for another	Precambrian or Phanerozoic rocks in subvolcanic basement

[a] Not necessarily muscovite, may be illite, etc.
[b] Limited data.

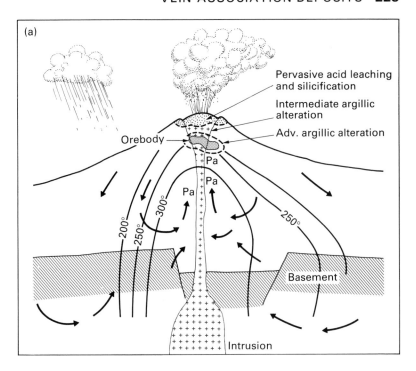

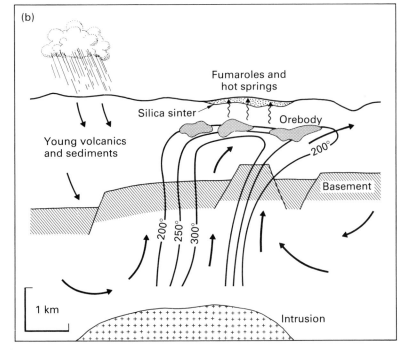

Fig. 16.6 Schemata for the formation of two types of epithermal precious metal deposits in volcanic terranes, based on Heald *et al.* (1987) with modifications from Henley (1991). (a) Acid sulphate type. Pa = propylitic alteration. Note that the mineralization occurs within the heat source. (b) Adularia–sericite type. The upwelling plume of hydrothermal fluid is outlined by the 200°C isotherm. The mushroom-shaped top reflects fluid flow in the plane of major fracture systems, a much narrower thermal anomaly would be present perpendicular to such structures. The heat source responsible for the buoyancy of the plume is shown as an intrusion several kilometres below the mineralized zone. In both schemata the arrows indicate the circulation of meteoric water. (a) is drawn at the same scale as (b).

have been reported (Henley 1991). This deposit and the ores of Goldfield, Nevada have been mined using selective methods, but those of the Pueblo Viejo district, Dominican Republic are being mined successfully using bulk mining methods (Muntean *et al.* 1990).

Proximity in time and space to the volcanism probably explains the high sulphur fugacity (much SO$_2$ gas!), which produced the extensive alunitization and the enargite–pyrite–covellite assemblage of these deposits. On the other hand the adularia-sericite deposits having mineralization at least 1 Ma later than the volcanism, being spatially unrelated to intrusive volcanics and with distinctive alteration assemblages, appear to have formed in the upper parts of geothermal systems above a deeper heat source (Fig. 16.6b). This accounts for the variety of volcanic host rocks and the broader upper part of the hydrothermal cell, which led to the development of more extensive mineralization districts. In this environment much more hydrothermal fluid was

available and much less sulphur thus producing different wall rock alteration mineral assemblages. If it is accepted that these deposits are fossil equivalents of present day geothermal systems then their gold contents may have accumulated very rapidly. Brown (1986) has calculated that it would require only 1500 years for such a system to transport 31.1 t (= one million troy ounces) of gold.

Although the formative pressure and temperature ranges for the two classes of deposit are similar, the activities of sulphur, oxygen and hydrogen will differ. Figure 16.7 shows the different geochemical environments that produced the different mineral and alteration assemblages. The highly sulphidized mineralogy of the acid sulphate deposits contrasts strongly with the less sulphidized adularia–sericite deposits. In acid sulphate deposits the assemblage pyrite–galena–enargite and the close association with alunite indicate an environment with relatively high sulphur fugacity, and the fluid responsible for this type of mineralization must have a pH of

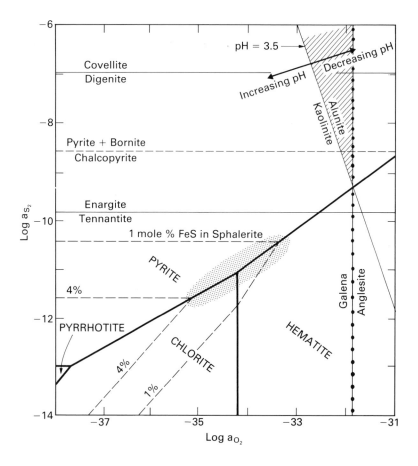

Fig. 16.7 Log a_{S_2}–a_{O_2} diagram showing the stability fields for the significant minerals in epithermal systems at 250°C. Diagonal lines indicate the probable conditions obtaining during the genesis of acid sulphate deposits and stippling those for adularia–sericite deposits. The position of the kaolinite–alunite line changes progressively with changes in sulphur concentration, so for the sake of simplicity one position only is shown. The field of magnetite, a mineral rarely found in these deposits, is masked by the field of iron-rich chlorite. Quartz is present throughout the diagram. Further details of the data used to construct this diagram will be found in Heald *et al.* (1987) from which this figure has been adapted.

less than three at the sulphur concentrations that probably obtained. With relatively high sulphur concentrations native sulphur might be precipitated—it occurs in some acid sulphate deposits and active geothermal systems.

The pyrite–chlorite–hematite assemblage of adularia–sericite deposits indicates lower sulphur and oxygen activities during their formation, as does the presence of tennantite, the absence of enargite and the common compositional range of 1–4 mole % FeS in sphalerite. These deposits also form at a higher pH than does acid sulphate mineralization (Heald *et al.* 1987).

Disseminated gold deposits

The range of deposit subtypes that can be described under this title is considerable. The best comprehensive reference is Boyle (1979, pp. 295–331) who recognized four subtypes but wrote that: 'These deposits cover a wide spectrum.' I will follow his classification.

Disseminated and stockwork gold–silver deposits in igneous intrusive bodies

Some of these are referred to as porphyry gold deposits and some are clearly copper-deficient, gold-rich porphyry copper deposits. Orebodies of this general type are commonly in the range 5–15 Mt with grades around 8–16 ppm, but there are many larger deposits of lower grade and some of these are reviewed in Sillitoe (1991b). The mineralization occurs in highly fractured zones of irregular outline which have been healed by veinlets, veins and stringers of auriferous quartz. These zones are marked by considerable hydrothermal alteration of the host rock, which varies with the rock type. In granitic rocks the common alteration is potassic alteration, sericitization, silicification, feldspathizatrion and pyritization; in intermediate and basic rocks, carbonatization, sericitization, serpentinization and pyritization.

Deposits of this subtype occur in orogenic belts in all the continents and range from Archaean to Phanerozoic in age. Good potted descriptions of eleven deposits are given by Boyle, who points out that there is a marked relationship in many gold belts between albitites, albite-porphyries, quartz–feldspar porphyries and gold. The exact nature of this relationship is uncertain and may vary from deposit to deposit. A good summary of these relationships as seen in Ontario can be found in Marmont (1983).

Disseminated gold–silver occurrences in volcanic flows and associated volcaniclastic rocks

Orebodies of this subtype have only recently been outlined and any marked increase in the price of gold in real terms will bring many more of these deposits into the class of potential orebodies. The mineralization occurs in large, diffuse volumes of alteration in rhyolites, andesites or basalts. Some are closely associated with known gold deposits. Reported gold values are mainly in the range of 0.02–0.03 ppm but some are much richer, e.g. Round Mountain, Nevada at 66 ppm. They are probably related to Carlin-type deposits, which are described below.

Disseminated deposits in tuffaceous rocks and iron formations

Gold deposits in tuffs and other pyroclastic rocks are common in the greenstone belts of the Precambrian (Woodall 1979). A well described example is the Madsen Mine, Ontario (Durocher 1983). The orebodies took the form of echelon ore zones hosted by heterogeneous, sheared and highly altered tuffs occurring in the lower parts of a tholeiitic–komatiitic sequence (Fig. 16.8). The orebodies, which were delineated by assay boundaries and had an average grade of about 8 ppm, were localized by rolls (open folds) in the hanging wall and footwall contacts. Orebodies at the Triton Gold Mine, Western Australia (Campbell 1953) occurred in a similar geological environment, with the same structural controls governing their location, and had a similar grade.

The spatial association of many Archaean gold deposits with ultramafic lavas suggests that komatiites may be the ultimate source rock of much gold, a point of importance in designing exploration programmes and one which has been well aired by Keays (1984).

A common factor of the greenstone belts of Archaean cratons is the presence of several large gold deposits (> 50 t Au) in areas containing numerous smaller deposits. These large and many smaller deposits occur in various facies of iron formation of which lean sulphide-bearing carbonate and silicate facies are perhaps the most important. Sawkins & Rye (1974) termed these 'Homestake-type' deposits

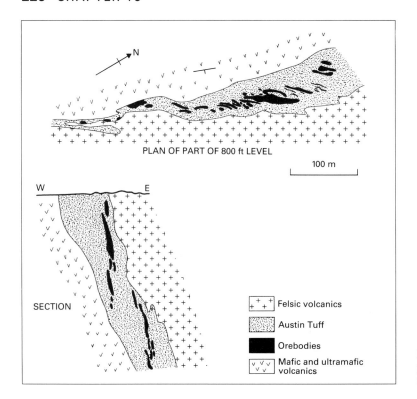

PLAN OF PART OF 800 ft LEVEL

100 m

W   E

SECTION

	Felsic volcanics
	Austin Tuff
	Orebodies
	Mafic and ultramafic volcanics

Fig. 16.8 Generalized geology of part of the Madsen Mine, Ontario. (After Durocher 1983.)

after the famous large and long lived mine in South Dakota, and included in this group Morro Velho, Brazil (Gaál & Teìxeìra 1988) and the Kolar Goldfield of India, where Natarajan & Mukerjee (1986) and Siva Siddaiah & Rajamani (1988) have described similar facies and mineral associations to those at Homestake. Many of the Zimbabwean deposits are of this type, as well as many in the Murchison Province, Western Australia.

The Homestake Mine has produced about 1100 t Au since it began production in 1877, and is the biggest gold mine in the USA. The orebodies occur only in the Homestake Formation, a thin (> 100 m) auriferous, quartz–sideroplessite schist unit within a thick sequence of metasediments that only contains minor metabasaltic rocks. The deposit has suffered complex polyphase folding and low to medium grade metamorphism that gave rise to considerable redistribution and concentration of the gold (and sulphides), such that the orebodies are now spindle-shaped zones which appear to be localized in part by dilatant zones formed by superposition of F_2 on F_1 folds (Fig. 16.9). The orebodies consist of quartz, chlorite and ankerite with pyrrhotite, arsenopyrite and gold. Most of the chlorite appears to be altered

cummingtonite or sideroplessite and much of the quartz takes the form of metamorphic segregations, veins and stringers, such that for many decades the deposit was considered to be epigenetic.

Stable isotope studies have shown that the sulphide sulphur and the oxygen of the quartz are indigenous to the Homestake Formation and the inference is that the gold too is syngenetic. A model for the formation of these deposits was proposed by Fripp (1976), involving circulating sea water, similar to that shown in Fig. 4.17 but with the saline solutions debouching into basins in which banded iron formation is being precipitated and Saager *et al.* (1987) published further studies, supporting Fripp's work, on deposits hosted by iron formation in Zimbabwe. Kerrich & Fryer (1979) have suggested that the fluids were of metamorphic origin but again postulate venting into a basin in which BIF (banded iron formations) was forming, cf. Fig. 4.10. On the other hand, Phillips *et al.* (1984) have suggested that some, perhaps all, of these deposits are epigenetic; but they side with Kerrich & Fryer in advocating a metamorphic source for the mineralizing fluids. These contrasting views lead to rather different exploration philosophies. A possible

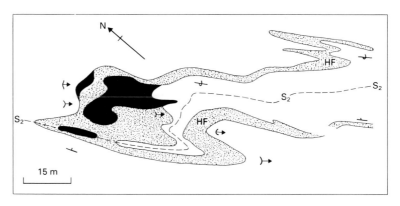

Fig. 16.9 Superposed folds, 2300 ft (701 m) level, Homestake Mine, South Dakota. An early fold with axial surface S_2 has smaller folds superposed on its limbs. Note development of orebodies in thickened hinge zones of later folds. Ore-bearing zones in black. HF = Homestake Formation. (Compiled from several sources.)

sequence of events, which ignores the origin of the mineralizing solutions, is shown in Fig. 16.10.

Carlin-type deposits

There are at least nine different names for this deposit type in the English literature. For this writer Carlin-type is the least wanting as far as the likelihood of misleading the reader is concerned, and it has the virtue of using the name of the first deposit of this type to be found. (The Carlin Mine, Nevada, came into production in 1965.) Nevertheless, we must not put these deposits into a straitjacket and expect all deposits to reproduce the exact features of the type deposit. This would constrain our exploration and scientific study of these deposits

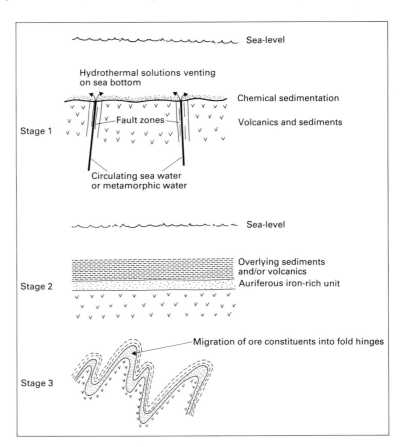

Fig. 16.10 Generalized model for the formation of Homestake-type deposits. (Modified from Sawkins & Rye 1974.)

and they are, at the time of writing, an important exploration target and one which is receiving much scientific attention. As a result of such studies Berger & Bagby (1991) have defined two deposit types—Carlin-type and Carlin-like—the latter are carbonate-replacement gold deposits associated with alkaline igneous intrusions in Central Montana and the Black Hills, South Dakota. Only the Carlin-type deposits will be considered here.

The deposits at present being exploited or proved up are mainly located in Nevada, Utah, Idaho, California and south-eastern China. Their grade varies from 1 to 13.4 ppm Au and mercury and silver are often won as by-products, although in some Au : Ag can be very high. Orebody tonnages are commonly in the range 5–80 Mt and resources can be very large, e.g. Round Mountain, Nevada, with 186 Mt grading 1.7 ppm Au (Tooker 1985). Orebodies generally have assay boundaries and their shapes vary from broadly tabular to quite irregular with faults and favourable horizons often exerting a control on their shape and position; for example, at the Getchell Mine the orebodies tend to be irregularly shaped pods elongated along favourable beds adjacent to and within the faults that acted as fluid conduits (Berger 1985). They are bulk mineable deposits. New finds are being made all the time (see Burger 1985) and there are many regions which could host such deposits; the young island arcs of the Western Pacific being favoured by Sawkins (1984) for exploration. An excellent summary of the principal characteristics of these deposits can be found in Berger & Bagby (1991) from which much of the following is taken.

The most favourable host rocks are fine-grained, finely laminated, carbonaceous, silty carbonates and carbonate-bearing siltstones and shales of marine origin. In the western USA these are Cambrian to Mississippian in age and occur within or adjacent to overthrust terranes related to continental margin tectonics. Doming of the mineralized rocks occurred and associated normal faults channelled hydrothermal solutions into favourable beds. The faulting often increased rock permeability by creating breccias, which host the ore in several deposits. Replacement orebodies never occur far from fractures. Igneous rocks, commonly granites, *sensu lato*, occur within or close to nearly all deposits. Great difficulties have been encountered in attempting to obtain radiometric dates on the wall rock alteration. Results to date suggest that the Nevadan deposits are not all the same age. Many appear to be Cretaceous and some Oligocene.

Wall rock alteration is marked by the development of jasperoid which has been much used as an exploration guide. It was formed before, during and after gold mineralization and usually replaces carbonate rocks and grains, as does introduced pyrite. Some detrital minerals were altered to new clay minerals (dickite, illite, kaolinite) during the ore mineralization stage and, following this, post-ore calcite veins were commonly developed. These may contain stibnite, cinnabar, orpiment, realgar, baryte and complex thallium minerals (Radtke, 1985).

Normally only free gold is present (i.e. no combined gold as tellurides) and the grain size is usually < 1 µm. Gold forms films on pyrite and amorphous carbon, is enclosed in silica, occurs within arsenical pyrite and in arsenian rims around pyrite grains. (Tellurides do occur at the Mercur Mine, Utah.) Native silver and electrum also occur. Pyrite, the most abundant sulphide, may be accompanied by marcasite and arsenopyrite. Base metal sulphides are generally uncommon. The mineralization persists in depth without any apparent zonation in the ore mineralogy, suggesting that these deposits are better classified as leptothermal or mesothermal deposits and should not be likened to the shallow, rapidly bottoming epithermal deposits of volcanic terranes. This conclusion is supported by fluid inclusion work, which indicates ore formation at depths of 2–4 km with main stage mineralization at 200–300°C.

A schema for the formation of Carlin deposits is given in Fig. 16.11. Sawkins (1984), Berger & Bagby (1991) and other writers favour intrusion-driven hydrothermal cells as the mineralizing mechanism. Berger & Bagby suggest that metalliferous, magmatic fluids from the intrusions became diluted by meteoric water in the contact zone and rose as an upwelling plume to deposit gold in the favourable zones described above. Fluid inclusion work suggests that boiling was not an important precipitation process, and presumably mixing with cooler meteoric water having a different chemistry was the mechanism of ore deposition.

From the regional tectonic point of view these deposits occur in complex terrane. Central and western Nevada is a case in point, consisting as it does of several geological terranes of distinct stratigraphical and structural domains that have been accreted throughout much of the Phanerozoic to the

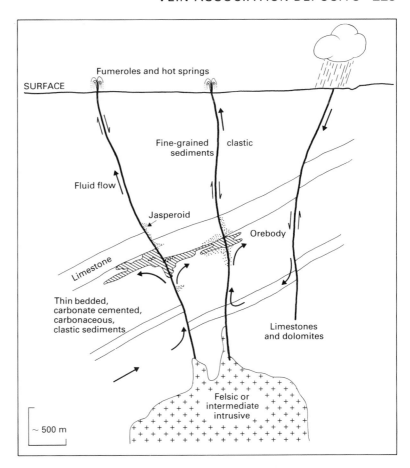

Fig. 16.11 Schema for the formation of Carlin-type deposits. (After Sawkins 1984.)

North American Craton. Sawkins (1984) suggested that at their time of formation the environment was either that of rear-arc rifting or a principal arc setting, and he favours the latter on the basis of age determinations of early to mid Tertiary. Silberman (1985) has shown that many geothermal systems in the general area of Nevada are only 0.5–3.0 Ma old, which favours the rear-arc rifting regime, such as the Taupo Volcanic Zone of North Island, New Zealand or an older analogue like the Caledonian marginal basin of Wales. Ivosevic (1984) also favours this environment. Even more difficult to place in its tectonic environment is the Cinola deposit of the Queen Charlotte Islands, British Columbia (Champigny & Sinclair 1982) which occurs in the suspect terrane of Wrangellia (Windley 1984).

Finally, it seems likely that carbon may have played a role in precipitating the gold, judging by its presence, or former presence, in many deposits;

Springer (1983) has speculated on the potential of Archaean carbonaceous schists as hosts of large low grade deposits of gold.

Unconformity-associated uranium deposits

Deposits of this type have come into prominence only during the last 15 years or so. The first to be discovered was the Rabbit Lake deposit in Saskatchewan in 1968, and the initial discoveries in the East Alligator River Field of Australia were made in 1970. As a result there is no agreed name for this deposit type, and in addition to the above title from Eckstrand (1984) they have been called *inter alia* 'unconformity vein-type deposits' and 'Proterozoic vein-like deposits.' Some are in no way vein-like, e.g. the McClean deposits (Figs 2.8, 2.9), and a name omitting the word vein is to be preferred. Since 1970 more deposits have been found and are now known to include (after Olympic Dam, South Australia) the largest (Jabiluka, Northern Territory, 200 000 t contained U_3O_8) and highest grade (Cigar Lake, Saskatchewan, 9.04% U) uranium deposits in the world. The Athabasca Basin in Saskatchewan and the Northern Territory of Australia host the main deposits so far found and these contain about half the western world's low cost uranium reserves (Dahlkamp 1984). The Key Lake Mine started up in October 1983 at a capital cost of £282 million, and, at its planned production level of 5443 t U_3O_8 p.a., its output is 12% of the total from the western world. Large deposits have also been found in the North-west Territories of Canada and in west Africa.

Orebodies of this deposit type range from very small up to more than 50 Mt (Jabiluka II) and grade from 0.3 to over 9%. They may also have many metal by-products, e.g. Jabiluka II has 15 ppm Au, Cluff Lake produces by-product gold and Key Lake contains large quantities of nickel, but when smelted the nickel contains too much uranium to be market-able. Other elements of note are Co, As, Se, Ag and Mo. The geological environment is that of middle Proterozoic, sandstone-dominated se-quences unconformably overlying older metamor-phosed Proterozoic basement rocks. Most of the mineralization occurs at or just below the uncon-formity, where it is intersected by faults passing through carbonaceous schists in the basement (Fig. 16.12). The Eagle Lake deposit is a partial exception to this generalization as its mineralization extends to about 500 m beneath the unconformity. The ore-bodies tend to be tubular (Figs 2.8, 2.9) to flattened cigar-shaped. The high grade orebodies grade out-wards into stratiform disseminations and fracture fillings and in Saskatchewan the mineralization may continue above the unconformity (Fig. 16.12). Important zones of wall rock alteration are devel-oped around the orebodies and these zones greatly broaden the exploration target. The dominant types of alteration are chloritization, argillization,

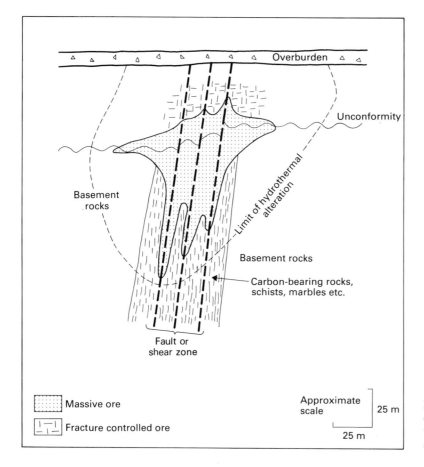

Fig. 16.12 Generalized diagram of an unconformity-associated uranium deposit. (After Clark *et al.* 1982.)

carbonitization (commonly dolomitization), silicification, pyritization and tourmalinization (Marmont 1989). A palaeoregolith is present at the unconformity below the Athabasca Group sandstone.

Although most of the Saskatchewan deposits are found along the basin rim (Fig. 16.13), the presence of the Cluff Lake Mine and other deposits in the Carswell Circular Structure, which brings the unconformity to the surface, and Cigar Lake at 400 m depth, indicates that other orebodies probably await discovery in deeper parts of the basin. The Carswell Circular Structure may have resulted from a metoritic impact or a diapiric cryptoexplosion in Cambro-Ordovician time (Pagel *et al.* 1986). The basin is now 1750 m deep at its centre but, on the basis of fluid inclusion studies, Pagel (1975) and Pagel & Jafferezic (1977) postulated that the basin sediments were once at a depth of 3500 m. These and other fluid inclusion studies indicate that both the uranium deposits and the Athabasca sandstones have been permeated by brines and CO_2-rich fluids at about 160–200°C. Stable isotopic H and O data (Pagel *et al.* 1980) suggest that these were one and the same fluid. (Similar fluids have been found in inclusions in the Australian deposits.) Isotopic studies by Bray *et al.* (1982) of the McClean Lake deposit permit a positive correlation of the $\delta^{13}C$

from graphite in the host metasediments and from siderite intergrown with uranium mineralization, suggesting that the carbon of the siderite has been derived from the graphite. Similarly, their work on sulphur isotopes showed that the sulphur in the ore zone sulphides could have been derived from the basement rocks. All the deposits have a largely epigenetic aspect, and radiometric dating at Key Lake (Trocki *et al.* 1984) shows that the mineralization is about 300 Ma younger than the Athabascan sediments and their major period of diagenesis.

All this evidence has led most geologists to postulate uranium deposition from low to medium temperature (100–300°C) geothermal systems driven by regional heating events (cf. Gustafson & Curtis 1983). The uranium is thought to have travelled as uranyl complexes [such as $UO_2(CO_3)_3^{4-}$] with reduction of U^{6+} as a result of carbonaceous material in the schists playing an important part in the mineralization process and involving reactions such as:

$$2UO_2(CO_3)_3^{4-} + 8H^+ + C \rightarrow 2UO_2 + 7CO_2 + 4H_2O.$$

The source of the uranium is still conjectural. For the Narbarlek deposit, Ewers *et al.* (1983) showed that it could have been derived from basement granite or metamorphics, or the overlying

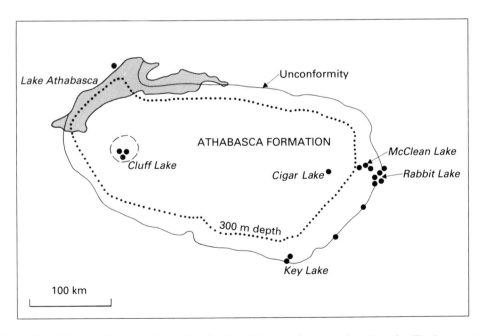

Fig. 16.13 Outline of the Athabasca Basin and distribution of the associated uranium deposits. The basement rocks outside the basin are older Proterozoic and Archaean in age. (After Clark *et al.* 1982.)

Kombolgie Formation. More recent studies embracing the Koongarra, Narbalek and Jabiluka deposits led Wilde *et al.* (1989) to conclude that the mineralizing solutions were terrestrial in origin and derived from oxidized, red bed sediments overlying the deposits. In both the Canadian and Australian fields there are uranium-rich rocks and uranium deposits present in the basement. Source rocks for the uranium are not therefore a problem in elucidating the genesis of these deposits.

With such characteristic features marking their geological environment, an exploration model for these deposits may be readily drawn up and favourable areas for exploration outlined. The student may care to try out such an exercise using the data given above and apply his model to the Canadian Shield using a 1 : 5 000 000 geological map. He can then compare his ideas with those of Clark *et al.* (1982). Of course, similar favourable areas can be outlined elsewhere and uranium deposits of this type have been found in France (George *et al.* 1983). An interesting summary of exploration for these deposits has been related by Ullmer (1985) and a useful summary of the geology of these deposits is provided by Marmont (1989).

17 / Strata-bound deposits

Introduction

The term strata-bound is applied, irrespective of their morphology, to those deposits which are restricted to a fairly limited stratigraphical range within the strata of a particular region. For example, the vein, flat and pipe lead–zinc–fluorite–baryte deposits of the Pennine orefields of Britain are restricted to the Lower Carboniferous and therefore spoken of as being strata-bound. To take a very different example, the stratiform deposits of the Zambian Copperbelt are all developed at about the same stratigraphical horizon in the Roan Series (Chapter 15) and these too may be described as strata-bound. Clearly, stratiform deposits can be strata-bound but strata-bound ores are not necessarily stratiform. Two important associations will be dealt with here: carbonate-hosted base metal deposits and sandstone–uranium–vanadium base metal deposits.

Carbonate-hosted base metal deposits

These are important producers of lead and zinc and also, sometimes principally, of fluorite and baryte. Copper is important in some fields, notably that of central Ireland.

Distribution in space and time

Most of the lead and zinc produced in Europe and the USA comes from this type of deposit. In Europe there are important fields in central Ireland, the Alps, southern Poland and the Pennines of England. In the USA there are the famous Appalachian, Tri-State (south-west Missouri, north-east Oklahoma and south-east Kansas), South-east Missouri and Upper Mississippi districts (the locations of some of these fields are shown on Fig. 23.15). There are also important fields in north Africa (Tunisia and Algeria) and Canada.

Few deposits of this type have as yet been located in the Precambrian. Minor Mississippi Valley-type mineralization, at least 2.3 Ga old, occurs in the Transvaal but the Gayna River deposit, Northwest Territory of Canada, has > 50 Mt running up to 10% combined metals. It is however, only 0.9 Ga old. Substantial deposits first appear in the Cambrian of south-east Missouri and important deposits occur in all systems, except the Silurian, up to the Cretaceous, as can be seen from the following examples.

Cambrian: South-east Missouri (Old and New Lead Belts); Ediacara, South Australia; Sardinia; Spain

Ordovician: Upper Mississippi

Devonian: Pine Point, Northwest Territories, Canada; Lennard Shelf area, Western Australia

Carboniferous: central Ireland; British Pennines; Tri-State; Illinois–Kentucky

Permian: Trento Valley, Italy

Triassic: Eastern Alps (Austria and northern Yugoslavia)

Jurassic: Southern Poland (also in Triassic and Devonian)

Cretaceous: Northern Algeria and Tunisia.

Environment

We are not concerned in this group of deposits with skarns and the skarn environment, but with deposits which occur mainly in dolomites and, to a lesser extent, in limestones. The most important feature is the presence of a thick carbonate sequence because thin carbonate layers in shales seldom contain important deposits of this type (Sangster 1976). The fauna and lithologies of the limestone and dolomite hosts show that they were mostly formed in shallow water, near shore environments of warm seas peripheral to intracratonic (or craton centre) basins, and a plot of major carbonate-hosted deposits on palaeolatitude maps shows a grouping of these deposits in low latitudes (Dunsmore & Shearman 1977). Some ore districts, e.g. Nanisivik, Canada and the Eastern Alpine District straddling Austria, Italy and Yugoslavia occur in former rift zones and others occur close to carbonate-shale facies changes, e.g. Pine Point, Canada, suggesting a near shelf-edge setting. The warmer climates of low latitudes encourage the development of reefs and so the frequent, but by no means universal, association of these deposits with reefs (e.g. South-east Missouri) and carbonate mudbanks (e.g. Ireland) is not surprising. The occurrence of reefs and carbonate mudbanks is related to ancient shorelines and sea bed topographies. Nowadays, along the shorelines where carbonate deposition is occurring we often find arid zones, as in the Persian Gulf, with desert-prograding supratidal flats or coastal sabkhas. There, gypsum and anhydrite are precipitated from marine-derived ground waters to form evaporites. These may be of considerable significance. The isotopic composition of the sulphur or sulphides from a number of carbonate-hosted deposits suggests origination from sea water sulphate, particularly sea water of the same age as the limestone country rocks. Sulphate evaporites are known to be interbedded with the limestones in relatively close regional proximity to many carbonate-hosted deposits (Dunsmore & Shearman 1977). Thus, as Stanton (1972) has emphasized, the primary regional control of such deposits is palaeogeographical.

Environments such as those described above developed in the past along the margins of marine basins that formed in stable cratonic areas, such as the Devonian Elk Point Basin of western Canada in which the Pine Point deposits occur. Carbonate-hosted lead–zinc deposits also occur in a very different environment—in the failed arms (aulacogens) of the triple junctions of rifted continental areas, as in the Benue Trough and the Amazon Rift Zone (Fig. 23.8), and on the flanks of embryonic oceans, as down the flanks of the Red Sea and the Tethys Ocean (Alps). In this environment too there is an important development of evaporites. A negative but important point is that these lead–zinc deposits are remote from post-host rock, igneous intrusives which might be the source for mineralizing solutions.

Returning to the cratonic basin–shelf sea environment we may note other regional controls, which are well exemplified by the Mississippi Valley region and the British deposits. In these regions, the orefields are present in positive areas of shallow water sedimentation and separated from each other by shale-rich basins. Such positive areas in the British Isles are often underlain by older granitic masses, and Evans & Maroof (1976) suggested that these very competent rocks fractured easily to produce channelways for uprising solutions which, on reaching the overlying limestones, gave rise to the mineralization. In addition, a large number of deposits are clearly related spatially to faults, some-times of a regional character (Fig. 5.7) up which the ore solutions may have passed.

Classification

Unfortunately there is no general agreement concerning the terminology to be used in describing these low temperature, carbonate-hosted deposits. Some authors include them all under the term Mississippi Valley-type (MVT) deposits. In 1976 Sangster divided carbonate-hosted, base metal deposits into two major types: (a) Mississippi Valley, and (b) Alpine. More recently (Sangster 1990) he appears to have abandoned this division but at the same time to have taken such a restrictive stand in deciding which deposits should be designated as Mississippi Valley-type that he wrote, 'Mississippi Valley-type deposits are, for the most part, a North American phenomenon'. Thus although he included the Silesian and Alpine Fields, he excluded the fluorite-rich deposits of Illinois and Kentucky, and the English Pennines, on account of their high fluorite and baryte content and the dominance of fracture control of the orebody locations. The present writer prefers to use the term Mississippi Valley-type to include all low temperature deposits containing one or more of the minerals baryte, fluorite, galena and sphalerite, that are carbonate-hosted *and epigenetic* in origin. Thus, as many authors do, I keep the epigenetic, fracture-controlled deposits, such as those of the English Pennines, in the Mississippi Valley class, but assign the syngenetic, carbonate-hosted deposits, such as Navan, Silvermines and Tynagh in Ireland, to the sedex class (pp. 197–202). An important feature that may help in deciding whether a deposit is of MVT or epithermal type is the nature of the fluid inclusions (Roedder & Howard 1988). In MVT deposits these most commonly display both homogenization temperatures in the range 100–150°C (although in some districts temperatures range up to about 200°C, e.g. the Alston Block of the English Pennine orefields, but not in the other Pennine orefields—Atkinson *et al.* 1982) *and* high salinities, 15–25 wt % NaCl equiv. In addition methane gas and droplets of immiscible oil are frequently present.

Russell & Skauli (1991) have proposed a different classification based on the enthalpy involved in the mineralizing systems that they believe are involved in the formation of most if not all carbonate-hosted, base metal deposits. (Enthalpy is a thermodynamic state function defined as $H = U + PV$, where U is the internal energy of a system and *P* and *V* are the pressure and volume respectively.) Thus their classification is very much dependent upon the particular genetic system that is considered to have given rise to the deposit or deposits under consideration. They suggested three classes. (a) A low enthalpy or Mississippi Valley-type, exemplified by most of the central North American Pb–Zn fields and formed from an evolved basinal brine which has been forced towards a basin margin by compaction and a hydraulically controlled head of water (as in Fig. 4.7b). (b) A medium enthalpy or Irish-type, exemplified in particular by the large Pb–Zn deposits of central Ireland and generated by modified, saline sea water convected within the upper crust as shown in Fig. 15.10. (c) A high enthalpy or Rosiclare–Pennine-type, in which the mineralizing fluid is basinal brine augmented by meteoric water that is circulating in convection cells driven by a nearby magmatic intrusion. In this type, magmatic fluorine is postulated as the source of the large quantities of this element present in Rosiclare–Pennine type orefields. (Rosiclare is another name for the fluorine-rich Illinois–Kentucky Field.) As there is some fluorite in the Alpine deposits, Russell & Skauli are inclined to include the Alpine-type in this third class. So, like Sangster, they exclude this last class of deposits from those they would designate Mississippi Valley-type *sensu stricto*. The reader will have to make his own decision as to which deposits he regards as Mississippi Valley-type; let us hope he does not go to the extreme to which one worker went, of proposing that only those deposits that contain J-type lead (see below) should be called Mississippi Valley-type!

The term Irish-type must be remarked on as it can be misleading if used without some knowledge of Irish mineralization. For in Ireland there is a complete spectrum from thoroughly epigenetic types in fracture-controlled Pb–Zn orebodies, some carrying baryte, through typical breccia-hosted bodies, like many in the Mississippi Valley itself (e.g. Harberton Bridge), to the big syngenetic, sedex-type deposits (Navan, Silvermines, Tynagh, etc.) as pointed out in Evans (1976a).

A number of workers, e.g. Rickard *et al.* (1975) and Sawkins (1984) have suggested that the sandstone-hosted, lead deposit type exemplified by Laisvall, Sweden and Largentière, France are Mississippi Valley-type deposits. Bjrlykke & Sangster (1981), however, argued that these, and similar deposits in other parts of the world, are sufficiently different to be considered as constituting a distinct and separate deposit type. Nevertheless Sangster (1990) wrote that the distinction is rather blurred, as in the South-east Missouri Field where the lower portions of several carbonate-hosted orebodies lie within the underlying Lamotte Sandstone.

There is thus a great deal of confusion as to what are the common properties of Mississippi Valley-type deposits and Sangster (1983b) suggested that this is because we are trying to classify the unclassifiable, that the differences between individual ore districts outweigh the similarities, both numerically and in significance.

Orebody types and situations

As has been made clear by the above discussion, the orebodies are very variable in type. In the British Pennines, vein orebodies with ribbon ore shoots occupying normal faults are the main deposit type in the northern field (Fig. 2.4). In the southern Pennines, veins are again the most important orebodies but there they occupy tear (wrench) faults. The orebodies in the Tri-State Field are in solution and collapse structures, caves and underground channelways connected with karst topography and these are common features in many ore districts. They arise from the fact that the host rocks formed in very shallow waters, such that even minor uplifts led to elevation above sea level, resulting in the common development of disconformities, nonconformities and the associated subaerial weathering that produced the solution-collapse breccias that host so many orebodies. At Pine Point, the ores are in interconnected small-scale solution cavities, which Dunsmore (1973) has suggested are the result of the dissolution of carbonate rocks by corrosive fluids generated by a reaction between petroleum and sulphate ion. Some of the geological situations in which these deposits occur are shown in Fig. 17.1. They may be listed as follows.

1 Above unconformities in environments, such as permeable reefs and facies changes (A-1); supratenuous folds (A-2); above the pinch-outs of permeable channelway horizons (A-3); above or in mudbank complexes (A-4).

2 Below unconformities in solution-formed open spaces (caves, etc.) related to a karst topography predating the unconformity (B-1); or in collapse structures formed by the dissolution of underlying beds by subsurface drainage (B-2 and B-3).

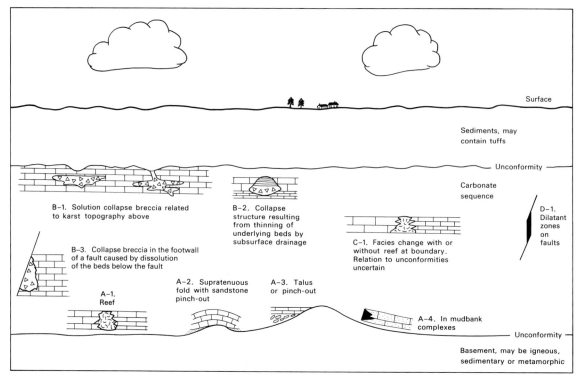

Fig. 17.1 Idealized vertical section illustrating the range of geological situations in which carbonate-hosted base metal deposits are known to occur. (Modified from Callahan 1967.)

3 At a facies change in a formation, or between basins of deposition (C-1).

4 In regional fracture systems (D-1).

Grade, tonnage and mineralogy

Average ore grades range mainly from 3 to 15% combined Pb + Zn with individual orebodies running up to 50%. Tonnages generally range from a few tens of thousands up to 20 Mt—see Fig. 15.8 for a grade–tonnage plot for these deposits. Lead and/or zinc are the elements that commonly determine economic viability. In a few mines silver and copper are important by-products, as are cadmium and germanium. Fluorite and/or baryte may be important by-products or the prime products that are mined. From Fig. 15.8 it can be seen that the majority of MVT deposits are < 10 Mt in size, but orefield totals, made up of a number of deposits, often lie in the range 50–500 Mt, e.g. Pine Point with over eight separate orebodies and the Upper Mississippi Valley with nearly 300. Most MVT orefields have a high Zn/Pb ratio and a few produce very little or no lead. Lead-dominated fields are rare and limited largely to the very productive Old and New Lead Belts of the South-east Missouri District (the New Lead Belt is now referred to more commonly as the Viburnum Trend), unless one includes the Laisvall–Largentière type deposits as suggested by Rickard, Sawkins and other workers.

The characteristic minerals of this ore association are galena, sphalerite, fluorite and baryte in different ratios to one another varying from field to field. Pyrite and especially marcasite may be common and chalcopyrite is important in a few deposits. Calcite, dolomite, other carbonates and various forms of silica usually constitute the main gangue material. Colloform textures are common in some ores, otherwise these ores are commonly, but not invariably, coarse-grained. High trace amounts of nickel seem to be characteristic of these ores (Ixer & Townley 1979).

Isotopic characteristics

Sulphur

Sulphur isotopic abundances have been studied in both the sulphide and sulphate minerals. These tend to vary from field to field (Fig. 17.2) but generally show a range of positive $\delta^{34}S$ values. This range may be explained in terms of fractionation as a function of mineral species, temperature or chemical environment or by the mixing of sulphur from different sources (Heyl *et al.* 1974). These authors have suggested that a comparison with values for crustal rocks (Fig. 17.2) indicates a crustal source for the sulphur of these deposits. Thus, the Pine Point values suggest that the sulphur was derived from marine evaporites (Rye & Ohmoto 1974) but the low values of the Cave-in-Rock District were thought by Richardson *et al.* (1988) to indicate a significant contribution during mineralization of H_2S from petroleum and possibly igneous or crustal sulphur from the basement.

Lead

Like sulphur, lead isotopic abundances are variable in nature. This variation results from the addition to existing lead of varying amounts of ^{206}Pb, ^{207}Pb and ^{208}Pb produced by the radioactive decay of uranium and thorium over geological time. ^{204}Pb is a stable isotope, possibly the decay product of a now unknown primordial radioelement. As ^{204}Pb amounts have remained unchanged, isotopic ratios are usually stated by reference to it, e.g. $^{206/204}Pb$, $^{207/204}Pb$ and $^{208/204}Pb$. Two distinct categories of lead, ordinary lead and anomalous lead, have been recognized. Ordinary lead has isotopic ratios that increase steadily with time so long as it remains in uniform source rocks (probably the mantle) in contact with constant amounts of uranium and thorium. Once it has been removed from its source and separated from the elements producing radiogenic lead ($^{206-208}Pb$), its isotopic composition is fixed and if a mathematical model is assumed for the rate of addition of radiogenic leads in the source region then a model lead age can be calculated. Generally, such ages are in reasonable agreement with other radiometric age determinations, but sometimes they are grossly incorrect and such leads are defined as anomalous. Some give negative ages, i.e. the model lead age says that they have not yet been formed! These leads must have had a more complicated history, presumably within the crust, during which they acquired extra amounts of

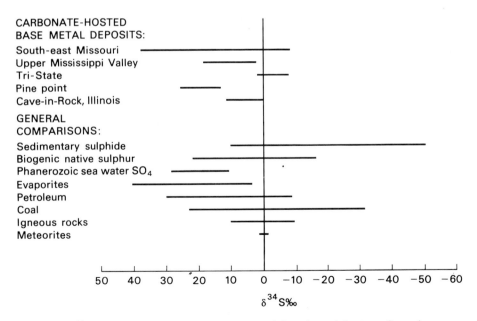

Fig. 17.2 The range of $\delta^{34}S$ for some carbonate-hosted, base metal deposits and the range for major sources of sulphur that could have contributed to the ore deposits. (Modified from Heyl *et al.* 1974.)

radiogenic lead. They are sometimes called J-type leads after Joplin, Missouri from where the first examples to be found were collected. Other leads, on a simple mathematical model, are older than the rocks in which they occur. These could be leads which were first removed from the mantle, then 'stored' in some older rocks before being remobilized and redeposited in younger rocks, e.g. the Silesian and Alpine deposits.

Leads from the various fields of the Mississippi Valley have been found to be notably enriched in radiogenic lead (i.e. having $^{206}Pb/^{204}Pb$ ratios of 20 or greater) compared to ordinary lead. They are all J-type leads. The strange thing is that although J-type lead is ubiquitous in the Mississippi Valley fields, most similar carbonate-hosted deposits, such as those of Pine Point, the British Pennines, central Ireland and southern Poland, contain ordinary lead. These facts suggest that the Mississippi Valley lead was derived from a crustal source relatively high in uranium and thorium, which could have provided it with anomalous amounts of radiogenic lead. A highly probable source of this nature would be the Precambrian basement (Heyl *et al.* 1974) because of the conspicuous similarity between the source age of the lead, as determined from the slope of the lead–lead isochron, and the accepted age of the basement rocks (1450 Ma) in the Upper Mississippi Valley district. On the other hand, work in the South-east Missouri district suggests that all the lead may have been derived by hydrothermal solutions passing through the permeable Cambrian Lamotte Sandstone.

The Upper Mississippi Valley district not only contains the most radiogenic lead yet known from a carbonate-hosted deposit anywhere in the world, it also shows an isotopic zoning (Fig. 17.3). The lead becomes progressively more radiogenic from west to east. Heyl *et al.* (1966) have pointed out that this could be produced by addition of highly radiogenic lead leached from the basement to more normal lead carried by basinal brines moving up dip into the district from the neighbouring Forest City and Illinois Basins (Fig. 17.3).

The fact that the lead in other carbonate-hosted ore districts (Pine Point, etc.) is ordinary lead has caused most workers to postulate a deep-seated uniform source, and it has been pointed out that there are a number of profound faults in the nearby Canadian Shield, which, when projected south-westwards, pass under the ore zone at Pine Point.

Apropos the Heyl *et al.* suggestion of basinal brine

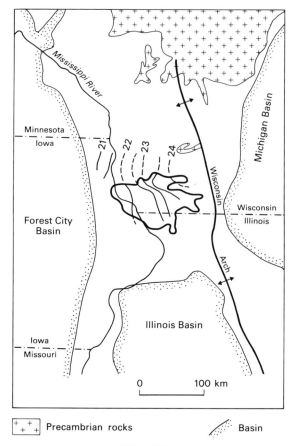

Precambrian rocks   **Basin**

Fig. 17.3 Variation of $^{206}Pb/^{204}Pb$ ratios in the Upper Mississippi Valley Mining District and the location of neighbouring sedimentary basins. (After Heyl *et al.* 1966.)

mineralizers, it should be noted that further south in the Mississippi Valley region, palaeomagnetic evidence shows that the mineralization in the fields 1–6 on Fig. 23.15 all occurred during a short interval of time from the late Pennsylvanian to early Permian (Symons & Sangster 1991). This provides strong evidence for the suggestion of Leach & Rowan (1986) that hot brines migrated northwards out of the Arkoma Basin of Arkansas as uplift took place during the Pennsylvanian to Permian Ouachita Orogeny. This was such a relatively brief time span as to indicate that the expulsion of brines from the basin was not a continual process induced by sediment loading and compaction (Fig. 4.7a), nor a continual flushing of the source rocks by gravity driven meteoric water flowing from a hydraulic head in an adjacent highland (Fig. 4.7b), unless the mineralization resulted only from the first flushing of the Arkoma Basin when the initial uplift and formation of the hydraulic head occurred (Symons & Sangster 1991).

Origin

There is little doubt that the majority of deposits of this class have been formed from epigenetic hydrothermal solutions. Some of the mechanisms of transportation and deposition of lead and zinc in hydrothermal solutions have been discussed in Chapter 4. The source(s) of these solutions and their metallic constituents is very problematical. As there is generally no obvious spatial association with igneous intrusions that could be regarded as ore fluid sources, and the work of Möller (1985) and Rankin & Graham (1988) indicates temperatures of below 270°C for the source regions of these fluids, then the mineralizing fluids must be either formation water or heated circulating surface water. For many fields outside the USA, lead isotopic studies suggest a deep source for the metals but in some cases basinal brines may have played an important role. Jackson & Beales (1967) and many others (see Chapter 4) have argued strongly for such a mechanism and this may have been the case for the Mississippi Valley districts. Again, the source of sulphur may have been partially deep-seated (Cave-in-Rock?) or from marine evaporites (Pine Point?) or from sea water.

Clearly Sangster (1983b) is right in recommending that we look at the differences between these deposits and do not try to force them into a straitjacket or look for one unique model for their formation. As the reader can discern from Chapter 4, we have today four general models for the genesis of carbonate-hosted lead–zinc deposits:

1 transport of the metals as bisulphide complexes, one fluid carrying metals and sulphur with precipitation by boiling, cooling by contact with ground water, etc.;
2 transport of metals as chloride complexes (more favoured hypothesis) and precipitation when this solution meets one carrying H_2S—mixing model;
3 transport of metals as chloride complexes and sulphur as sulphate in the same solution, precipitation when the sulphate is reduced by encountering organic material;
4 organometallic complexes as carriers of the metals, H_2S in the same solution with precipitation by cooling.

Much more research will have to be carried out before we can decide whether one general model is applicable or whether we have a number of processes at work with different combinations being operative during the genesis of different deposits.

Sandstone uranium–vanadium-base metal deposits

These deposits are found in terrestrial sediments, frequently fluviatile, which were generally laid down under arid conditions. As a result, the host rocks are often red in colour and for this reason copper deposits of this type are commonly referred to as 'red bed coppers'. Uranium-rich examples are called Colorado Plateau type, carnotite type, Wyoming roll-front type, Wyoming geochemical-cell uranium type, western states type, or sandstone-uranium type. The last term is now in common use.

In deposits of this type one or two metals are present in economic amounts, whilst the others may be present in minor or trace quantities. Thus copper mineralization (with chalcocite, bornite and covellite) is widespread in red bed successions though it is not often up to ore grade (pp. 35–36) and the same applies to silver and lead–zinc mineralization. Uranium mineralization ($\pm$vanadium) may be accompanied by trace amounts of the above metals but usually occurs as separate deposits.

Sandstone–uranium-type deposits

Uranium deposits of this general type are widespread in the USA and they have provided over 95% of its domestic production of uranium and vanadium. They can be divided into four classes (Adams 1991) but there is only space for a generalized view in this chapter. From the global point of view they probably constitute a quarter of the non-communist world's reserves, and they are now known in many parts of the world (Dahlkamp 1978). In the USA, these deposits are well developed in the Colorado Plateau region and in Wyoming (Fig. 17.4).

Metals occurring in these deposits in significant quantities are: uranium, vanadium, copper, silver, selenium and molybdenum. A deposit may contain any one or more of these metals in almost any combination except vanadium and copper, which are usually mutually exclusive. Amounts of uranium, vanadium and copper vary enormously within and between deposits and many orebodies fluctuate so much in grade that a single overall

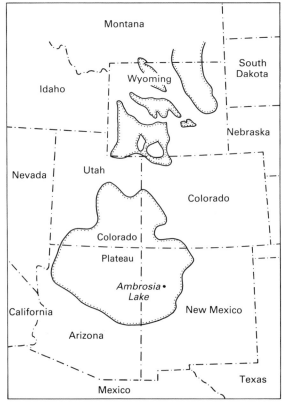

Fig. 17.4 Map showing the Colorado Plateau and Wyoming Basins. (After Rackley 1976.)

average figure is not informative. Generally, grades vary from 0.1 to 1% U_3O_8, but locally can be much higher, with such phenomena as whole tree trunks entirely replaced by uranium minerals. The usual

range in mineral deposits is 1000 to 10 000 t of contained U in ores grading 0.1–0.2% U. Many deposits contain less than 1000 t U, but some contain more than 30 000 t U. Some American deposits carry up to 1.5% V_2O_5 and some up to 0.2% Mo, but Mo in many cases is deleterious (Eckstrand 1984).

Most of the orebodies are similar. Small irregular pods are common and are sporadically distributed within a favourable rock unit. The larger deposits form mantos hundreds of metres long, about 100 m wide and a few metres thick; these may also be referred to as tabular bodies. They can be mined by open pit or underground methods whilst the smaller bodies can be exploited by *in situ* leaching. The elongate orebodies follow buried stream courses or lenses of conglomeratic material. The most common forms of deposit are termed (a) roll-front (Fig. 17.9); (b) blanket or peneconcordant (Fig. 17.5a); and (c) stack or tectolithological (Fig. 17.5b), which are often related to permeable fault zones. These different morphologies can be related to the flow of mineralizing fluids through the host rock. The deposits are epigenetic, in the sense that they were formed in their present position after the host sediment was deposited—how much later is very debatable. The typical orebody represents an addition of less than 1% of ore minerals, which are accommodated in pore spaces where they form thin coatings on the detrital grains, whilst, in the case of high grade deposits, they may entirely fill the pore spaces. The disseminated form and microscopic size of the ore minerals increase the susceptibility to subsequent oxidation and remobilization by both alteration and weathering. The principal primary

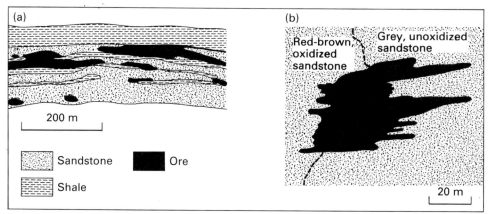

Fig. 17.5 Cross sections through two forms of sandstone uranium-type deposits in the Ambrosia Lake Field, New Mexico. (a) Blanket or peneconcordant. (After Dahlkamp 1978.) (b) Stack or tectolithological. (After Dixon 1979.)

uranium minerals are pitchblende and coffinite $((USiO_4)_{1-x}(CH)_{4x})$. Vanadium, if present, is generally in the form of roscoelite (vanadium mica) and montroseite $(VO(OH))$.

Sedimentological studies have shown that the usual immediate host rocks of these ores are fossil stream channel deposits. These consist of linear formations of permeable sandstone and conglomeratic sandstone enclosed by relatively impermeable rocks—shales, mudstones, etc. (Figs 17.6–17.8). During deposition, climatic conditions were warm to hot and seasonably humid. Abundant vegetation grew in the depositional area and animals burrowed and mixed dead vegetation with the sediment. Frequent reworking by the streams incorporated

sufficient organic material into the sediment to produce reducing conditions when it decayed. The sands deposited under these conditions were organic-rich, pyritic, light to dark grey, and the associated clays were light to medium grey or green, pyritic and commonly carbonaceous. Petrographic studies show that the sediment was often derived from a granitic source area and during weathering of the granite its trace content of uranium would be oxidized to the hexavalent state and taken into solution. This uranium would migrate through the basin of deposition to be lost in the sea unless it came into contact with reducing conditions in the organic-rich sediment a short distance beneath the sediment–water interface, in which case it would

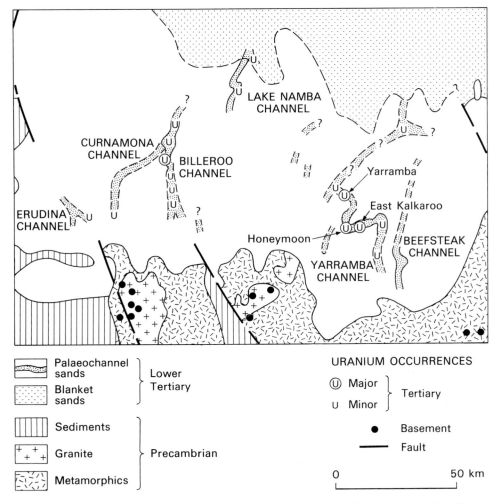

Fig. 17.6 Fossil stream channels with uranium mineralization in the Tertiary of South Australia. Note the many uranium occurrences in the basement which may have been the source of the uranium. (After Brunt 1978.)

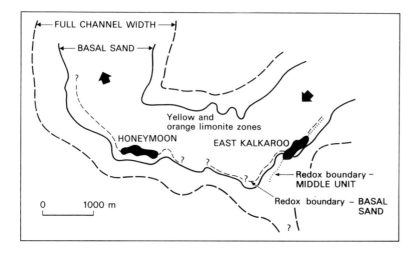

Fig. 17.7 Part of the Yarramba channel showing the position of two major uranium occurrences. Both deposits occur in embayments along the same channel bank. Patchy ore grade mineralization is also present along the redox boundaries away from the deposits. The opposite bank is barren as the yellow limonitic oxidation zone extends up to the pinchout of the sand units. (After Brunt 1978.)

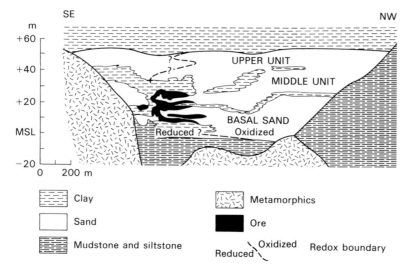

Fig. 17.8 Cross section across the Yarramba channel showing the East Kalkaroo deposit. Note the roll-front configuration of the ore. (After Brunt 1978.)

enrich the sediment (Rackley 1976). In some areas acid tuffs rather than granites appear to have been the source rocks for the uranium, and the devitrification of tuffaceous sediments may well be the major source of the uranium in the US deposits (Adams 1991). As the oxygen-rich waters encroached upon the reducing environment an irregular tongue-shaped zone of oxidized rock was formed. The interface, or redox boundary, between the oxidized and reduced rocks has, in cross section, the shape of a crudely crescentic envelope or roll, the leading edge of which cuts across the host strata (Fig. 17.9) and points down dip towards reduced ground that still contains authigenic iron disulphides. The

reduced ground is generally grey in colour, the oxidized ground is drab yellow to orange or red owing to the development of limonite and hematite by the alteration of the sulphides. Upon encountering reducing conditions, the uranium became reduced to the insoluble tetravalent state and was precipitated. Continuous or episodic introduction of oxygenated ground water resulted in continuous or episodic solution and redeposition of uranium and migration of the redox interface down the palaeoslope. This process can lead to ore grade accumulation at or near the concave edge of the roll and, to a lesser extent, in reduced rock near the upper and lower limbs of the roll. Later reduction or

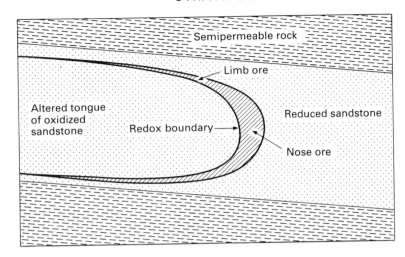

Fig. 17.9 Idealized cross section of a roll-front uranium deposit. (After Reynolds & Goldhaber 1978.)

oxidation of the ore beds may materially alter the form and mineralogy of the orebodies and obscure the primary redox relationships. The small amount of fluid inclusion work that has been carried out on these deposits suggests that the associated calcite and baryte were formed at temperatures a little under 100°C (Poty & Pagel 1988).

Useful summary discussions of this deposit type can be found in Tilsley (1981), Nash *et al.* (1981), Northrop & Goldhaber (1990) and Adams (1991).

18 / Sedimentary deposits

Broadly speaking, sediments can be divided into two large groups, allochthonous deposits and autochthonous deposits. The allochthonous deposits are those which were transported into the environment in which they are deposited and they include the terrigenous (clastic) and pyroclastic classes. The autochthonous sediments are those which form within the environment in which they are deposited. They include the chemical, organic and residual classes. Table 18.1 shows the relationship between these various terms.

Some sediments are sufficiently rich in elements of economic interest to form orebodies and examples from both sedimentary groups will be described in this chapter. The student must realize, however, that the total range of sedimentary material of economic importance is much greater than can be included in this small volume although some more will be covered in Chapters 20 and 21. The residual deposits will be dealt with in the next chapter together with other ores in whose formation weathering has played an important role.

Allochthonous deposits

Allochthonous sediments of economic interest are usually referred to by ore geologists as mechanical accumulations or placer deposits. They belong to the terrigenous class formed by the ordinary sedimentary processes that concentrate heavy minerals. Usually this natural gravity separation is accomplished by moving water, though concentration in solid and gaseous mediums also may occur. The dense or heavy minerals so concentrated must first be freed from their source rock and must possess a high density, chemical resistance to weathering and mechanical durability. Placer minerals having these properties in varying degrees include: cassiterite, chromite, columbite, copper, diamond, garnet, gold, ilmenite, magnetite, monazite, platinum, ruby, rutile, sapphire, xenotime and zircon. Since sulphides break up readily and decompose they are rarely concentrated into placers. There are, however, some notable Precambrian exceptions (perhaps due to a non-oxidizing atmosphere) and a few small recent examples.

Placer deposits have formed throughout geological time, but most are of Tertiary and Recent age. The majority of placer deposits are small and often ephemeral as they form on the earth's surface usually at or above the local base level, so that many are removed by erosion before they can be buried. Most placer deposits are low grade, but can be exploited because they are loose, easily worked materials which require no crushing and for which relatively cheap semi-mobile separating or hydraulic mining plants can be used. Mining usually takes the form of dredging, about the cheapest of all mining methods. Older placers are likely to be lithified, tilted and partially or wholly buried beneath other lithified rocks. This means that exploitation costs are much higher and then the deposits, to be economic, must be of high grade or contain unusually valuable minerals such as gold, as in the Precambrian Witwatersrand of RSA described below, and diamonds; examples of which are the Cretaceous

Table 18.1 A classification of sedimentary rocks

Group	Class
I. Allochthonous sediments	*Terrigenous deposits*—clays, siliclastic sands and conglomerates *Pyroclastic deposits*—tuffs, lapillituffs, agglomerates, volcanic breccias
II. Autochthonous sediments	*Chemical precipitates*—carbonates, evaporites, cherts, banded iron formation, ironstones, phosphorites *Organic deposits*—coal, lignite, oil shales *Residual deposits*—laterites, bauxites

diamondiferous conglomerate near Estrela Do Sul, Brazil (Anon. 1988a) and the Max Resources' prospect at Nullagine, 200 km south-west of Port Hedland, Western Australia. The latter consists of a Tertiary conglomerate 2–3 m thick and 2–3 m beneath the surface, that has a recoverable grade of 0.23 ct m^{-3} with approximately 60% of the stone being of gem quality (Anon. 1989a).

The world wide distribution of placer deposits is very much a product of variation (both at present and in the recent geological past) of geomorphological processes acting at the earth's surface, provided primary sources are present. For example, in semi-arid to arid morphogenetic zones, fluvial processes may be more effective in liberating and transporting heavy minerals than in reworking and concentrating them to economic grades. These processes must therefore be carefully analysed before choosing areas for exploration (Sutherland 1985).

Placers can be classified in various ways but in this book the simple, traditional, genetic classification shown in Table 18.2 will be used. The traditional usage is to be found in Lindgren (1922), McKinstry (1948), Bateman (1950), Routhier (1963), Lamey (1966), Jensen & Bateman (1979) and many other textbooks. The different usage introduced by Macdonald (1983), which is very likely to be an influential book, has been adopted by Edwards & Atkinson (1986); so we are probably entering a period of confusion and it remains to be seen which set of terms becomes the standard usage of the future. Colluvial is not altogether a happy choice of term as for many writers it implies accumulation at the *base* of a cliff or slope, and it is often used as a synonym for talus. Anyone who has consulted the Oxford English Dictionary would eschew its usage forthwith!

Residual placers

These accumulate immediately above a bedrock source (e.g. gold or cassiterite vein) by the chemical decay and removal of the lighter rock materials and they may grade downwards into weathered veins as in some tin areas of Shaba. In residual placers chemically resistant light minerals (e.g. beryl) may also occur.

Residual placers only form where the ground surface is fairly flat; when a slope is present, creep will occur and eluvial placers will be generated (Fig. 18.1). Residual placers formed over carbonatites are important as producers of apatite, e.g. at Jacupiranga, Brazil; Sokli, Finland (Notholt 1979) and Sukulu, Uganda (Reedman 1984). They are sources and potential sources of niobium, zircon, baddeleyite, magnetite and other minerals. These residual placers have often formed on carbonatites which are themselves subeconomic.

Eluvial placers

These are formed upon hill slopes from minerals released from a nearby source rock. The heavy minerals collect above and just downslope of the source and the lighter non-resistant minerals are dissolved or swept downhill by rain wash or are blown away by the wind. This produces a partial concentration by reduction in volume, a process which continues with further downslope creep. Obviously, to yield a workable deposit this incomplete process of concentration requires a rich source. In some areas with eluvial placers, the economic material has accumulated in pockets in the bedrock surface, e.g. cassiterite in potholes and sinkholes in marble in Malaysia.

Table 18.2 A classification of placer deposits

Mode of origin	Class (traditional usage)	Usage in Macdonald (1983)
Accumulation *in situ* during weathering	Residual placers	Eluvial
Concentration in a moving solid medium	Eluvial placers	Colluvial
Concentration in a moving liquid medium (water)	Stream or alluvial placers Beach placers Offshore placers	Fluvial Strandline Marine placers
Concentration in a moving gaseous medium (air)	Aeolian placers	Desert or coastal aeolian

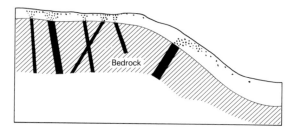

Fig. 18.1 The formation of residual (left) and eluvial (right) placer deposits by the weathering of cassiterite veins.

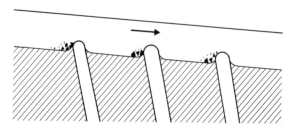

Fig. 18.2 Quartzite ribs interbedded with slate serving as natural riffles for the collection of placer gold.

Stream or alluvial placers

These were once the most important type of placer deposit and primitive mining made great use of such deposits. The ease of extraction made them eagerly sought after in early as well as in recent times and they have been the cause of some of the world's greatest gold and diamond rushes.

Our understanding of the exact mechanisms by which concentrations of heavy minerals are formed in stream channels is still incomplete. Rubey (1933) considered fall velocity to be the most important segregating mechanism and Rittenhouse (1943) used the concept of hydraulic equivalence to explain heavy mineral concentrations. Brady & Jobson (1973), however, showed that fall velocities are of little importance and they found that bed configuration and grain density are the most important factors.

It is well known that the heavy mineral fraction of a sediment is much finer grained than the light fraction (Selley 1976). There are several reasons for this. Firstly, many heavy minerals occur in much smaller grains than do quartz and feldspar in the igneous and metamorphic rocks from which they are derived. Secondly, the sorting and composition of a sediment is controlled by both the density and size of the particles, known as their hydraulic ratio. Thus a large quartz grain requires the same current velocity to move it as a small heavy mineral. Clearly, if we have a very rapid flow all grains of sand grade will be in motion, but with a slackening of velocity, the first materials to be deposited will be large heavy minerals, then smaller heavy minerals, plus large grains of lighter minerals. If the velocity of the transporting current does not drop any further then a heavy mineral concentration will be built up. For this reason such concentrations are developed when we have irregular flow and this may occur in a

number of situations—always provided a source rock is present in the catchment area.

The first example is that of emergence from a canyon. In the canyon itself net deposition is zero. As the stream widens and the gradient decreases at the canyon exit, any heavy minerals will tend to be deposited and lighter minerals will be winnowed away. Again, where we have fast-moving water passing over projections in the stream bed, the progress of heavy minerals may be arrested (Fig. 18.2). Waterfalls and potholes form other sites of accumulation (Fig. 18.3) and the confluence of a swift tributary with a slower master stream is often another site of concentration (Fig. 18.4) (Best & Brayshaw 1985). Most important of all, however, is deposition in rapidly flowing meandering streams. The faster water is on the outside curve of meanders and slack water is opposite. The junction of the two, where point bars form, is a favourable site for deposition of heavy minerals. With lateral migration of the meander (Fig. 18.5), a pay streak is built up which becomes covered with alluvium and eventually lies at some distance from the present stream channel.

Obviously, placer deposits do not form in the meanders of old age rivers because current flow is too sluggish to transport heavy minerals. In the upper reaches, current flow may be too rapid and

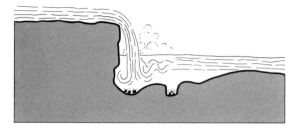

Fig. 18.3 Plunge pools at the foot of waterfalls and potholes can be sites of heavy mineral accumulations.

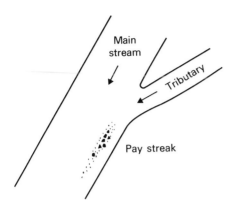

Fig. 18.4 A pay steak may be formed where a fast-flowing tributary enters a master stream.

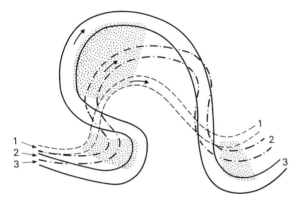

Fig. 18.5 Formation of pay streaks (dotted) in a rapidly flowing meandering stream with migrating meanders. 1 = original position of stream; 2 = intermediate position; 3 = present position. Note that pay streaks are extended laterally and downstream. Arrows indicate direction of water flow.

there may also be a lack of source material. The middle reaches are most likely to contain placer deposits where we have graded streams in which a balance has been achieved between erosion, transportation and deposition. Gradients measured on a number of placer gold and tin deposits average out at a little under 1 in 175.

Much of the above discussion hinges on the concept of hydraulic equivalence in situations where the mineral grains are being transported either in a series of saltation leaps or in suspension. Reid & Frostick (1985) have pointed out, however, that in most fluvial *and littoral marine* situations the transport of sand and larger particles is largely as part of

a traction carpet, in which case settling equivalent is unimportant. They consider that the important processes are (a) entrainment equivalence—larger light mineral grains stand proud on the carpet and are subject to a greater lift and drag and are entrained, and (b) interstice entrapment—the movement of smaller heavy mineral grains into the interstices of coarser sediment, as a result of which gravels will be better traps than sand. This results in fine grains commonly lagging behind coarse grains when both are being transported over a coarse bed and dynamic lag enrichments of heavy minerals are built up (Slingerland 1984).

Much of the world's tin is won from alluvial placers in Brazil and Malaysia—but see Chapter 12. A good summary description of the principal Malaysian deposits is given in Dixon (1979).

Beach placers

The most important minerals of beach placers are: cassiterite, diamond, gold, ilmenite, magnetite, monazite (an economically important carrier of REE), rutile, xenotime and zircon. Examples include the gold placer of Nome, Alaska, diamonds of Namibia, ilmenite–monazite–rutile sands of Travencore and Quilon, India, rutile–zircon–ilmenite sands of eastern and western Australia, and magnetite sands of North Island, New Zealand. Of course a source or sources of the heavy minerals must be present. These may be coastal rocks, or veins cropping out along the coast or in the sea bed, or rivers or older deposits being reworked by the sea. Recent marine placers occur at different topographical levels owing to Pleistocene sea level changes (Figs 18.6, 18.9). The optimum zone for heavy mineral separation to take place is the tidal zone of an unsheltered beach but concentration may also occur on wave-cut terraces. Some raised beaches formed during high Pleistocene sea levels contain placer ores, such as at Nome, Alaska.

Important beach placers stretch for about 900 km along the eastern coast of Australia and these are particularly important for their rutile and zircon production. They occur in Quaternary sediments that form a coastal strip up to 13 km wide and usually 30–40 m thick. Placer deposits occur along the present day beaches and in the Pleistocene sands behind them (Fig. 2.17). These stabilized sands are characterized by low arcuate ridges which probably outline the shape of former bays. As the thickest heavy mineral accumulations are usually adjacent to

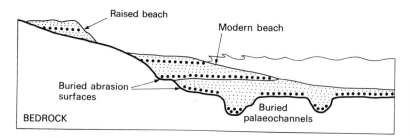

Fig. 18.6 Sketch section to illustrate some sites of beach placer deposits. Placers shown by heavy stipple. (After Selley 1976.)

the southern side of headlands, a reconstruction of the palaeogeography is important during exploration. Thus at Crowdy Head (see Fig. 2.17) the deposit between A and B appears to be related to an old headland, now a bedrock outcrop at B. The point A also appears to mark the site of a former headland and further mineralization may be present to the south-west of this point. Similar relationships exist in the beach placer deposits of Oregon, USA (Peterson *et al.* 1986.)

Beach placers are formed along shorelines by the concentrating agency of waves and shore currents. Waves throw up materials on to the beach, the backwash carrying out the larger, lighter mineral grains, which are moved away by longshore drift, the larger and heavier particles thus being concentrated on the beaches. Heavy mineral accumulations can be seen on present day beaches to have sharp bases and to form discrete laminae. They are especially well developed during storm wave action. Inverse grading is present in these laminae. At the base there is a fine-grained and/or heavy mineral-rich layer which grades upwards into coarser and/or heavymineral-poor sands. These laminations develop during the backwash phase of wave action (Winward 1975). The breaking wave moves sand into suspension and carries it beachward and as its velocity drops, its load is deposited. Then the water flow reverses and a surface layer of sand is disturbed, becoming a high particle density bed flow. During such a bed flow the smaller and denser particles sink to the bottom of the flow, producing the reserve grading and also helping to concentrate the heavy minerals. The heavy mineral-poor sand is thus closest to the surface, waiting to be removed by the next wave. A considerable tidal variation is also important in that it exposes a wider strip of beach to wave action, which may lead to the abandonment of heavy mineral accumulations at the high water mark, where they may be covered and preserved from erosion by seaward advancing aeolian deposits.

Thus beaches on which heavy mineral accumula-

tions are forming today include many upon which trade winds impinge obliquely, and ocean currents parallel the coast, these two factors favouring longshore drift. In addition, these beaches face large areas of ocean and so are subjected to fierce storms and large waves. Such situations are found along the eastern and western coasts of Africa and Australia where various important heavy mineral concentrations occur. These concentrations, as noted above, are often sited close to a headland and on the side facing longshore currents and drift. In the case of one example of the Oregon deposits described by Peterson *et al.* (1986) the site of heavy mineral accumulation is centred on a narrow region 0.75–1.25 km south of Otter Rock Headland (Fig. 18.7), which coincides with a major change in coastline orientation. Here the increasing shoreline curvature produces a decreasing velocity of northward flow as a result of changing shoreline orientation relative to oblique wave approach, and less easily entrained heavy minerals are deposited whilst easily entrained light minerals, such as quartz and feldspar, are carried further north. Peterson *et al.* showed that these heavy mineral concentrations near Oregon headlands occur in an area of maximum shoreline curvature, at a distance south of the north-bounding headland that is proportional to the seaward extent of the headland.

But just how do such ephemeral deposits become preserved? The answer is still the subject of considerable debate. Along the present day coast of New South Wales, heavy mineral accumulations formed by storm action are rarely preserved. They are reworked and redeposited in diluted form, possibly because these beaches are now in a stable or slightly erosive stage. In the mining operations inland from the foredune of this area the Holocene and Pleistocene deposits are seen to form overlapping layers separated by heavy mineral-poor quartz sand. These layers dip south-eastwards towards the prevailing winds (Fig. 18.8). This suggests that for the preservation of heavy mineral deposits either the

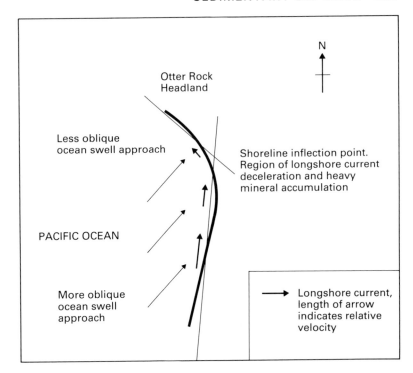

Fig. 18.7 Diagram illustrating the decrease in longshore current velocity with change of coastline orientation that leads to the formation of heavy mineral accumulations along the Oregon coastline. (Modified from Peterson *et al.* 1986.)

shoreline must prograde because of a previously more abundant sediment supply than at present, or the sea level must fall to remove the heavy minerals from the sphere of wave activity.

Beach placers played an important role in the early days of the titanium industry during the late 1940s and their importance increased with the development of the chloride process, which is preferred to the older sulphate process for the production of titania and titanium as it causes less pollution. Initially the chloride process required a feedstock with $TiO_2 > 70\%$, but now some plants can use feedstock down to 60% TiO_2. This means that any ilmenite in the feedstock must be largely weathered to leucoxene. (Pure unaltered ilmenite is about 52% TiO_2.) Only beach placer deposits can

supply such highly altered ilmenite and in addition can supply rutile concentrates, which are even more valuable. Economic beach placers with titanium oxide minerals are typically 10 m thick, 1 km wide and over 5 km long. Closely spaced smaller bodies may be workable. Sets of orebodies that are parallel but not contiguous often mark strandlines on a series of marine terraces (Fig. 18.9). The strong weathering necessary to release, separate and enrich titanium oxide minerals results in large Quaternary deposits being limited to latitudes below 35°. Intense weathering reduces the source rock mass by leaching, especially reducing the volume of feldspathic, mafic and calcareous materials; produces monominerallic grains and leaches iron from ilmenite, in some cases raising the TiO_2 content to 80% (Force 1991).

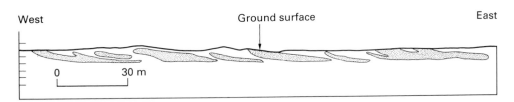

Fig. 18.8 Cross section of heavy mineral concentrations in Quaternary sands at Cudgen, New South Wales. Note seaward (easterly) dip. (After Force 1991.)

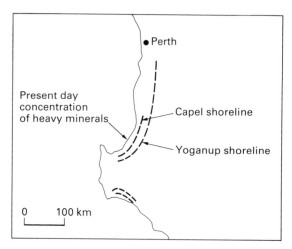

Fig. 18.9 Palaeocoastlines and present day coastline in south-western Australia. The heavy mineral concentrates occur along three former shorelines of Pliocene to early Pleistocene age. The oldest, Yoganup, has six shorelines at 66–26 m elevation, the Capel has 10 shoreline deposits at 6–4 m elevation and the youngest, Minninup, is at or slightly above present day sea level. (From Edwards & Atkinson 1986.)

Offshore placers

These occur on the continental shelf usually within a few kilometres of the coast. They have been formed principally by the submergence of alluvial and/or beach placers (drowned placers). Well studied examples are the tin placers of Indonesia (Batchelor 1979). Offshore placers are becoming increasingly important as heavy mineral producers and, with the development of more efficient dredges that are capable of working along storm affected coastlines, they will help in prolonging the life of this deposit type. For example, it is estimated that the life of the TiO_2 offshore sands of Hainan, China would increase from 28 to 56 years with the introduction of more efficient dredging. Such dredges also would enable the exploitation of the newly found monazite-rich sands off the north and south-western coasts of Sri Lanka. These occur at a depth of 10–15 m some 8 km offshore (Wickremeratne 1986, Anon. 1989b). Many other offshore placers await exploitation and more will undoubtedly be discovered.

Aeolian placers

The most important of these have been formed by the reworking of beach placers by winds, e.g. the large titanomagnetite iron sand deposits of North Island, New Zealand that are estimated to contain more than 1000 Mt of titanomagnetite, 300 Mt of which are present in the Taharoa deposit (Anon. 1983c). The titanomagnetite of these black sands is derived from poorly consolidated, Quaternary, andesitic lahars which crop out in coastal cliffs and stream banks around volcanic cones.

Coastal aeolian dunes have great volume and homogeneity and those with significant contents of heavy minerals can be economic even at low grades. Coastal dunes are of three main types: foredunes, transgressive dunes and stationary dunes. Foredunes are immediately adjacent to present or fossil beaches and they may be interbedded and mined in conjunction with beach deposits. Transgressive dunes have migrated inland from beach deposits and consist of longwall types, still parallel to but now detached from the shoreline, and cliff-top types which are parabolic accumulations on cliffed or other steep shorelines. The Jennings Eneabba deposit of Western Australia (about 350 km north of Perth) appears to have formed as a cliff-top dune. Stationary dunes are those tied to bedrock features.

High grade aeolian deposits occur at Richards Bay, RSA on the Natal coast facing the Indian Ocean. They occur in a belt along the modern coast up to about 1 km inland and 5–25 km north of Richards Bay. The dunes are locally over 100 m high and rest on a platform at 20–40 m elevation. The deposits average 20 m in thickness and carry 10–14% heavy minerals of which ilmenite, zircon, leucoxene and rutile constitute about a half. By-product monazite is produced.

Fossil placer deposits

The most outstanding examples are the gold–uranium-bearing conglomerates of late Archaean to middle Proterozoic age. The principal deposits occur in the Witwatersrand Goldfield of South Africa, the Blind River area along the north shore of Lake Huron in Canada (only trace gold) and at Serra de Jacobina, Bahia, Brazil. Other occurrences are known in many of the other shield areas (Hutchinson 1987, Minter 1991). The host rocks are oligomictic conglomerates (vein quartz pebbles) having a matrix rich in pyrite (or more rarely hematite), sericite and quartz. The gold and uranium minerals (principally uraninite) occur in the matrix together with a host of other detrital minerals.

In the Witwatersrand Goldfield (Fig. 18.10) the orebodies appear to have been formed around the periphery of an intermontane, intracratonic lake or shallow water inland sea at and near entry points where sediment was introduced into the basin. Deposition took place along the interface between river systems, which brought the sediments and heavy minerals from source areas to the north and west, and a lacustrine littoral system that reworked the material (Pretorius 1975, 1981). The individual mineralized areas formed as fluvial fans (Fig. 2.18), which were built up at the entry points. Placer concentrations usually occur in very mature scour-based pebble lag and gravel bar deposits, overlain by trough cross-bedded quartz arenites. Each fan was the result of sediment deposition at a river mouth that discharged through a canyon and flowed across a relatively narrow piedmont plain before entering the basin. The formation of a deltaic fan is shown in Fig. 18.11. Continual uplift of the land along basin-margin faults produced frequent resorting of the sediments, which were further winnowed by

longshore currents. Carbon bands associated with the conglomerates were once algal mats that fringed the deltas. They contain gold and uranium that presumably was taken into sea water solution and precipitated by the mats. These processes led to the formation of the world's greatest goldfield, which between its discovery in 1886 and 1983, produced over 35 000 t of gold from ores with an average grade of $10 \, \mathrm{g \, t^{-1}}$. Average mined grades are now below this figure.

The problem of the provenance of so much gold and pyrite has received much attention. The work of Klemd & Hallbauer (1987) and Robb & Meyer (1990) has revealed the presence of many peraluminous granites in the probable source area that have suffered extensive hydrothermal alteration accompanied by uranium–gold–pyrite mineralization. Erosion of these altered granites could have provided a significant proportion of the minerals in the placer deposits, but mass balance considerations indicate that it is still necessary to infer the existence of other gold deposits in the source area whose

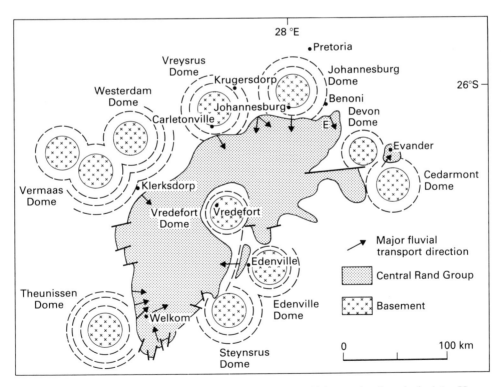

Fig. 18.10 Map showing the distribution of the Central Rand Group, which contains the principal Au–U mineralization, within the Witwatersrand Goldfields together with adjacent granite domes and sites of major fluvial influx. E indicates the position of the East Rand Goldfield (Fig. 2.18). (After Minter *et al.* 1988).

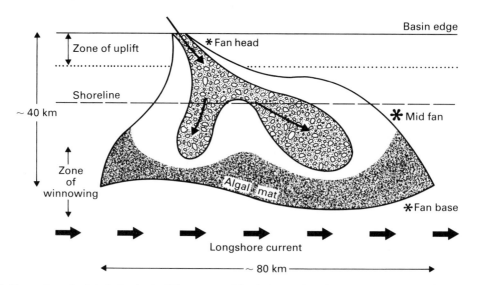

Fig. 18.11 Formation of a deltaic fan in the Witwatersrand Basin (Pretorius 1975). The coarse material is shown by the pebble pattern, the algal mat by the finer pattern. Asterisk sizes indicate the relative gold values. (From Barnes, 1988, *Ores and Minerals,* Open University Press, with permission.)

erosion could make up the deficit. The work of Shepherd (1977), however, showed that much of the gold and uranium came from different sources. He studied the fluid inclusions in the quartz pebbles associated with uranium-rich compared with gold-rich conglomerates and found that the inclusions in pebbles from uranium-rich units were marked by the presence of liquid CO_2, whereas the converse is the case in gold-rich areas.

Different sources for the gold and uranium are also postulated by Hutchinson & Viljoen (1988). They claim that the gold was not introduced into the Witwatersrand Basin with detritus from surrounding highlands but was derived along the margin of the basin by the reworking of proximally deposited, gold-rich, pyritic, sulphide-facies exhalites. These they believe were formed near hydrothermal discharge vents generated by extensive, shallow marine volcanism. The uraninite they do consider to have been transported fluvially from the eroding hinterland. A recent discussion of these and the history of ideas about the genesis of the gold–uranium mineralization can be found in Pretorius (1991).

The Blind River area has also been closely studied and here the uranium deposits (Fig. 2.19) appear to have been laid down in a fluviatile or deltaic environment, perhaps during a wet period preceding an ice age. Unlike the Witwatersrand, the host conglomerates are now at the base of the enclosing arenaceous succession and they appear to occupy valleys eroded in the softer greenstones of the metamorphic basement (Roscoe 1968).

These deposits mark an important metallogenic event in the late Archaean and early Proterozoic, as this type of metal concentration is rare in the older Archaean and essentially absent from younger rocks. The presence of large amounts of apparently detrital pyrite and uraninite in the rocks is a problem. At the present day detrital pyrite is uncommon but known, whilst detrital uraninite is very rare but has been reported. Uraninite is not stable in water in equilibrium with atmospheres containing today's oxygen levels, and this has been taken as supporting evidence for an oxygen deficient, CO_2-rich atmosphere during much of the Precambrian. Thus an increase in oxygen content with time would be the control limiting this type of mineralization to the Archaean and early Proterozoic. This model of an anoxic Precambrian atmosphere has been challenged by a number of workers (cf. Windley 1984) and crucial evidence was presented by Simpson & Bowles (1977). They described heavy mineral concentrations in recent sediments of the Indus that contain pyrite and uraninite, the latter being strikingly similar to that of the Blind River conglomerates. Apparently, under conditions of rapid uplift, erosion, transport and burial, uraninite can persist as a detrital mineral. A useful recent discussion of the problem was given by Robinson & Spooner (1984).

Autochthonous deposits

In this section we will be concerned with bedded iron and manganese deposits. The iron deposits can be divided conveniently into the Precambrian Banded Iron Formations (BIF) and the Phanerozoic ironstones.

Banded iron formations (BIF)

These form one of the earth's mineral treasures. Besides the term BIF, these rocks are known in different continents under the terms itabirite, jaspillite, hematite-quartzite and specularite. They occur in stratigraphical units hundreds of metres thick and hundreds or even thousands of kilometres in lateral extent. Substantial parts of these iron formations are usable directly as a low grade iron ore (e.g. taconite) and other parts have been the protores for higher grade deposits (Chapter 19). Compared with the present enormous demand for iron ore, now approaching 10^9 t p.a., the reserves of mineable ore in the banded iron formations are very large indeed (James & Sims 1973). An extraordinary fact emerging from recent studies is that the great bulk of iron formations of the world was laid down in the very short time interval of 2500–1900 Ma ago (Goldich 1973, James & Trendall 1982). The amount of iron laid down during this period, and still preserved, is enormous—at least 10^{14} t and possibly 10^{15} t, i.e. 90% or more of the total BIF in the Precambrian (Gross 1991). Band iron formations are not restricted to this period, older and younger examples being known, but the total amount of iron in these is far outweighed by that deposited during the former short time interval and now represented by the BIF of Labrador, the Lake Superior region of North America, Krivoi Rog and Kursk, USSR and the Hamersley Group of Western Australia.

Banded iron formation is characterized by its fine layering. The layers are generally 0.5–3 cm thick and in turn they are commonly laminated on a scale of millimetres or fractions of a millimetre. The layering consists of silica layers (in the form of chert or better crystallized silica) alternating with layers of iron minerals. The simplest and commonest BIF consists of alternating hematite and chert layers. Note that the content of alumina is less than 1% contrasting with Phanerozoic ironstones, which normally carry

several per cent of this oxide. James (1954) identified four important facies of BIF.

1 *Oxide facies.* This is the most important facies and it can be divided into the hematite and magnetite subfacies according to which iron oxide is dominant. There is a complete gradation between the two subfacies. Hematite in least altered BIF takes the form of fine-grained grey or bluish specularite. An oolitic texture is common in some examples, suggesting a shallow water origin, but in others the hematite may have the form of structureless granules. Carbonates (calcite, dolomite and ankerite rather than siderite) may be present. The 'chert' varies from fine-grained cryptocrystalline material to mosaics of intergrown quartz grains. In the much less common magnetite subfacies, layers of magnetite alternate with iron silicate or carbonate and cherty layers. Oxide facies BIF typically averages 30–35% Fe and these rocks are mineable provided they are amenable to beneficiation by magnetic or gravity separation of the iron minerals.

2 *Carbonate facies.* This commonly consists of interbanded chert and siderite in about equal proportions. It may grade through magnetite–siderite–quartz rock into the oxide facies, or, by the addition of pyrite, into the sulphide facies. The siderite lacks oolitic or granular texture and appears to have accumulated as a fine mud below the level of wave action.

3 *Silicate facies.* Iron silicate minerals are generally associated with magnetite, siderite and chert which form layers alternating with each other. This mineralogy suggests that the silicate facies formed in an environment common to parts of the oxide and carbonate facies. However, of all the facies of BIF, the depositional environment for iron silicates is least understood. This is principally because of the number and complexity of these minerals and the fact that primary iron silicates are difficult to distinguish from low rank metamorphic silicates. Probable primary iron silicates include greenalite, chamosite and glauconite, some minnesotaite and probably stilpnomelane. Most of the iron in these minerals is in the ferrous rather than the ferric state, which, like the presence of siderite, suggests a reducing environment. P_{CO_2} may be important, a high value leading to siderite deposition, a lower one to iron silicate formation (Gross 1970). Carbonate and silicate facies BIF typically run 25–30% Fe, which is too low to be of economic interest. They also present beneficiation problems.

4 *Sulphide facies.* This consists of pyritic carbonaceous argillites—thinly banded rocks with organic matter plus carbon making up 7–8%. The main sulphide is pyrite which can be so fine-grained that its presence may be overlooked in hand specimens unless the rock is polished. The normal pyrite content is around 37%, and the banding results from the concentration of pyrite into certain layers. This facies clearly formed under anaerobic conditions. Its high sulphur content precludes its exploitation as an iron ore; however, it has been mined until recently for its sulphur content at Chvaletice in Czechoslovakia.

Precambrian BIF can be divided into two principal types (Gross 1970, 1980).

1 *Algoma type.* This type is characteristic of the Archaean greenstone belts where it finds its most widespread development but it also occurs in younger rocks including the Phanerozoic. It shows a greywacke–volcanic association suggesting a geosynclinal environment and the oxide, carbonate and sulphide facies are present, with iron silicates often appearing in the carbonate facies. Algoma-type BIF generally ranges from a few centimetres to a hundred or so metres in thickness and is rarely more than a few kilometres in strike length. Exceptions to this observation occur in Western Australia where late Archaean deposits of economic importance are found. Oolitic and granular textures are absent or inconspicuous and the typical texture is a streaky lamination. A close relationship in time and space to volcanic rocks hints at a volcanic source of the iron

and many regard deposits of this type as being exhalative in origin, e.g. Fralick *et al.* (1989). Goodwin (1973) in a study of this deposit type in the Canadian Shield showed that facies analysis was a powerful tool in elucidating the palaeogeography and could be used to outline a large number of Archaean basins. His section across the Michipicotin Basin is shown in Fig. 18.12. Large deposits of middle Archaean age occur in the Guyanan and Liberian Shields and prior to the break-up of Gondwanaland these iron formations occupied an area of 250 000 km^2 (James 1983).

2 *Superior type.* These are thinly banded rocks mostly belonging to the oxide, carbonate and silicate facies. They are usually free of clastic material. The rhythmical banding of iron-rich and iron-poor cherty layers, which normally range in thickness from a centimetre or so up to a metre, is a prominent feature and this distinctive feature allows correlation of BIF over considerable distances. Individual parts of the main Dales Gorge Member of the Hamersley Brockman BIF of Western Australia can be correlated at the 2.5 cm scale over about 50 000 km^2 (Trendall & Blockley 1970), and correlations of varves within chert bands can be made on a microscopic scale over 300 km (Trendall 1968).

Superior BIF is stratigraphically closely associated with quartzite and black carbonaceous shale and usually also with conglomerate, dolomite, massive chert, chert breccia and argillite. Volcanic rocks are not always directly associated with this BIF, but

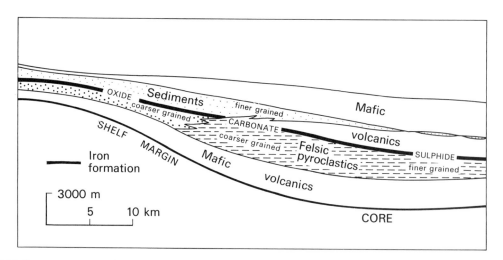

Fig. 18.12 Reconstructed stratigraphical section of the Michipicotin Basin showing the relationship of the oxide, carbonate and sulphide facies of banded iron formation to the configuration of the basin and the associated rock types. (After Goodwin 1973.)

they are nearly always present somewhere in the stratigraphical column. Superior type BIF may extend for hundreds of kilometres along strike and thicken from a few tens of metres to several hundred metres. The successions in which these BIF occur usually lie unconformably on highly metamorphosed basement rocks with the BIF, as a rule, in the lower part of the succession. In some places they are separated from the basement rocks by only a metre or so of quartzite, grit and shale and in some parts of the Gunflint Range, Minnesota, they rest directly on the basement rocks.

The development of Superior BIF reached its acme during the early Proterozoic, and Ronov (1964) has calculated that BIF accounts for 15% of the total thickness of sedimentary rocks of this age. Stratigraphical studies show that BIF frequently extended right around early Proterozoic sedimentary basins and Gross (1965) suggested that BIF was once present around the entire shoreline of the Ungava craton for a distance of more than 3200 km.

The associated rock sequences and sedimentary structures indicate that Superior BIF formed in fairly shallow water on continental shelves (Gross 1980, Goodwin 1982), in evaporitic barred basins (Button 1976), on flat prograding coastlines (Dimroth 1977) or in intracratonic basins (Eriksson & Truswell 1978). Trendall (1973) has suggested that the rhythmic microbands of the Hamersley Group so closely resemble evaporitic varves that a common origin is probable. He suggested that the banding originated by the annual accumulation of iron-rich precipitates whose deposition was triggered by evaporation from a partially enclosed basin with an average water depth of about 200 m.

There is general agreement that BIF are chemically or biochemically precipitated. Blue-green algae and fungi have been identified in the Gunflint Iron Formation of Ontario (Awramik & Barghoorn 1977) and some of these resemble present day iron-precipitating bacteria which can grow and precipitate ferric hydroxide under reducing conditions. However, there is no agreement on the source of the iron. One school considers that this was derived by erosion from nearby landmasses, another that it is of volcanic exhalative origin, e.g. Gross (1986). One major drawback to the terrestrial derivation is that if large amounts of iron and silica were transported from the continents, large quantities of aluminous material must either have been left behind or transported and dispersed in the sea with, or not far from, the iron deposits. No such residual bauxites or aluminous sediments have been discovered. On the other hand, Miller & O'Nions (1985) gave an estimate for submarine hydrothermal supply of iron to the present oceans of $\leq 10^{10}$ kg p.a., yet it is claimed that the Hamersley BIF of Western Australia alone required $\geq 10^{10}$ kg p.a. They concluded that a major contribution of iron from the continents did occur unless the hydrothermal iron input during the Proterozoic was overwhelmingly greater than the present day one. On the other hand, from their study of the isotopic composition of neodymium in the Hamersley and Michipicoten (Ontario) BIF, Jacobsen & Pimentel-Klose (1988) concluded that much of the iron in BIF was leached by hydrothermal water circulating through Archaean mid-ocean ridge systems.

The mechanism of transport of the iron is also hotly debated. Today with an oxygen-rich atmosphere very little iron is transported to the oceans in ionic solution and most travels in colloidal solution or as particulate matter and is deposited mainly in muds. We have no true Phanerozoic analogues of the BIF of the early Proterozoic. For those who advocate a CO_2-rich, oxygen-poor early atmosphere the explanation is simple—iron could then travel as the bicarbonate in ionic solution and, since aluminium does not form a bicarbonate, the two would be separated and another genetic problem solved. With the significant development of oxygen in the atmosphere, large scale formation of BIF would cease. Garrels (1987, 1988) assumed transport of iron in the Fe(II) state and developed a quantitative model for the deposition of iron in large basins to which it was carried in stream waters and precipitated by evaporation to produce iron mineral–chert varves like those described from BIF in many parts of the Precambrian, but especially well developed in the Hamersley Group of Western Australia (Trendall 1973). This is an extremely well thought out model and worthy of careful examination and testing.

Conversely, after a careful and lengthy appraisal of the genesis of BIF, Kimberley (1989) concluded that all iron formations are the result of exhalative activity. Cherty iron formation (the principal facies of Superior type BIF) he postulated as forming from low temperature ($< 300°C$) hydration of newly formed igneous crust by circulating sea water in convection cells. Thus he links the marked production of cherty iron formations in the early Proterozoic and the much smaller but significant development in the late Proterozoic to particularly rapid crustal accumulation in opening rifts, followed by abrupt failure of the rift and hydration of the new crust. Kimberley cites evidence of Phanerozoic deposition of small cherty iron formations within rifts, a possible modern analogue in the Red Sea, evidence of recent exhalative iron deposition in Venezuela and much other evidence in favour of this theory. Non-cherty iron formation he attributes to seismic pumping of sea water along strike-slip or transform faults that pass through ophiolite-bearing sedimentary successions along continental margins, where the iron-dissolving waters have become hypersaline because they were pumped through evaporites or cooling plutons. These fluids are believed to have been modified by ascent through argillaceous sediment leading to cooling and loss of silica.

The Lake Superior region

For an example of BIF, we can look briefly at the deposits in the USA to the west and south of Lake Superior (Bayley & James 1973). This was one of the greatest iron ore districts of the world. The western part, which is shown in Fig. 18.13, can be divided into three major units: a basement complex (> 2600 Ma old); a thick sequence of weakly to strongly metamorphosed sedimentary and volcanic rocks (the Marquette Range Supergroup); and later Precambrian (Keweenawan) volcanics and sediments.

Banded iron formation is mainly developed in the Marquette Range Supergroup, but in the Vermilion district it is present in the basement. In this district there is a great thickness of mafic to intermediate volcanic rocks and sediments. Banded iron formation, mainly of oxide facies, occurs at many horizons, generally as thin units rarely more than 10 m thick, but one iron formation (the Soudan) is much thicker and has been mined extensively.

The remaining iron ore of this region comes from the Menominee Group of the Marquette Range Supergroup. All the iron formations of this group in the different districts are of approximately the same age. The Marquette Range Supergroup shows a complete transition from a stable craton to deep water conditions. Clastic rocks were first laid down on the bevelled basement but most of these were removed by later erosion and in many places the Menominee Group rests directly on the basement. Iron formation is the principal rock type of this group. Despite the approximate stratigraphical equivalence of the major iron formations, they differ greatly from one district to another in thickness, stratigraphical detail and facies type. They appear to have been deposited either in separate basins or in isolated parts of the same basin or shelf area. The only evidence of contemporaneous volcanism is the occurrence of small lava flows in the Gunflint and Gogebic districts. It has been suggested *apropos* this region that volcanism appears to have been detrimental rather than conducive to iron concentration.

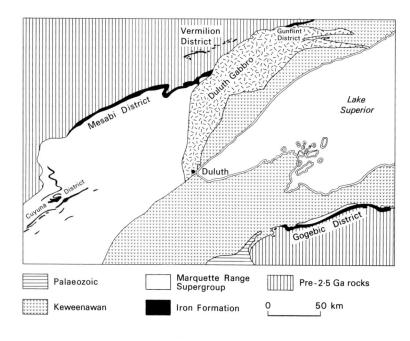

Fig. 18.13 Distribution of iron formation in Minnesota and northern Wisconsin. (After Bayley & James 1973.)

The same iron formation appears in the Mesabi and Gunflint districts. It is 100–270 m thick and consists of alternating units of dark, non-granular, laminated rock and cherty, granule-bearing irregularly to thickly bedded rock. The granules are mineralogically complex containing widely different proportions of iron silicates, chert and magnetite; some are rimmed with hematite. The iron formation of the Cuyuna district consists principally of two facies, thin bedded and thick bedded, which differ in mineralogy and texture. The first is evenly layered and laminated, the layers carrying varying proportions of chert, siderite, magnetite, stilpnomelane, minnesotaite and chlorite, while the second contains evenly bedded and wavy bedded rock in which chert and iron minerals alternately dominate in layers 2–30 cm thick. Granules and oolites are present. In the Gogebic district the iron formation is 150–310 m thick and consists of an alternation of wavy to irregularly bedded rocks characterized by granule and oolitic textures. The iron in the irregularly bedded rocks is principally in the form of magnetite and iron silicates, and granule textures are common. The evenly bedded iron formation is mineralogically complex, consisting of chert, siderite, iron silicates and magnetite. Each mineral may dominate a given layer and may be accompanied by one or more of the other minerals.

Phanerozoic ironstones

These are usually classified into two types, Clinton and Minette, but both are now of very diminished economic importance as they are of low grade and impossible to beneficiate economically on account of their silicate mineralogy. Mining of ironstones in the UK, once very important, has now ceased, and exploitation of these ores within the EEC will probably come to an end in a few years. They are moving, as it were, from the category of reserve to that of resource even though there are still many megatonnes in the ground.

1 *Clinton type.* This forms massive beds of oolitic hematite–chamosite–siderite rock. The iron content is about 40–50% and they are higher in Al and P than BIF. They also differ from BIF in the absence of chert bands, the silica being mainly present in iron silicate minerals with small amounts as clastic quartz grains. Clinton ironstones form lenticular beds usually 2–3 m thick and never greater than 13 m. This type of ironstone appears to have formed in shallow water along the margins of continents, on continental shelves or in shallow parts of miogeoclines. It is common in rocks of Cambrian to Devonian age. One of the best examples is the Ordovician Wabana ore of Newfoundland (Gross 1970).

2 *Minette type.* These are the most common and widespread ironstones. The principal minerals are siderite and chamosite or another iron chlorite, the chamosite often being oolitic; the iron content is around 30% while lime runs 5–20% and silica is usually above 20%. The high lime content forms one contrast with BIF and often results in these ironstones being self-fluxing ores.

Minette ironstones are particularly widespread and important in the Mesozoic of Europe, examples being the ironstones of the English Midlands, the minette ores of Lorraine and Luxembourg and the Salzgitter ores of Saxony. Unlike the BIF, neither the Minette nor the Clinton ironstones show a separation into oxide, carbonate and silicate facies. Instead the minerals are intimately mixed, often in the same oolite.

Sedimentary manganese deposits

Sedimentary manganese deposits and their metamorphosed equivalents produce the bulk of the world's output of manganese. Residual deposits are the other main source. The USSR is the world leader in manganese ore production, and in 1989 produced 9.1 Mt, i.e. about 41% of world production. Approximately 75% of this came from the Nikopol Basin in the Ukraine and much of the remainder from the Chiatura Basin in Georgia. The other important producers are the Republic of South Africa (3.62 Mt), Gabon (2.45 Mt), Australia (2 Mt), Brazil (1.8 Mt) and India (1.1 Mt). Total world production is about 22.1 Mt and about 95% of this is consumed by the steel industry. The remaining 1 Mt or so is used for a multitude of purposes. Weiss (1977) wrote that manganese is 'used in feed, food, fertilizer, fungicides, facebricks, frits, flux, fragrances, flavors, foundries, ferrites, fluorescent tubes, fine chemicals, ferric leaching—and ferroalloys. It is an oxidant, deoxidant, colorant, bleach, insecticide, bactericide, algicide, lubricant, nutrient, catalyst, drier, scavenger—and much much more'. Manganese for these various purposes must be of a specific type in a particular form and certain producers have become expert in supplying the special ores required and the end products derived from them.

An important use is in dry cell batteries in which

ground manganese oxide acts as an oxidizing agent or depolarizer. Battery grade manganese oxide requires at least 80% MnO_2, < 0.05% metals which are electronegative to zinc and no nitrates. Battery grade ore from Gabon is 83–84% Mn, from Ghana 78% and Greece 75%.

The geochemistry of iron and manganese is very similar and the two elements might be expected to move and be precipitated together. This is indeed the case in *some* Precambrian deposits; for example, in the Cuyuna District (Fig. 18.13), manganese is abundant in some of the iron formations and forms over 20% of some of the ores. In other areas there seems to have been a complete separation of manganese and iron during weathering, transport and deposition so that many iron ores are virtually free of manganese and many manganese ores contain no more than a trace of iron. The mechanism of this separation is still unknown. Stanton (1972) and Roy (1976) discuss a number of possibilities. Firstly, there is the possibility of segregation at source owing to manganese being leached more readily from source rocks because of its relatively low ionic potential. This is, however, unlikely to produce more than a few per cent difference in the Mn/Fe of the extracted material compared with the ratio in the source rocks. A second possibility arises from the observation that many hot springs produce more manganese than iron, suggesting that iron has been precipitated preferentially from these hydrothermal solutions before they reached the surface. The third possibility is segregation by differential precipitation. Chemical considerations suggest that a limited increase in pH in *some* natural situations may lead to the selective elimination of iron from iron–manganese solutions. A fourth possibility is that separation occurs during diagenesis. The development of reducing conditions will cause both the iron and the manganese to be reduced and to go into solution and move laterally and upward. When the solutions reach an oxidizing environment the two are precipitated. However, since iron will always be the last to be reduced and hence mobilized, and the first to be oxidized and hence immobilized, manganese will tend to be progressively separated in diagenetic solutions. Recently a number of workers have proposed that anoxia plays a role in the formation of the giant deposits of shallow marine environments in intracratonic basins (Glasby 1988). Anoxia leads to an increase in the dissolved manganese in the deeper sea water of adjacent shale basins and,

following a marine transgression on to a nearby platform, the dissolved manganese can be precipitated in the shallow oxic zone of the platform. Shallow marine manganese deposits are thus considered to be lateral facies of black shales formed in anoxic basins. This process is very clearly set forth by Force & Cannon (1988).

Another enigma of manganese deposits is the almost universal restriction of fossil deposits to shallow water environments of deposition, whilst in modern deep sea areas there is an unprecedented concentration of manganese, Roy (1988) and Glasby (1988) have estimated that, whereas the total terrestrial manganese deposits amount to about 6.4×10^9 t, deep sea manganese nodules formed since the lower Miocene (12 Ma ago) hold about 10^{11} t Mn, i.e. 16 times the manganese in terrestial deposits. Almost equally puzzling is the time distribution of economic deposits, the first of which appear in rocks about 3000 Ma old, i.e. towards the end of the Archaean, but these are economically insignificant particularly when compared with BIF deposits of this age. Large deposits do not appear until the Proterozoic within which the world's largest known resource of manganese occurs in the Transvaal Supergroup (2500–2100 Ma) of the Kalahari Field. Glasby (1988) has estimated this as containing 5026.3 Mt of Mn compared with the world's total resources of 6376.1 Mt!

Classification of manganese deposits

A number of classifications have been put forward and a useful one for working explorationists and mining geologists is that put forward by Machamer (1987); see Table 18.3. Many primary manganese deposits are not economically mineable, but they may form valuable protores in areas where supergene enrichment has occurred. In his table Machamer has added the supergene equivalents of the primary deposits and this greatly enhances its usefulness over the more common classifications, which usually only deal with primary deposits. It should be noted that each primary type, when subjected to secondary alteration, gives rise to supergene deposits which also have distinct characteristics.

Type I. These concentrations are rocks composed in the main of manganese carbonates (queluzites), manganese silicates (gondites) or a mixture of both and there often is a close association with andesitic to basaltic metavolcanics. Tectonic thickening or isoclinal folding may produce mineable thicknesses. Their frequent volcanic association and the analogy with Archaean BIF has led many authors to postulate an exhalative origin. Roy (1988), however, contends that these deposits are all concentrated in shallow water, intracratonic basins more of Proterozoic rather than Archaean type and, unlike the Archaean BIF, there is little evidence of an exhalative origin.

When exposed to tropical and subtropical weathering, oxidation and supergene enrichment can give rise to high grade ores, especially when the protore is carbonate.

Type II. These are economically very important deposits. They consist of discrete beds of manganese oxide or carbonate intercalated with BIF. In the Brazilian, Bolivian and RSA examples three distinct manganese-rich units are separated by BIF and thin beds of ferruginous arkose and the stratigraphically lowest bed is usually the thickest and most valuable. These are platform-type deposits but some, e.g. many of those in West Africa, have a volcanic setting and a number of workers have suggested an exhalative origin for these and have included the Kalahari Field in this category. However, other workers, e.g. Roy (1988), dispute this conclusion most strongly.

Type III. These are Mn-rich graphitic or carbonaceous shales of no economic importance unless they have undergone lateritization, when important deposits may be formed, such as those in the Francevillian Series of Gabon. A point to watch in their economic assessment is the level of aluminium, phosphorus or other components that may be deleterious in metallurgical processes, as these elements tend to be concentrated during the laterization.

Type IV. These are beds of manganese carbonates of little worldwide significance. When of high enough grade and thickness the Mn carbonate is extracted and calcined to produce a saleable Mn oxide sinter.

Type V. These deposits formed in a shelf environment under estuarine and shallow marine conditions (Varentsov & Rakhmanov 1977). On one side they may pass into a non-ore-bearing coarse clastic succession, sometimes with coal seams, that lies between the manganese orefields and the source area for the sediments. On the other side they pass into an argillaceous sequence that marks deeper water deposition.

The largest manganese ore basin of this type is the South Ukrainian Oligocene Basin and its deposits include about 70% of the world's reserves of manganese ores. It forms a part of the vast South European Oligocene Basin, which also contains the deposit of Chiatura and Mangyshlak in the USSR and Varnentsi in Bulgaria. The distribution of ore deposits in the southern Ukraine is shown in Fig. 18.14.

The manganese ore forms a layer interstratified with sands, silts and clays. It is 0–4.5 m thick, averages 2–3.5 m and extends for over 250 km. There are intermittent breaks owing to post-Oligocene erosion. A glauconitic sand is frequently present at the base of the ore layer which consists of irregular concretions, nodules and rounded earthy masses of manganese oxides and/or carbonate in a silty or clayey matrix. A shoreward to deeper water zoning is present (Figs 18.14, 18.15). The dominant minerals of the oxide zone are pyrolusite and psilomelane. The principal carbonate minerals are manganocalcite and rhodochrosite. Progressing into a deeper water environment, the ore layer in the carbonate zone grades into green-blue clays with occasional manganese nodules. The Chiatura deposit of Georgia shows a similar zonal pattern. The average ore grade in the Nikopol deposits is 15–25% Mn and at Chiatura it varies from a few to 35%.

The lack of associated volcanic rocks suggested to older workers that these deposits were formed by weathering and erosion from the nearby Precambrian shield, which contains a number of rock types (spilites, etc.) that could have supplied an abundance of manganese. Palaeobotanical research has shown that the deposition of the ores coincided with a marked climatic change from humid subtropical to cold temperate. This change could have affected the pattern of weathering and transportation of the manganese. An older but similar deposit is the huge Groote Eylandt Cretaceous deposit, Gulf of Carpentaria, Australia, which was also formed in shallow marine conditions just above a basal unconformity during a marine transgression (Ostwald 1981).

More recently it has been suggested that these deposits are related to periods of high sea level stands linked to simultaneous anoxic events in adjacent black shale basins (Thein 1990). At Imini, Morocco it appears that the ores were formed in a zone of mixing near the Upper Cretaceous coastline where oxidized ground water of meteoric origin

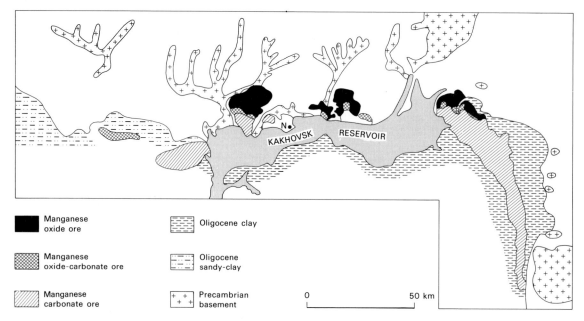

Fig. 18.14 Distribution of manganese ore in the South Ukrainian Basin. The northern and eastern parts of the map area with outcrops of Precambrian basement are largely covered by Quaternary sediments. N = Nikopol. (Modified from Varentsov & Rakhmanov 1977.)

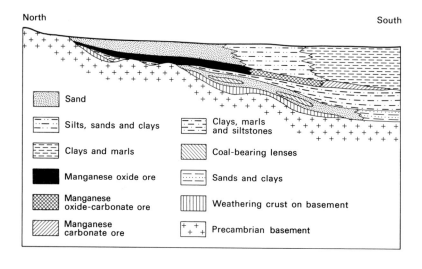

Fig. 18.15 Diagrammatic cross section through the Nikopol manganese deposits showing the zonation of the manganese ores and the transgressive nature of the sedimentary sequence with its overlap on to the Precambrian basement of the Ukrainian Platform. (After Varentsov 1964.)

bearing Mn (and other metals) entered reducing sea water, itself enriched in Mn derived from the mobilization of Mn from the upper sediment layers in adjoining shale basins (Thein 1990).

Weathering and oxidation of such primary ores can produce high grade, high quality supergene ores, such as those of Groote Eylandt, Northern Territory, Australia (Pracejus *et al.* 1988).

Volcanic-associated exhalative deposits

Whereas many workers consider the deposits of types I–V to be of non-volcanic origin, there is a further type that many would agree is volcanogenic-exhalative. This is geologically the most widespread manganese deposit type, but economically of little or minor importance compared with the deposits

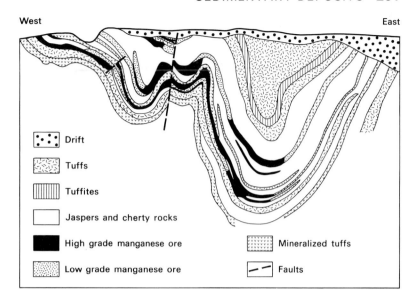

West

East

Drift

Tuffs

Tuffites

Jaspers and cherty rocks

High grade manganese ore

Low grade manganese ore

Mineralized tuffs

Faults

Fig. 18.16 Geological section through the Kusimovo deposits, Southern Urals. (After Varentsov & Rakhmanov 1977.)

described above. This type was divided by Shatsky (see Varentsov & Rakhmanov 1977) into a greenstone association with spilite-keratophyre volcanism and a porphyry association with trachyrhyolite volcanism. The greenstone association is the important one. It occurs in geosynclinal-type terrane and consists mainly of a silicate facies; sometimes carbonates are dominant but rarely oxides. A good example is the Devonian Kusimovo deposit (Fig. 18.16), which occurs in a group of deposits near Magnetogorsk in the Southern Urals. There, an intimate association exists in strongly folded rocks with basic and intermediate volcanics and the orebodies are layers or lenses that rest conformably on the underlying pyroclastics. They range from a few centimetres to 5 m in thickness and are bedded, with bands of braunite alternating with cherty jasperoid layers. The manganese mineralogy of these deposits varies with the grade of metamor-

phism in a complicated manner and the tenor of the ores is commonly 15–25% Mn.

The evidence for a volcanic-exhalative origin of manganese ores of this class is fairly compelling. Not only are they intimately associated with volcanic activity in many different settings including modern island arcs (Stanton 1972), but they frequently carry volcanic fragments including shards and they are enriched in elements, such as As, Ba, Cu, Mo, Pb, Sr and Zn, that appear to be volcanic contributions (Zantop 1981). Colley & Walsh (1987) have described typical deposits of this type in Fiji. Here the small bodies of Fe–Mn rock are associated with hemipelagic sediments and basaltic rocks and are similar to hydrothermal mounds that are forming at the present day on the ocean floor at sites such as the Galapagos Ridge. Some secondary enrichment has occurred to produce minor orebodies, which over the period 1949–71 yielded 200 kt of ore.

19 / Residual deposits and supergene enrichment

In the previous chapter we considered the concentration into orebodies of sedimentary material removed by mechanical or chemical processes and redeposited elsewhere. Sometimes the material left behind has been sufficiently concentrated by weathering processes and ground water action to form residual ore deposits. For the formation of extensive deposits, intense chemical weathering, such as in tropical climates having a high rainfall, is necessary. In such situations, most rocks yield a soil from which all soluble material has been dissolved and these soils are called laterites. As iron and aluminium hydroxides are amongst the most insoluble of natural substances, laterites are composed mainly of these materials and are, therefore, of no value as a source of either metal. Sometimes, however, residual deposits can be high grade deposits of one metal only. A recent discussion of the development of residual and supergene deposits can be found in Brimhall & Crerar (1987).

Residual deposits of aluminium

When laterite consists of almost pure aluminium hydroxide, it is called bauxite and is the chief ore of aluminium—the metal used in everything from beer cans to jumbo jets. Total world production of bauxite in 1989 was about 92 Mt and most of this was used for the production of aluminium metal. About 4–5 Mt, however, was absorbed by non-metallurgical operations for the manufacture of refractories, abrasives, alumina chemicals and in the cement industry. There are certain chemical requirements that bauxites must meet to be economic (Table 19.1). For refractories the iron content must be low since its presence tends to lower their fusion temperatures. The content of titania, alkalis and alkali earths must also be low and many bauxites do not meet these requirements. High alumina ores from China and Guyana are among the best for this end use. In the manufacture of high alumina cement, bauxite, which replaces the clay or shale in Portland cement, is fused with limestone. This cement resists corrosion by sea water or sulphate-bearing water. An excellent review of the non-metallurgical uses of alumina can be found in Benbow (1988).

An important by-product of bauxite treatment is gallium and among non-ferrous metals aluminium is unusual in having only one technology for extraction in commercial use in market economy countries—the Hall–Héroult Process discovered in 1886. Before this was developed aluminium was a precious metal used only in jewelry!

Bauxite will develop on any rock with a low iron content or one from which the iron is removed during weathering. As with placer deposits, bauxites

Table 19.1 Bauxite grades by end use. (From Edwards & Atkinson 1986)

		Metallurgical	Chemical	Cement	Refractory (calcined)	Abrasive (calcined)	Proppants
Requirements		High alumina, low iron, low silica, low titania	Must be gibbsitic, very low iron, silica low but not critical	Moderately high alumina, low silica, iron may be high, diaspore preferred	High alumina, low iron, low silica, very low alkalis	High alumina, moderately low iron, low silica, low titania	High alumina, low silica, low clays, iron content not critical
Typical analyses	Al_2O_3	50–55%	>55%	45–55%	>84.5%	80–88%	None available
	SiO_2	0–15%	5–18%	<6%	<7.5%	4–8%	
	Fe_2O_3	5–30%	<2%	20–30%	<2.5%	2–5%	
	TiO_2	0–6%	0–6%	3%	<4%	2–5%	

are vulnerable to erosion and most deposits are therefore post-Mesozoic. Older deposits, however, are known, for example those in the Palaeozoic of the USSR. Some eroded bauxite has been redeposited to form what are called transported or sedimentary bauxites.

After oxygen and silicon, aluminium is the third most common element in the earth's crust, of which it forms 8.1%. Aluminium displays a marked affinity for oxygen and is not found in the native state. In weathered materials it accumulates in clay minerals or in purely aluminous ones, such as gibbsite, boehmite and diaspore, which are the principal minerals of bauxite. The mineralogy of bauxites depends on their age. Young bauxites are gibbsitic. With age, gibbsite gives way to boehmite and diaspore. Bauxite deposits are usually large deposits worked in open pits. The largest deposit is that of Sangaredi in Guinea, where there is at least 180 Mt forming a plateau up to 30 m thick and averaging 60% alumina, and this country is, after Australia (37.85 Mt), the western world's largest producer— 17 Mt in 1989. Next in line in that year were Jamaica (9.4 Mt), Brazil (8.25 Mt), India (3.9 Mt) and Surinam and Yugoslavia, both with about 3.4 Mt.

Classification of bauxites

Bauxite deposits are extremely variable in their nature and geological situations. As a result, many different classifications have been put forward although here we have space to look at only four, those of Harder, Hose, Grubb and Hutchison (Table 19.2). Grubb's simple scheme is based on the topographical levels at which these deposits were formed and

it involved assigning some karst bauxites to the high level class and some to the low level—an opinion not supported by Bárdossy (1982). Hutchison (1983) combined Grubb's two classes into one, which he terms lateritic crusts, whilst recognizing karst and sedimentary bauxites as separate entities—a scheme adopted by Evans (1980). A useful discussion of the mineralogy and geochemistry of bauxites can be found in Maynard (1983).

High level or upland bauxites

These generally occur on volcanic or igneous source rocks forming thick blankets of up to 30 m which cap plateaux in tropical to subtropical climates. Examples occur in the Deccan Traps of India, southern Queensland, Ghana and Guinea. These bauxites are porous and friable, show a remarkable retention of parent rock textures, and are dominantly gibbsitic. They rest directly on the parent rock with little or no intervening underclay. Bauxitization is controlled largely by joint patterns in the parent rock, with the result that chimneys and walls of bauxite often extend deep into the footwall.

Low level peneplain-type bauxites

These occur at low levels along tropical coastlines, such as those of South America, Australia and Malaysia. They are distinguished by the development of pisolitic textures and are often boehmitic in composition. Peneplain deposits are generally less than 9 m thick and are usually separated from their parent rock by a kaolinitic underclay. They are frequently associated with detrital bauxite horizons produced by fluvial or marine activity.

Table 19.2 Classification of bauxite deposits

Harder & Greig (1960)	Hose (1960)	Grubb (1973)	Hutchison (1983)
Surface blanket deposits	Bauxites formed on peneplains	High level or upland bauxites	Lateritic crusts
Interlayered beds or lenses in stratigraphical sequences	Bauxites formed on volcanic domes or plateaux		
Pocket deposits in limestones, clays or igneous rocks	Bauxites formed on limestone or karstic plateaux	Low level peneplain-type bauxites	Karst bauxites
Detrital bauxites	Sedimentary reworked bauxites	Sedimentary bauxites	

Karst bauxites

These include the oldest known bauxites—those in the lands just north of the Mediterranean, which range from Devonian to mid-Miocene. Other major deposits are the Tertiary ones of Jamaica and Hispaniola. These bauxites overlie a highly irregular karstified limestone or dolomite surface. Texturally, karst bauxites are quite variable. The West Indian examples are gibbsitic ores with a structureless earthy, sparsely concretionary texture. European karst bauxites, on the other hand, are generally lithified and texturally pisolitic, oolitic, fragmental or even bedded. Mineralogically they are predominantly boehmitic ores. From these and other facts Grubb contended that the West Indian bauxites have strong affinities with upland deposits, whilst the European karst bauxites are more reminiscent of peneplain deposits.

Transported or sedimentary bauxites

This is a small class of non-residual bauxites formed by the erosion and redeposition of bauxitic materials.

Genesis of bauxites

Traditionally three processes have been put forward for the formation of bauxite: (a) weathering and leaching *in situ* of bedrock to produce aluminium and iron-rich residues; (b) enrichment of sediments or weathered rocks in aluminium by ground water leaching; and (c) erosion and redeposition of bauxitic materials. The first process has been generally considered to have been the mode of formation of many bauxite deposits and Raman (1986) summed it up as follows. The genesis of lateritic bauxite deposits is governed largely by the cumulative effect of several controls and important among them are (a) a favourable parent rock with easily soluble minerals whose leaching will leave a residual enrichment of aluminium and/or iron, (b) effective rock porosity allowing free circulation of water, (c) high rainfall with intermittent dry spells, (d) good drainage, (e) tropical warm climate, (f) low to moderate topographical relief, (g) prolonged stability, and (h) presence of vegetation including bacteria. An in-depth discussion of these processes and many economic deposits can be found in Bárdossy & Aleva (1990).

The work of Brimhall *et al.* (1988) has introduced a novel idea into the discussion of bauxite genesis.

These workers made a quantitative study of the bauxite near Jarrahdale in the Darling Downs area of Western Australia; the Darling Downs contain about 900 Mt of bauxite reserves which make it one of the largest bauxite regions in the world. Using a mass balance approach they showed that the enrichment of aluminium to ore grade is the result of addition of large quantities of aeolian dust from the arid continental interior. Fine-grained dust that has been transported over great distances is enriched in aluminium and iron as it has a high content of clay minerals and amorphous alumino-silicate material. Pye (1988) has published an interesting map showing how many of the known bauxite deposits lie on the main trajectories of continentally derived, wind blown dust and it may well be that this process has played an important part in the formation of all bauxite deposits.

Iron-rich laterites

Most iron-bearing laterites are too low in iron to be of economic interest. Occasionally, however, laterites derived from basic or ultrabasic rock may be sufficiently rich in iron to be workable, though in some cases other metals, such as cobalt and nickel, may also have been enriched to such an extent as to poison the ore. These deposits, which may be as much as 20 m thick but are usually less than 6 m, consist of nodular red, yellow or brown hematite and goethite which may carry up to 20% alumina. Deposits of this type form mantles on plateaux and are worked in Guyana, Indonesia, Cuba, the Philippines, etc.

A good example is the Conakry deposit in Guinea which is developed on dunite, the change from laterite to dunite being sharp. Most of the laterite consists of a hard crust, usually about 6 m thick. The ore as shipped contains 52% Fe, 12% alumina, 1.8% silica, 0.25% phosphorus, 0.14% S, 1.8% Cr, 0.15% Ni, 0.5% TiO_2 and 11% combined water. This last figure illustrates one of the drawbacks of these ores—their high water content, which may range up to 30%. This has to be transported and then removed during smelting.

Auriferous bauxites and laterites

An exciting development in the exploration of bauxites and laterites was the discovery of gold in the early 1980s in a bauxite near Boddington, 130 km south-east of Perth, Western Australia. By 1988 mineable reserves of 45 Mt averaging 1.8 ppm Au, at a cut-off of 1 ppm had been defined. The deposit was set to produce 10.885 t p.a. Au by the end of 1989, which would make it the largest gold mine in Australia (Bird 1988).

Secondary enrichment of gold is well known in Australian gold deposits, and many 5–20 oz nuggets have been found recently, using metal detectors, in laterites of the Coolgardie district, Western Australia. Secondary gold concentrations in Australia, such as the nugget-bearing bauxite of the Cloncurry region of Queensland, are often found where little or no basement gold mineralization is known (Wilson 1983). Recently gold-bearing laterites from the Mato Grosso, Brazil have been described by Michel (1987). Clearly we have here a new deposit type of worldwide significance, particularly for the shield areas of Gondwanaland, and residual deposits developed over greenstone belts will be the first exploration target in the search for further deposits. Comparable gold concentrations are also known from older rocks, e.g. Proterozoic palaeosols of the Transvaal, RSA (Martini 1986). It is possible that economic platiniferous laterites may be found in the future (Bowles 1985).

Residual deposits of nickel

The first major nickel production in the world came from nickeliferous laterites in New Caledonia where mining commenced in about 1876. It has been calculated that there are about 64 Mt of economically recoverable nickel in land-based deposits. Of this, about 70% occurs in lateritic deposits, although less than a half of current nickel production comes from these ores. Golightly (1981) has given a good review of this deposit type.

Residual nickel deposits are formed by the intense tropical weathering of rocks rich in trace amounts of nickel, such as peridotites and serpentinites, running about 0.25% Ni. During the lateritization of such rocks, nickel passes (temporarily) into solution but is generally quickly reprecipitated either on to iron oxide minerals in the laterite or as garnierite and other nickeliferous phyllosilicates in the weathered rock below the laterite. Cobalt too may be concentrated, but usually it is fixed in wad. Grades of potentially economic deposits range from 1 to 3% Ni + Co and tonnages from about 10 to 100 Mt as can be seen in Fig. 19.1. Besides by-product cobalt there may be recoverable copper, as in the Buhinda prospect, Burundi, which at a cut-off of 0.8% Ni contains 28.6 Mt running 1.62% Ni, 0.12% Co and 0.31% Cu (Derkmann & Jung 1986).

Nickel deposits of New Caledonia

Much of New Caledonia is underlain by ultrabasic rocks, many of which are strongly serpentinized. A typical environment of nickel mineralization is shown in Fig. 19.2 and a more detailed profile in Fig. 19.3. The nickel occurs in both the laterite and the

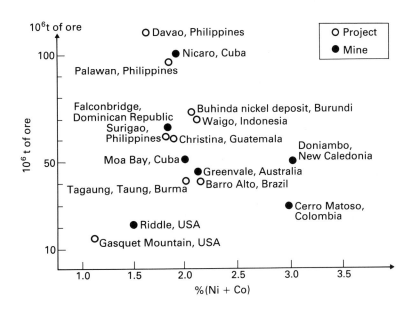

Fig. 19.1 Grade–tonnage diagram for nickeliferous laterite deposits. (After Derkmann & Jung 1986.)

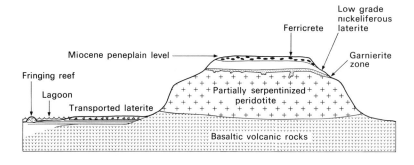

Fig. 19.2 Diagrammatic profile of a peridotite occurrence in New Caledonia showing the development of a residual nickel deposit. (After Dixon 1979.)

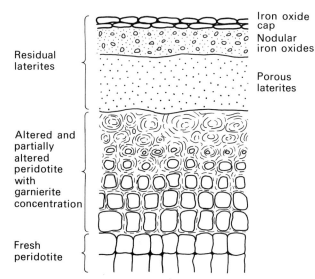

Fig. 19.3 Section through nickeliferous laterite deposits, New Caledonia. (After Chételat 1947.)

weathered rock zone. In the latter it forms distinct masses, veins, veinlets or pockets rich in garnierite which occur around residual blocks of unweathered ultrabasic rock and in fissures running down into the underlying rock. The material mined is generally a mixture of the lower parts of the laterite and the weathered rock zone. Above the nickel-rich zone there are pockets of wad containing significant quantities of cobalt. Grades of up to 10% Ni were worked in the past, but today the grade is around 3% Ni. It has been estimated that there are 1.5 Gt of material on the island assaying a little over 1% Ni. All laterites take time to develop and it is thought that those on New Caledonia began to form in the Miocene.

The Greenvale Nickel Laterite, north Queensland

This deposit was discovered in 1966 as a result of the comparison of the geological environment with that of New Caledonia (Fletcher & Couper 1975). The section above the fresh serpentinite (which runs 0.28% NiO) is similar to that of the New Caledonian occurrences (Fig. 19.4). Nickel and cobalt are concentrated to ore grade in a laterite mantle covering about two thirds of the serpentinite.

Erosion has removed the ore zone from the rest of the peridotite. Ore reserves run to 40 Mt averaging 1.57% Ni and 0.12% Co. The ore zone occurs mainly in the weathered serpentinite, often towards the top, and partially in the overlying limonitic laterite.

Residual deposits of chromium

It is possible that in a few years time we will see a chromium-bearing laterite at Range Well in Western Australia brought into production (Anon. 1989c). The 31 Mt of reserves that have been outlined grade 3.64% Cr and occur in laterite overlying ultramafic rocks. The chromium that might be produced would

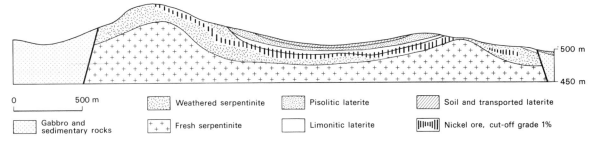

Fig. 19.4 Diagrammatic section of the Greenvale Serpentinite, Queensland, Australia, showing the position of the nickel orebody. (After Fletcher & Couper 1975.)

be suitable for the chemical industry or the manufacture of a 12–14% chromium pig iron for stainless steel production. The latter possibility arises from the fact that the chromium 'occurs predominantly . . . in hematite and goethite'.

Residual deposits of titanium

Anatase deposits in the weathered overburden of alkalic rocks in Brazil form some of the world's largest titania resources and these may in time replace the orthomagmatic and placer sources. Known resources are put at 300 Mt grading 20% or more anatase; see references in Force (1991). These deposits have been formed over three of the 44 known alkalic stocks disposed around the Brazilian Parana Basin. Weathered mantles on some of the other stocks contain important resources of REE, phosphate and baddeleyite. The mantles over the titaniferous rocks are up to 200 m thick and have a basal phosphate-rich zone.

Residual deposits of REE

Mention has been made in Chapter 9 of the strange occurrence of a tropical-type regolith over part of the Sokli Carbonatite that lies within the Arctic Circle. A similar regolith, but much richer in REE, overlies a carbonatite at Mount Weld near Laverton, Western Australia—a more likely setting for the development of such a weathering product! Inferred reserves (equivalent to probable ore) include 7.4 Mt with REO grading 9.4%, 6.1 Mt with Y_2O_3 grading 0.33%, 22.8 Mt with Nb_2O_5 grading 1.86% and 2.2 Mt with Ta_2O_5 grading 0.099%. These reserves lie within a larger lower grade resource with 39 Mt of REO. The mineralization is complex, more diverse than comparable Brazilian deposits and generally occurs from 30 to 75 m depth with thicker sections

reaching 100 m (Anon. 1988b). Some details of the Brazilian occurrences are given by Mariano (1989).

Supergene enrichment

Although it is applied more commonly to the enrichment of sulphide deposits, the term supergene enrichment has been extended by many workers to include similar processes affecting oxide or carbonate ores and rocks such as those of iron and manganese. In supergene sulphide enrichment the metals of economic interest are carried down into hypogene (primary) ore where they are precipitated with a resultant increase in metal content, whereas in the case of iron and manganese ores it is chiefly the gangue material that is mobilized and carried away to leave behind a purer metal deposit.

Supergene sulphide enrichment

Surface waters percolating down the outcrops of sulphide orebodies oxidize many ore minerals and yield solvents that dissolve other minerals. Pyrite is almost ubiquitous in sulphide deposits and this breaks down to produce insoluble iron hydroxides (limonite) and sulphuric acid:

$$2FeS_2 + 15O + 8H_2O + CO_2 \rightarrow 2Fe(OH)_3 + 4H_2SO_4 + H_2CO_3 \text{ and}$$
$$2CuFeS_2 + 17O + 6H_2O + CO_2 \rightarrow 2Fe(OH)_3 + 2CuSO_4 + 2H_2SO_4 + H_2CO_3.$$

Copper, zinc and silver sulphides are soluble and thus the upper part of a sulphide orebody may be oxidized and generally leached of many of its valuable elements right down to the water table. This is called the zone of oxidation. The ferric hydroxide is left behind to form a residual deposit at the surface and this is known as a gossan or iron hat—such features are eagerly sought by prospectors. As the water percolates downwards through the zone of oxidation, it may, because it is still carbonated and still has oxidizing properties, precipitate secondary minerals, such as malachite and azurite (Fig. 19.5).

Often, however, the bulk of the dissolved metals stays in solution until it reaches the water table below which conditions are usually reducing. This leads to various reactions that precipitate the dissolved metals and result in the replacement of primary by secondary sulphides. At the same time, the grade is increased and in this way spectacularly rich bonanzas can be formed. Typical reactions may be as follows:

$PbS + CuSO_4 \rightarrow CuS + PbSO_4$ (Covellite + anglesite),

$5FeS_2 + 14CuSO_4 + 12H_2O \rightarrow 7Cu_2S + 5FeSO_4 + 12H_2SO_4$ (Chalcocite),

$CuFeS_2 + CuSO_4 \rightarrow 2CuS + FeSO_4$ (Covellite).

This zone of supergene enrichment usually overlies primary mineralization, which may or may not be of ore grade. It is thus imperative to ascertain whether newly discovered near surface mineralization has undergone supergene enrichment, for, if this is the case, a drastic reduction in grade may be encountered when the supergene enrichment zone is bottomed. For this purpose a careful polished section study is often necessary.

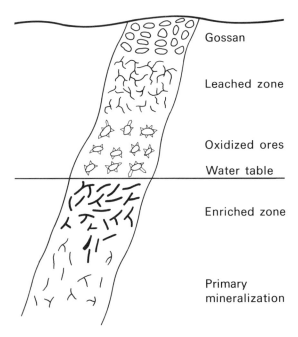

Fig. 19.5 Generalized section through a sulphide-bearing vein showing supergene enrichment. (Modified from Bateman 1950.)

Clearly, such processes require a considerable time for the evolution of significant secondary mineralization. They also require that the water table be fairly deep and that ground level is slowly lowered by erosion—this usually means that such phenomena are restricted to non-glaciated land areas.

Supergene enrichment has been important in the development of many porphyry copper deposits and a good example occurs in the Inspiration orebody of the Miami district, Arizona (Fig. 19.6). Primary ores of this district are developed along a granite–schist contact with most of the ore being developed in the schist. The unenriched ore averages about 1% Cu (Ransome 1919) and consists of pyrite, chalcopyrite and molybdenite. Supergene enrichment increased the grade up to as much as 5% in some places. The schist is more permeable than the granite and more supergene enrichment occurred within it. The enrichment shows a marked correlation with the water table (Fig. 19.6), where it starts abruptly. Downwards, it tapers off in intensity and dies out in primary mineralization. Chalcocite is the main secondary sulphide, having replaced both pyrite and chalcopyrite.

The nature of the primary (hypogene) mineralization can have a significant influence on the development of supergene copper mineralization and Anderson (1982) and Titley (1982b) have discussed this effect. Highly pyritic primary mineralization can produce more sulphuric acid during weathering giving rise to more efficient leaching of copper than in less pyritous deposits and, in suitable circumstances, rich blankets of supergene ore may then develop. A very useful recent study of supergene copper mineralization can be found in Cook (1988).

The leached and oxidized zones may not be without economic importance as ores. One of the world's largest open pit gold–silver mines, Pueblo Viejo, Dominican Republic (27 Mt grading 4.23 g t^{-1} Au and 21.6 g t^{-1} Ag plus another body of 14 Mt) is developed in the oxidized zone of sulphide protore (Russell *et al.* 1981). Another and fascinating example is the world's only germanium–gallium mine, the Apex Mine, Utah (Bernstein 1986). At this locality germanium and gallium have been concentrated in the secondary iron oxides—material which is often examined only for the exploration data it may yield. It is very probable that there are more such orebodies awaiting discovery by mineral sleuths who have in mind the needs of the high tech. industries.

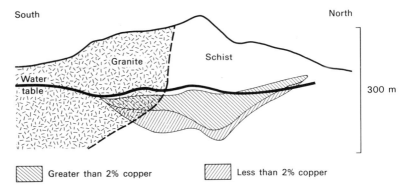

Fig. 19.6 Cross section through the Inspiration orebody at Miami, Arizona, showing the relationship between the high and low grade ores and the position of the supergene enriched zone (Cu over 2%) relative to the water table. (Modified from Ransome 1919.)

Supergene enrichment of banded iron formation

Most of the world's iron ore is won from orebodies formed by the natural enrichment of banded iron formation (BIF). Through the removal of silica from the BIF the grade of iron may be increased by a factor of two to three times. Thus, for example, the Brookman Iron Formation of the Hamersley Basin in Western Australia averages 20–35% Fe but in the orebodies of Mount Tom Price it has been upgraded to dark blue hematite ore running 64–66% Fe.

The agent of this leaching is generally considered to be descending ground water, though a minority school in the past has argued the case for leaching by ascending hydrothermal water. In general, these orebodies show such a marked relationship to the present (or a past) land surface that there is little doubt that we are dealing with a process akin to

lateritization. This relationship is exemplified clearly by the orebodies of Cerro Bolivar, Venezuela (Ruckmick 1963), which are shown in Fig. 19.7. These orebodies are developed in a tropical area having considerable relief and thus the ground water passing through them can be sampled in springs emerging from their flanks. This water carries 10.5 ppm silica and 0.05 ppm iron. Its pH averages 6.1. Clearly, the rate of removal of silica is about 200 times that of iron so that the iron tends to be left behind whilst the silica is removed. The iron is not entirely immobile and this probably accounts for the fact that many orebodies of this type consist of compact high grade material of low porosity. Obviously, the voids created by the removal of silica have been filled by iron minerals, normally hematite. This process seems to be taking place at Cerro Bolivar at the present time and Ruckmick has calculated that

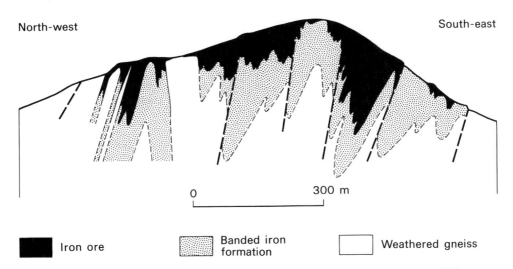

Fig. 19.7 Cross section through the iron orebodies of Cerro Bolivar, Venezuela. (After Ruckmick 1963.)

South   North

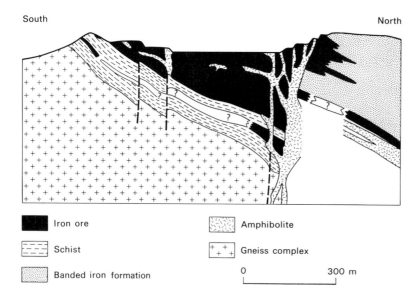

	Iron ore		Amphibolite
	Schist		Gneiss complex
	Banded iron formation		0 300 m

Fig. 19.8 Longitudinal section through the Iron Monarch orebody, South Australia. (Modified from Owen & Whitehead 1965.)

if the leaching process occurred in the past at the same rate as it is proceeding today then the present orebodies would have required 24 Ma for their development.

That downward-moving waters were the agent of leaching is attested to by the frequency with which enriched zones of BIF occur in synclinal structures into which downward-moving ground water has been concentrated. The leaching is, of course, intensified if the BIF is underlain by impervious formations, such as slate as in the Western Menominee District of Michigan and schist in the Middleback Ranges of South Australia. The BIF of the Middleback Ranges occurs in the Middleback Group. The rocks of this group have suffered folding along northerly axes followed by easterly cross-folding. This produced a number of domes and basins. The domes have largely been removed by erosion leaving isolated cross-folded synclinal areas scattered across an older basement. Downward-moving ground waters have produced orebodies by leaching the BIF in the keels of these synclines. The evidence for this mechanism of formation is particularly good at the Iron Monarch body. This orebody occupies a northward plunging syncline. The main northern down plunge termination of the orebody is a north-west trending dyke complex which appears to have acted as a dam to the downward-moving water preventing extensive leaching from proceeding further down the keel of the syncline (Fig. 19.8). For this reason, very little enrichment has occurred on the down plunge side. Total production from this

and similar orebodies in the Middleback Ranges up to 1989 was 185 Mt of high grade ore. Present production is 2 Mt p.a. of ore grading 64+% Fe.

Enormous orebodies of this type continue to be found, such as the N4E Mine in the Carajas region of Brazil. This monster contains at least 1251 Mt of ore averaging 66.13% Fe and only 1% SiO_2 and 0.038% P.

Supergene enrichment of manganese deposits

Deep weathering processes akin to lateritization can also give rise to the formation of high grade manganese deposits. Although not comparable in geographical area with the previously described sedimentary deposits (Chapter 18), they nevertheless form important accumulations of manganese. Large deposits of this type occur at Postmasburg in South Africa, Groote Eylandt, Australia and in Gabon, India, Brazil and Ghana.

Since manganese is more mobile than iron, the problem arises as to why manganese is retained in the weathering profile. The reason is probably because in most cases the residual manganese deposits are formed on low grade manganiferous limestones and dolomites. These rocks are low in iron and silica, and manganese takes the place of iron in the laterite profile. The carbonates are easily dissolved (by comparison with silica) and the higher pH that probably prevailed in these environments would have tended to immobilize the manganese.

Residual and other manganese deposits are described in detail by Roy (1981), an excellent description of the supergene ores at Groote Eylandt is given by Pracejus *et al.* (1988) and a summary of supergene manganese deposits is given in Table 18.3. In view of the great economic importance of the vast deposits of the Postmasburg Field it is pertinent to point out that not all workers accept a supergene origin, and De Villiers (1983) has detailed evidence for a hydrothermal origin.

Supergene enrichment of uranium deposits

An important example of this enrichment is the Rössing Mine (Chapter 9), where 40% of the uranium is present in secondary minerals.

20 / Industrial minerals

Introduction

As noted in Chapter 1 industrial minerals have been defined as including any rock, mineral or other naturally occurring substance of economic value but excluding metallic ores, mineral fuels and gemstones. Nevertheless, as was demonstrated using bauxite as an example, many metallic ores have non-metallurgical end uses and these may be remarkably diverse and important. Industrial minerals dominate metals in tonnage produced and total product value (Tables 1.2, 1.3) and have more rapid growth rates. Recycling and substitution are of little general importance in the industrial minerals field compared with that of metals and prices are more stable.

For some industrial minerals the unit value is so low that transport costs over any appreciable distance can make them non-competitive and an efficient bulk transport infrastructure is essential; consequently deposits of these minerals must be exploited close to a market. For many industrial minerals, particularly those of low unit value, the first essential to be discovered is a market rather than a deposit. This is the converse of the situation for metals and industrial minerals of high unit value, as these can be traded internationally.

A good example of the value of exploring for a market first is given by Harris (1986). The point he made is that industrial minerals companies should search out areas where thriving industries are consuming bulky industrial minerals that have to be transported long distances. An example is western North America where large volume extender and filler minerals used in industries such as paper manufacture, e.g. kaolin, talc and $CaCO_3$ were, at the start of the 1980s, being brought in from far afield—all the kaolin right across the continent from Georgia and all the talc from Ontario. Examination of the consumption of these minerals in the five major markets of adhesives and sealants, paint, paper, plastics and rubber suggested an average growth rate of 5% p.a. in this region. This has greatly stimulated the development of new sources in the western part of the continent. Thus

kaolin deposits in Saskatchewan and Idaho were being evaluated or developed in the 1980s, as were talc deposits in Montana and $CaCO_3$ in Washington and California. Such developments will be well placed for exports to expanding markets of other countries of the Pacific Rim. Examples of the value of market creation will be given in the section on olivine.

All the above and other important points, such as the definition of ore, grade (including deposit homogeneity and physical properties), mineralogical form, grain size and environmental considerations are discussed in Chapter 1 and these matters will not be reconsidered here.

Some recent trends should be noted. Product purity, grain size and other specifications have become, and will continue to become, more rigid, making stringent process and quality control practices necessary. For example, in the manufacture of high quality paper only the finest grades of filler and coating material can be used. The mineral processing techniques for industrial minerals continue to assume more elaborate forms, e.g. the processing of kaolin may now involve grinding and airfloating, washing, delamination, calcining, magnetic separation, etc.; mica and talc may be micronized (ultra-fine grinding) and so on. New applications are frequently developed, often as a result of research by mineral producers themselves.

Although industrial minerals are generally thought of as having a wide occurrence, a small number are rare. Thus for boron, garnet, iodine, lithium minerals, natural sodium carbonates, nitrates and vermiculite, about 90% of world production is concentrated in only three countries. In addition, for a total of 13 industrial minerals, more than 95% of world production comes from only five countries. Besides the commodities listed above, these include bromine, industrial diamonds, REE, rutile, wollastonite and zirconium and hafnium. Much of the supply of another 12 commodities is mined in only five countries (Noetstaller 1988). There is thus room for much exploration and competition, particularly if new markets can be developed!

Many different classifications of industrial minerals have been proposed and a valuable discussion of these can be found in Noetstaller (1988). It is felt that a consideration of these would not be a worthwhile use of space in this short work. However, I would draw attention to Noetstaller's 'Profile of Industrial Minerals by End-Use Classes' as this may assist geologists in identifying potential uses for any materials found in their concession areas. Another important source of information in this respect is the specialist articles in the journal *Industrial Minerals*. The basic reference work is Lefond (1983) but for background reading for this chapter I would recommend the short review by Scott (1987) and the book by Harben & Bates (1990).

The variety of industrial minerals and their uses is so vast that only a book like Harben & Bates can begin to do justice to their diversity. I have therefore chosen nine topics or minerals in an attempt to illustrate this great variation and the importance of physical or chemical properties, or a combination of both, in the uses to which these resources can be put. These topics are of course supplemented by the industrial minerals and the descriptions of non-metallurgical uses of metallic ores covered in previous chapters. The chosen topics are aggregates and constructional materials, clays, evaporites, graphite, limestone and dolomite, magnesite, olivine, perlite and phosphate. Subjects touched upon already and the relevant pages are: bauxite (pp. 3–4, 262–264), baryte (pp. 211–213), beryl (p. 124), chromite (pp. 128–134), diamonds (pp. 104–113), fluorite and baryte (pp. 233–239), graphite (p. 167), lithium (p. 124), manganese (pp. 257–261), phosphate (pp. 114–120), talc (pp. 165–167), titania (pp. 136–138, 247–250, 267), and vermiculite (pp. 114–120).

Aggregates and constructional materials

This is such an enormous subject that whole books have been devoted to it, e.g. Collis & Fox (1985) and Prentice (1990). Here it is possible only to skim through the topic, which includes coarse aggregates, fine aggregates, structural clay products, cement and concrete, glass and insulation materials and lightweight aggregates.

Coarse aggregates

Industrial usage separates coarse and fine aggregates since they are used for different purposes. Coarse aggregates have rock particles > 5 mm diameter. In the USA most of the particles should be retained on an ASTM No. 4 (3/8 in, 9.5 mm) sieve; in the UK aggregate grading must comply with BS regulations using BS sieves. Of course the volume and size of particles retained by a particular sieve are governed by the particles' shapes. This means that, although coarse aggregates are sold in nominal size grades (e.g. 14, 20 mm, etc.) produced by screening processes, undersize and oversize particles are present in each size grade. Limits on the permitted amounts of undersized and oversized particles are set in most countries.

Above 5 mm size particles may occur naturally in glacial, alluvial, beach or marine gravels or may be produced by crushing igneous, sedimentary or metamorphic rocks. In desert regions it may be necessary to use carbonate sands of coastal origin or material in active wadis in mountainous areas. In tropical regions near surface rocks may be weakened by weathering but occasional hills of fresh rock may be present.

Aggregates must be tested carefully to assess their suitability for various functions. If they are to be embedded in bitumen or cement they must react favourably with them. Resistance to heavy loads, high impacts and severe abrasion, together with durability are all important. Many properties have to be tested.

Compressive strength is measured by applying uniaxial pressure to a cube of rock and recording the pressure when failure occurs; this is the uniaxial or unconfined compressive strength (UCS). Weak to moderately strong rocks, e.g. some limestones and sandstones, fail at 5–50 MPa; granites are strong to extremely strong and fail at 90–260 MPa.

Water absorption by aggregates during concrete production can lead to shrinkage and cracking; some basalts and mudstones have high absorption, granites and limestones low absorption.

Materials for road surfacing must score well in the following four tests. *Aggregate crushing value* (ACV) is a measure of the load-bearing capacity, which cannot always be correlated with the UCS value. *Aggregate impact value* (AIV) is measured by dropping a hammer of standard weight from a standard height 15 times on to a sample. *Aggregate abrasion value* (AAV) is a measure of the weight loss during abrasion of the sample by sand or by rotation within a steel cyclinder. *The polished stone value* (PSV) is a measure of the ability of aggregates to inhibit skidding.

Flakiness is a measure of the particle shape. Few rocks on breaking are devoid of any tendency to develop elongate or flattish rather than cuboidal fragments. The above test values tend to decrease with an increase of flakiness. *Resistance to weathering* is difficult to assess but there are tests that attempt to measure this property. Quarry operators have to take great care that weathered rock is not included in their aggregates, otherwise unsound aggregate may be delivered to the consumer.

Normally, coarse aggregate producers are sited very close to their markets, but in special circumstances they can export over considerable distances, e.g. the Glensanda Quarry, Scotland where shallow draft, bulk carrier ships can load alongside the granite quarry. In this case the benefits of large scale production and cheap transport permit profitable export of aggregate to markets as far removed as Texas.

Fine aggregates

With an upper size limit of 5 mm these can be much coarser than the geologist's sand grade. They may be derived from natural loose materials, beach sands, dune sands, etc., by crushing rock or by crushing artificial materials, such as slag. Fine aggregates may be used in concrete making; road construction; precast concrete products, such as beams and other large sections for buildings and bridges; drainage pipes; tiles; mortars; plasters; trench fillings on which to lay cables, water pipes, etc.; filter materials; glass manufacture; foundry sand and many other purposes. With such a variety of uses many different specifications and tests must be met and passed by producers.

In many of the above uses the fine aggregate acts as a filler, cutting down the amount of binding material (cement, bitumen) that is required or imparting stiffness to a mixture of materials. Thus, because the packing behaviour of the grains is important, grain size analysis and shape assessment (roundness and sphericity) are essential in determining the suitability of fine aggregates for particular uses.

Sources of coarse and fine aggregates

Naturally occurring aggregates—sand and gravel—are common in many parts of the world. Difficulties in locating such deposits may obtain in lower latitudes where chemical weathering is dominant and laterization removes silica from many source rocks, so that quartz sands may be uncommon. In higher latitudes, especially those which have been glaciated, sand and gravel deposits are common. The major mineral in these sands is usually quartz. Some of the locations in which sand and gravel deposits may be found are shown in Fig. 20.1. *River terrace deposits,* like those of river flood plains, are commonly gravelly below and more silty and clayey above. Older terraces have suffered more weathering and erosion and are therefore more likely to be contaminated by clay and iron oxides. In *alluvial cones,* built up on plains by rivers carrying debris from an upland, the coarse material is proximally deposited. *Kames and eskers* are gravel-rich deposits of glacial origin. *Beach deposits* can be very variable and just one of many possible configurations is sketched.

In all these deposits deleterious matter may be present and the commonest in fine aggregates are clay, coal, carbonates, plant material and iron oxides. Clay can be eliminated by scrubbing (washing). Appreciable amounts of coal fragments may render a sand deposit useless; smaller amounts may be removed, especially from gravels, by jigging. Carbonates may occur in quartz sands as shells, concretions or thin limestone bands. Their presence may greatly reduce the sand's value, e.g. they will cause shrinking and cracking in concrete. Likewise the presence of iron oxide coatings on the grains may restrict the uses to which a sand may be put. Plant material can be removed by washing and, if necessary, by calcining (expensive!).

Structural clay products

Besides bricks, important clay products used by the building industry are vitrified clay pipes, floor, and wall tiles. Most brickclays are won from unconsolidated or soft sediments. Quartz may form up to 90% of the clay, but satisfactory bricks can be made from quartz-poor clay. The clay minerals are usually mixtures of kaolinite, illite, smectite and chlorite. A high proportion of smectite will impart a higher drying shrinkage than the other clay minerals and may give rise to other difficulties. After being ground and moistened with water, a brickclay must be capable of taking a good shape by moulding or pressure and of retaining that shape without detrimental shrinkage, warping or cracking when the bricks are dried and then fired. A high clay content will make the material too plastic and 20% is about right. Iron minerals are converted to hematite on

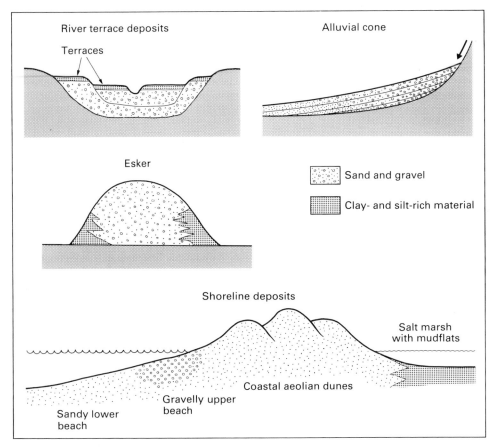

Fig. 20.1 Sketch sections (at different scales) of selected types of sand and gravel deposits. For discussions of these see text.

firing and give bricks their characteristic colours. Ochres, manganese oxides and other colorants may be added to the mix. There are many other factors, such as calcite and carbon content, firing conditions, etc., that contribute to the finished colour. Carbon can be a valuable constituent in that it lowers fuel costs.

In Europe and North America brickworks are now large sophisticated units which require thick, consistent sources of supply. Suitable materials are won from deposits of Upper Palaeozoic to Holocene age. In central England Carboniferous and Triassic marls are much used, but about half the British brick output comes from the basal 70 m of the widespread Upper Jurassic Oxford Clay. This excellent brick-making clay has illite as the dominant clay mineral with a smaller amount of kaolinite; chlorites, smectite and vermiculite are also present. Calcite is low

and the appreciable content of finely divided carbon reduces the firing cost by as much as 75%.

Clay specifications for vitrified clay pipe manufacture are much more rigid than for brick making and imported clays may have to be mixed with local materials to satisfy these requirements.

Decorative floor and wall tiles are made from specific mixes of different types of clay; ball clay, china clay, limestone, quartz sand, etc. of various origins rather than from the product of a single clay pit.

Cement and concrete

Portland cement is manufactured by calcining a mixture of limestone and clay or shale. About 18–25% of clay is added, less if the limestone is argillaceous. The limestone does not have to be chemically very pure but pure limestones are preferred, and the magnesia content must be low. After calcining, the resultant clinker is ground to produce cement and about 5% of gypsum or anhydrite is added to act as a setting retardant. A useful discussion of raw material requirements for cement manufacture can be found in Dunn (1989).

Basic *concrete* mixes are around 60–80% of aggregate (about one-third of this is fine aggregate), 6–18% cement and 14–22% water. Additives are used to improve the fluidity, aid the setting, waterproof and colour the concrete. Quality control of all the constituents is most important.

A number of recent structural failures in concrete dams, flyovers and buildings have apparently been caused by aggregate–cement reactions. They are due to the presence of alkalis released by cement setting reactions into the residual fluids. These fluids will react with cryptocrystalline silica (in chert and flint, silicified limestone, greywacke, etc.) to form an expansive silica gel which can exert pressures as high as 14 MPa and shatter the concrete. Careful selection of aggregates to exclude these reactive materials is at present the only answer to this problem. A useful discussion of deleterious substances in aggregates can be found in Mather & Mather (1991).

Building stones

Stone is used in a variety of ways in the building industry. Roughly broken stone may be used for rock-fill and large blocks for coastal defences. Better shaped blocks are used as armourstone on breakwaters and other harbour works.

Building stone that is quarried, cut and dressed into regularly shaped blocks is termed dimension stone (ashlar). For use as a building stone a rock should possess mineral purity, that is, be free from decomposition products produced by secondary alteration of the rock. It should have the necessary mechanical strength to sustain loads and stresses in service (i.e. a high crushing strength). The durability (resistance to weathering) should be high if the stone is to be used in the polluted atmospheres of cities, sulphur dioxide being particularly deleterious. The hardness, workability and directional properties are likewise important. The hardness ranges from that of soft coquina or limestone to that of granite. Workability depends partially on hardness. Limestone is easy and cheap to dress but granite is expensive. Workability also depends on the presence or absence of planes or incipient planes of splitting

(joints). Porosity and permeability determine the water content and therefore the susceptibility to frost action. The texture affects the workability: fine-grained rocks split and dress more readily than coarse ones. Lastly the colour should be attractive and permanent and not affected by weathering, nor should there be accessory minerals such as pyrite, that will oxidize and produce unsightly stains.

Glass

The Crystal Palace, built for the Great Exhibition in London in 1851, was the first building in which glass was a major feature. Glass is now an important material in buildings and large sheets are produced by rolling or drawing, or by floating on molten metal. As glass will not be covered elsewhere in this book other uses of glass will be touched on here.

Over 90% of manufactured glass is of the soda–lime–silica type for which the raw materials are high purity quartz sand (glass sand), limestone and soda ash (sodium carbonate). Some dolomite and/or feldspar (or aplite or nepheline syenite) is also added to improve chemical durability and resistance to devitrification. To obtain certain characteristics (e.g. heat resistance) other mineral products, such as borates, fluorite, arsenic and selenium, may be added.

Loosely consolidated sands and sandstones may make suitable glass sands. In these the silica content ranges from 95 to 99.8%. Alumina up to 4% is acceptable for common glass production but it must be < 0.1% for optical glass, although a small amount does help to prevent devitrification. Iron oxides can be tolerated only for green and brown bottle glass; otherwise iron as Fe_2O_3 must be < 0.05% or even < 0.015% for some purposes. The sand grains should be even-sized (i.e. well sorted) and less than 20 but more than 100 mesh (ASTM). Few sands are sufficiently pure that no processing is necessary. Refractory grains, such as chromite and zircon, must be removed as these will not dissolve in the melting furnace but will persist to form black and colourless specks in the glass.

Gypsum

This mineral has been long used by man. It was known and prized by the Assyrians and Egyptians and the latter used it to make plaster for the pyramids. In the form of alabaster it has been long used as an ornamental stone and in mediaeval times alabaster ornaments, sculptures and altarpieces were exported from the Nottingham area of England to many other parts of Europe.

When heated to 107°C gypsum becomes the hemihydrate $CaSO_4.\frac{1}{2}H_2O$—plaster of Paris. It then can be mixed with water, and sometimes with fine aggregate, for rendering walls and ceilings and the manufacture of plasterboard. This accounts for more than 70% of world consumption. Other uses are in cement and fertilizers and as a filler in paper, paint and toothpaste and in the manufacture of gypsum muds for oilwell drilling.

Gypsum is an evaporite and as such accumulated in basins or on supratidal salt flats under arid conditions (sabkhas); see below. It is now quarried or mined from seams in rocks or the caprocks of salt domes, where ages range from Cambrian to Tertiary, but the largest accumulations in Europe and North America are in the Permo-Triassic. Gypsum is a low cost product, about £6–7 t^{-1} CIF (carriage, insurance and freight) in the UK in October 1991, and only in favourable circumstances can it be transported over long distances. Nevertheless, Spain exports nearly 1 Mt p.a. to the eastern USA for plasterboard manufacture. The future of gypsum mining lies under a cloud as considerable substitution of natural gypsum by desulphogypsum from coal-fired power stations can be expected in the future. Japan is now a leading country in this trend (Ellison & Makansi 1990).

Insulators and lightweight aggregates

A number of rocks and minerals are used for noise and heat insulation in buildings. *Pumice* is mixed with cement and moulded to form lightweight insulating blocks. *Perlite* (considered later) has a similar use as does vermiculite (pp. 114–120). Pumice and perlite being rhyolitic, volcanic glass are obtained only from Tertiary and younger rocks. Other minerals used for this purpose are diatomite and expanded micas.

Clays

Clay minerals are layered silicates with sheets of SiO_4 tetrahedra combined in various ways with sheets containing $Al(O,OH)_6$ and/or $Mg(O,OH)_6$ octahedra. Disorder in the stacking of the layers results in the growth of only very small grains (usually < 0.002 mm diameter) with unusual surface properties. Water and other molecules are attracted to their surfaces resulting in the development of plasticity. In some clay minerals exchangeable cations (Na^+, K^+, H^+, Ca^{2+}, or Mg^{2+}) are essential to their structure and give them important chemical properties with industrial applications. Clay deposits normally consist of one or more clay minerals with varying proportions of non-clay minerals, the most common being quartz, feldspar, carbonate minerals, gypsum and organic material. The types of clay raw materials and some of their uses are given in Table 20.1. Most commercial clays have (a) kaolinite, (b) montmorillonite (smectite), or (c) attapulgite as the dominant clay material. Halloysite is chemically similar to kaolinite and sepiolite is similar to attapulgite. They therefore will be considered here under these three headings.

Kaolinite-bearing clays

Kaolin

This is a soft white plastic clay with a low iron content that is often called china clay. Commercial deposits contain 10–95% kaolinite. Generally kaolin deposits should have kaolinite with at least 55% of particles finer than 2 μm, and should be extremely white with a high reflectance.

Industrial uses of kaolin are based mainly on its physical properties and different applications demand distinct combinations of properties so that individual grades are rarely suitable for every use. Kaolins are assessed for brightness, particle size distribution, viscosity (important for the valuable paper-coating clays) and for pottery clays the iron and potash contents, fired brightness, casting behaviour and other ceramic properties. Kaolin which is to be fired should be free of sulphur; this can be present in the form of alunite as well as pyrite. The quantity and quality of the kaolin can be improved by processing. Firstly the non-kaolin minerals are largely removed, then flotation is used to remove the remaining iron and titanium minerals, and magnetic separation completes this process and removes micas. Bleaching may be used to increase the whiteness and delamination to break down any kaolinite into individual platelets.

Total world production in 1988 was 23.2 Mt of which the USA produced about 39%, the UK 14% and the USSR 13%. The remaining production came from many countries spread all over the world. Kaolin has a medium to high unit value and is traded internationally. In October 1991 prices were as

Table 20.1 Clay raw materials, their mineralogy and uses. (Modified from Scott 1987)

Kaolin (sometimes known as china clay): kaolinite of well ordered crystal structure plus very minor amounts of quartz, mica and sometimes anatase. Highest quality material will be almost pure kaolinite and contain > 90% of particles < 0.002 mm. Low to moderate plasticity. Uses: paper filler and coating; porcelain and other ceramics; refractories; pigment/extender in paint; filler in rubber and plastics; cosmetics; inks; insecticides; filter aids

Ball clay (plastic kaolin): kaolinite of poorly ordered crystal structure, with varying amounts of quartz, feldspar, mica (or illite) and sometimes organic matter. Kaolinite particles usually finer grained than in kaolin. Moderate to high plasticity. Uses: pottery and other ceramics; filler in plastics and rubber; refractories; insecticide and fungicide carrier

Halloysitic clay: halloysite with varying amounts of quartz, feldspar, mica, carbonate minerals and others. Similar particle size distribution to ball clay. Moderately plastic. Uses: pottery and other ceramics; filler in plastics and rubber

Refractory clay or fireclay: structurally disordered kaolinite with some mica or illite and some quartz. Very plastic. Uses: refractories (firebricks); pottery, vitrified clay pipes

Flint clay: kaolinite and diaspore. Plastic only after extended grinding. Uses: refractories

Common clay and shale: kaolinite, and/or illite and/or chlorite, sometimes with minor montmorillonite, quartz, feldspar, calcite, dolomite and anatase. Moderately to very plastic. Uses: bricks, tiles and sewer pipes

Bentonite and Fullers' earth: montmorillonite (smectite) with either Ca, Na or Mg as dominant exchangeable cation. Minor amounts of quartz, feldspar and other clay minerals. Forms thixotropic suspension in water. Forms a very plastic sticky mass. Uses: iron ore pelletizing; foundry sand binder; clarification of oils; oil-well drilling fluids; suspending agent for paints; adhesives, etc.; absorbent

Sepiolite and attapulgite: sepiolite or attapulgite with very minor amounts of montmorillonite, quartz, mica, feldspar. Thixotropic and plastic as montmorillonite clays. Uses: absorbent; clarification of oils; some oil well drilling fluids; special papers

---

follows: coating clays £75–120 t^{-1}, filler clays £45–65 t^{-1} and pottery clays £30–90 t^{-1} (all FOB European ports).

Primary kaolin deposits are formed by the alteration of feldspar-rich rocks, such as granites, gneisses, arkoses and feldspathic conglomerates, by weathering or hydrothermal processes or a combination of both. Rocks low in iron form the best protoliths, e.g. leucogranites. A high iron content may give rise to non-economic kaolins unless the protolith lay beneath peat and was leached by humic acids; in this way fireclays lying beneath coals have been formed. Tectonized, faulted or well jointed rocks in which weathering can penetrate to considerable depths are particularly favourable parent rocks. The traditional view is that kaolinization results from intense weathering in hot humid climates and therefore a number of workers have suggested that palaeoclimatic criteria are important for prospecting for kaolin deposits. Thus by the use of plate tectonic considerations, it may be possible to decide whether certain rocks in particular regions might have been exposed to tropical weathering (Kužvart & Böhmer 1986). The traditional view has been challenged recently by Bird & Chivas (1988) who pointed out that the Lower Permian kaolin deposits of eastern Australia were formed during cold or glacial conditions arising from the poleward drift of Australia during the late Palaeozoic. Their isotopic investigation of kaolinite samples indicated that they formed from cold waters—a conclusion that is difficult to reconcile with the classical interpretation of kaolinite weathering as a tropical or subtropical phenomenon.

Hydrothermal alteration also may form kaolin, but in general this appears to be far less important as a genetic process than weathering.

Kaolin deposits formed by the above processes may remain *in situ* or may be eroded and deposited to form sedimentary deposits, such as the important Cretaceous and Tertiary deposits of Central Georgia and South Carolina. Here the deposits occur randomly in quartz sands as elongate or curved lenses up to 13 m thick and 1.3 km long. In such deposits the sand may be an important by-product. In south-west England the primary kaolin deposits are famous for their size and quality and they have yielded some 128 Mt since production began in the mid-eighteenth century (Highley 1990).

The English deposits were formed in Hercynian granites in which sodic plagioclase was preferentially kaolinized, only very intense kaolinization has altered the K-feldspar (Fig. 20.2). The remaining minerals are largely unaltered and the orebodies vary from hard unaltered granite to soft kaolinite.

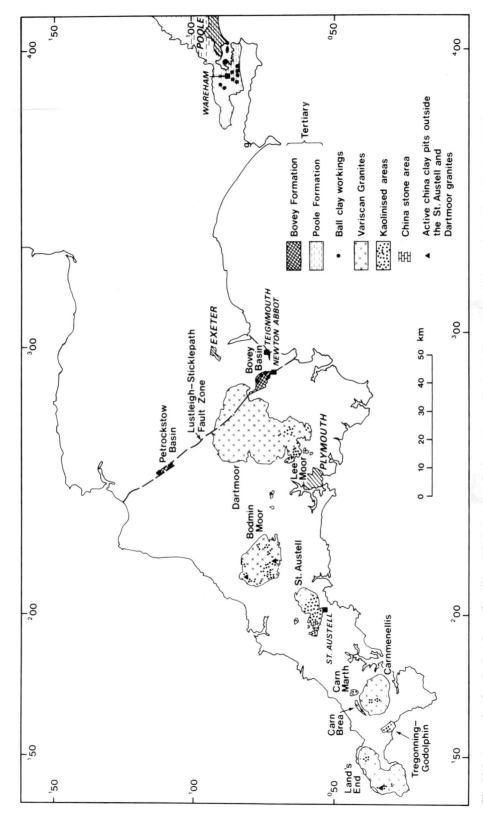

Fig. 20.2 A map showing the locations of kaolin and ball clay deposits in south-west England. (Reproduced with permission from Highley 1990.)

Typical deposits are funnel- or trough-shaped, narrowing downwards and extending as deeply as 230 m. Deposits in all the plutons have been exploited but production from the St Austell Granite now accounts for about 75% of the region's output, with much of the remainder coming from the south-western margin of the Dartmoor Granite.

Sheppard (1977) suggested that the English deposits resulted from the action of meteoric water during the tropical weathering known to have affected south-west England in the early Tertiary. He cited as evidence the isotopic values for δD and $\delta^{18}O$ obtained from the kaolinites and the general relationship of the deposits to present or recent erosion surfaces. On the other hand Bristow (1987b) suggested two phases of hydrothermal alteration, the first probably being associated with granite emplacement and the second being a much later convective flow induced by radiogenic heat from these high heat flow granites. Bristow cited the following points in favour of hydrothermal kaolinization: (a) unaltered granite overlies kaolinization at depths > 250 m; and (b) the kaolinization is associated with greisen-bordered, quartz–tourmaline veins and the crystallinity index of the kaolinite increases towards these veins. A third possibility is that hydrothermal processes prepared the granites for later weathering to complete the process. Or is the isotopic evidence simply the effect of a later overprint?

Ball clay

This is an old English term that refers back to the days when these clays were rolled into balls for transporting. Ball clays are plastic, sedimentary clays usually about 70% kaolinite but also containing illite, quartz, montmorillonite, chlorite and 2–3% of carbonaceous material. This organic matter helps to make the clay highly plastic and increases its green strength, i.e. its pre-firing strength. It burns on firing which aids in producing a white or near white body. High quality ball clays are used to make wall tiles, vitreous china, sanitary ware and stoneware, the lower quality for pipes, bricks and tiles—for other uses see Table 20.1. Prices in October 1991 ranged from £15–£110 t^{-1} FOB according to the grade. Ball clays are produced in significant commercial quantities in four countries only. In 1987 the USA was the largest producer with about 841 000 t, of which about 70% came from Tennessee. The UK is the largest exporter of ball clays, the other two main sources being the Cheb Basin, Czechoslovakia

and the Don Basin, USSR (Russell 1988a). Owing to the relative scarcity of ball clays, other types of clay, generally with lower white-firing characteristics, are used for much ceramic production—these are known as *plastic clays* and the main producers are Germany and France.

All major deposits of ball clay are sedimentary and their common content of carbonaceous material indicates a swampy depositional environment. Those in the UK (Fig. 20.2) were laid down in three basins during the early Tertiary. The deposits are interlayered with sands, lignites and lower quality clays. Within the Bovey Basin there are more than 40 workable, ball clay beds ranging in thickness from 1 to over 5 m. They lie within an area of about 46 km². The deposits in all three basins appear to have developed in shallow lakes fed by rivers carrying sediment from an upland area of granites, slates and greywackes that was undergoing surface kaolinization.

Refractory clay or fireclay

These clays are composed principally of kaolinite; they do not burn white and can withstand temperatures > 1500°C. Their use for the manufacture of refractories has suffered in recent years from competition by magnesite-, bauxite- and olivine-based products. Besides kaolinite they often contain gibbsite, boehmite and/or diaspore. For commercial use they must be low in mica and iron, which can produce low melting point glasses. In Europe they are largely Carboniferous and more indurated than ball clays. Their production is principally as a by-product of opencast coal working.

Flint clay

This is a compact clay, having a conchoidal fracture, composed mainly of kaolinite. It fires to a high density body having low permeability.

Smectite (montmorillonite) and attapulgite clays

These clays are composed principally of montmorillonite or attapulgite (palygorskite), which have many variants but several properties in common, such as medium to high surface area and sorptive properties and excellent decolorizing, binding and thickening powers. Attapulgite is virtually unaffected by electrolytes but the smectites are very reactive as they have a layer lattice structure in which internal substitution of Si^{4+} and Al^{3+} by lower valency cations has left unsatisfied negative charges. These are balanced by loosely held exchangeable cations, usually Ca^{2+}, Mg^{2+} and Na^+, on the interlayer surfaces. Their properties vary with the dominant cation but, by a simple sodium-exchange process, Ca-montmorillonite can be converted to the Na variety. Na-montmorillonite swells in water, is often known as swelling bentonite and is in great demand. Calcium bentonite does not have this property. In Britain it is known as fuller's earth and has been used since Roman times for fulling woollen cloth (i.e. removing the grease). In the USA material referred to as fuller's earth, which has similar properties to calcium bentonite, is attapulgite or sepiolite.

Smectite clays have a unique combination of physico-chemical properties leading to very varied industrial applications. These include small crystal size, high plasticity, sorption and cation-exchange capacity, large chemically active surface areas, swelling characteristics and important rheological, water-sealing and bonding properties (Highley 1990). The major use for bentonites is in oil well drilling muds. Attapulgite and sepiolite are also used in these muds, but sepiolite, being the only clay mineral stable at high temperatures, is favoured for muds used in drilling geothermal wells. These clays have a phenomenal number of other uses, e.g. acid-activated calcium bentonite is a high value-added chemical product used as a bleaching and refining agent in the processing of edible oils and fats; certain grades of bentonite are used as fillers in paper and yet others to make absorbent granules for pet litter!

Workable deposits of bentonite usually occur as beds of Mesozoic or Tertiary age. They may be 1–3 m thick and persist over many kilometres. The majority are believed to have been formed by the alteration *in situ* of volcanic ash of rhyolitic composition. The famous Miocene attapulgite–sepiolite deposits of Florida and Georgia are thought to have formed as sediments in a restricted, shallow marine environment. It is difficult to arrive at production figures for the minerals in this group because the nomenclature is very confused. About half the world's bentonite production of around 6 Mt is mined in the USA and a large proportion of the world total of attapulgite and sepiolite ($\sim$ 3.5 Mt). Seventeen other countries are significant producers of these minerals.

Evaporites

The principal products of these widespread deposits are gypsum, common salt and potash. Concentrations of borates, nitrates and other salts are relatively uncommon.

Nature, occurrence and genesis

Wherever evaporation exceeds rainfall evaporites may form. This may occur in the supratidal zone (sabkhas) or within a restricted body of water, which may occupy a small or large basin. Modern examples of these evaporite-forming environments are well known. There are also thick and extensive Phanerozoic evaporitic deposits, the so-called saline giants, of which no modern equivalents have been found. Some of these may have formed in deep restricted waters, others have clear shoreline and shallow-water features—see Schreiber (1986) in which a useful summary of ideas on evaporite formation can be found.

Quaternary and Recent evaporites occur mostly in subtropical zones between 15–35° latitude, in equatorial elevated plateaux, arctic deserts and rain shadows of high mountain ranges and continental regions, as in Central Asia. Ancient evaporites may have been formed in any of these environments. The oldest date back well into the Archaean but their best development was during the Phanerozoic in which their distribution is quite irregular in both time and space. Most evaporites are of marine origin but terrestrially formed deposits can be of economic importance.

Gypsum, anhydrite and halite are the most common evaporitic minerals but 39 other minerals—too many to discuss here—are normally listed as common (Schreiber 1986). Although many of these minerals may occur in the same deposit, many evaporites consist only of gypsum–anhydrite, whilst others consist of halite with little calcium sulphate. Potassium and magnesian salts are much less common but important reserves are present in the Devonian of the Williston and Elk Point Basins of western North America and in Byelorussia of the USSR, in the Carboniferous of the Paradox Basin, Utah, the Permo-Carboniferous of the Upper Kama Basin along the western edge of the Urals, USSR, the Permian Zechstein of northwestern Europe and the Permian of New Mexico and Texas, the Tertiary of Italy and Spain and other regions.

The evaporation of sea water leads first to the precipitation of calcium carbonate, followed by calcium sulphate, halite, potassium and magnesium salts. The natural deposition of minerals from concentrated sea water in brine pools and other restricted bodies of water follows this pattern. For the formation of deposits of any size a great volume of sea water must evaporate, e.g. if a basin of sea water 1000 m deep was evaporated completely, only about 14 m of evaporite would be deposited, of which 12.9 m would be halite, 1.5 m $MgCl_2$, 1 m $MgSO_4$, 0.7 m $CaSO_4 + CaCO_3 + CaMg(CO_3)_2$ and 0.4 m KCl reflecting the ratio of these salts in normal sea water. Clearly for the development of thick evaporite deposits considerable replenishment of the body of evaporating water by new sea water must take place. Relative humidity is an important control on the precipitation of evaporitic minerals in arid climates. For the development of brines from which halite can be precipitated the mean relative humidity must be $< 76\%$ and for potassium salts $< 67\%$; but most low-latitude coastal regions have relative humidities of 70–80%; however, low values do occur over land masses. Thus along many arid coastlines $CaSO_4$ will be the only precipitate and the best situation for halite and potassium salt precipitation will be a marine basin nearly surrounded by land (Schreiber 1986).

Evaporites may suffer a number of post-depositional changes. They may be replaced by other minerals, dissolved away to produce collapsed strata and dissolution residues or deformed during or after burial to produce contorted and brecciated beds and diapiric structures, including salt domes.

Sabkha deposits

The relative importance of these compared with basinal deposits is a moot point, particularly as they may flank evaporitic basins and one deposit type may pass laterally into the other. In addition, if sabkhas sink to form a deeper water lagoon, a vertical transition may develop. We therefore must not expect that individual evaporite deposits fall simply into one or another depositional model. The gypsum deposits of the Purbeckian (Upper Jurassic) of south-eastern England (Fig. 20.3) lie in the first ancient sediments to be recognized as being of sabkha origin (Shearman 1966) and the Lower Clear Fork Formation (Permian) of Texas, with its anhydrite and halite deposits, appears to have accumulated in a sabkha–brine pool environment.

Ancient basinal evaporites

These are believed to have developed in shallow and deep water environments. Shallow water evaporites tend to occur as massive beds with considerable lateral continuity (tens of kilometres) and the associated sediments yield abundant evidence of their shallow water origin. Deep water evaporites also have great horizontal continuity but now great vertical continuity as well. They are thin-bedded to laminar and may contain sulphate and halite sections tens to hundreds of metres in thickness. The majority of these saline giants were formed within interior basins (see Chapter 23), but others occur in rift valley settings, e.g. the Gulf of Mexico, Gabon and Red Sea deposits.

Two principal models have been put forward for the genesis of these basinal saline giants. In the *desiccated basin model* several basin-filling and drying-out events are postulated, since one drying-out event would produce only a thin evaporitic section. Each evaporitic event gives rise to concentric zones of carbonate, sulphate and chloride (Fig. 20.4a). The *brine-filled basin model* is the one used most commonly to explain saline giants (Fig. 20.4b). The evaporating water is replenished by continuous inflow of normal sea water, the more saline water forming within the basin is denser, sinks below the fresh inflow and is trapped within the basin by the bar at the inlet. Minerals may be precipitated evenly over the basin but with increasing salinity there will be a mineralogical change and a layer cake sequence, as in Fig. 20.4a, may be produced. On the other hand, in a variation of this model it is postulated that precipitation of less soluble evaporitic minerals occurs close to the inflow and that of more soluble minerals takes place from more concentrated brines situated further away, giving the tear drop distribution shown in Fig. 20.4b.

The Zechstein evaporites

Some workers have suggested that during the formation of certain saline giants both models operated at different times. This idea has been put forward to explain the complications seen in the Permian Zechstein evaporites of north-western Europe. The water level in the Zechstein Sea (Fig. 15.1) is believed to have been controlled by worldwide eustatic changes, such that at low sea level stand it became a completely enclosed basin leading to a high degree of desiccation.

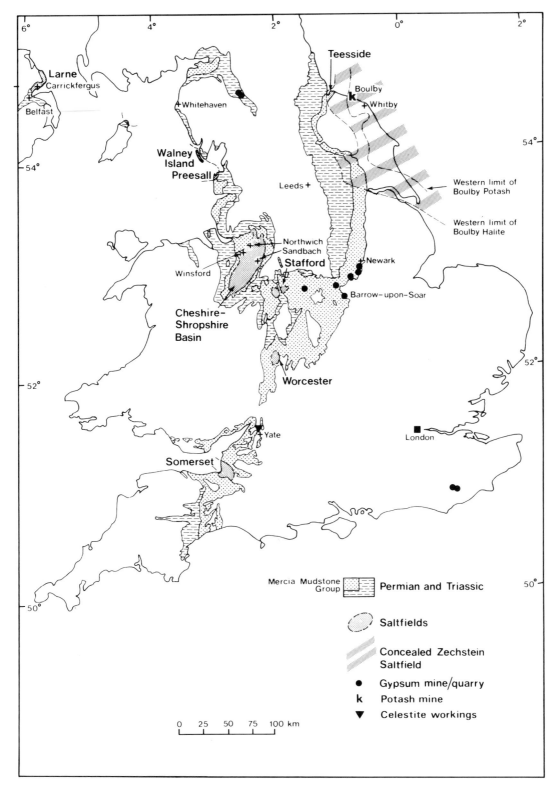

Fig. 20.3 The distribution of Permo-Triassic rocks in the UK that contain evaporite workings. (Reproduced with permission from Highley 1990.)

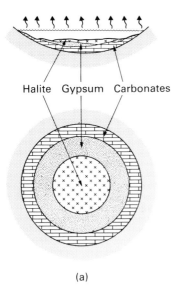 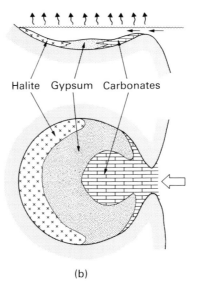

Halite Gypsum Carbonates

Halite Gypsum Carbonates

(a)

(b)

Fig. 20.4 Postulated patterns of evaporite distribution. (a) The bull's eye pattern developed in completely enclosed basins. (b) The tear drop pattern resulting from deposition in a partially open basin. (Reproduced with permission from Schreiber (1986) *Sedimentary Environments and Facies,* Blackwell Scientific Publications.)

The Zechstein succession is up to 3000 m thick and has been divided into five depositional cycles. The ideal cycle starts with a thin clastic unit and passes upwards through limestone, dolomite, anhydrite, halite to potassium and magnesian salts as evaporitic concentration increased the salinity (Schreiber 1986). These evaporites are exploited right across the site of the former sea from the UK, through The Netherlands and Germany to Poland. In many places salt pillars and domes have formed and mining occurs in these and in less disturbed strata. In the UK, sylvinite ore (a mixture of sylvite and halite) was discovered in north-western England in 1939 during drilling for oil (Fig. 20.3). Here the Boulby Mine accounts for about 50% of UK potash consumption. The ore is high grade containing about 25% K_2O (Highley 1990). The Zechstein evaporites are also important as cap rocks in the gas fields of The Netherlands and the southern North Sea. The reservoir rocks here include the Rotliegendes sandstone (Fig. 15.2) and the gas originated in the underlying Carboniferous coals.

Salt

The multiplicity of end uses of this mineral was referred to in Chapter 1. About 50–60% of world production goes to the chemical industry for the manufacture of chlorine and sodium chemicals, and salt is used to produce soda ash (sodium carbonate) in the Solvay process. Large amounts are used in

northern countries as a de-icing agent on roads to the detriment of the environment and concrete structures in which it may set off, or accelerate, aggregate–cement reactions. Salt retails at £20 t^{-1} delivered in the UK. World production in 1988 was 184.9 Mt of which the USA produced 35.3 Mt (but was still a net importer!), China 22 Mt, Germany 16.7 Mt, the USSR 15.5 Mt, Canada 10.6 Mt, India 8.4 Mt, France 7.9 Mt and the UK 7.1 Mt. A large number of other countries produce significant tonnages of salt. Much of the production in western Europe comes from Triassic deposits and all the saltfields marked on Fig. 20.3 win salt from the Trias, which in England is also a useful celestite producer.

Potash

About 95% of potash production goes into the manufacture of fertilizers; the remainder into the manufacture of potassium hydroxide, which is used in soaps, detergents, glass, ceramics, dyes and drugs. World production of potash in 1988 was 31.9 Mt of which 93% came from only six countries. Only another six countries have outputs > 40 000 t —a very different picture from salt production. The leading producers in 1988 were the USSR 11.3 Mt, Canada 8.3 Mt, Germany 5.8 Mt, the USA 1.5 Mt, France 1.5 Mt and Israel 1.3 Mt. The price of muriate of potash (KCl) with ≮ 60% K_2O is £71– 74 t^{-1} CIF a British port.

Borates

The main uses of borates in the USA in 1989 were insulation and textile fibreglass 42%, borosilicate glass 9% and detergents 9%. In western Europe the detergent use is about 3% and fibreglass only 18%. The other principal uses are in agriculture, fire retardants and enamel glazes. World demand in 1989 was about 960 000 t B_2O_3 equivalent, whereas world production capacity is about 1.3 Mt. The USA and Turkey produce about 85% of world output. The American production is from Boron, Death Valley and Searles Lake in the desert region of California and the Turkish from three districts about 200 km south of Istanbul. World reserves have been estimated at 322 Mt.

There are 11 important borate minerals, mainly sodium and calcium borates of which borax is but one. These were deposited in lakes and playas in desert regions and their deposits are of variable size. The Boron deposit in the Mojave Desert, about 160 km north of Los Angeles, is a lenticular bed 3.2×1.6 km in plan which ranges in thickness from 1 to 90 m and is of Miocene age. The boron is believed to have been introduced into the lake water by thermal springs.

Nitrates

Although the world acquires most of its nitrogen from the atmosphere, natural nitrates have been produced in Chile since about 1830. Approximately 750 000 t of sodium and potassium nitrates are produced each year forming < 0.3% of the world's nitrogen needs. The ores occur in a belt about 30 km wide and extending for about 700 km through the Atacama Desert. They are largely in the form of surface caliche (calcrete) and the grade is around 7–8% nitrate.

Graphite

Price and production

This is a valuable mineral whose price in October 1991 varied according to its grade and purity from US\$325–360 t^{-1} for powder running 80–85% C and US\$1000–1300 t^{-1} for 97–99% purity, to US\$750–1500 t^{-1} for lump graphite grading 92–95% C. World production is difficult to assess for a number of reasons but the 1986 British Geological Survey figure was 645 000 t with China estimated at

185 000 t. The other top producers were South Korea 99 218 t, USSR 83 000 t, Brazil 47 000 t, India 40 483 t, Mexico 37 780 t, Austria 36 167 t, Czechoslovakia 25 254 t, and North Korea 25 000 t. Since then two new mines have opened in Canada, three more are under development and a number of possible orebodies are being investigated. It appears at the time of writing that the market may be moving into an oversupply situation.

Graphite is marketed in a number of forms. The two basic divisions are into amorphous and crystalline. Amorphous is a misuse of the term as this graphite is really microcrystalline. It has a black earthy appearance, is graded primarily on its carbon content and commercial ores of this type, which come mainly from China, Mexico and South Korea, contain 50–90% C. Its main geological source is metamorphosed beds of coal or carbonaceous material. Crystalline graphite may be subdivided into flake, vein and powder. Flake graphite consists of lustrous, mica-like grains disseminated through metamorphosed carbonaceous rocks (e.g. black shales, marbles) that have suffered at least garnet grade metamorphism. These rocks may grade from 1 to 90% C. Many countries supply it but the major producers are China and the Malagasy Republic (Madagascar). Within a year or two Canada will be an important producer too. Vein graphite (sold in lump, chip and dust forms) is relatively rare, the best deposits being found in Sri Lanka. Some of these veins may consist almost entirely of graphite; others may run 80–90%. The graphite is often coarse grained; fine-grained material is sold in powder form as 'chip' and 'dust' graphite.

Synthetic graphite is manufactured by graphitizing petroleum, coke or anthracite in an electric furnace. In general it has different properties, and therefore uses, from natural graphite, so there is little competition between the two, except in the process of recarburizing steel where it competes with amorphous graphite (Russell 1988b). The world annual production of synthetic graphite is about 1.5 Mt.

Uses

Graphite is important for one or a combination of the following properties:
1 high lubricity;
2 low coefficient of thermal expansion;
3 good electrical and heat conductivity;
4 flexibility and sectility over a wide temperature range;

5 chemically inert and non-toxic;
6 generally not wetted by metals.

In 1987 the major uses to which graphite was put in the USA were refractories 30.5%, foundries 15%, lubricants 14%, brake linings 13%, pencils 7%, steel making 5% and batteries 3% (Russell 1988b). For these various uses different specifications are laid down by the consumers and these may be found in Russell's article, which covers much more ground concerning graphite than its title suggests. Many of the refractories are used in the steel industry in the form of bricks (magnesia-graphite), crucibles, retorts, stoppers, sleeves and nozzles. It is in foundry uses that graphite's non-wettability property is important. It is mixed with a bonding agent (refractory clay, sand, talc, etc.) for use as a coating within moulds to prevent metal castings from adhering to the mould. With the replacement of asbestos in brake linings over recent years the graphite component has risen from 1–2% to 15%. Graphite has long been used in dry cell zinc–carbon batteries. For use in the long life alkaline type the graphite must have a *minimum* carbon content of 98%, be free of metallic impurities and have a grain size of 5 μm. Norwegian flake graphite is particularly suitable for this use. However the trend towards rechargeable Ni–Cd batteries may mean that battery usage is no longer a developing market for graphite.

Graphite deposits

Most graphite is of metamorphic origin and only subordinate amounts are won from vein and residual deposits.

Metamorphic graphite

This results largely from the contact or regional metamorphism of organic matter in sediments. The degree of graphitization depends principally on temperature but pressure and the nature of the original organic matter also play a part. True graphite probably forms above 400°C. Metamorphosed coal seams are important graphite producers in several countries. A number of mines around Jixxi in Heilongjiang Province, China work this type of deposit, contributing about 19% of China's output from rock grading 17–21% flake graphite. In Mexico, Triassic coals have been metamorphosed to mixtures of graphite, anthracite and coke by granite intrusions. The main beds are around 3 m thick but folding has led to thicknesses of up to 7 m. By con-

trast, an example of graphite produced by regional metamorphism can be seen in Austria. Here in the Grauwacken Zone of the Eastern Alps, about 75 km south-west of Vienna, two underground mines work microcrystalline graphite ore running 40–90% C. Production is 15 000–20 000 t p.a. and the reserves total approximately 1.6 Mt (Holzer 1986).

A large tonnage of metamorphic graphite comes from metamorphosed black shales, carbonaceous limestone and other carbonaceous rocks. Sometimes the metamorphism is of contact type, as at Kuldzhuktan in Uzbekistan, USSR where a gabbroic intrusion into limestone has given rise to pyroxene–garnet–wollastonite–graphite hornfels or skarn. Another graphitic skarn deposit is Norway's only graphite producer (see p. 167). However, graphite produced by regional metamorphism is an increasingly important source of this mineral and many mines work this type of deposit including the new Canadian ones. As mentioned above, grades vary considerably and sometimes a single orebody of, say, pelitic gneiss, which runs only 3–4% C, may contain layers or lenses many metres thick and perhaps a kilometre long, where grades may range up to 90%. As an example of low grade material we may note the new prospect being investigated by Calgraphite Mines in Canada. This has about 50 Mt of proved and probable reserves running 3% C from which a 94–96% C product has been obtained in preliminary mineral dressing tests.

Vein graphite

The best examples of this deposit type are found in Sri Lanka where they occur in a Precambrian high grade metamorphic terrane dominated by granulite facies rocks (Fig. 20.5). Most of the deposits are found in the south-western sector of the island, which is a region characterized by the presence of wollastonite-bearing calciphyres and cordierite gneisses of low to medium granulite facies grade (Katz 1987). At the Bogala Mine the immediate host rocks are quartzo-feldspathic biotite and hornblende gneisses with garnet gneissses and a few small lenses of calciphyres. The graphite veins are mineralized fractures whose thickness varies from a few millimetres to about 1 m. Only those more than 5 cm thick are worked. Four productive veins are present at depth but each divides into a number of branches near the surface. The average ore contains over 90% C. Rock fragments, quartz, feldspar, calcite, magnetite, pyrite and chalcopyrite form the

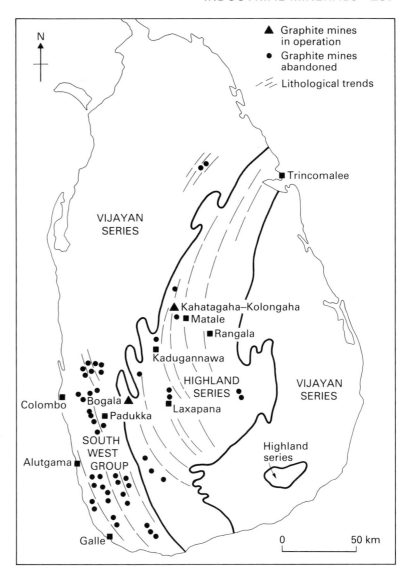

Fig. 20.5 Map showing the location of graphite mines in Sri Lanka. (After Katz 1987.)

principal gangue (Silva 1987). The veins have been traced to a depth of 250 m and along strike for up to 500 m.

The genesis of vein graphite has been much debated over the last century or more. In a recent study of the Bogala orebodies Katz (1987) suggested that the CO_2-rich fluids characteristic of granulite facies terranes became sufficiently focused to produce hydraulic fracturing and precipitate graphite. He supported this hypothesis with temperature estimates based on X-ray diffraction and stable carbon isotopic data. After an investigation of graphite in veins and microscopic veinlets in silli-

manite gneisses and plutonic rocks of New Hampshire, Rumble *et al.* (1986) postulated that carbon was mobilized as CO_2 and CH_4 during metamorphic devolatilization reactions and transported in aqueous fluids through hydraulic fractures to sites of precipitation.

Residual deposits

Because of its unreactive nature graphite can be concentrated in deposits formed by the weathering of graphitic gneisses and schists. An example occurs in the Manampotsy District in east central Malagasy

Republic. Here the graphite is distributed through a loose residue of kaolin and mica. The workable zone of weathering is about 3–30 m thick and carries 3–10% graphite. Similar deposits are mined in northern Shanxi Province, China.

Limestone and dolomite

Limestone is probably the most important industrial rock or mineral used by man and, together with dolomite, it is quarried at thousands of localities throughout the world. Limestone contains at least 50% of the minerals calcite and dolomite, with calcite predominating. In dolomites (dolostones) the mineral dolomite is the major carbonate. There is a complete gradation from impure to high calcium limestone (> 95% $CaCO_3$). High purity dolomites are defined as carrying > 87% $CaMg(CO_3)_2$. Three components occur in the majority of limestones: carbonate grains, micrite and sparite. Many limestones consist of sand-sized calcite grains in a sparry cement (these are usually high-calcium limestones), while others are so full of micrite that they are simply lithified carbonate muds. Modern carbonate sediments are composed of aragonite, high Mg calcite and low Mg calcite. Ancient limestones, composed of low Mg calcite, formed through diagenetic replacement of aragonite grains and loss of Mg from originally high Mg calcite. Other diagenetic changes may involve dolomitization—a replacement process that may greatly increase the porosity—and silicification. Common impurities are clay, sand, chert and organic matter. Limestones and dolomites possess a wide range of colour, grain size and bed thickness. All these features can have an influence on their utilization.

The most commonly exploited limestone is compact and lithified, but softer porous varieties may be used for certain applications (e.g. Cretaceous Chalk in the UK and France). Other naturally occurring forms of calcium carbonate, e.g. aragonite and shell sands, marble (much commercial 'marble' is polished limestone), carbonatite, calcareous tufa, vein calcite and cave onyx, may compete with, or be an alternative source of, $CaCO_3$ where limestone is absent. Vein calcite may come from veins composed almost entirely of that mineral or may be a by-product of fluorspar and baryte operations.

Uses

Table 20.2 gives some of the more important end uses of limestone according to product size, and just a few additional remarks will be made here. Limestones and marbles have been used by man as building stones for many centuries. So few are suitable for use as dimension stone that their products may be exported over considerable distance, e.g. in mediaeval times Caen stone was imported from Normandy to build many of England's great cathedrals and monasteries. A short discussion on building stones will be found earlier in this chapter.

Lime is a manufactured chemical resulting from the calcination of limestone or dolomite. The term should not be used for other materials, such as ground limestone used for agricultural purposes (Freas 1989). The primary lime products are quicklime, pebble lime, hard-burn dolomite or, when water is added, calcium hydroxide or slaked lime. High purity limestone is preferred because on calcination it loses 44% of its weight and any impurities will be approximately doubled in amount in the lime. The many types of kiln are reviewed by Freas. The end uses of lime are formidable in number. The principal consumer is the steel industry where it is used as a flux. Other important uses are for water treatment and purification; flue gas desulphurization, for which limestone is also used; chemical processes, such as the manufacture of caustic soda, carbide, bleaches and various inorganic chemicals; in wood pulp and paper production and sewage treatment. An excellent review of the markets and uses for lime can be found in O'Driscoll (1988b).

The important requirements for coarse aggregate usage will be found earlier in this chapter.

Calcium carbonate is making great strides as a filler in a field where it has to compete with ground mica, kaolin, nepheline syenite, talc and wollastonite. The high value, high quality, fine calcium carbonate or dolomite fillers are defined in the USA as those having a minimum fineness of 97% passing an ASTM 325 mesh (45 µm) sieve. Anything from 6 to 10 Mt p.a. of such fillers are produced in the USA alone. There are two forms of carbonate filler: carbonates obtained by grinding natural rock and precipitated calcium carbonate (PCC). Precipitated calcium carbonate is made by slaking lime with water and reacting it with CO_2 to form orthorhombic or trigonal crystals and grains. The whole process is one requiring much technical skill to produce the right product for the right market.

Table 20.2 Uses of limestone in terms of product size (excluding cement production). Important assessment requirements are given in parenthesis for each size[a]. (Modified from Scott 1987)

>1 m	Cut and polished stone ('marble'). (Mineable as large blocks containing no planes of weakness. Consistent white or attractive patterns of coloration. Low porosity and frost resistant)	3–8 cm	Filter bed stone. (Compressive strength. Chemical purity. Moisture absorption. Abrasion resistance. Crust formation)
>30 cm	Building stone. (Wide spacing of bedding and jointing. Low porosity and frost resistant. Consistent appearance. High compressive strength)	3–8 mm	Poultry grit. (Chemical purity. Shape of grains)
		< 4 mm	Agriculture. (Chemical purity. Organic matter)
>30 cm	Rip-rap or armour stone. (High compressive and impact strength. High density. Low porosity and frost resistant. Wide spacing of bedding and jointing)	< 3 mm	Iron ore sinter, foundry fluxstone and non-ferrous metal fluxstone, self-fluxing iron ore pellets. (Chemical purity)
1–30 cm	Kiln feed stone for lime kilns. (Chemical purity. Degree of decrepitation during calcination and subsequent handling. Strength in relation to design of kiln and crushing plant. Burning characteristics of stone)	<0.2 mm	Filler and extender in plastics, rubber, paint, paper, putty. (Chemical purity. Whiteness and reflectance. Oil, ink and pigment absorption. pH. Nature of crushing and grinding circuits in relation to compressive strength)
		<0.2 mm	Asphalt filler. (Most pulverized limestone including dust collector discharge can be used)
1–20 cm	Aggregate, including concrete (cement and bituminous), roadstone, railway ballast, roofing, granules, terrazzo and stucco. (Impact strength. Crushing strength. Abrasion resistance. Resistance to polishing. Soluble salts. Alkali reactivity with cement. Tendency to form particles of particular shape)	<0.2 mm	Mild abrasive. (Near white colour. Low quartz content)
		<0.2 mm	Glazes and enamels. Mine dust. Fungicide and insecticide carrier. (Chemical purity. Near white colour. Organic matter)
		<0.1 mm	Flue gas desulphurization. (Chemical purity. Surface area. Microporosity)
0.2–5 cm	Chemicals and glass. (Chemical purity. Organic matter. Abrasion resistance)	Various sizes	Bulk fill. (Depends on customer requirements. Size gradation important)

[a] Elements usually determined in chemical analysis are Si, Al, Fe, Mn, Mg, Ca, Na, K, S, P and loss on ignition. Other elements may be important for some uses, e.g. toxic elements in filler for plastic toys.

Fillers in general no longer merely substitute for a more expensive ingredient—they have become functional fillers, i.e. they add colour, stiffness, opacity, increase electrical conductivity or resistance to heat. Some fillers are surface treated with chemicals to change their properties; this applies in particular to fine carbonate fillers, which are comprehensively reviewed in O'Driscoll (1990).

A refractory material, $CaO.MgO$, is manufactured from dolomite by calcining it at about 1500° C. This is used for lining metallurgical furnaces and other high temperature purposes.

Assessment and mining

Many of the important requirements for the different uses of limestone and dolomite are given in Table 20.2 and these govern the assessment of a limestone body or guide the exploration geologist in his search for a suitable deposit. It should be borne in mind that different beds in a deposit can be mined for different purposes, e.g. a high-calcium limestone for lime production and a dolomite section for crushed stone. Overburden is generally present and must be stripped off; its nature and thickness can greatly affect the economics of the operation. It may also have an environmental importance, as sufficient acreage must be acquired for its disposal. The amenability of the rock to mining must be ascertained. The joint pattern must be determined as this will guide the engineers in planning the method of extraction. The presence of reefs may introduce new rock types and a knowledge of reef trends may enable the geologist to predict their probable presence or absence in a particular limestone body. If a high purity product is required a considerable amount of drilling and core analysis will be necessary.

Magnesite and magnesia

Magnesia is one of the world's much used and versatile chemicals. It uses range from refractories through fertilizers to lining upset stomachs and it is obtained from three sources: magnesite, sea water and natural brines. A further magnesite product is magnesium metal. Magnesia for the refractive industry is prepared from magnesite by calcining at 1600–2000°C to produce dead-burned magnesia:

$$MgCO_3 \rightarrow MgO + CO_2\uparrow$$

For other industrial users who require an active magnesia, calcination is at 800–1000°C to produce caustic calcined or light-burned magnesia. In the sea water process a more complicated procedure must be followed. First lime must be produced and then slaked to prepare calcium hydroxide, which is added to the sea water; a simple ion exchange reaction results in magnesium hydroxide sludge, which is then calcined to produce magnesia:

$$CaCO_3 \rightarrow CaO + CO_2\uparrow$$

$$CaO + H_2O \rightarrow Ca(OH)_2$$

$$Ca(OH)_2 + MgCl_2 \rightarrow CaCl_2 + Mg(OH)_2$$

$$Mg(OH)_2 \rightarrow MgO + H_2O\uparrow$$

A significant contaminant of sea water magnesia is boron oxide and an ancillary process for reducing the boron content may be used during the calcination stage. Magnesium brines are treated in a similar manner to sea water although slaked calcined dolomite rather than lime may be used as in some sea water plants; or their contained $MgCl_2$ is decomposed at 600–800°C in a reactor vessel:

$$MgCl_2 + 2H_2O \rightarrow Mg(OH)_2 + 2HCl\uparrow$$

Production and prices of magnesite and magnesium

Drawing up production figures for magnesite production involves a lot of guesswork, as the figures for the three leading producers, the USSR 2.5 Mt?, China 2 Mt? and North Korea 1.9 Mt?, have to be estimated. Assuming the above figures, world production in 1987 was 12.3 Mt; the other leading producers being Turkey 1.19 Mt, Austria 0.95 Mt, Greece 0.9 Mt and Czechoslovakia 0.67 Mt. In October 1991 the price of calcined magnesite for agricultural purposes was £80–90 t⁻¹ CIF and for industrial purposes £125–270 t⁻¹ CIF according to type and purity.

Magnesium is the lightest structural metal (density 1740 versus 2700 kg m⁻³ for aluminium) and its alloys have very high strength to mass ratios and are used in the transport, particularly aerospace, industries. Non-structural uses of the metal include desulphurization of iron and steel, metallothermic reduction of titanium and zirconium, explosives, dry batteries and anodic corrosion rods. The demand for magnesium metal in the late 1980s was increasing steadily at about 5% p.a. and the price in November 1991 was £1550–1750 t⁻¹.

Uses of magnesia

About four-fifths of world production goes into the manufacture of magnesian refractories which, because of their inertness and high melting points, are used in lining steel and non-ferrous metal furnaces, cement kilns, etc. The refractories industry has rigid specifications: $MgO \nleq 95\%$, $Fe_2O_3 < 1\%$, a lime : silica ratio of 2 : 1, bulk density of 3400 kg m⁻³ and a low boron content. The manufacture of magnesia–carbon refractories calls for such pure magnesia that only a few magnesite mines can satisfy this market and much of the magnesia used comes from sea water processing plants.

The remaining one-fifth or so of magnesia production goes into such diverse uses as animal feedstuffs, fertilizers, special cements, gas scrubbing equipment, paper, pulp, rubber, plastics, fire-proof boards and Milk of Magnesia!

Magnesite deposits

The mineral magnesite is rarely found pure as it is the end member of a complete solid solution series to siderite ($FeCO_3$) and also Mn and Ca can substitute for Mg to a limited extent. Thus the purity of dead-burned magnesia produced from different deposits is very variable and variations in composition within a single orebody may create production difficulties.

Magnesite of industrial interest exists in two principal natural forms (Schmid 1987, Pohl 1989): macrocrystalline (sparry) magnesite and cryptocrystalline magnesite. The former is coarse-grained and largely confined to deposits of metasomatic or sedimentary origin hosted by ancient marine

platform carbonate suites; the latter has a porcelain-
or bone-like appearance and is extremely fine-
grained, and occurs as lenticular masses, veins and
stockworks in serpentinites. These are usually
smaller deposits than the carbonate-hosted ones.
However, very large sedimentary deposits of cryptocrystalline magnesite have been discovered recently
in Queensland, Australia and their exploitation will
put that country into the forefront of magnesia
production.

Deposits of macrocrystalline (sparry) magnesite

These are lenticular, stratiform and often strata-
bound deposits that can attain considerable thicknesses; the Asturreta orebody, Spain is up to 150 m
thick. Tonnages range up to 50 Mt or higher. The
immediate host rock is frequently dolomite, the
enclosing carbonate sequence being most commonly
Proterozoic or Palaeozoic in age. This deposit type
occurs worldwide. The coarse grain size of some of
these deposits may be primary but many deposits
have suffered regional metamorphism and their
fabric is largely or entirely metamorphic. The
coarse-grained magnesite typically displays replacement relationships towards the country rocks; these
are usually dolomite, but magnesite can be seen replacing metapelite, quartzite and chert (Pohl 1989).

In addition to the magnesite in these deposits,
there may be minor amounts of dolomite—
sometimes with evidence of two generations: the
first fine grained, grey and pre-magnesite, the second
coarse-grained, white and post-magnesite. Quartz,
talc, chlorite, pyrite and organic material may also
occur.

The Asturreta Mine, described by Velasco et al.
(1987), is a typical example of this deposit type and
its geological setting is shown in Fig. 20.6. Magnesite
occurs within the Namurian dolomitic section at a
number of localities where it is well stratified and
concordant with the enclosing rocks. At Asturreta,
coarse-grained, white magnesite occurs in beds
3–7 cm thick, which are thought to be diagenetically
crystallized rhythmites. Stromatolytic relicts are
present. Chert, quartz grains, aragonite, calcite,
chalcopyrite, malachite, iron oxides, baryte, fluorite
and albite occur in very minor amounts. Both
dead-burned and caustic calcined magnesia are
produced.

The genesis of this deposit type is hotly debated,
with hypotheses of formation covering a complete
spectrum from primary sedimentary, through dia-

genetic alteration, low temperature epigenetic
metasomatism to high temperature metamorphic–
metasomatic action. There is no room in this
volume to discuss these different ideas adequately,
the reader being therefore recommended to examine
a number of the papers in Möller (1989). However,
a small part of this debate may be seen in the
discussion on Velasco et al. (1987) by Chaye
d'Albissin et al. (1988) concerning the Asturreta
Mine.

Velasco et al. had found that most primary fluid
inclusions in much of the coarse-grained dolomite
and in sparry magnesite from the Asturreta deposit
yielded homogenization temperatures (uncorrected
for pressure) in the range 160–175°C. Studies of the
crystallinity of illite and the degree of graphitization
of carbonaceous material in the orebodies indicate
that they were raised to these temperatures or
slightly higher ones. Work on oxygen and carbon
isotopes in the carbonates indicates similar temperatures. From this data, careful textural studies and
other work, Velasco et al. concluded that metasomatic replacement of syngenetic dolomite by
magnesite during advanced diagenesis gave rise to
the deposit. Chaye d'Albissin et al. disputed this
conclusion on the grounds that all magnesite deposits of this type are of primary sedimentary origin and
they quote evidence in support of their argument
from a number of similar deposits in Spain and
other parts of the world.

Deposits of cryptocrystalline magnesite

These can be divided into those developed within
ultramafic complexes and undoubted sedimentary
deposits. In the first category the magnesite is
associated with ophiolites, which have often suffered
considerable deformation and hydrothermal alteration, and it occurs in lenses, veins, stockworks and
irregularly shaped bodies where fracture zones cut
these altered rocks. The deposits usually have a few
hundred metres vertical spread and workable orebodies vary from several hundred thousand to a few
million tonnes. Grades are 20% $MgCO_3$ upwards.
The vein orebodies range from several to 45 m in
thickness and can be as much as 4 km long and
300 m deep. Wall rock contacts are often sharp. The
magnesite of these orebodies may contain small
amounts of Ca but generally the Fe content is
strikingly low compared with the macrocrystalline
magnesite. Associated minor minerals are dolomite,
chalcedony, quartz, talc, sepiolite and serpentine.

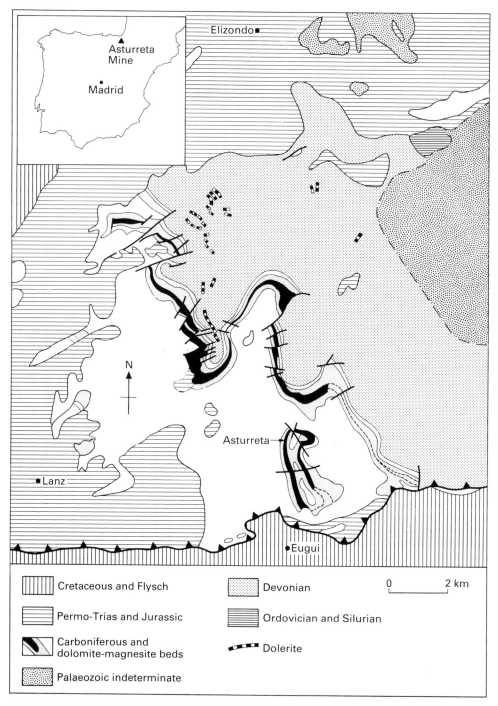

Fig. 20.6 Map showing the location and geological setting of the Asturreta Mine, northern Navarra, Spain. (Modified from Velasco *et al.* 1987.)

Deposits of this type occur in the alpine orogenic belts of Austria, Yugoslavia, Greece and Turkey and many other parts of the world (Pohl 1989). Again various modes of genesis have been suggested, including alteration by descending, meteoric waters enriched in Mg^{2+} and biogenic CO_2, CO_2 metasomatism during regional metamorphism, circulating hydrothermal solutions, etc. These mechanisms are invoked for some of the deposits discussed in papers in Möller (1989).

Primary sedimentary deposits of dolomite and magnesite ought to be much more common in the geological column than they are, in view of the fact that sea water is supersaturated with respect to dolomite by almost two orders of magnitude (Blatt *et al.* 1980). The reason for their scarcity is probably because Mg^{2+} is a much smaller ion than Ca^{2+} and therefore binds water molecules to itself very tightly. Unless these can be stripped off by unusual chemical conditions the magnesium ions cannot enter a nucleating crystal structure.

Nevertheless there are now many well-documented examples of what must be primary sedimentary deposits formed by chemical precipitation *in situ* and mechanical (clastic) sedimentation. Many of these appear to have formed in inland lakes and Schmid (1987) has given us a graphic description of magnesite deposition in Salda Lake—a present day lake in southern Turkey that is approximately 60 km² in extent. Near the lake is highly weathered serpentinite containing a magnesite stockwork and lenses of magnesite. Much magnesite from the serpentinite has been deposited as a mud in this lake and in a wide, swampy, former river mouth to the south of the lake. Considerable amounts of the mud have been recrystallized to form cryptocrystalline lumps and nodules. The iron content of this magnesite is very low. Schmid suggested that the newly found Australian deposits in Queensland had a similar origin. These are about 65 km north-west of Rockhampton and form the largest known, low-iron magnesite deposits in the world. Here the Kunwarara deposit is reported to contain at least 800 Mt of ore containing an estimated 400 Mt of high purity, cryptocrystalline magnesite (Anon. 1989d). Another prospect in the area is said to contain at least 188 Mt of magnesite. Two others are being explored. The Kunwarara orebody has an average thickness of 7.75 m and is covered by, on average, 4.2 m of soft black soil overburden. It consists of magnesite lumps and nodules within a soft magnesite matrix.

The world's largest magnesite deposits occur in the Liaodong Peninsula, China (Fig. 20.7). They and their host rocks are of early Proterozoic age (Schmid 1984, Qiusheng 1988). The magnesite-bearing zone has been outlined along approximately 60 km and has an outcrop width of 1–2 km, but is known to continue further to the north-east. There are 17 moderate to large deposits and many smaller ones. Reserves of magnesite for this district have been put at 2.5 Gt! The magnesite bodies are stratiform, concordant with the bedding in the host dolomites and have an average thickness of 100 m. Minor amounts of talc, tremolite, quartz, dolomite, calcite and pyrite accompany the magnesite. The magnesia content is very uniform. This, and the excellently preserved sedimentary structures, such as hail pits on bedding surfaces and slump folds, that have been found, and stable isotopic investigations, suggested to Qiusheng (1988) that the deposits are of primary sedimentary origin.

Another and very interesting aspect of this region is the occurrence of marble-hosted, metamorphosed boron orebodies; these show a spatial relationship to volcanic centres. There are 77 known deposits of which 12 large deposits make up 98% of the boron reserves in the peninsula, the largest being about 2 km long and up to 100 m thick!

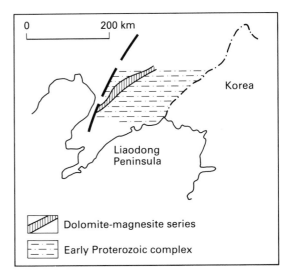

Fig. 20.7 Sketch map showing the location of magnesite deposits in the Liaodong Peninsula, China. (Based on maps in Schmid 1984 and Qiusheng 1988.)

Olivine

Olivine is a solid solution series having two end members—forsterite (Mg_2SiO_4) and fayalite (Fe_2SiO_4). Fayalite melts at 1205°C, forsterite at 1890°C; thus magnesian-rich olivine is a highly refractory material. Substitution by other cations, including Ca^{2+} is very low. Most olivine in workable deposits runs 48–50% MgO, 41–43% SiO_2, 6.1–6.6% FeO and 0.2–0.8% CaO.

Production and price

World production increased dramatically in the last decade and in 1989 was about 4 Mt; but as a number of firms do not release production figures this is very much an estimate based on information in Griffiths (1989). Norway is the leading producer with an output of about 2.1 Mt (but has a production capacity of 4 Mt). Spain also has a high production capacity, about 1.5 Mt p.a., and enormous reserves (> 100 Mt), but inferior transport facilities mean that Spanish output is mainly marketed within the EEC. Olivine prices are given in Table 20.3.

Table 20.3 Olivine prices in October 1991

Bulk, crushed for blast furnace	£8–14 t^{-1} CIF
Bulk, dry, graded refractory aggregate	£21–25 t^{-1} CIF
Foundry sand, bagged, delivered UK	£40–65 t^{-1}
Foundry sand ex-plant or mine USA	$53–85 t^{-1} FOB

Uses

The large increase in production in recent years has been in response to olivine's use as a slag conditioner in iron and steel making. The usual slag forming agents in iron smelting are dolomite or limestone with fluorspar to lower the melting point. Silica may be added too, according to the iron ore composition, its presence being an aid in lowering the melting point of the charge. Because olivine contains both magnesia and silica it is an excellent substitute for dolomite and silica when treating low silica ores and, as it has a higher magnesia content than dolomite, fewer tonnes are required and less slag produced.

A traditional use of olivine has been as a foundry sand thanks to its high melting point, good thermal conductivity, high heat capacity (giving it resistance to thermal shock), low thermal coefficient of expansion and high green strength. It has the additional advantage that it is not, like silica sand, a health hazard and there is no risk of workers developing silicosis.

Another traditional use is as a blast cleaning agent —again there are no health risks involved. This use is likely to increase as more and more countries prohibit the use of quartz sand for sand blasting. Many European countries already have done so.

Among the other uses to which olivine is put are the manufacture of refractory bricks, ladles, torpedo tubes and heat storage units (night storage heaters). In this last use olivine is facing strong competition from magnetite ('FeOlite'), which has better storage properties. Olivine's high density (3300 kg m^{-3}) has led to its being used in the North Sea oilfields as ballast for oil platforms and for covering undersea pipelines.

Specifications for some of the above uses are given in Table 20.4

Table 20.4 Specifications for olivine for various uses. (After Dunham 1986)

Steel slag conditioner
MgO : SiO_2 molar ratio > 1.75 : 1
MgO : SiO_2 wt % ratio approximately 1 : 1
CaO, Na_2O, K_2O all low

Size of particles:
  lump olivine—1–5 cm diameter
  sinter feed—all < 6 mm, 65% > 0.85 mm
  pellet feed—all < 2 mm

Foundry sand
High refractoriness; i.e. low alkalis and CaO, high MgO
  content
97% > 90 µm and < 325 µm

Refractory bricks
Mole ratio of MgO : SiO_2 as high as possible. Ratio
  made up to > 2 : 1 by the addition of MgO
Cr_2O_3 and Al_2O_3 both useful
Low Na_2O, K_2O and CaO
Size range 1–8 mm

Abrasive
No silica minerals present
Normal grade in range 1–1.5 mm, special types
  < 0.25 mm

Geology

Olivine is an essential mineral of peridotite and rocks composed almost exclusively of this mineral are known as dunite; this is the only rock of economic interest as a source of olivine. Most dunites of commercial interest are probably of alpine type (see Chapter 10), but some deposits are not sufficiently well described for one to be dogmatic on this point.

Norway's production comes largely from the Åheim district. This is in the Proterozoic North-western Gneiss Complex of Southern Norway. The Åheim Mine is situated close to a deep water harbour about 300 km south-west of Trondheim. The dunite runs 90–95% olivine with minor quantities of pyroxene, chlorite and other minerals, and the quality of the rock is very consistent. The deposit occupies about 6.5 km^2 and is surrounded by gneiss of the Fjordane Complex. The dunites and associated eclogites of the district are believed to be mantle-derived and to have undergone granulite and eclogite facies metamorphism. Gravity anomalies over the Åheim body indicate an eastward extension at depth and the total volume is estimated to be 25–30 km^3 (Bugge 1978). It is no wonder that the reserves have been described as limitless!

A/S Olivin, which is wholly government owned, began operations in 1948 and slowly increased production to meet the demands of the traditional blast cleaning and foundry sand markets. However, at the low output of about 100 000 t p.a. the operation was barely economic and it was realized that a really profitable operation could only be achieved by increasing production and lowering unit costs in order to expand existing markets and create new ones. A unique example of the latter is the use of olivine in North Sea oilfield operations (mentioned above under 'Uses'), which are on A/S Olivin's doorstep. To cut mining costs, ripping, rather than drilling and blasting, is employed to break up much of the dunite. The six open pits worked separately before 1980 were consolidated into a single operation. A new processing plant was installed and the capacity is 4 Mt p.a. although present output is around 2 Mt p.a. Loading facilities at the harbour can accommodate vessels of up to 80 000 t. The processing plant produces three blast furnace grades, two refractory grades and five foundry sand grades. A refractories plant produces bricks and other refractories. Sixty per cent of sales are in Europe but bulk carriers take foundry grade sands to the USA, South America, New Zealand and Iran.

Another Norwegian company, Franzefoss Bruk A/S, is shipping crude ore to the Eastern Seaboard of the USA for procesing at Aurora, Indiana, where the finished product can compete successfully with that from Washington State, which is handicapped by the high overland transport costs.

Spain has dunite in abundance in Galicia. The Landoy Mine has reserves of about 100 Mt and a production capacity of 1–1.5 Mt p.a. However, its prices are dependent on a number of factors, including the port used. Puerto de Carino, 12 km from the mine, can only handle vessels of up to 6 000 t . The second port used is 52 km away involving higher overland transport costs.

In the *USA* olivine production is largely concentrated in Washington State where, although there are large reserves, production is only around 150 000 t p.a. At Twin Sisters Mountain, reserves of approximately 200 000 Mt of good quality olivine have been outlined. This is worked by the Olivine Corporation, which is carving out a market for itself by developing and producing refractory incinerators for waste disposal. Waste disposal is a large problem in the USA and, as each incinerator requires 20–100 t of olivine per unit, the company is establishing a useful new market.

Other countries producing olivine include Austria, Italy, Japan and Pakistan.

Perlite

The name perlstein was given by some nineteenth century German petrologists to certain rhyolitic, glassy rocks with numerous concentric cracks which, on fragmentation, yielded pieces vaguely resembling pearls. The more modern name perlite is now universally used. The important property of perlites is that they are hydrated rocks carrying 2–5% water and on heating by flash roasting to temperatures close to their melting point, the contained water is converted to steam and the grains swell into light, fluffy, cellular particles. This results in a volume increase of 10–20 times and produces a material with low thermal conductivity, considerable heat resistance and high sound absorption.

Uses

Over half the world's production goes into the construction industry as aggregate for insulation boards, plaster and concrete in which weight reduction and special acoustic or thermal insulation properties are required. It is used for loose-fill insulation of cavity walls and for the thermal insulation of storage tanks for liquified gases.

Horticultural applications include use as a rooting medium and soil conditioner and as a carrier for herbicides, insecticides and chemical fertilizers. It is used for filtering water and other liquids, in food processing, in pharmaceuticals and as a filler in paints, plastics and other products. Its uses for animal feedstuffs, poultry litter and crop farming is of growing importance (Lin 1989). For some of these uses pumice is a competitor, it has been expanded for us by nature and therefore does not require furnace treatment. The crushing strength of pumice is higher than that of perlite so that its use makes for a stronger concrete. Pumice, however, has the economic disadvantage that it must be shipped to processing plants in the bulky expanded form whereas perlite can be shipped as the crude, unexpanded rock. Vermiculite is a much more important competitor with many properties in common with perlite. Its occurrence and uses have been discussed on p. 119.

Production and price

World production of perlite in 1989 was over 2.5 Mt from which about 1.8 Mt of processed perlite was prepared. The USA produced about 29%, the USSR about 26%, Greece about 24%, Turkey about 6% and Hungary about 5%. Other producers with significant output are Italy, Japan, Czechoslovakia and Mexico. The UK could join this group if test work on the Antrim Perlite project in Northern Ireland is a success.

In recent years prices have remained very stable; they are given in Table 20.5.

Table 20.5 Perlite prices

Raw, crushed, graded, loose in bulk	£40–45 t^{-1} CIF
Aggregate, expanded, ex-works, UK	£200–260 t^{-1}
Filter aids, expanded, milled, delivered UK	£312–335 t^{-1}

Geology

Perlite, like other glasses, devitrifies with time so that commercial deposits are mainly restricted to areas of Tertiary and Quaternary volcanism. Perlite occurs as lava flows, dykes, sills and circular or elongate domes. The domes are the largest and most commercially important bodies and they can be as much as 8 km across and 270 m in vertical extent. Many of these lava domes cooled quickly in their outer parts to obsidian but the interiors remained hot and formed fine-grained, crystalline rock. In certain instances the obsidian has been hydrated as a result of penetration by ground water forming perlite. Remnants of unaltered obsidian may remain in the perlite, which may also contain phenocrysts of quartz, feldspar and other minerals.

As a result of this mode of formation a particular perlite body may be spherulitic, pumiceous, obsidian-carrying, contain much breccia and show a great variation in volatile content. A sound knowledge of the distribution of these variables within a working quarry is essential in order to preserve a consistent finished product and this calls for much advance drilling and bench sampling. For example, in Hungary's one producing quarry at Palhaza, in the extreme north-east of that country, 4 t of waste are mined for each tonne of usable perlite, because the deposit is so heterogeneous. The variation in volatile content results in the production of three grades each giving different expansion results: $100\,g\,l^{-1}$, $70–100\,g\,l^{-1}$ and $70\,g\,l^{-1}$. Eight size grades are made of each of these three qualities of perlite (Shackley 1990).

In the *USA*, about 85% of production comes from New Mexico, where there is a wide distribution of perlitic domes and flows. The No Agua Peaks, a group of four rhyolitic hills, are believed to contain one of the largest reserves of perlite in the world. Other perlite-producing states are: Arizona, California, Colorado, Idaho and Nevada.

The *USSR* possesses a number of important perlite deposits in Armenia. These include some highly porous varieties that are used after simple crushing and classifying as lightweight aggregate. Reserves run to many millions of tonnes. Another important producing area lies in eastern Kazakhstan and eastern parts of central Asia.

The *Greek* deposits occur in Milos—an island in the Aegean Sea. Here four companies work a number of quarries. At Vouthia Bay there are loading facilities capable of handling vessels up to 30 000 d.w.t alongside a 250 000 t p.a. processing plant. These advantages cut handling costs considerably for the company involved—Silver & Baryte Ores Mining. Other deposits occur on nearby islands but are not worked.

Phosphate rock

Phosphorus is a fundamental element in life. The bones of mammals consist of phosphate, phosphorus is an important constituent of the genetic material DNA and a primary nutrient in the growth of crops. For modern man to feed himself he must have access to huge quantities of artificial fertilizer. Consequently a large tonnage of phosphate is mined each year to form, with potash, nitrates and sulphur, the basis of a vast fertilizer industry which consumes over 90% of the phosphate that is mined annually.

Phosphorus is present in most rocks in minor to trace amounts, but it is only in phosphate rock that the P_2O_5 content is high enough for it to constitute a phosphate ore, where values as high as 40% P_2O_5 may be attained. With rocks of a suitable type and up-to-date processing plants, ores as low as 4% P_2O_5 can be worked. It should be noted that the phosphate content of phosphate rock is generally quoted as %P_2O_5. Other forms quoted include %BPL (bone phosphate of lime—bones were once the main source of phosphate) and %P. Conversion factors are: %P_2O_5 × 2.1853 = %BPL and %P_2O_5 × 0.4364 = %P.

The phosphate in commercial deposits is almost invariably in the form of apatite—commonly fluorapatite $Ca_5(PO_4)_3F$ or carbonate fluorapatite which has the approximate formula $Ca_{10}(PO_4)_{6-x}(CO_3)_x(F,OH)_{2+x}$. Phosphate deposits are of two main types—igneous and marine sedimentary. The former deposits, which supply about 16% of the world's production, are discussed in Chapter 8 and it is the sedimentary deposits that will be discussed in this section. A third source—guano deposits—is rapidly declining as a commercial ore; whilst a fourth source—apatite-rich iron ores at Kiruna and Grängesberg in Sweden—is of very minor importance.

Uses

Most phosphate rock needs beneficiation to increase the BPL content to between 60 and 80%. This is done by crushing, sizing and flotation to reduce the content of impurities such as quartz, chert, clay or shale, mica and carbonates. A certain amount of crushed rock is applied directly to soils, if they are acidic, as a fertilizer and to increase the pH. Electric furnace treatment is used on some lower grade ores (24–30% P_2O_5) to produce elemental phosphorus, which may be used for fertilizer manufacture, or for the preparation of pure phosphoric acid or food-grade acid (Russell 1987). However, the bulk of crushed ore is treated with sulphuric acid to produce phosphoric acid which can be shipped economically

by tanker to fertilizer plants around the world. This phosphoric acid is also used for the manufacture of calcium phosphate animal feedstuffs and for certain chemical processes. Various grades of superphosphate are also prepared from ores by reaction with sulphuric acid. About 90% of world phosphate production goes for fertilizer manufacture, the remaining 10% for the manufacture of animal feedstuffs, detergents, food and drink products, fire extinguishers, dental products and the surface treatment of metals.

For every tonne of phosphoric acid produced by sulphuric acid treatment about 3 t of waste phosphogypsum result and vast stockpiles of this material have been built up. Some has been used as a soil conditioner and minor amounts in plasterboard and cement manufacture. A pilot plant was set up in Florida recently to test the possibility of producing

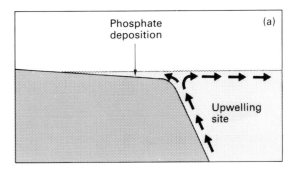

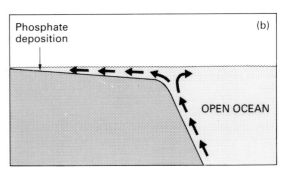

Fig. 20.8 Possible model for phosphorite formation beside an upwelling site initiated in a deep ocean. (a) Nutrient-rich, deep ocean water floods on to the continental shelf and low grade phosphorite deposits form. (b) A sea level rise leads to a major marine transgression resulting in reworking of phosphatic shelf sediments and the shoreward transport of phosphatic grains to form major deposits in the coastal zone and in marginal embayments. (After Cook 1984.)

sulphuric acid and aggregate from this waste material. Another by-product of the mining of some deposits is uranium and the IMC Fertilizer Company produces about 500 t p.a. of yellow cake at its New Wales plant in Florida. In addition, some fertilizer plants already recover fluorine and others are installing the necessary equipment to do so. This fluorine is being used to make artificial cryolite.

Production and price

World production of phosphate climbed steadily from 22 Mt in 1950 to 142 Mt in 1980 in response to world demand, but since then there has been little growth in demand and production in 1989 was about 153 Mt. Demand in many regions has decreased in recent years because of the high level of grain stocks in North America and western Europe. This has resulted in low grain prices and the compulsory set-aside of land, two factors that tend to depress fertilizer usage. For example, phosphate rock usage peaked in western Europe at 23 Mt in 1980 but was down to 17.1 Mt by 1989. This declining demand and an increase in the world's production capacity has led to an oversupply of phosphate and the price for phosphate rock has declined from about US\$50 t^{-1} in 1980 to \$35 t^{-1} for 60–66% BPL ore.

The top four producers in 1989 were the USA (48.9 Mt), the USSR (34.5 Mt), Morocco (18 Mt) and China (11.4 Mt). These countries have for some years produced about 75% of world output. The remaining 25% came from 26 other countries.

Geology of phosphorites

Sedimentary phosphate deposits are known as phosphorites and world reserves of this material are

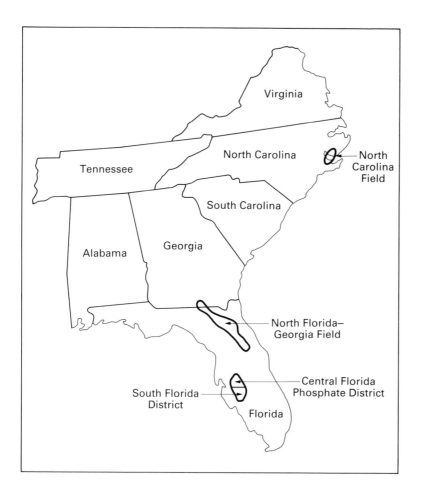

Fig. 20.9 Location of the phosphate districts of economic importance in the south-eastern USA.

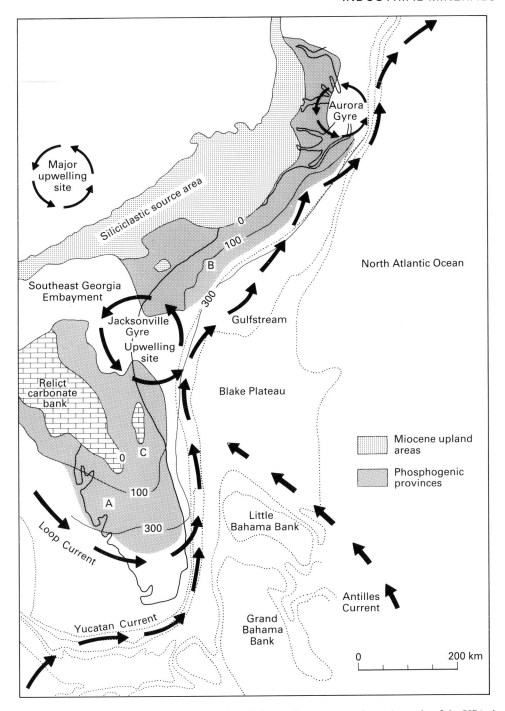

Fig. 20.10 A Miocene palaeogeographical reconstruction of the south-eastern continental margin of the USA showing oceanic currents, major topographically induced upwelling sites and areas of phosphate deposition. (After Compton *et al.* 1990, with some modifications based on van Kauwenbergh & McClellan 1990.)

thought to be in excess of 200 000 Mt (Cook 1984). Phosphorites occur on every continent and range in age from Precambrian to Recent, but almost all the commercially exploited deposits are Phanerozoic. However, some extensive continental areas have no commercial phosphorites, e.g. western Europe, Canada, Greenland and the north-eastern USSR.

Phosphorites generally form beds a few centimetres to tens of metres thick that are composed of grains (frequently termed pellets) of cryptocrystalline carbonate fluorapatite, which is often conveniently referred to, particularly in the field, as collophane. Most grains are well rounded and up to about 2 mm across. They may contain grains of quartz, clay, framboidal pyrite and carbonaceous material. In some deposits it is clear that many of these phosphatic grains were originally carbonate that has been phosphatized diagenetically. Oolites occur in many deposits and some may be primary but others clearly are phosphatized calcite oolites. Other forms of collophane include nodules, which may be up to tens of centimetres across, and mudstone phosphorites. Phosphorites often show much evidence of reworking, perhaps with the addition of new phosphatic material and later *in situ* leaching, which removes carbonate to form enriched residual deposits. Phosphate deposits commonly have interbeds of shale and chert.

At the present day phosphorites are forming in low latitudes off the coasts of continents where there is known to be marked upwelling bringing deep oceanic waters on to continental shelves, e.g. off the coasts of Chile, Namibia and eastern Australia. Major upwelling sites develop where oceanic boundary currents—such as the Gulf Stream off the coast of Florida and the Carolinas (see Fig. 20.10)—steer around bathymetric highs producing gyres within which Ekman pumping occurs (spiral upward movement of water). The deep ocean is a vast reservoir of phosphate, which increases in amount when the rate of deep oceanic circulation slows down. If continental drift gives rise to the development of new seaways or the drift of a continent into low latitudes (pp. 325–326, then this phosphate may be recirculated into shallow oceanic areas and on to continental shelves where a great increase (around 500 times) will develop in the biomass, leading to the formation of widespread, low grade phosphate deposits (Fig. 20.8). However, these will be on the deeper parts of shelves and it seems that economic deposits only develop when a rise in sea level results in a marked marine transgression that leads to reworking of the shelf deposits, and the shoreward transport of phosphate grains to form major deposits in coastal traps, such as bays and other marginal embayments, and around structural highs.

The deposits of Florida and North Carolina appear to be a good example of this process. Figure 20.9 shows the distribution of the economic phosphorite fields. These fields are responsible for much of the present day phosphate rock production in the USA. The Miocene phosphatic sediments are usually coarser with siliceous impurities on the landward side and they grade seawards to finer grained deposits and carbonate. Sedimentation took place on the shallow water coastal and nearshore platform adjacent to structural highs. The deposits in the Central District have only a moderate overburden thickness, good ore thicknesses, a BPL content averaging 70% and a significant uranium content. A reconstruction of the Miocene palaeogeography with the positions of major upwelling sites which fed phosphate on to the shallow marine platforms is shown in Fig. 20.10.

21 / Short notes on selected industrial minerals

The subject of industrial minerals and their uses is so vast that it is impossible to do justice to this topic in an introductory text. In this chapter some brief comments will be made on minerals, rocks and mineral groups that have not received much attention in previous chapters. More detailed treatments can be found in Harben & Bates (1990) and Lefond (1983).

Abrasives

The most important points in choosing an abrasive (grit) is (a) that it must be harder than the work piece, and (b) that it is imperative to match the abrasive to the job to be done. As an example of the second point we can look at the household cleaning of brass and silver. Brass cleaners often carry very fine-grained quartz and mica. The quartz will leave very fine but largely imperceptible scratches on brass ornaments; however, such scratches cannot be tolerated on silver and only mica is added to the cleaning fluid to act as an abrasive.

Properties required

Hardness. This is measured on Mohs Scale or by determining the microindentation hardness using Vickers, Knoop or Rockwell hardness numbers.
Strength and toughness. The grit must be strong enough to bear the forces applied to it, i.e. it must not fracture under the polishing loads.
Friction. The frictional coefficient must not be so high that the grit sticks to the work piece.
Chemical inertness. No chemical action is allowable.
Thermal shock resistance. This must be considerable and must be measured, as grits are often subjected to repeated high and low temperatures during working.
Refractoriness. The grit must not melt or dissociate during use.
Shape. This affects the cutting rate. Sharp splinters may grind quickly but will break more easily than equant shaped grains.
Size. This can be critical depending on the job in hand.
Endurance. Working life of the grit.

Some abrasives

Quartz will be present in a sand or crushed sandstone in various sizes and can be sieved to obtain the grade required; it is, however, only about as hard as steel. *Garnet* is a little harder followed by emery which is a mixture of corundum + magnetite ± hematite ± spinel ± plagioclase, but the use of these minerals is largely superseded by artificial abrasives—carborundum, etc. However, the bulk of *diamond* production, both natural and artificial, is used in the abrasives industry. Softer materials include *diatomite, pumice,* various *metal oxides* and *whiting* (ground and washed chalk).

Asbestos

This material is now considered to be the cause of asbestosis (scarring of the lung tissue) and many types of cancer. Certain forms, e.g. crocidolite (blue asbestiform riebeckite), have been banned in a number of European countries and there is a virtual ban on all forms of asbestos for most uses in the USA. Natural and artificial substitutes are now used whenever possible.

About 95% of world production is of chrysotile (asbestiform serpentine) which is less dangerous than crocidolite. It occurs in veinlets and veins forming stockworks in serpentinites. The asbestos fibres grew perpendicular to the vein walls and hence their fibre length and value is governed by vein thickness. Deposits are mined in bulk and the asbestos separated in the mill. The leading producers and their output in 1988 were USSR 2.6 Mt (estimated), Canada 710 kt, Brazil 208 kt and Zimbabwe 187 kt. World production has declined from 4.5 Mt in 1983 to about 4 Mt today.

Fluorspar

Some writers use this name for the mineral fluorite, others for the raw ore containing fluorite and gangue and yet others, particularly in industry, for the finished product, which comes in three grades: acid, ceramic and metallurgical. About half the fluorite produced is consumed in iron and steel making, but this demand is decreasing as new technology is adopted by smelters. Acid grade (minimum of 97% CaF_2) is used for the manufacture of hydrofluoric acid (HF), much of which is consumed in the production of fluorocarbons and of artificial cryolite (Na_3AlF_6), the latter being essential for the smelting of aluminium. Ceramic grade, No. 1 (95–96% CaF_2) and No. 2 (80–92% CaF_2), is used in the manufacture of special glasses, enamels and glass fibres as well as in other industrial applications. Metallurgical grade must run 60% CaF_2 and have $< 0.3\%$ sulphide and $< 0.5\%$ Pb. World production of fluorite in 1989 has been put at 5.66 Mt—a record output for this mineral. China is the leading producer followed in order by Mexico, Mongolia, the USSR and RSA.

Much fluorite is won from Mississippi Valley-type deposits (Chapter 17) although the USA is now a minor producer. It also occurs in pipes (mineralized diatremes), cryptovolcanic structures, alkaline igneous complexes, carbonatites and other deposit types.

Nepheline syenite

This is an alkaline, plutonic igneous rock with high sodium and aluminium contents and low silica—a chemical make-up that enables it to compete with feldspar and aplite as a source of sodium and aluminium in the manufacture of container and sheet glass. Its lower fusion temperature cuts energy costs, but feldspar has the edge on it in price. The sodium acts as a flux and the aluminium reduces the tendency of glass to devitrify and improves its chemical durability. Nepheline syenite is also used in the manufacture of some vitreous whiteware, glazes and enamels and as a filler in paint, rubber and plastics.

For commercial purposes nepheline syenite must contain $< 2\%$ coloured minerals so that its iron content is very low. Outside the USSR two deposits supply MEC with this rock: the Blue Mountain deposit in southern Ontario and the Stjernoy Island deposit of Norway, which is 400 km within the Arctic Circle. Their 1987 production was 506 kt and 240 kt respectively. Phonolite, the volcanic equivalent of nepheline syenite, is mined on a minor scale in France and Germany.

Pyrophyllite

This is a soft, white mineral with a pearly lustre that resembles talc and is used for many similar purposes. The principal markets for pyrophyllite are refractories, ceramics and steel making. However, in the last-named market it seems that the advent of continuous casting techniques is leading to a decrease in the use of pyrophyllite in zircon ladle bricks, traditionally one of pyrophyllite's important end uses. The Australian prices for bulk purchases ex-store are US\$25–35 t^{-1} for refractory grade and \$35–45 t^{-1} for ceramic and filler grade. Unfortunately pyrophyllite production is lumped statistically with talc, steatite and soapstone. It has been estimated that world production lies in the 0.9–1.1 Mt p.a. range of which approximately half is in Japan and a quarter in South Korea. China is believed to be a substantial supplier and large new reserves have recently been found in that country.

Pyrophyllite is an occasional product of high grade hydrothermal alteration (Chapter 3) and economic deposits have been formed in a number of orogenic belts as the result of the hydrothermal alteration of acid and intermediate volcanics. In the deposits of the Carolina Slate Belt in the USA, heated meteoric water appears to have been the mineralizing agent (Klein & Criss 1988), but Bryndzia (1988) considers that at the Foxtrap Pyrophyllite Mine, Newfoundland, Canada's only pyrophyllite producer, the fluids responsible for the alteration of the rhyolite flows and pyroclastics emanated from the nearby Holyrood Batholith, a granitoid intrusion within whose aureole there are many zones of pyrophyllitization which are traceable over 7 km.

Sillimanite minerals

Under this heading, industry groups andalusite, kyanite and sillimanite which are polymorphs of Al_2SiO_5. These are highly refractory minerals and this is their major use. Minor uses include abrasives, glazes and non-slip flooring production. Commercial grades must run 56% Al_2O_3 or better and 42% SiO_2; acid soluble Fe_2O_3 must be $< 1\%$, TiO_2 $< 1.3\%$ and CaO and MgO must not exceed 0.1% each. Sillimanite minerals command a good price, £90–155 t^{-1}, according to grain size and Al_2O_3 content. However, as they have different physical properties, their markets are different and they are traded as separate mineral concentrates. World production is about 0.5 Mt, of which in 1987 RSA produced 39%, mainly in the form of andalusite. The USSR came next with about 20% as kyanite, and the USA third with 18%, as kyanite and related minerals.

India produces all three polymorphs and came fourth with 10%.

Economic deposits of kyanite and sillimanite are generally of regional metamorphic origin and andalusite of contact metamorphic origin. The large South African output comes mainly from the metamorphosed Daspoort Shales, which lie in the contact aureole of the Bushveld Complex (Chapter 10). These shales are over 60 m thick and run 5–20% andalusite. Known reserves will last for 30 years at the present production rate and it has been suggested that the total resources are sufficient to keep the world supplied at present consumption levels for 500 years! The andalusite hornfelses are soft and no blasting is required. The ore is crushed and the andalusite separated in a heavy media plant. France is the other big andalusite producer—about 70 000 t p.a. from the Damrec Mine, Brittany.

The USA is by far the largest producer of kyanite, much of which comes from kyanite quartzites at Willis Mountain, Virginia. India and Sweden are also important kyanite producers, the latter from a mine in Precambrian quartzites running 30% kyanite. Virtually all world sillimanite production, about 30 000 t p.a., comes from India; most of this used to come from immense boulders of massive sillimanite (with minor corundum) but by-product sillimanite from placer operations is now the principal source.

Slate

Many readers will be familiar with the use of this rock as a roofing material and ornamental stone. Those guilty of a 'misspent youth' will have devoted much time to propelling billiard balls on slate-bedded tables! However, in recent years slate has been put to other uses. Rapid heating of crushed slate to about 1200°C produces a porous, slag-like material suitable as lightweight concrete aggregate. In North Wales finely powdered slate (fullersite) is produced for use as a filler in bituminous compounds such as automobile underseal, plastics and industrial adhesives. It is used also in insecticides and fertilizers.

Sulphur

Sulphuric acid is the most important inorganic chemical produced in terms both of volume and use. As has been indicated earlier, it is derived from a number of sources, such as the smelting of sulphide ores and the sulphur recovered from crude oil (approximately 50% of world sulphur production comes from oil refineries). Rather less than 40% of world sulphur output comes from working beds of native sulphur and deposits in the caprocks of salt domes. Much of this sulphur is produced by the Frasch process in which the sulphur is melted by pumping hot water into it and then pumping the resultant liquid to the surface. Poland is the world's major producer of native sulphur with an output of nearly 5 Mt p.a. using the Frasch process; the USA is next with 3.2 Mt. Other important Frasch producers include Mexico and the USSR. The sulphur in these deposits is believed to have resulted from the biogenic alteration of gypsum to sulphur and calcite. In the Polish Miocene deposits this reduction appears to have occurred at places where the gypsum was upfaulted, covered by impervious clay and saturated with saline water and hydrocarbons from lower strata.

Trona

The bulk of the world's soda ash (sodium carbonate) is produced by the Solvay process, with halite as the starting point. However, there is important production of natural sodium carbonate in the form of trona ($Na_2CO_3.NaHCO_3.2H_2O$) in Wyoming, USA, Kenya and Mexico. Botswana and China are expected to become producers in the near future and in due course some Turkish deposits may be exploited. Output in the USA in 1989 was almost 9 Mt, in Kenya 235 kt and Mexico 145 kt. The remaining 22.34 Mt was produced by Solvay plants scattered across the world. Most soda ash is consumed by the glass industry. Other users include the chemical, textile, paper and fertilizer industries.

The Wyoming deposits belong to the Eocene Green River Formation of lacustrine origin. At least 42 beds of trona are present in the Wilkins Peak Member interbedded with marl, oil shale, halite and clastic sediments. Twenty-five beds range from 1 to 12 m in thickness and extend over 250 km or more. The trona is probably ultimately of volcanic origin. Searles Lake in California is another non-marine evaporite deposit from which trona is won. The Kenyan deposit is in Lake Magadi in the Eastern Rift Valley. In the lake, which floods and dries out annually, trona beds up to 35 m thick are present and these are still being formed by brines entering the lake from hot springs. Again the ultimate source is thought to be volcanic. The trona is extracted by a floating dredger, then crushed, slurried and pumped to the treatment plant.

Wollastonite

This mineral, not mined at all until the 1930s, is now in considerable demand as it can substitute for asbestos in some of the uses of that mineral. This is particularly true of long fibre wollastonite, which is used in fibre boards and panels. Another major use is as a filler in plastics and ceramics. In the USA the price ex-works in October 1991 was US$260–265 *per short ton* for best grade acicular, – 200 mesh material.

Wollastonite has the ideal formula $CaSiO_3$ but Fe, Mg or Mn can substitute for small amounts of Ca and thereby reduce the whiteness. In 1987 recorded world output was of the order of 120 000 t; 72 kt were produced in the USA, 31 kt in India and nearly 16 kt in Finland. The Magata Mine, RSA was commissioned in 1990 to produce 2500 t of ore per month and the search is on for more deposits, explorationists being encouraged by the high price and demand. It was reported recently that a 100 Mt deposit has been found in China and a number of encouraging prospects are being investigated in Canada.

The majority of, if not all, wollastonite production is from contact metamorphosed impure limestones. Wollastonite also occurs in a number of alkaline igneous rocks and it is conceivable that economic deposits could exist in some of them. The principal producer in the USA is the Fox Knoll Mine in the Adirondack Mountains of New York State. Here impure Proterozoic limestone in the contact aureole of an anorthosite has been metamorphosed to a wollastonite–garnet hornfels. This contains bands rich in one or the other mineral and the ore averages 60% wollastonite and 40% garnet plus impurities. The ore is crushed and magnetic separation used to free the non-magnetic wollastonite from the feebly magnetic garnet and diopside. Diopside is separated from the garnet electrostatically.

22 / The metamorphism of ore deposits

It is all too often forgotten that many ore deposits occur in metamorphosed host rocks. Among these deposits, the substantial proportion that are syn-depositional or diagenetic in age must have been involved in the metamorphic episode(s) and consequently their texture and mineralogy may be modified considerably, often to the advantage of mineral separation processes. In addition, parts of the orebody may be mobilized and epigenetic features, such as cross-cutting veins, may be imposed on a syngenetic deposit. Ores in metamorphic rocks may be metamorphic in the sense that their economic minerals have been concentrated largely by metamorphic processes; these include the pyrometasomatic (skarn) deposits and some deposits of lateral segregation origin. On the other hand they may be metamorph*osed*, in the sense that they predate the metamorphism and have therefore been affected by a considerable change in pressure–temperature conditions, which may have modified their texture, mineralogy, grade, shape and size—structural deformation often accompanies metamorphism. This chapter will be concerned entirely with metamorphosed ores.

It is important from the economic as well as the academic viewpoint to be able to recognize when an ore has been metamorphosed. Strongly metamorphosed ores may develop many similar features to high temperature epigenetic ores for which they may be mistaken. This can be very important to the exploration geologist, for in the case of an epigenetic ore, particularly one localized by an obvious structural control, the search for further orebodies may well be concentrated on a search for similar structures anywhere in the stratigraphical column, whereas repetitions of a metamorphosed syngenetic ore should be first sought for along the same or similar stratigraphical horizons. Notable examples of orebodies long thought to be high temperature epigenetic deposits, but now considered to be metamorphosed syndepositional ores, are those of Broken Hill, New South Wales and the Horne Orebody, Quebec. In both areas recent exploration having a stratigraphical basis has been successful in locating further mineralization.

Certain types of deposit, by virtue of their structural level or development in geological environment, are rarely, if ever, seen in a metamorphosed state. Some examples are porphyry copper, molybdenum and tin deposits, ores of the carbonatite association and placer deposits. On the other hand, certain deposits, such as the volcanic-associated massive sulphide class have generally suffered some degree or other of metamorphism such that their original textures are commonly much modified.

Three principal types of metamorphism are normally recognized on the basis of field occurrence. These are usually referred to as contact, dynamic and regional metamorphism. Contact metamorphic rocks crop out at or near the contacts of igneous intrusions and in some cases the degreee of metamorphic change can be seen to increase as the contact is approached. This suggests that the main agent of metamorphism in these rocks is the heat supplied by the intrusion. As a result, this type is sometimes referred to as thermal metamorphism. It may take the form of an entirely static heating of the host rocks without the development of any secondary structures, such as schistosity or foliation, though these are developed in some metamorphic aureoles.

Dynamically metamorphosed rocks are typically developed in narrow zones, such as major faults and thrusts, where particularly strong deformation has occurred. This type of metamorphism is accompanied by high strain rates and a range of *PT* conditions depending on the structural level and other factors, such as proximity of igneous intrusions and penetration by hot hydrothermal solutions of deep seated origin, which may cause what is sometimes described as hydrothermal metamorphism. Epigenetic ores, developed in dilatant zones along faults, often show signs of dynamic effects (brecciation, plastic flowage, etc.) owing to fault movements during and after mineralization.

Regionally metamorphosed rocks occur over large tracts of the earth's surface. They are not necessarily associated with either igneous intrusions or thrust belts, but these features may be present. Research has shown that regionally metamorphosed rocks generally suffered metamorphism at about the time they were intensely deformed. Consequently they contain characteristic structures, such as cleavage, schistosity, foliation or lineation, which can be seen on both the macroscopic and microscopic scales, producing a distinctive fabric in rocks so affected. Whereas regional metamorphism is normally associated with major orogenic activity, thick basinal sequences not involved in orogenesis may show incipient regional metamorphism; this may be referred to as burial metamorphism. Metamorphism developed at mid-oceanic ridges, although it may be regional in extent, is referred to as ocean-floor metamorphism.

How can we recognize when ores have been metamorphosed? One way is to compare their general behaviour with that of carbonate and silicate rocks that have undergone metamorphism. In contact and regionally metamorphosed areas these rocks generally show:

1 the development of metamorphic textures;
2 a change of grain size—usually an increase;
3 the progressive development of new minerals. In addition, as noted above, regionally metamorphosed rocks usually develop certain secondary structures. Let us examine each of these effects in turn.

Development of metamorphic textures

Deformation

All three types of deformation, elastic, plastic and brittle, are important—elastic deformation largely because it raises the internal free energy of the grains and renders them more susceptible to recrystallization and grain growth. Plastic deformation occurs by primary (translation) gliding or by secondary (twin) gliding. In translation gliding, movement occurs along glide planes inside a grain without any rotation of the crystal lattice. Translation gliding can be readily induced in the laboratory in galena, which has glide planes parallel to {100}. In twin gliding, rotation of part of the crystal lattice occurs so that it takes up a twinned position. Rotation is initially on the molecular scale, but with continued applied stress it spreads across the whole grain by rotation of one layer after another. The normal result is the development of polysynthetic twins. One example familiar to geologists is the secondary twinning of carbonates in deformed marbles. Most of the soft opaque minerals contain many potential glide

planes, there being eight in galena, and as a result, the stresses which may affect these minerals during dynamic and regional metamorphism can lead to plastic flowage. Elevated temperatures facilitate such flowage, but flow can be induced at lower temperatures by increasing the applied pressure.

Brittle deformation takes the form of rupturing or shearing and both generally follow lines of weakness in grains, such as cleavages, twin planes, etc. Ruptures are often healed by new growth of the same or another mineral in the space created, but on the other hand, softer minerals may flow into this space thus isolating the fragments of the ruptured mineral. This is particularly the case in polyphase grain aggregates made up of strong and weak minerals. At a given degree of deformation the stronger minerals, such as pyrite and arsenopyrite, may fail by rupturing whilst the softer minerals, such as galena and the sulphosalts, may flow. These processes can give rise to grain elongation and the development of schistose textures. Instructive drawings and photomicrographs illustrating these textures can be found in Pesquera & Velasco (1989).

Recrystallization

The deformed state is one of high potential energy and, if the temperature is high enough, annealing will occur leading to a reduction of this high energy level. Of the various processes that take place during annealing, recrystallization is the most important. It consists of the replacement of strained grains by strain-free grains followed by grain growth, the new grains meeting in growth impingement boundaries, which represent an unstable configuration. The presence of these grain boundaries leads to an increase in the free energy of the system over that which it would possess if no grain boundaries were present. This extra free energy is that of unsatisfied bonds at or near the surface of the grains; it is called the interfacial free energy.

The shapes of grains in a polycrystalline aggregate are governed by two main factors: firstly, the need to reduce the overall free energy level to a minimum, and secondly, the requirement to fill space. The first requirement would be fulfilled by spherical grains but these clearly would not fill space. The microscopic examination of polished sections of artificially annealed metals and minerals shows that most grains meet three at a time at a point. Separation of these grains shows that they are bounded by a number of flat surfaces identical with those in a soap froth (Smith 1964). Inspection of such froths (which can be done by shaking up some soap solution in a plain tumbler) shows that in three dimensions the great majority of bubbles (and, by inference, grains too) meet in threes along lines. These lines are called triple junctions. When intersected by a surface (e.g. that of a polished section) these junctions appear as points (Fig. 22.1a). In a monominerallic aggregate the angles between the grain boundaries around such points are equal to or close to 120°, provided the section is normal to the triple junction. Such equilibrium grain configurations will normally be present in annealed (i.e. metamorphosed) ores. They are present also in some autoannealed ores, such as chromite deposits in large lopoliths, and at least one case of their occurrence in an epigenetic ore has been recorded

(Burn 1971). In investigating such grain configurations in ores, it is usual to measure a number of these angles and to use a frequency plot to determine whether they peak at 120° (Stanton 1972). Calculation of the standard deviation will produce a measure of the degree of perfection of the annealing. Very beautifully annealed mineral aggregates can be seen in the gold–copper ores of Mount Morgan (Lawrence 1972).

So far we have been concerned with monominerallic aggregates. In the case where a phase is in contact with two grains of a different phase the dihedral angle (θ) is no longer 120° (Fig. 22.1b) but some other constant angle, which varies according to the composition of the phases. The values of a number of these dihedral angles are known, mainly from the work of Stanton, and again measurements can be made to see if the grain aggregate has been annealed.

Increase in grain size

In an annealed aggregate, grain boundary energy is the dominant force leading to structural modification. An influx of heat will raise the energy level and, if the temperature is high enough to permit diffusion, grain boundary adjustments will occur. Since the total grain boundary free energy is proportional to the grain boundary area, grain growth will occur in order to reduce the number of grains per unit volume. Grain growth does not go on indefinitely and it has been shown empirically by metallurgists that the stable average grain diameter is given by:

$$De^2 = C(T - T_0)$$

where De is the mean grain diameter, C and T_0 are constants for a given metal, and T is the absolute temperature. Thus, with increasing grades of metamorphism, monominerallic aggregates in particular, and indeed all grain aggregates should show an increase in grain size. Vokes (1968) has demonstrated that this is the case for metamorphosed volcanic-associated massive sulphide deposits in Norway.

When recrystallization occurs under a directed pressure then the effects known as Riecke's Principle may become important. This is the phenomenon of pressure solution causing parts of grains to go into solution at points of contact, with the dissolved material being redeposited on those parts of grains that are not under such a high stress. This produces

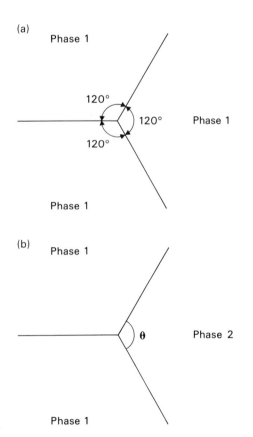

Fig. 22.1 (a) Grain boundary configuration about a triple junction point in a monominerallic aggregate. (b) The same configuration when a second phase is present. θ is the dihedral angle and it can be less than or greater than 120°.

flat elongate grains, which give the ore a schistose or gneissose appearance.

Textural adjustments and the development of new minerals

As Vokes (1968) has pointed out, because of the considerable ranges of pressure and temperature over which sulphide minerals are stable, we do not find the progressive development of new minerals with increasing grade of metamorphism that are present in many classes of silicate rocks. Banded iron formations on the other hand do show the abundant development of new mineral phases (James 1955, Gole 1981, Haase 1982), as do many manganese deposits. Sulphide assemblages by comparison only show some minor mineralogical adjustments.

Textural adjustments

Two commonly noted phenomena are the reorganization of exsolution bodies and the destruction of colloform textures. Though solid solutions may be created during metamorphism by the diffusion of one or more elements into a mineral, the general tendency is for exsolution bodies to migrate to grain boundaries to form intergranular films and grains during the very slow cooling period that follows regional metamorphism.

A general decrease in the incidence of colloform textures with an increasing grade of metamorphism has been reported from pyritic deposits in the Urals, Japanese sulphide deposits and Canadian Precambrian volcanic-associated massive sulphide deposits. Both these adjustments generally improve the ores from the point of view of the mineral processor. Exsolution bodies do not pass with their host grains into the wrong concentrate, and the replacement of colloform textures by granular intergrowths leads to better mineral separation during grinding.

Mineralogical adjustments

In contact metamorphism the main effects so far studied are those seen next to basic dykes cutting sulphide orebodies. These include the following reactions and changes:

1 pyrite + chalcocite $\rightarrow$ chalcopyrite + bornite + S;
2 pyrite + chalcocite + enargite $\rightarrow$ chalcopyrite + bornite + tennantite + 4S;
3 the iron content of sphalerite increases;
4 copper diffuses into sphalerite;

5 marginal alteration of pyrite to produce pyrrhotite + magnetite.

The principal mineralogical change reported from regionally metamorphosed sulphide ores is an increase in the pyrrhotite : pyrite ratio with the appearance of magnetite at higher grades. The sulphur given off by the reactions involved in pyrrhotite and magnetite production may diffuse into the wall rocks, promoting the extraction of metals from them and the formation of additional sulphide minerals.

Some effects on orebodies and implications for exploration and exploitation

Some results of fold deformation of orebodies have been touched on in various places in earlier chapters. For example, in Fig. 15.6, the effective thickening of the orebody in part of the Luanshya deposit is evident, and in Fig. 16.10, the migration of ore components into fold hinges during deformation and metamorphism is shown. These orebodies at the Homestake Mine were further deformed during later phases of folding (Fig. 16.9), and became spindle-shaped.

Many massive sulphide deposits occur in regionally metamorphosed terrane of all grades of metamorphism; some are very severely deformed and there are many excellent studies of their metamorphism, e.g. Selkman (1984) and Frater (1985) as well as of their deformation (van Staal & Williams 1984) and of the reconstruction of original features through the use of 'unfolding techniques' (Zachrisson 1984). Sangster & Scott (1976) discussed some of the implications for exploration arising from the metamorphism of these bodies and their country rocks. They pointed out that a massive sulphide deposit after undergoing even mild metamorphism is often radically different in many ways from its original form, and that metamorphic changes can directly influence the exploration for, and the development of, these orebodies. Metamorphism of the host rocks can so change them that it is difficult for the geologist to recognize and trace favourable rock environments. Because many of these orebodies are associated with acid volcanics, it is imperative that the distinction be made between metamorphosed silicic volcanics and metaquartzites, meta-arkoses or granite-gneisses. Similarly, the recognition of a metamorphosed cherty tuff layer can be of great assistance as a marker horizon, in

Table 22.1 Some metamorphic equivalents of primary rock types accompanying volcanic-associated massive sulphide deposits

Primary rock type	Medium grade metamorphism	High grade metamorphism
Chert	Quartzite	Quartzite
Pyritic BIF	Pyrite–pyrrhotite–mica schist	Pyrite-pyrrhotite–mica gneiss
Rhyolite Rhyolite-tuff Rhyolite-breccia Rhyolite-agglomerate	Muscovite–feldspar–quartz schist	Granitic gneiss
Andesite Andesitic volcaniclastics	(Biotite)–hornblende–plagioclase schist	Amphibolite with clinopyroxene $\pm$ biotite $\pm$ garnet
Basalt	Oligoclase or andesine amphibolites $\pm$ epidote $\pm$ garnet	Andesine or labradorite amphibolite $\pm$ garnet

structural interpretation and because it may represent a former exhalite with associated massive sulphide deposits (see Chapter 15). For example, at one of the world's oldest known Besshi deposits in the Lower Proterozoic of Gairloch, Scotland, which has suffered amphibolite facies grade regional metamorphism and polyphase deformation, geochemical studies have revealed the presence of exhalative components in the footwall rocks and the quartz-magnetite hanging wall schists are clearly of this origin (Jones *et al.* 1987). The explorationist has to 'see through' metamorphosed suites of rocks to

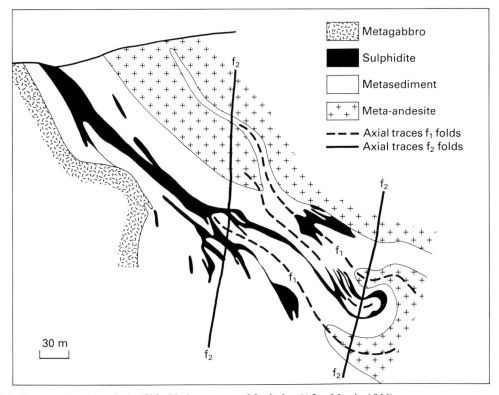

Fig. 22.2 Cross section through the Chisel Lake ore zone, Manitoba. (After Martin 1966).

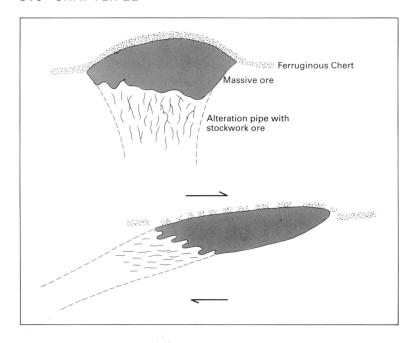

Ferruginous Chert

Massive ore

Alteration pipe with
stockwork ore

Fig. 22.3 Schema showing the
effect of shearing (below) on an
undeformed massive sulphide
deposit (above). (After Sangster
& Scott 1976.)

reconstruct the original stratigraphy and environment in the search for favourable areas and horizons. Some of these problems will be clearly grasped from a perusal of Table 22.1.

Metamorphism and deformation of a volcanic-associated massive sulphide orebody may also give rise to a different mineralogy, grain size and shape of the deposit, which can affect its geophysical and even its geochemical response. Such deposits were probably originally roughly circular or oval in plan and lenticular in cross section. Deformation may change them to blade-like, rod-shaped or amoeba-like orebodies, which may even be wrapped up and stood on end, and then be mistaken for replacement pipes, e.g. the Horne Orebody of Noranda, Quebec! A good example of what can happen to what was possibly originally one single massive sulphide body is shown in Fig. 22.2. Martin's paper is well worth studying as it is a good, but succinctly described, example of the problems that such deformation can pose to the mine geologists and engineers.

When considerable shearing is involved in the deformation, a flattening effect by transposition along shear or schistosity planes may transpose the alteration pipe and the stockwork beneath the massive sulphide lens into a position nearly parallel to the schistosity and the lens (Fig. 22.3). In extreme cases, the sulphide lens may be sheared off the stockwork zone, separating the two. Metamorphism may also have a profound effect on the rocks of the alteration zone. Chloritized andesites in alteration pipes have frequently been metasomatized to a bulk composition that will produce cordierite–anthophyllite hornfelses—the 'spotted dog rock' or 'dalmatianite' of the Noranda district—in the thermal aureoles of intrusives. The deformational styles of four sediment-hosted, lead–zinc deposits have been documented and compared by McClay (1983), who showed that the tectonic style developed in any one deposit depends upon the conditions of deformation, the sulphide and host rock compositions and the nature of the layering within the deposits.

Finally it must be recorded that epigenetic mineralization formed before, or early in the deformational history of an area, may also develop metamorphic textures and structures similar or identical to some of those described above. An example of just such a case has been carefully documented by Brill (1989).

Part 3
Mineralization in space and time

23 / The global distribution of ore deposits: metallogenic provinces and epochs, plate tectonic controls

Introduction

It has long been recognized that specific regions of the world possess a notable concentration of deposits of a certain metal or metals and these regions are known as metallogenic provinces. Such provinces can be delineated by reference to a single metal (Figs 23.1–23.3) or to several metals or metal associations. In the latter case, the metallogenic province may show a zonal distribution of the various metallic deposits (Fig. 5.5). The recognition of metallogenic provinces has usually been by reference to epigenetic hydrothermal deposits, but there is no reason why the concept should not be used to describe the regional development of other types of deposit provided they show a geochemical similarly. For example, the volcanic-exhalative antimony–tungsten–mercury deposits in the Lower Palaeozoic inliers of the eastern Alps form a metallogenic province stretching from eastern Switzerland through Austria to the Hungarian border. The establishment of an international commission to prepare metallogenic maps of the world has led to the publication of many such maps. These are often based on metallogenic provinces and are a common tool of exploration geologists seeking new mineral deposits (Feiss 1989, Noble 1980).

By contrast with the simple definition in the above paragraph many writers have extended the regional concentration of certain elements, element associations and deposit types to imply a genetic relationship (Turneaure 1955). In the 1920s J.E. Spurr wrote of deposits in these provinces as having 'a blood relationship' and Feiss (1989) wrote, 'The issue is ultimately to establish their [the constituent deposits] common origin.' The present writer prefers, if it is possible, not to use genetic implications as a method of classification and advocates the simpler definition of metallogenic provinces. With regard to their size, Petrascheck (1965) suggested that at least one dimension of such a province should be 1000 km long. Turneaure on the other hand was happy to allow areas as small as mineral districts to be designated as metallogenic provinces. Most writers appear to think in terms of considerably

larger areas than this without being quite as rigid as Petrascheck.

Within a metallogenic province there may have been periods of time during which the deposition of a metal or a certain group of metals was most pronounced. These periods are called metallogenic epochs. Some epochs are close in time to orogenic maxima, others may occur later. Thus in the Variscan orogenic belt of north-western Europe and its northern foreland, which form the metallogenic province shown in Fig. 5.5, the principal epochs of epigenetic mineralization were Hercynian (end Carboniferous to early Permian) and Saxonian (middle Triassic to Jurassic). The orogenic events in this belt culminated about the end of the Carboniferous and the Saxonian mineralization and associated vulcanicity is post-orogenic. Important provinces and epochs of syngenetic mineralization are also present in this orogen, notably the Iberian Pyrite Belt with its volcanic-associated massive sulphide deposits of Carboniferous age developed over an exposed strike length of 200 km. A comparable but longer province is that of the Japanese Miocene Kuroko deposits—1200 km, and another comparable but older metallogenic province is the Proterozoic (1800–1730 Ma) volcanic belt of Arizona, which stretches over 300 km (Anderson & Guilbert 1979).

Metallogenic provinces and epochs of tin mineralization

Tin deposits are an excellent example of an element restricted almost entirely from the economic point of view to a few metallogenic provinces. Those outside the USSR are shown in Figs 23.1–23.3. Even more striking is the fact that most tin mineralization is post-Precambrian and confined to certain well-marked epochs. Equally striking is the strong association of these deposits with post-tectonic granites. Among tin deposits of the whole world, 63.1% are associated with Mesozoic granites, 18.1% with Hercynian (late Palaeozoic) granites, 6.6% with Caledonian (mid-Palaeozoic) granites and 3.3% with Precambrian granites. In some belts,

313

Fig. 23.1 Tin belts on continents around the Atlantic Ocean. Dotted areas indicate concentrations of workable deposits. (Modified from Schuiling 1967.)

tin mineralization of different types and ages occurs. For example, in the Erzgebirge of Germany, Lower Palaeozoic or Precambrian stratiform deposits of tin are developed, which are probably volcanic-exhalative in origin (Baumann 1965), together with epigenetic deposits associated with Hercynian granites. It is possible that these granites represent anatectic material from the Lower Palaeozoic or Precambrian of the region and that some of the

stratiform tin was remobilized when the granitic magmas were formed by partial melting.

The figures quoted above suggest that increasing amounts of tin have been added to the crust with decreasing antiquity. The puzzling feature in this connexion is that there does not appear to be any concomitant increase in the level of trace element tin with decrease in geological age in granites inside or outside tin belts. As epigenetic tin deposits are

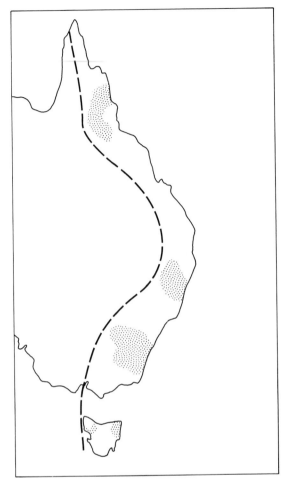

Fig. 23.2 The Palaeozoic tin belt of eastern Australia. The principal fields are shown with dotted ornamentation. (Modified from Hills 1953.)

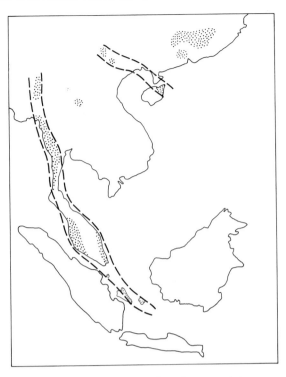

Fig. 23.3 Tin belts and fields of south-eastern Asia. These form the so-called tin girdle of south-eastern Asia, but it should be noted that if lines were drawn to join all the fields together, as is done by some authors, e.g. Lericolais *et al.* (1987), then the lines would cross tectonic trends at a high angle.

developed in the cupolas and ridges on the tops of granitic batholiths, it has been suggested that the paucity of such deposits in geologically older terranes is a function of their deeper levels of erosion. If this is the case, one might ask why no Precambrian tin placers are known containing cassiterite derived from such deposits, since Precambrian gold and uranium placers are present in a number of continents. The ultimate source of tin, as for other metals, must have been the mantle—this is shown by the recent tin mineralization discovered along the Mid-Atlantic Ridge. The more immediate source, however, must have been the crust or upper mantle, as Schuiling (1967) has argued from a consideration of the Namibian–Nigerian belt, where

pre-continental drift, Precambrian mineralization, syn-drift, Jurassic mineralization and post-drift, Eocene mineralization occur. If the source of the tin belonging to these different epochs had been below the African plate, then the tin deposits of different ages should occur in parallel belts. That they are not so distributed suggests derivation of the tin from within the crust or upper mantle of the African plate. The existence of tin metallogenic provinces suggests that parts of the upper mantle were (and are?) relatively richer in tin, and that this tin has been progressively added to the crust, where it has been recycled and further concentrated by magmatic and hydrothermal processes. The recognition of the development of tin provinces is of fundamental importance to the mineral exploration geologist searching for this metal. It is probable that any further discoveries of important deposits of this metal will be made within these provinces or their continuations. Continental drift reconstructions suggest that a continuation of the eastern Australian tin belt may be present in Antarctica.

Some other examples of metallogenic epochs and provinces

Banded iron formation

These rocks, of which the commonest facies is a rock consisting of alternating quartz and hematite-rich layers, are virtually restricted to the Precambrian.

They occur in the oldest (> 3760 Ma old) Isua sediments of western Greenland and in most Archaean greenstone belts. Their best development took place in the interval 2500–1900 Ma ago, in early Proterozoic basins or geosynclines situated near the boundaries of the Archaean cratons. There was comparatively little development of BIF after this period, and Phanerozoic ferruginous sediments are of the hematite–chamosite–siderite type. The latter also tend to be grouped into narrow stratigraphical intervals, for example the Minette ores of this class, which are well developed in the Jurassic of western Europe.

Nickel sulphide deposits

These deposits are almost entirely Precambrian in age and are mainly restricted to a few Archaean greenstone belts in Ontario, Manitoba and Ungava (Canada); the Western Australian Shield; the northern part of the Baltic Shield; and in Zimbabwe. They show, therefore, a good development of metallogenic provinces, but there is considerable evidence that those with mantle-derived sulphur are confined to the Archaean and early Proterozoic (see Chapter 11).

Titanium oxide ores of the anorthosite association

The mid Proterozoic mobile belts (1800–1000 Ma old) of Laurasia and Gondwanaland are characterized by post-tectonic andesine–labradorite anorthosites containing titanium–iron oxide deposits. Examples occur at Bergen, Egersund and Lofoten (Norway); St Urbain and Allard Lake (Quebec); Iron Mountain (Wyoming), and Sanford Lake (New York). According to Windley (1984), most of these Proterozoic anorthosites crystallized in the interval 1700–1200 Ma with a peak at 1400 Ma, which indicates an important metallogenic epoch. These anorthosites are confined to two linear belts in the northern and southern hemispheres when plotted on a pre-Permian continental drift reconstruction and thus they also outline two metallogenic provinces for titanium.

Trace element delineated provinces

Burnham (1959) showed that the major ore deposits of the south-western USA lie in provinces outlined by greater than average trace element content in the crystal lattices of chalcopyrite and sphalerite. The most useful trace elements were found to be cobalt, gallium, germanium, indium, nickel, silver and tin. These exhibit well-defined geographical distributions (Fig. 23.4). Burnham suggested that the variations in trace element content are probably due to variations in the compositions of the fluids during crystal growth. The metallogenic belts appear to be independent of time, wall rock type or intrusions, and Burnham consequently suggested that they are of deep-seated origin and related to deep-seated compositional heterogeneities in the mantle.

Plate tectonics

Mining geologists have for many decades attempted to relate various types of mineralization to large scale crustal structures. Prior to the development of plate tectonic theories of crustal evolution and deformation, the Hall–Dana theory (of geosynclines and their deformation and uplift to form mountain

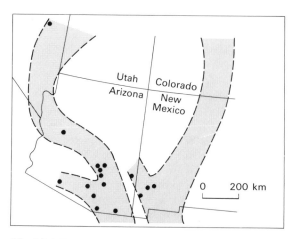

Fig. 23.4 Trace element delineated metallogenic provinces in the south-western USA. The positions of some major porphyry copper deposits are indicated. (Modified from Burnham 1959.)

chains) dominated geological thought. Various schemes of pre-orogenic, syn-orogenic and post-orogenic stages of magmatism were proposed as integral parts of this orogenic cycle. It was noted by mining geologists that many ore deposits occur in geosynclinal regions, and the hypothesis of distinct stages of magmatism linked to the evolution of geosynclines naturally led to the concept of accompanying stages of metallogenesis, since in the 1940s and 1950s much emphasis was still placed on the magmatic–hydrothermal theory for the origin of the majority of metallic deposits.

When relating the genesis of orebodies to major tectonic features, such as island arcs and continental mountain chains, it is essential to know the age relations between the mineral deposits and their host rocks; that is, whether the ores are syngenetic and thus part of the stratigraphical sequence, or whether they are epigenetic and therefore younger, perhaps much younger, than their host rocks. One of the reasons why a knowledge of age relations is important is that in the classical geosynclinal theory it was held that all the rocks of the geosyncline were formed within a subsiding linear trough. The plate tectonic theory tells us that some components of orogenic belts may have been transported for hundreds or even thousands of kilometres from their place of origin. If they contain syngenetic ore deposits then these must have evolved in a very different environment from that in which we now find them. The increasing trend towards syngenetic interpretations of many orebodies is therefore very important.

Since this chapter was originally written in 1979, three books have been published on this subject alone: Hutchison (1983), Mitchell & Garson (1981) and Sawkins (1984, 1990). In what follows I have drawn on all three for data and ideas but, for reasons of space, this chapter must be very selective and for broader treatments of this subject the reader is referred to these three works.

We now know a great deal concerning plate tectonic development during the Phanerozoic; evidence for similar activity during the Proterozoic is also abundant and, for many, convincing (Windley 1984). A number of workers have recently applied plate tectonic principles to explain Archaean geology, particularly the genesis of greenstone belts. These belts are generally considered to be restricted to the Archaean and early Proterozoic, but Tarney *et al.* (1976) argued persuasively that the Rocas Verdes Marginal Basin in southern Chile represents a young example of greenstone belt formation.

Plate convergence and spreading centres are among the important features that control the global location of mineral deposits. These features are depicted in Fig. 23.5 Plate convergence owing to subduction can occur entirely within oceanic areas, adjacent to continental margins or within such margins, and this process is accompanied by extensive activity involving materials derived from the mantle wedge above the Benioff Zone, and also from the subducted plate. Other important features to which mineralization appears to be related include hot spots (mantle plumes), rifting and other extensional tectonics, and collision tectonics. The various rock and tectonic environments involved are reviewed in the above books, and a good discussion of the sedimentation in these different tectonic environments was given by Mitchell & Reading (1986). In this chapter I will adopt the six tectonic settings discussed by Mitchell & Reading, which are:

1 interior basins, intracontinental rifts and aulacogens;
2 oceanic basins and rises;
3 passive continental margins;
4 subduction-related settings;
5 strike-slip settings;
6 collision-related settings.

Interior basins, intracontinental rifts and aulacogens

There are two types of sedimentary basin within continental interiors: large basins often over 1000 km across, and relatively narrow, fault-bounded rift valleys.

Continental interior basins (intracratonic basins)

These may contain entirely continental sediments, much of which may have been deposited in large lakes, e.g. the Chad Basin in Africa in which the palaeo-lake Mega Chad extended over at least 300 000 km^2. The Eyre Basin, Australia, is another good example. The Chad Basin has an area of about 600 000 km^2 and contains up to 2 km of Mesozoic and Tertiary sediments. Some interior basins, e.g. Hudson Bay, have been inundated by the sea and may contain mainly marine sediments. In some of these the marine transgression is accompanied by mineralization, e.g. the Permian Kupferschiefer of northern Europe and the late Proterozoic Central African Copperbelt (see Chapter 15), although it must be mentioned that recent workers have

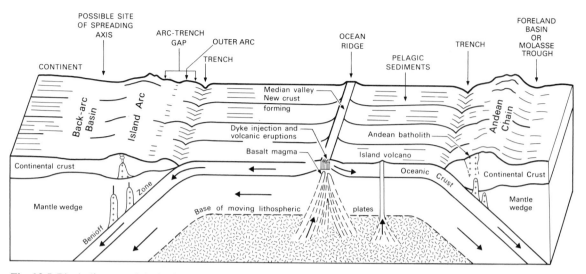

Fig. 23.5 Block diagram of the basic elements of oceanic crust formation and plate movements. On the right is the site of the Andean-type arc with a molasse trough on the continental side, its deposits overlapping on to Precambrian crust, and an arc–trench gap and trench on the oceanic side. On the left is the island arc setting with its trench, troughs of deposition in the arc–trench gap, volcanic arc and (not shown) the back-arc wedges of clastic sedimentation. The back-arc basin represents the site of the Japan Sea-type setting which may or may not have its own spreading axis.

suggested a rift setting for the latter. The first continental basins were developed as soon as sufficient craton was present, and this seems to have been about 3000 Ma ago in southern Africa (Windley 1984). Here the Dominion Reef and Witwatersrand Supergroups with their important gold mineralization (Chapter 18) were laid down. Proterozoic basins with associated mineralization include the one containing the Huronian Supergroup with the uraniferous conglomerates of Blind River (Elliot Lake), and those of the Athabasca (Canada) and Alligator River (Australia) successions with their unconformity-associated uranium deposits.

Phanerozoic intracontinental basins are often important for their evaporite deposits, such as the Permian Zechstein evaporites of Europe, the Devonian evaporites of the Elk Point Basin of western Canada, the Silurian evaporites of the Michigan Basin and others listed in Chapter 20. These and similar evaporites are important for potash and soda production. In some basins, e.g. the Paris Basin (France) and the Triassic basins of the British Isles, important gypsum deposits occur. Platform carbonates are present around some basins and these may host lead–zinc mineralization, e.g. the coarse, dolomitic Presqu'ile Formation that contains many of the orebodies of the important Pine Point Field on

the margin of the Elk Point Basin. This plate tectonic setting is of course transitional to the shelf environment of passive continental margins, and carbonate-hosted base metal deposits occur in both settings.

In this section, mention must be made of the sandstone uranium-type deposits developed in the Wyoming and Colorado Basins (Fig. 17.4), the Mississippi Embayment, and clastic basins in other continents (Chapter 17). Lastly, though not strictly pertinent to the theme of this book, the importance of these basins as gas, oil and coal producers should not be forgotten!

Domes, rifts (graben) and aulacogens

These are initiated by the doming of continental areas which, due to stretching, develop three rift valleys that meet at a 120° triple junction (Burke & Dewey 1973). As is shown in Fig. 23.6, two of the rift valleys may combine to form a divergent plate boundary leading through graben development to ocean spreading, whilst the third arm may show only partial development. This third arm may develop a considerable thickness of sediments, with some volcanics and igneous intrusions. All these may be structurally deformed but the geological history of such zones is relatively simple. They do not often

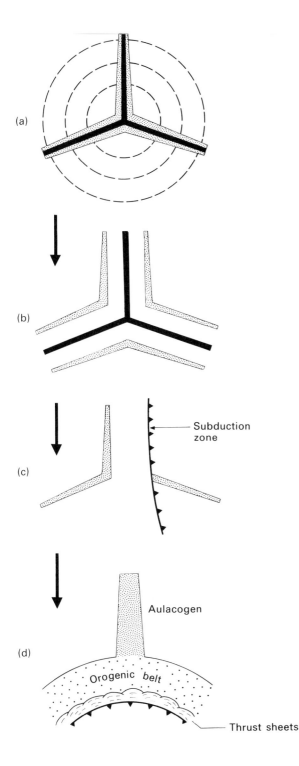

progress much beyond the graben stage and are called failed arms or aulacogens. The associated domes are believed to have formed over mantle plumes (hot spots).

Hot spot-associated mineralization

The triple junction by the shoulder of Brazil is thought to have been generated by the Niger Mantle Plume (Fig. 23.8). Plumes may cause melting of the continental crust forming granite intrusions, e.g. the Cabo Granite of Brazil. Now the large tin–tungsten provinces are associated with subduction zones throughout the world, but there are some tin provinces that do not fit this pattern, e.g. those of Nigeria, Rondônia (Brazil), Saudi Arabia and the Sudan. Tin in these provinces is often associated with sodic granites in ring complexes and the specific association in Nigeria is with A-type hypersolvus biotite granites (Bowden 1982). Mitchell & Garson (1976) suggested that mantle plumes were responsible for the generation of these granites and their associated mineralization. Both the granite magma and the associated metals were probably derived from the crust. The unroofing of these A-type tin granites has given rise to many eluvial and alluvial deposits of cassiterite and columbite. The richest alluvial concentrations are found in the vicinity of granite cupolas that have undergone only shallow erosion (Kinnaird & Bowden 1991).

Tin deposits are also associated with the granites of the Bushveld Complex, RSA, as are also the bulk of South Africa's fluorite resources, which occur as pipes, veins and mineralized breccia bodies in granite. This huge igneous intrusion and that of the Stillwater Complex, USA (Chapter 10) have been attributed to hot spot activity in an anorogenic setting. The huge Dufek Complex in Antarctica may also be related to a mantle hot spot, and in the future it could become a hunting ground for

Fig. 23.6 Schematic diagrams showing the development of an aulacogen. (a) Development of three rift valleys with axial dykes in a regional dome. (b) The three rifts develop into accreting plate margins (e.g. early Cretaceous history of Atlantic Ocean–Benue Trough relationship). (c) One arm of the system begins to close by marginal subduction causing deformation of sediments and volcanics. (d) Continental margin collides with subduction zone forming orogenic belt, and the failed arm is preserved as an aulacogen. (After Burke & Dewey 1973.)

platinum (de Wit 1985). The metamorphic aureoles of these vast intrusions may host industrial minerals of various sorts, e.g. the andalusite deposits in the aureole of the Bushveld Complex (Chapter 21).

The Sudbury Structure, Canada, with its copper–nickel ores, could also have been developed over a hot spot, but much current opinion favours a meteorite impact origin (Chapter 11), and this has also been suggested as the origin of the Bushveld Complex. The Palabora Complex with its copper-bearing carbonatites is about the same age as the Bushveld Complex and may be related to the same hot spot activity. This and other hot spot activity, rifting and associated mineralization in Africa have been reviewed by Olade (1980). The alkaline igneous activity and the associated carbonatites of the Kola Peninsula (Chapter 8) may also be related to hot spot activity.

The mid Proterozoic (1700–1000 Ma) bimodal association of anorthosite and granite, which gave rise to the enormous anorthositic massifs with their associated titanium mineralization (p. 316 and Chapter 10), probably represents a major but abortive attempt worldwide to develop continental fragmentation. A possible modern analogue is the line of Phanerozoic intrusions in a belt 200 km wide and 1600 km long stretching from Hoggar in Niger down to the Benue Trough (Windley 1984).

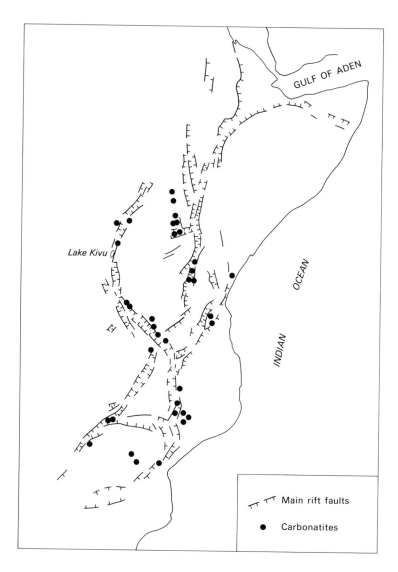

Fig. 23.7 Distribution of carbonatite intrusions relative to the East African Rift Valleys. (After Mitchell & Garson 1976.)

Mineralization associated with continental rifting

The initial stage of graben development is marked by deeply penetrating faults forming pathways to the mantle and giving rise to volcanism. This is usually of alkaline type, sometimes with the development of carbonate lavas and intrusives and occasionally kimberlites (Fig. 23.7). Erosion of these may lead to the formation of soda deposits (e.g. Lakes Natron and Magadi in East Africa) and the intrusive carbonatites may carry a number of metals of economic interest (Chapter 8) as well as being a source of phosphorus and lime.

In the grabens themselves, sediments with or without volcanics may accumulate. In the East African rift valleys, red bed type deposits are common, with conglomerate fans along the escarpments, and playa-lakes have produced evaporites.

The water of Lake Kivu is rich in zinc and precipitates sphalerite. The deposition of this and other metals is believed to be due to hydrothermal solutions, which represent ground water that has been cycled through volcanics and sediments at times of high rainfall (Degens & Ross 1976).

Recently formed aulacogens occur at both ends of the Red Sea and the Gulf of Suez is one of these. It contains a succession of Neogene salt, limestone and clastic sediments 4 km thick, and in it lie the Ras Morgan Oilfields. Other young aulacogens also contain salt deposits and oilfields, e.g. the North Sea. Slightly older aulacogens are present on both sides of the Atlantic (Fig. 23.8), and one occurs where the shoulder of Brazil fits into Africa. At this point, spreading occurred on all three arms of a triple junction 120–80 Ma ago; then the Benue Trough closed with, Burke & Dewey suggested, a short-lived period

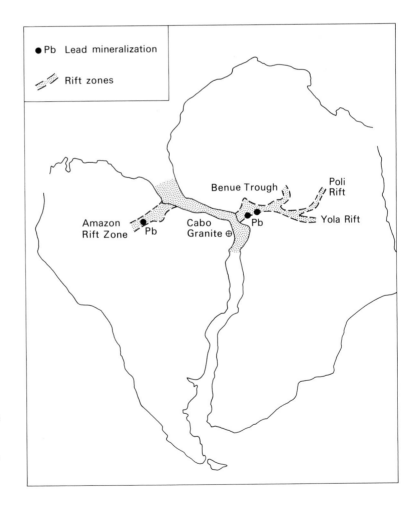

Fig. 23.8 Sketch map showing the locations of the Benue Trough, the Amazon Rift Zone and the lead mineralization within these aulacogens. (After Burke & Dewey 1973 and Mitchell & Garson 1976.)

of subduction, whilst the South Atlantic and Gulf of Guinea arms continued to spread. The Benue Trough is an aulacogen about 560 km long having a central zone of high Bouguer anomalies flanked on either side by elongate negative anomalies. This pattern is believed to be due to the presence of near surface masses of crystalline basement and intermediate igneous rocks below the central zone. Lead–zinc–fluorite–baryte mineralization occurs in fractures in Lower Cretaceous limestone, and similar lead mineralization is present in the Amazon Rift Zone. Tholeiitic igneous activity also took place and intrusions of diorite, gabbro and pyroxenite are present. The origin of the mineralization in both aulacogens is controversial as it may be associated with basic magmatism or with circulating brines which have passed through evaporite deposits. Akande *et al.* (1989) have recently published compelling evidence for a basinal brine source for the ore fluid, which reacted with evaporites and clastic and carbonate rocks, from which it leached some of its constituents, before mineralizing fractures at a higher level.

A number of Proterozoic aulacogens that carry more important mineralization than that in the above examples have been recognized and Burke & Dewey (1973) have identified a number of these in North America (Fig. 23.9). Those running perpendicular to the western margin of the continent carry great thicknesses of rocks of the Belt Series. The northernmost trough contains the Coppermine River Group consisting of more than 3 km of basalt flows with native copper mineralization overlain by greater than 4 km of sediments with evaporites near the top. Epigenetic deposits associated with the rifting include the Great Bear Lake Uranium Field. To the south is the Athapuscow Aulacogen with red beds, sedimentary uranium deposits, evidence of the former presence of evaporites (Stanworth & Badham 1984), and alkali syenite and granite intrusions with associated beryllium–REE greisen deposits (Trueman *et al.* 1988).

Further south is the Alberta Rift, which passes into British Columbia and contains about 15–20 km of late Precambrian sediments in which important

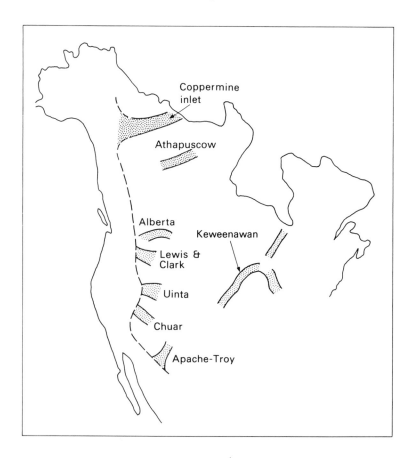

Fig. 23.9 Some of the Proterozoic aulacogens of North America. (Modified from Burke & Dewey 1973.)

stratiform and epigenetic lead–zinc mineralization occur, including the famous Sullivan Mine at Kimberley, British Columbia. Phosphorite deposits are also present. Hoffman (1989), however, viewed this region as a remnant back-arc basin trapped within a continent rather than as an intracratonic rift and he disputed the existence of aulacogens in Utah and California. To the east, the Keweenawan Aulacogen, which is the same age as the Coppermine Aulacogen, also carries a considerable thickness of basalt and clastic sediments, and again native copper mineralization is present (Elmore 1984). Other mining camps that are believed to lie within aulacogens include the copper and lead–zinc ores of Mount Isa, Queensland, and the McArthur River lead–zinc deposit (Dunnet 1971), the Bergslagen base metal deposits of Sweden and the lead–zinc–silver ores of Broken Hill, New South Wales (Plimer 1988). Red bed coppers are well developed in upper Palaeozoic aulacogens of the USSR, and uranium mineralization is also present in some of these.

Among orthomagmatic ore deposits in aulacogens, mention must be made of the Great Dyke of Zimbabwe (Fig. 10.3), the copper–nickel ores of the Noril'sk-Talnakh region of the USSR (Fig. 11.11), and the Duluth Complex, USA.

A number of porphyry molybdenum deposits have been discovered recently in aulacogens, e.g. in the felsic rocks of the Oslo Graben (Ihlen *et al.* 1982) and the Malmbjerg deposit of Greenland (Sawkins 1990).

Ocean basins and rises

We must now pass to the later stages of rifting that lead on to the development of embryonic oceans, such as the Red Sea. With further extension of the crust and the commencement of continental drift, deep crustal flowage and tensional faulting will combine to thin the crust along the graben. At some stage during this process an opening to the sea may initiate marine conditions. Observations from the Rhine Graben, Mesozoic deposits along the Atlantic coastlines and the Miocene of the Red Sea region indicate that evaporite series of great thickness may form at this time. These evaporites contain halite as well as gypsum and therefore have a double economic importance.

The Red Sea is known to be floored by basalt and there is evidence that new oceanic crust is being formed along its median zone. On either side, pelagic sediments are forming, but the depths are not yet sufficiently great nor are the other factors present that lead to phosphorite development, as in areas of upwelling oceanic currents along some Atlantic and Pacific coasts where phosphorite deposits are forming at the present day.

The possible mode of development of oceanic crust along the median zone of the Red Sea is shown in Fig. 23.10. As this new crustal material moves away from the median zone, layer 1 of the oceanic crust, in the form of pelagic sediment, is added to it. As Sillitoe (1972a) has suggested, there is strong evidence that many Cyprus-type sulphide deposits are formed during this process of crustal birth, and black smoker deposits have been reported from the median zone of the Red Sea (Blum & Puchelt 1991). If slices of the oceanic crust are thrust into mélanges at convergent junctures or preserved in some other way then sections similar to that shown in Fig. 23.10 may be expected. This is the case where these cupriferous pyrite orebodies are found in Cyprus, Newfoundland, Turkey, Oman, etc. Deposits of massive sulphides have been discovered on a number of spreading ridges (pp. 72–73) and they show a number of similarities to the Cyprus deposits. Similar situations to the oceanic spreading ridges are found in back-arc basins and indeed many workers hold that the Troodos Massif of Cyprus was developed in such a milieu. In this massif, non-economic nickel–copper mineralization and economic chromite deposits occur in the basic plutonic rocks.

Hydrothermal mineralization with the development of copper, zinc, silver and mercury has been reported from oceanic ridges in the Atlantic and Indian Oceans by Dmitriev *et al.* (1971). Surprisingly, tin mineralization was found in both ridges, accompanied in one case by typical hydrothermal minerals normally associated with granite environments, such as tourmaline, topaz, fluorite and baryte. Normally tin in continental areas is associated with granitic activity; however, important cassiterite orebodies occur in or at the contact of a Proterozoic ultramafic intrusion in northern Guangxi, China. This ultramafic association is marked by a trace element assemblage in the cassiterite of Cr, Ni, Co and Ti rather than the Nb–Ta of granite-associated cassiterite (Wenkui 1988) and tin values of 17–18 ppm in the ultramafic intrusion (Kequin & Jinchu 1988). Cupriferous pyrite mineralization of stockwork-type, which occurs in metabasalts, has been reported by Bonatti *et al.* (1976) from the Mid-Atlantic Ridge. They put

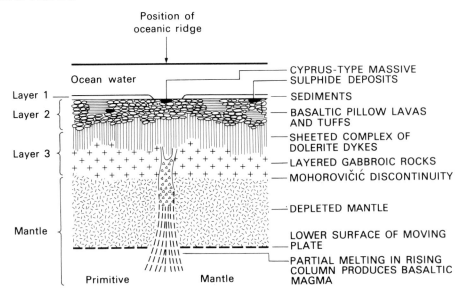

Fig. 23.10 Schematic representation of the development of oceanic crust along a spreading axis. The crustal layering and the possible locations of Cyprus-type massive sulphide deposits are shown. (After Cann 1970 and Sillitoe 1972a.)

forward evidence that it was generated by sea water solutions circulating through the oceanic crust in a manner similar to that described for the Cyprus deposits in Chapter 4. As is the case with a number of ridges in the Pacific Ocean, black smokers have now been found on the Mid-Atlantic Ridge (Rona *et al.* 1986), indicating that the hydrothermal processes that produce them probably operate along most active sea floor spreading centres.

The median zone of the Red Sea is notable for the occurrence of hot brines and metal-rich muds in some of the deep basins. Those studied by Degens & Ross (1969, 1976) can be defined by the 2000 m contour. The largest of these deeps is the Atlantis II which is 14×5 km and 170 m deep. At the base, multicoloured sediments usually 20 m thick, but up to 100 m, rest on basalt, and piston coring has demonstrated the presence of stratification and various sedimentary facies high in manganites, iron oxides and iron, copper and zinc sulphides. This segment of the Red Sea is a slow spreading centre that is opening at the rate of 15 mm p.a. The Atlantis II deposit is the largest sulphide accumulation that has been found on the sea floor. It contains about 227 Mt of metalliferous sediment of which 90 Mt grades 2.06% Zn, 0.45% Cu and 38.4 g t^{-1} Ag and it has formed within the last 15 000 years (Zierenberg & Shanks 1988). Feasibil-

ity studies are under way for their recovery. Above the sediments, the deeps contain brine solutions; these and the sediments run up to about 60°C and 30% sodium chloride. It has been suggested that the sediments and brines have originated either from the ascent of juvenile solutions, or from the recirculation of sea water, which has leached salt from the evaporites flanking the Red Sea and passed through hot basalt from which it has leached metals. Dupré *et al.* (1988), however, using lead isotopes showed that the lead was not derived from a basaltic source but rather from Recent biogenic terrigenous sediments and other unexplored sediments. Whatever the source of the metals, if this second mechanism is the sole or principal one, then metal-rich muds of this origin will form only when evaporites are sufficiently near to supply the required salt. This being the case, it is possible to outline areas of the present oceans which can be said, from the distribution of known evaporites, to have high or low probabilities of containing these deposits. A study of this nature has been made for the North Atlantic by Blissenbach & Fellerer (1973). However, as similar metal-rich muds have now been found on both the Atlantic and East Pacific Rises, this latter hypothesis may be suspect. The possible economic development of deposits of this type has been discussed by Amann (1985).

Passive continental margins

As ocean spreading gradually forces two continents apart, both sides of the original rift become passive margins (also termed trailing, inactive or Atlantic-type), with the development of a continental shelf, bounded oceanwards by a slope and landwards by a shoreline or an epicontinental sea. The type of sediment forming on shelves today depends on latitude and climate, on the facing of a shelf relative to the major wind belts and on tidal range (Mitchell & Reading 1986). In the past, shelves in low latitudes were often covered with substantial platform carbonate successions, and these can be hosts for base metal deposits of both epigenetic and syngenetic nature (so-called Mississippi Valley-type, Irish-type and Alpine-type deposits).

A number of small, stratiform, sandstone-hosted copper and lead–zinc deposits occur in the Cretaceous along the western edge of Africa from Nigeria to Namibia, and further north continue inland to the margin of the Ahaggar Massif. Similar deposits are known along the northern margin of the High Atlas Mountains on the margin of the Morocco Rift, and older ones, which are related to Infracambrian rifting and the development of a Palaeotethys Ocean, are found along the northern edge of Gondwanaland (Olade 1980, Sillitoe 1980b). In many ways these copper deposits resemble those described in the first part of Chapter 15 and may be a variant of them. The lead ores of Laisvall in Sweden and Largenitière and Les Malines in France appear to have been developed in a similar tectonic setting (Ramboz & Charef 1988).

A number of the world's most important sedimentary manganese deposits occur just above an unconformity and were formed under shallow marine conditions on shelf areas e.g. Nikopol and Chiatura, USSR and Groote Eylandt in Australia (Chapter 18).

Passive continental margins that have suffered marine transgressions are also important for phosphorite deposits. Palaeomagnetic research has revealed that the majority of phosphorites formed at low latitudes although a few appear to have been deposited 30–50° north or south of the Equator. Faunal evidence suggests that many Cambrian phosphorite deposits were formed in an east–west seaway extending from Australia into Asia and perhaps into Europe, in a similar fashion to the later Mesozoic Tethyan seaway (Cook & McElhinny 1979). Both these seaways, and the east–west seaway in which the Cretaceous–Eocene phosphorites were laid down, occurred in low to intermediate latitudes. The fact that a coastline is in low latitudes will not necessarily produce the strong upwelling required for phosphogenesis (pp. 298–300). A narrow east–west seaway at a low latitude will probably be marked by considerable upwelling because of the strong westerly directed flow through it (Fig. 23.11a). In a seaway at higher latitudes any upwelling will be much weaker and sea water temperatures lower, resulting in a much lower degree of biogenic activity. A newly formed, narrow, north–south seaway will probably have a complex pattern of interfering currents with little or no associated upwelling (Fig. 23.11c). Only when a longitudinally aligned ocean is 2000 km or so wide will ocean-wide currents sweep along continental coasts creating upwelling gyres at points of deflection along the shelf (Fig. 20.10) and giving rise to extensive and significant phosphogenesis (Fig. 23.11d) (Cook & McElhinny 1979, Cook 1984). Thus, as indicated in Fig. 23.11, strong upwelling and associated phosphogenesis can occur in a latitudinally aligned seaway during its juvenile and mature stages but only at a mature stage in a longitudinally directed seaway.

Many workers now favour a continental shelf environment for the deposition of the Proterozoic Superior-type BIF (Windley 1984). McConchie (1984), in a comprehensive assessment of the evidence from the Hamersley Group, which contains the most extensive accumulation of sedimentary iron deposits known, makes a compelling case for a mid to outer shelf environment.

Finally, we must note the important beach placer deposits that are developed along the trailing edges of many continents, particularly those where trade winds blow in obliquely to the shoreline combining with ocean currents to give rise to marked longshore drift. Among these we can list the diamond placers of the Namibian coast, the rutile–zircon–monazite–ilmenite deposits of the eastern and western coasts of Australia, and the similar deposits of Florida and the eastern coasts of Africa and South America.

Subduction-related settings

In these settings oceanic lithosphere is subducted beneath an arc system on the overriding plate. The subduction zone is usually marked by a deep sea *trench* and, between this and the active *volcanic arc*, is the *arc–trench gap*. The arc–trench gap in some arcs is comprised of an *outer arc* and a *fore-arc basin*.

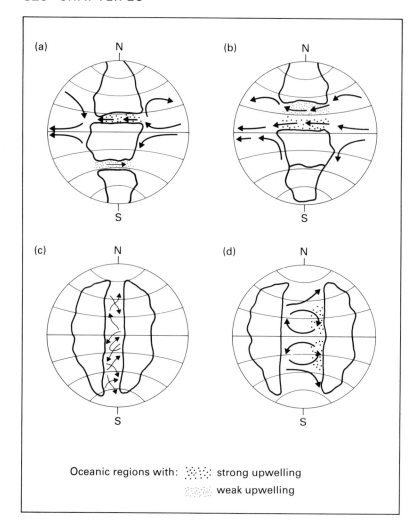

Oceanic regions with: ⬚ strong upwelling
weak upwelling

Fig. 23.11 Schematic representation of the probable current patterns and areas of upwelling consequent upon the rifting and drifting apart of continents. (a) A juvenile stage in which phosphogenesis occurs in a low latitude seaway but not in the equivalent high latitude seaway. In the mature stage (b) phosphogenesis is limited to the coast, which has remained in low latitudes. With a longitudinal seaway in the juvenile stage (c) a network of small currents interfere with each other, but in the mature stage (d) large scale oceanic currents can produce upwelling. (After Cook & McElhinny 1979.)

The outer arc, which may rise above sea level, is usually built of oceanic floor or trench sediments scraped off the subducting plate above low angle thrusts and tectonically accreted to the overlying plate to form an *accretionary prism* or *tectonic mélange*. The fore-arc basin contains flat lying, undeformed sediments that may reach 5 km in thickness and which reflect progressive shallowing as the basin fills up: turbidites, shallow marine sediments and fluviatile–deltaic–shoreline complexes. The sediments are derived mainly from the volcanic arc or by erosion of uplifted basement (Windley 1984). Arc systems may be either *island arcs* or *continental margin arcs*. Island arcs are of several kinds: some are clearly intra-oceanic (Tonga, Scotia), some are separated from sialic continent by small semi-ocean basins (Japan, Kurile), some pass laterally into a continental margin fold belt (Aleutian) and some are built against continental crust (Sumatra–Java). In continental margin arcs, the volcanic arc is situated landward of the oceanic crust–continental crust boundary, as in the Andes (Fig. 23.5). Behind volcanic arcs there is the back-arc area usually referred to when behind island arcs as the *back-arc basin* or *marginal basin* or when behind continental margin arcs as a *foreland basin* or *molasse trough,* e.g. the Amazon Basin behind the Andes.

Mineralization in island arcs

It is convenient to divide the mineral deposits in these arcs according to whether they were formed outside the arc–trench environment and transported to it by plate motion (allochthonous deposits), or whether they originated within the arc complex (autochthonous deposits) (Evans 1976b).

Allochthonous deposits

Rocks and mineral deposits formed during the development of oceanic crust may by various trains of circumstances arrive at the trench at the top of a subduction zone (Fig. 23.12). This material is largely subducted, when some of it may be recycled; some, however, is obducted to form accreted terrane. Examples of what appear to be Cyprus-type sulphide orebodies thrust into a mélange occur in north-western California, where we find the Island Mountain deposit and some smaller occurrences in the Franciscan Mélange. Further north in the Klamath Mountains of Oregon the obducted Josephine Ophiolite contains the important Turner Albright deposit (Kuhns & Baitis 1987). Cyprus-type massive sulphides are also likely to be present at the base of island arc sequences, whether these are the initial succession or a second or later one formed by the migration of the Benioff Zone, because in most cases

the arc basement will have originated at an oceanic ridge.

Clearly, any other deposits formed in new oceanic crust may also eventually be mechanically incorporated into island arcs and the most likely victims will be the chromite deposits of alpine-type peridotites and gabbros (Thayer 1964, 1967). Although some writers have proposed that these deposits were intruded partly as fluid magma and partly as crystal mush, Thayer (1969a,b) has cited much evidence in favour of their being regarded as cumulates that originated in the upper mantle, often at mid-ocean ridges. Recent evidence from the Papuan ultramafic belt is particularly compelling in this respect (Thayer 1971). Here, in a block apparently composed of obducted oceanic crust and upper mantle, rock units of pillowed basalts, gabbro and peridotite rest upon each other in thicknesses exceeding 2000 m for each unit. The plutonic rocks all show well-developed cumulate textures. Podiform chromitites are the most important magmatic deposits in the allochthonous peridotites (Thayer 1971). Platinum metal deposits are known in some of these, but they are more important as the source rocks for placer deposits of these metals. Economic

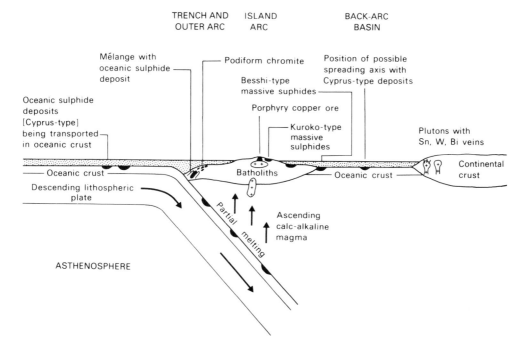

Fig. 23.12 Diagram showing the development and emplacement of some mineral deposits in an island arc and its adjacent regions. (Modified from Sillitoe 1972a,b.)

deposits of podiform chromitite occur in present day arcs in Cuba, where they are found in dunite pods surrounded by peridotite, and on Luzon in the Philippines, again in dunite in a layered ultramafic complex. Numerous occurrences are present in ancient island arc successions (cf., e.g. Thayer 1964). According to Thayer (1971), the allochthonous 'high temperature' lherzolitic peridotites and associated gabbroic 'granulites' of Green and the garnet peridotites of O'Hara contain no significant deposits of oxides or sulphides. A recent development of great importance in this context, and for exploration programmes, is the conclusion of Pearce *et al.* (1984) that all major podiform chromite deposits were formed in marginal basins and not at mid-ocean ridges.

If we follow Thayer (1971) in excluding the Precambrian ultramafic sheets from the class of alpine-type plutonic rocks of the ophiolite suite, then it must be recorded that at present remarkably few magmatic sulphide segregations are known in these rocks. Uneconomic pyrrhotite–pentlandite accumulations are known in Cyprus but the only deposits so far being exploited are those of the Acoje Mine in the Philippines, where nickel and platinum sulphides occur. The accelerated pace of mineral exploration in present day arcs will indicate eventually whether such deposits are truly rare or whether many are still awaiting discovery. At the moment, the latter possibility appears to be the more plausible.

Autochthonous deposits

In considering these deposits it is convenient to divide the arc development into three stages: initial or tholeiitic, main calc-alkaline stage and waning calc-alkaline stage.

Initial or tholeiitic stage. The probable hydrous nature of such magmas and their derivation by partial melting of mantle material (Ringwood 1974) implies that cogenetic massive sulphide deposits may well be expected to form during this stage. However, no massive sulphide deposits have as yet been detected in the earliest tholeiitic sequences of island arcs, but such sequences must be considered as promising grounds for their discovery. Stanton (1972) has surmised that the early basic lavas of the Solomon Islands, with sulphide amygdules and disseminations, belong to this stage.

Main calc-alkaline stage. This is the main period of arc development. Considerable sedimentation and subaerial volcanicity may have occurred during the tholeiitic stage, but the main island arc building and plutonic igneous activity belong to this stage. Baker (1968) showed that whilst the early stage of island arc volcanism is dominated by basalt and basaltic andesite, the more evolved arcs have andesite as the dominant volcanic rock. Ringwood (1974) has summarized the investigations, which have shown that the early tholeiitic stage is succeeded by magmas having a calc-alkaline trend and which are probably ultimately derived from two sources—subducted oceanic crust, and partial melting of the mantle wedge overlying the Benioff Zone. This will result in the diapiric uprise of wet pyroxenite from just above the Benioff Zone. The pyroxenite will undergo partial melting as it rises and the magmas so produced will fractionate as they rise to produce a wide range of magmas possessing calc-alkaline characteristics.

The exposure of much more rock to subaerial erosion will greatly increase the volume of sediment that reaches inter-island gaps, the arc–trench gap and the back-arc area. Reef limestones will then be common. Rapid erosion on the flanks of subaerial volcanoes, ignimbritic activity (related to the increase in silica content of the magmas) giving rise to submarine lahars, submarine slumping and so on, would all contribute to produce a vast volume of material for local sedimentation and for transportation by turbidity currents to the arc–trench gap, along submarine canyons into the trench itself and in the reverse direction into the back-arc regions.

Volcanic massive sulphide, stockwork, skarn and vein deposits are formed in this stage. Conformable sulphide orebodies of Besshi-type develop at this stage in back-arc basins (Fox 1984, Jones *et al.* 1987, Vance & Condie 1987, Smith *et al.* 1990), accompanying the andesitic to dacitic volcanism. They occur in complex structural settings characterized by thick greywacke sequences, as in the late Palaeozoic terrane of Honshu where the volcanism is basaltic to andesitic, and the turbidites may be of arc or continental derivation as indicated in Fig. 23.13. The lack of ophiolitic components, the petrological association and the pyroclastic nature of the volcanics indicates a different environment from that of the Cyprus-type deposits. Workers investigating the exhalative tungsten mineralization in the Austrian Alps (pp. 210–211) are concluding that these de-

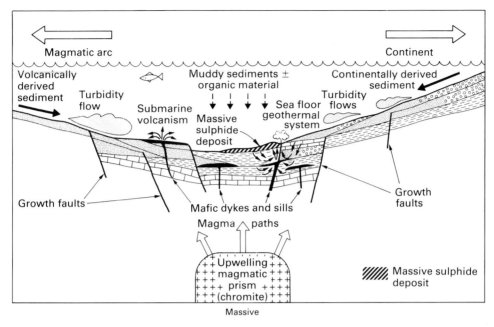

Fig. 23.13 Sketch section across a back-arc basin showing how a Besshi-type massive sulphide deposit could evolve in a basic volcanic and turbiditic environment in which both continentally and arc-derived sediment is present. The basin has formed by the development of rifting and growth faults in response to extensional stresses above a Benioff Zone (Fig. 23.12). (Modified from Mattinen & Bennett 1986.)

posits too were formed in a back-arc basin environment, with the initial magmatic concentration of tungsten taking place in boninitic melts (Thalhammer *et al.* 1989).

Sillitoe (1972a) suggested that the solutions which deposited some of these massive sulphide deposits might have come directly from subducted sulphide bodies (Fig. 23.12), with the metals in them being derived from the subducted bodies and layers 1 and 2 of the subducted ocean floor. Sillitoe (1972b) has further suggested that metal-bearing brines expelled from layers 1 and 2 would induce fusion in the wedge of mantle above the Benioff Zone, ascending with the resulting hydrous calc-alkaline magmas to form porphyry copper deposits when the volcanism is dominantly subaerial. Exploration in recent years has shown that many porphyry coppers are present in island arcs (Fig. 14.7) with, contrary to earlier ideas, both diorite and Lowell–Guilbert model types being present. There is no evidence that continental crust is required for their formation; therefore they can be sought for in any island arc with or without continental crust. The young age of the orebodies in some recent arcs indicates that they belong to the

late stages of island arc evolution and perhaps overlap into the waning stage. Sillitoe's theory that the ultimate source of the metals in porphyry deposits is to be sought in subduction zones is attractive as it can account for the relatively uniform nature of porphyry copper deposits despite the great variety of their host rocks. His theory is attractive on other grounds. It can account for the regional differences seen in these deposits (Chapter 14), relating them to differences in the content of copper, molybdenum and gold in the subducted oceanic crust.

Skarn deposits are common at igneous–carbonate country rock contacts in the older and more complex arcs, such as Indonesia, the Philippines and Japan. Many of the granites and granodiorites with which these deposits are associated also possess spatially related pegmatitic and vein mineralization, and the resulting metallic association can be quite complex, as in the Kitakami and Abukama Highlands of Japan. Skarn and vein deposits of tin are well known from parts of the Indonesian Arc. Less well known are the gold veins in the contact zones of diorites and granodiorites of the Solomon Islands. Other vein deposits are not common in the younger arcs, including those of mercury, which is surprising in view of their development in association with andesites in the Alpine orogenic belt, e.g. the eastern Carpathians. Cinnabar veins do occur in the more complex arcs of the Philippines, Japan and New Zealand.

Waning calc-alkaline stage. The massive sulphide deposits of Kuroko-type probably belong to the later stages of development of island arcs. They are associated with the more felsic stages of calc-alkaline magmatism and are marked by the distinctive features that have been described in Chapter 15. Kuroko-type ores in ancient island arcs include the Palaeozoic deposits of Captain's Flat, New South Wales; Buchans, Newfoundland; Parys Mountain, Wales; and Avoca, Ireland. Present opinion is that these deposits too are wholly or mainly developed in back-arc settings and Leat *et al.* (1986) pointed out that an extensional setting favours the development of long-lived, rhyolitic, high level magma chambers, often with bimodal basalt–rhyolite volcanic associations like those of North Wales, with which Kuroko-type deposits are frequently associated. The Kuroko-type deposits of Japan are thought by Horikoshi (1990) to have formed in a weakly extensional environment at the waning stage of the opening of the Sea of Japan about 15.3 Ma ago and Halbach *et al.* (1990) described a modern analogue of Kuroko-type deposits from the Okinawa Trough—a back-arc basin.

As Stanton (1972) has pointed out, manganese deposits of volcanic affiliation (pp. 260–261) are very common in eugeosynclinal successions of all ages from Archaean to Recent and island arcs are no exception, e.g. such deposits occur in the Bonin Arc (Usui *et al.* 1986). Large numbers of deposits of subeconomic size occur in the Tertiary volcanic areas of Japan, Indonesia and the West Indies and in the Recent volcanic sequences of the larger southwestern Pacific islands. These deposits have been formed *in situ* and are not tectonic slices of pelagic manganese deposits of the oceanic crust. They belong to this and the preceding stage of island arc evolution.

The nature of back-arc basins and their sedimentary and volcanic contents are too variable and complex to be discussed here. A good summary can be found in Mitchell & Reading (1986). There is a scarcity of observed mineralization in modern back-arc basins, partly because of the thick blanket of sediment and volcanic ash present in most of them

(Mitchell & Garson 1981), so we must turn to ancient marginal basins for more evidence of the deposits that formed in them. According to various workers Besshi-type sulphide ores form in this environment (see above), and a number of ophiolites with their associated sulphide and chromite deposits have been assigned to this setting. The Taupo Volcanic Zone of North Island, New Zealand, with its geothermal systems, demonstrates the mineralizing processes that can take place in marginal basins (Chapter 4) and the similar but older Welsh Basin, UK, has Kuroko-style and Cu–Pb–Zn vein mineralization (Ball & Bland 1985, Reedman *et al.* 1985) and, in addition, sedimentary and volcanic manganese deposits and oolitic ironstones.

Mineralization in continental margin arcs

These arcs are also referred to as Andean or Cordilleran Arcs, but both terms are inappropriate because parts of some continental margin arcs, e.g. Sumatra and Java in the Sunda Arc, are separated from the continental interior by marine basins (Mitchell & Garson 1981). The principal features of these arcs include a trench with turbidites, high heat flow and regularly arranged metamorphic zones, as well as monzonitic–granodioritic plutons and batholiths, foreland basins that are often molasse-filled, and widespread sulphide and oxide mineralization (including the extensive development of porphyry copper and molybdenum deposits). Active volcanoes are frequently present in mountain chains underlain by crust of greater than normal thickness due to either crustal shortening or underplating. Although these are regions of plate convergence with its concomitant compression, many graben structures are developed running parallel or obliquely to the arcs. Some magmatic activity and mineralization are related to this rifting.

The Andean Arcs lie along the western margin of South America where it abuts against the Nazca Plate, which is moving down a Benioff Zone beneath the present mountain chain (Fig. 14.7). This chain is immensely rich in orebodies of various types and metals, but on the whole these show a plutonic–epigenetic affiliation rather than a volcanic–syngenetic one. This is no doubt to some extent a function of erosion. The profusion of metals may be related to proximity to the East Pacific Rise and the abundance of metals that are being added to the oceanic crust at this spreading axis. Sawkins (1972) has listed the main features of Andean Cordilleran deposits, emphasizing their close relationship in time and space with calc-alkaline intrusives and their occurrence at high elevations (2000–4000 m above sea level) at relatively shallow levels in the upper crust. He reminded us that this implies that erosion of the Andes down to present sea level would remove this entire suite of ore deposits.

The Andes are characterized by linear belts of mineralization (Peterson 1970, Sillitoe 1976, Frutos 1982) that coincide with the morphotectonic belts which run down the mountain chain (Grant *et al.* 1980). The Coastal Belt, mainly consisting of Precambrian metamorphic rocks, is important for its iron–apatite deposits of skarn and other types. The Western Cordillera (made up mainly of Andean igneous rocks) is a copper province particularly important for porphyry-type deposits carrying copper, molybdenum and gold (Fig. 14.8); and the Altiplano, a Cretaceous–Tertiary intermontane basin filled with molasse sediments, is part of a larger province of vein and replacement Cu–Pb–Zn–Ag deposits that extends along the Eastern Cordillera. This belt consists of Palaeozoic sedimentary rocks and Palaeozoic and Andean igneous rocks, with the tin (+ W–Ag–Bi) belt (Fig. 16.4) along its eastern margin. The tin–silver belt south of Oruro (Chapter 16 and Fig. 16.4) has produced vast amounts of silver from Spanish colonial times to the present day and indeed the eastern Pacific continental margin arcs from British Columbia to Chile are well endowed with large silver deposits (deposits in which silver forms 50% of the value of the ore), but the western Pacific arcs have virtually none. Graybeal & Smith (1988) have used the variation of Ag/Au in porphyry copper deposits to estimate the relative silver content of magma generated along Benioff Zones. Many regions along the eastern Pacific Rim, where silver deposits are abundant, occur in porphyry copper provinces with high Ag/Au values, suggesting that the development of silver deposits is controlled by the relative abundance of silver in subducted oceanic crust. Oceanic crust subducted in the western Pacific is, in general, much older than that being subducted under the Americas. Graybeal & Smith suggested that enrichment of silver in young oceanic crust at spreading centres may reflect metallogenetic evolution over geological time. The Western Cordillera of the Andes has become, during the last 25 years, an important gold province (Sillitoe 1991a). Of the 18 principal deposits (those with about 10 Mt or more of gold), around half are inferred to be parts of porphyry copper sys-

tems. Acid sulphate epithermal deposits occur above porphyry mineralization and adularia–sericite and skarn deposits around porphyry-type mineralization, which itself may be either rich or poor in gold.

The overall metal zonation of the Andes has been related to the remobilization of metals, subducted in the oceanic crust, at different depths down the Benioff Zone (Wright & McCurry 1973), to migration of Benioff Zones, and to changes in their inclination (Mitchell & Garson 1972, Mitchell 1973, Sillitoe 1981b). The porphyry deposits are related to I-type granitoids probably generated along the Benioff Zone or in the mantle wedge above it, but the tin deposits are related to S-type granitoids, perhaps representing anatectic melts that derived their tin from continental crust. There is indeed good reason to believe that this is recycled tin (Dulski *et al.* 1982). Similar situations occur in Alaska and China with tin mineralization related to deeper sections of the Benioff Zone than the porphyry copper mineralization. Simpson *et al.* (1987) concluded that the Chinese tin mineralization developed in tensional settings along major fracture zones that guided the intrusion of post-tectonic granites.

Metallogenic variations are known from other parts of the continental margin arcs of the Americas and they have been tabulated by Windley (1984). The types of skarn deposit found in continental margin settings are given in Table 13.1.

The North American Cordillera is a vast storehouse of mineral treasures developed during a complicated tectonic history too involved for discussion here. A useful summary for the western USA can be found in Guild (1978), and Davies (1989) has written a succinct review of porphyry copper and molybdenum deposits in North America. One point, however, should be made, and that is the important connection between back-arc rifting and base and precious metal vein deposits (Ivosevic 1984). Extensional regimes in the back-arc region are also believed to have controlled the emplacement of Climax-type porphyry molybdenum deposits in the western States (Sawkins 1984).

Strike-slip settings

Two different types of fault are included here: the transform fault and the long known tear or wrench fault. Strike-slip faults vary in size from plate boundary faults, such as the San Andreas Fault of California and the Alpine Fault of New Zealand, through microplate boundaries and intraplate faults, such as those of Asia north of the Himalayas, down to small scale fractures with only a few metres offset. Whether these structures occur as single faults or as fault zones (and where complex patterns with some extension are present in the resulting sedimentary basins), they may be important in locating economic mineral deposits.

Well recognized transform faults in continental crust have little or no associated mineralization, although some potential is perhaps indicated by the location of the Salton Sea Geothermal System (which is a potential ore-forming fluid, p. 58) within the San Andreas Fault Zone, but it must be noted that this system is not underlain by normal continental crust. Transform faults are, however, loci of higher heat flow and could well be channelways for hydrothermal solutions, as certainly appears to be the case in the Red Sea. Here, the brine pools and metal-rich muds, described in 'Ocean basins and rises' above, appear to be located above transform faults (Mitchell & Garson 1981). Other possible examples of transform faults acting as structural controls of mineralization come from a study of the hypothetical continental continuations of transform faults. For example, many of the diamond-bearing kimberlites of West Africa lie along such lines (William & Williams 1977) and some onshore base metal deposits of the Red Sea region show a similar relationship. The Donnybrook–Bridgetown Shear Zone of Western Australia, which is considered to be analogous to the San Andreas Fault System, hosts the Greenbushes pegmatite group (Sn–Ta–Li producer) which is potentially one of the largest rare metal resources in the world (Partington 1990).

Modern strike-slip basins occur in oceanic, continental and continental margin settings (Mitchell & Reading 1986). They may also occur in back-arc basins. The classic onshore strike-slip basin is the Dead Sea with its important salt production, and the best known offshore basins are those of the Californian Continental Borderland, some of which host commercial oil pools. Ancient strike-slip basins are difficult to identify. However, two examples of economic importance are the Tertiary Bovey Basin of England (p. 280 and Fig. 20.2) with its economic deposits of sedimentary kaolin and minor lignite, and the late Mesozoic, coal-bearing, lacustrine basins of north-eastern China. A similar structural relationship was noted by Groves & Foster (1991) in the Yilgarn Block (Western Australia) where major Archaean gold deposits are commonly adjacent to

transcraton shear zones and on the district scale are usually present in geometrically related, shorter strike-length, smaller scale structures.

Henley & Adams (1979) studied giant placer deposits and their tectonic settings. They defined giant placer deposits as those where total gold production has exceeded 148 t and showed that the majority of these are situated on the Pacific margins, where they formed during the Tertiary in similar tectonic and sedimentary environments. The plate boundaries involved are both convergent and regional strike-slip faults, such as the Alpine Fault in New Zealand and the San Andreas Fault of California (Fig. 23.14a). Two factors of regional evolution were important in the development of each giant placer. First there was the rejuvenation of an older orogenic belt containing significant quantities of bedrock gold to provide a source, and second, continual high rates of uplift of this belt resulting in multiple episodes of sediment reworking and the development of large concentrations of placer gold (Fig. 23.14b).

Fundamental faults and lineaments often have orebodies and even orefields spatially related to them. Among the first to point to the importance of this relationship were Billingsley & Locke (1941) in their consideration of orebody distribution in the Cordillera of the USA and Wilson (1949) dealing with the Ontario–Quebec region of Canda. Lineament analyses have made great advances in the last 20 years as the result of the availability of data from remote sensing and photogeological techniques. Here we have space only to consider a few examples.

Heyl (1972, 1983) drew attention to the 38th Parallel Lineament of the central USA and its relationship to a number of orefields in the Mississippi Valley and elsewhere (Fig. 23.15). This lineament is a zone marked by wrench faults, lines of alkalic, gabbroic, ultramafic and kimberlitic intrusions, suggesting connections into the mantle, and changes in stratigraphy across it. Gravity anomaly patterns suggest an 80 km, Precambrian dextral movement, with a Phanerozoic offset of 8–10 km. Important orefields occur at the intersections of this lineament and other structures, for example the Central Kentucky Orefield occurs where the West Hickman Fault Zone and the Cincinnati Arch cross it and the very important South-east Missouri Field where it is crossed by the Sainte Genevieve Fault System and the Ozark Dome.

Many workers have commented on the close connection between gold deposits in the Archaean

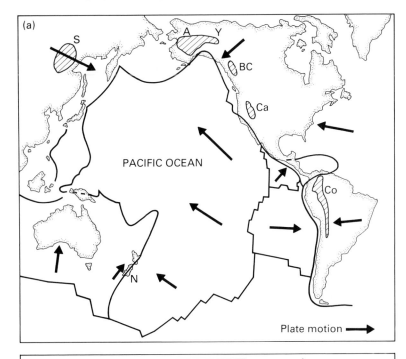

Fig. 23.14 (a) Distribution of giant placer goldfields around the Pacific rim. A = Alaska, Y = Yukon, BC = British Columbia, Ca = California, Co = Colombia and adjoining regions, N = New Zealand, S = eastern Siberia. (b) Stratigraphy and principal tectonic events in the evolution of Californian gold placers, with a schematic representation of the reworking of placers that led to the development of large, rich placer deposits. (After Henley & Adams 1979.)

Abitibi Greenstone Belt of Ontario–Quebec and regional fault zones and good recent discussions of this are in Kerrich (1986) and Osmani *et al.* (1989). About 18% of the world's production of antimony comes from the Archaean Murchison Schist Belt—a greenstone belt in the Kaapvaal Craton, RSA. The antimony–gold mines lie along the 35–40 km Antimony Line, which is considered to be a zone of shearing hosting most of the mineralization (Vearncombe *et al.* 1988). Finally we can glance at a small

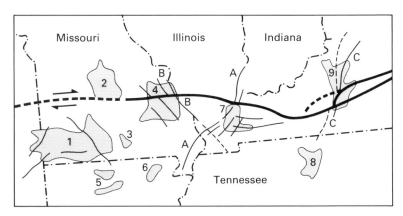

Fig. 23.15 Map showing about half the 38th Parallel Lineament as outlined by Heyl (1972) (broad dark lines), other faults (narrow lines) and some of the base metal–fluorite–baryte orefields of the Mississippi Valley. Key to orefields: 1 = Tri-State, 2 = Central Missouri, 3 = Seymour, 4 = South-east Missouri, 5 = Northern Arkansas, 6 = North-east Arkansas, 7 = Illinois–Kentucky, 8 = Central Tennessee, 9 = Central Kentucky; faults: A = New Madrid Fault Zone, B = Ste Genevieve Fault Zone, C = West Hickman Fault Zone.

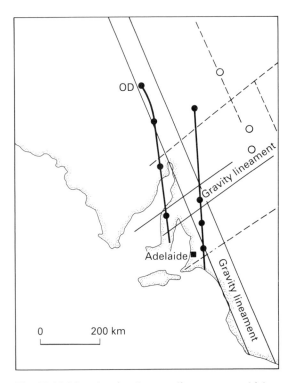

Fig. 23.16 Map showing the two alignments on which the largest of the copper deposits in this part of South Australia lie. The pecked lines represent geological lineaments, open circles are uranium deposits, OD = Olympic Dam. (After Lambert *et al.* 1987.)

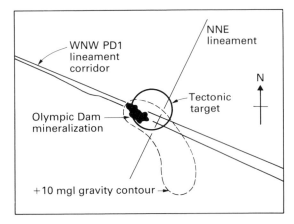

Fig. 23.17 Plan of the Olympic Dam orebody in relation to the west-north-west photolineament corridor PD1 and the north-north-west trending gravity anomaly, which has a sinistral flexure at its northern end bringing it into line with the photolineament. The location of Olympic Dam is shown on Fig. 23.16. (After O'Driscoll 1986.)

part of the important lineament analysis carried out in Australia by O'Driscoll which defined the Olympic Dam location as a priority drilling target (Woodall 1984, O'Driscoll 1986). The largest copper deposits of the Stuart Shelf–Adelaide Geosyncline region of South Australia lie on two alignments, (Fig. 23.16). The giant among these deposits, Olympic Dam, is close to the intersection of two lineaments and a sinistral flexure in the north-north-west trending gravity anomaly (Fig. 23.17).

Collision-related settings

Continental collision results from the closing of an oceanic or marginal basin whose previous existence is revealed by the presence of lines of obducted ophiolites. In some orogens interpreted as collision belts, such as the Damarides of southern Africa, this evidence is missing, and palaeomagnetic studies have not yielded evidence for significant separation at any stage of the cratonic blocks on either side of the orogen. Now collision can occur between two active arc systems, between an arc and an oceanic island chain or between an arc and a microcontinent, but the most extreme tectonic effects are produced when a continent on the subducting plate meets either a continental margin or island arc on the overriding plate. Subduction can then lead to melting in the heated continental slab and the production of S-type granites (Pitcher 1983)–Fig. 23.18. In many continental collision belts, highly differentiated granites of this suite are accompanied by tin and tungsten deposits of greisen and vein type (Chapters 12 and 16). The tin and tungsten deposits of the Hercynides of Europe (Cornwall, Erzegebirge, Portugal, etc.) are good examples. An excellent study of this activity is to be found in Beckinsale (1979) who discussed the generation of tin deposits in south-east Asia in the context of collision tectonics.

With continental collision of the type illustrated in Fig. 23.18, belts of older I-type granitoids, possibly with associated porphyry copper and molybdenum deposits, are related to the subduction of oceanic lithosphere, and belts of younger S-type granitoids, which may have associated tin (and tungsten) mineralization, are formed during continental collision. In south-eastern Asia (Fig. 23.19) there are three different granite provinces that run parallel to each other. The Eastern Granite Province contains hornblende–biotite and biotite granites of Permian age (265–230 Ma). They are I-type granitoids with initial $^{87}Sr/^{86}Sr$ ratios of 0.705–0.714 (Lehmann 1990), which are, relative to the Main Range Granites of the Central Province, poor in tin, and the ratio of historic tin output of Central Province to Eastern Province granites in Malaysia is 19 : 1. Further north in this province in Thailand there is a well developed Permian volcanic–plutonic arc with a porphyry copper deposit at Loei. The Central Province hosts the Main Range Granites of Malaysia. These are post-tectonic, peraluminous, S-type, biotite granites whose ages lie in the range 220–200 Ma and whose initial strontium ratios are 0.716–0.751. The famous Malaysian tinfields of the Kinta Valley and Kuala Lumpur occur in this province, but the same province in central and northern Thailand is much less well endowed with tin.

The Western Province is restricted to western Thailand and Burma and carries granitic intrusions of Cretaceous to Tertiary age (130–50 Ma). The older granites are largely I-type and the younger

Fig. 23.18 (a) Two continents on a collision course as oceanic crust between them is subducted. Here we have the Andean-type situation of Fig. 23.5 in which I-type granites are generated along the Benioff Zone. Some of these granites may host porphyry copper deposits. (b) Collision has taken place and the leading section of the underriding plate has been blanketed with thick thrust slices of sediment and mélange. Temperatures rise sufficiently in this continental crust to permit partial melting and the formation of S-type granites. If the crustal material contained above average amounts of tin, tungsten, etc., then these granites may have associated, epigenetic deposits of these metals. The I-type and S-type granites, with their associated mineralization, may occur in parallel belts as in Thailand and Malaysia. (After Beckinsale 1979.)

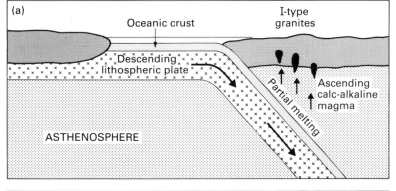

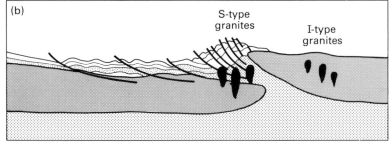

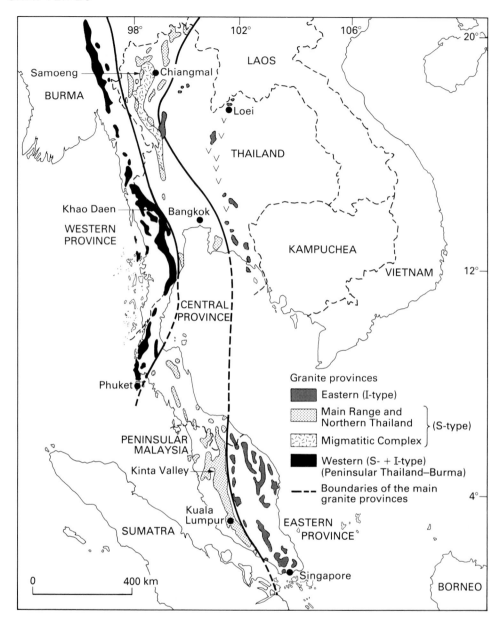

Fig. 23.19 The Granite Provinces of the South-east Asian Tin Belt. (Modified from Lehmann 1990.)

S-type. In the latter tin has been worked for over a century or so at Kao Daen, Phuket and Samoeng. As Beckinsale (1979) pointed out, the geochronological, geological and geochemical data can be explained by postulating the successive collisions of three microcontinental plates (Fig. 23.20). In this reconstruction it is postulated that in the Permian an ocean or, more probably, a marginal basin occupied

the area of the present central Thailand and eastward subduction of oceanic crust gave rise to the development of the Permian volcanic–plutonic Eastern Province with its porphyry copper mineralization (Fig. 23.20a). This marginal basin closed in the Triassic with a continental collision that resulted in the genesis of the S-type tin granites of the Central Province (Fig. 23.20b). Following this collision the

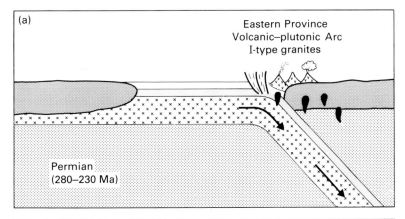

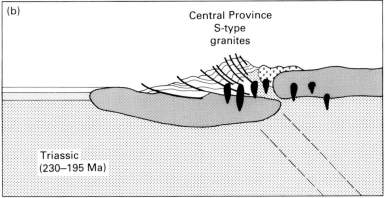

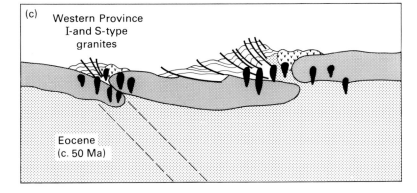

Fig. 23.20 Diagrams based on the work of Beckinsale (1979) and illustrating a possible plate tectonic reconstruction to account for the formation of the South-east Asian Tin Belt. For further explanation see text.

marginal basin to the west was gradually closed, again by eastward subduction of oceanic crust, which gave rise to I-type granite emplacement in western Thailand at about 130 Ma. A second continental collision began at about 100 Ma and produced a phase of S-type granite formation, but in this event considerable overlap of I- and S-type granites resulted, as in Phuket Island, perhaps as a result of more pronounced subduction of the leading

edge of the Burma Plate (Fig. 23.20c). Finally it may be remarked that belts of I-type granites and S-type tin granites similar to those in Malaysia appear to be present in the Hercynides of Europe.

Uranium deposits, particularly of vein type, may be associated with S-type granites as in the Hercynides (Fig. 5.5) and Damarides (Rössing deposit, Chapter 9). Unfortunately, not all S-type granites have associated mineralization, for example those found in a belt stretching from Idaho to Baja California (Miller & Bradfish 1980). Sawkins (1984) has suggested that the reason that some have associated tin mineralization, some uranium and others neither, presumably reflects the geochemistry of the protolith from which these granites were derived, and their subsequent magmatic history.

Of course, the various collision possibilities described above can result in deposits developed in arcs of various types being incorporated into collision belts. The other features, such as intermontane basins and foreland basins, which are present in arc orogens (described in previous sections), will have similar mineralization potential.

In some collision belts where there has been restricted development of oceanic crust, such as the European Alps (Hsü 1972, Dercourt & Paquet 1985), there was a plate margin setting where the plate motion was dominantly lateral. Subduction of oceanic crust was not, therefore, developed on any substantial scale, and this had important results as far as the evolution of the Alps is concerned. It accounts for several problems of Alpine geology: the distinctive tectonic style, the subordinate development of tectonic mélanges, the longevity of the Alpine flysch trenches and the very small amount of calc-alkaline igneous activity of either plutonic or volcanic type. This has resulted in a virtual absence of porphyry copper deposits and Besshi- and Kuroko-type massive sulphide deposits. Indeed, by comparison with the Andean belt or mature island arc terranes, the paucity of post-Hercynian mineral deposits in the Alps is most marked (Evans 1975).

Despite the fairly common occurrence of ophiolites in parts of the Alps, this remark also applies to chromite deposits, which are insignificant in size. The most important metalliferous mineralizations of the Alps are uranium–sulphide deposits in the Permian, which may have developed in a graben environment, and lead–zinc deposits in Triassic carbonate sequences of the eastern Alps. The latter appear to be syngenetic. The occurrence of baryte, celestine and anhydrite in the ores, the presence of evaporite beds in the host limestones and the general geological setting invite a comparison with the lead–zinc ores of the Red Sea region. Thus, as Evans (1975) has shown, by considering the possible nature of the evolving Alpine fold belt from a plate tectonic point of view it is possible to account for the ore deposits that are present and the absence of others. These, and other deposits likely to be developed in Mediterranean-type orogens, are the ones upon which the search for ore deposits should be concentrated in the post-Hercynian rocks of the Alpine region.

Space does not allow an extension of this chapter to deal with Precambrian deposits, but it is clear from the literature that plate tectonic interpretations are applicable, at least in part, to the problems of the geology of Precambrian shields; the genesis of many Precambrian orebodies in the light of this theory has already been considered by a number of workers. Some useful summaries are given in Hutchison (1983) and Windley (1984), with a shorter review in the next chapter.

24 / Ore mineralization through geological time

It is now well known to geologists that the earth, and its crust in particular, has passed through an evolutionary sequence of changes throughout geological time (Windley 1984). These changes have been so considerable that we must expect them to have had some influence on the nature and extent of mineralization. Reference has already been made in this book (Chapter 23) to the association of most of the world's tin mineralization with Mesozoic and late Palaeozoic granites, to the virtual restriction of banded iron formation to the Precambrian and the bulk of it to the interval 2500–1900 Ma ago, and to the importance of the Precambrian for nickel and orthomagmatic ilmenite deposits. In Chapter 11, it was noted that the lack of numerous Phanerozoic nickel sulphide deposits may be due to depletion of the mantle in sulphur during the Archaean, and in Chapter 15 attention was drawn to the fact that volcanic-associated massive sulphide deposits show important changes with time; these are discussed in detail in Hutchinson (1983). We will now examine such changes in the type and style of mineralization in a little more detail. These changes can be discussed conveniently in terms of the Archaean. Proterozoic and Phanerozoic intervals and the environments that prevailed during them.

First, however, it must be remarked that although the Precambrian occupies about eight-ninths of geological time, very little is known as to whether plate tectonic processes acted then and, if they did, whether their mode of operation differed from those known to us through Phanerozoic studies. Precambrian research suggests that the Phanerozoic plate tectonic model, somewhat modified, is applicable for developments over the last 2500 Ma. As yet there is no general consensus about the type of plate regime in the Archaean and even those indications we have are biased towards the late Archaean, and little is known about large scale structures prior to about 3000 Ma ago (Park 1988). On the other hand Ming-guo & Windley (1989) argued that the processes of accretion and crustal growth were similar in ancient and modern times.

The Archaean

This interval, 3800–2500 Ma ago, is notable both for the abundance of certain metals and the absence of others. Metals and metal associations developed in significant amounts include Au, Ag, Sb, Fe, Mn, Cr, Ni–Cu and Cu–Zn–Fe. Notable absentees are Pb, U, Th, Hg, Nb, Zr, REE and diamonds.

Two principal tectonic environments are found in the Archaean: the high grade regions and the greenstone belts. The former are not in general important for their mineral deposits, which include Ni–Cu in amphibolites, e.g. Pikwe, Botswana, and chromite in layered anorthositic complexes, e.g. Fiskenæsset (west Greenland) as well as chromitite seams in dunite lenses known to be at least 3800 Ma old (Chadwick & Crewe 1986). The greenstone belts on the other hand are very rich in mineral deposits whose diversity has been described by Watson (1976). The principal mineral deposits are related to the major rock groups of the greenstone belts and their adjoining granitic terranes as follows:

1 ultramafic flows and intrusions: Cr, Ni–Cu;
2 mafic to felsic volcanics: Au, Ag, Cu–Zn;
3 sediments: Fe, Mn and baryte;
4 granites and pegmatites: Li, Ta, Be, Sn, Mo, Bi.

Chromite

This is not common in greenstone belts but a very notable exception is present at Sherugwel (formerly Selukwe c. 3500 Ma) in Zimbabwe This is a very important occurrence of high grade chromite in serpentinites and talc-carbonate rocks intruded into schists which lie close to the Great Dyke. It resembles the podiform class of deposit, though it is in a very different tectonic environment from that in which Phanerozoic examples of this class are normally found (Stowe 1987).

Nickel–copper

These deposits are composed mainly of massive and disseminated ores developed in or near the base of komatiitic and tholeiitic lava flows and sills as described in Chapter 11. Only four important fields occur, those of south-western Australia (Kalgoorlie Belt), southern Canada (Abitibi Belt), Zimbabwe and the Baltic Shield of northern USSR. As these metals and their host rocks are probably mantle-derived, this suggests the existence of metallogenic provinces controlled by inhomogeneities in the mantle, though the anomaly may not consist of an excess of nickel but rather of a concentration of sulphur which led to the extraction of nickel from silicate minerals.

Gold

Gold has been won in smaller or larger amounts from every greenstone belt of any size and its occurrence is the principal reason for the early prospecting and mapping of these belts. The gold is principally in vein deposits cutting basic or inter-mediate igneous rocks—both intrusions and lava flows—but the more competent intrusives such as the Golden Mile Dolerite of Kalgoorlie, are more important. Some gold deposits show an association with banded iron formation and these appear to have been deposited from subaqueous brines to form exhalites. The greatest concentration of gold mineralization occurs in the marginal zones of the greenstone belts near the bordering granite plutons and it decreases towards the centre of the belts. This may suggest that it has been concen-trated from the ultrabasic–basic volcanics by the action of thermal gradients set up by the intrusive plutons. On the other hand, or as an additional factor, the spatial relationship with regional fault zones that have controlled the distribution of the volcanics, the synvolcanic sediments and the min-eralization, is clearly significant, and many gold camps in the Abitibi Greenstone Belt of Ontario and Quebec appear to be positioned where cross-lineaments intersect these regional faults (Hodgson 1986).

Silver is usually present with gold in the green-stone belts. In the Abitibi Belt, Canada it is found in a Au–Ag–Cu–Zn association at granite–mafic vol-canic contacts. These granites also have porphyry-style Cu–Mo mineralization (see references in Windley 1984).

Copper–zinc

Volcanic-associated massive sulphide deposits are very common in the Archaean, especially in the Abitibi Belt of southern Canada (Sangster & Scott 1976). These deposits are principally sources of copper, zinc and gold, but they are of Primitive-type and their lead content is normally very low. The virtual absence of lead mineralization from these greenstone belts may to some extent reflect the fact that during the Archaean there had been insufficient time for much lead to be generated by the decay of uranium and thorium in the mantle.

Iron and manganese

Banded iron formation is common throughout Archaean time but not in the quantities in which it appears in the Proterozoic. It is generally the Algoma type (Chapter 18) that is present. There is some production from Archaean iron ore in Western Australia and from the Michipicoten Greenstone Belt, Canada where, at the Helen Mine, there occurs one of the world's largest siderite deposits. Morton & Nebel (1984) have recently published evidence favouring an exhalative origin for this. However, in contrast to all other Precambrian areas, China has large and significant BIF deposits in the Archaean. This BIF development reached a peak in the late Archaean and the BIF occurs both in high grade gneiss terranes and greenstone belts, being more abundant *in the former*. This gneiss-hosted occur-rence is unique to China. Another intriguing point concerning these Chinese Archaean occurrences is that nearly all are of Superior type (oxide facies) (Ming-guo & Windley 1989). Further studies of these occurrences will throw much light on our ideas of early crustal growth and the development of BIF in time and space.

Manganese deposits of any significance, but still small, did not develop until 3000 Ma ago post-dating by at least 800 Ma the oldest known geolog-ical sequence containing BIF and base metal sul-phide ores (Roy 1988).

The early to mid Proterozoic

The late Archaean and early Proterozoic (the transition date is arbitrarily put at 2500 Ma) were marked by a great change in tectonic condi-tions. The first stable lithospheric plates developed, although these seem to have been of small size. Their appearance permitted the formation of sedi-mentary basins, the deposition of platform sedi-ments and the development of continental margin troughs.

Gold–uranium conglomerates

The establishment of sedimentary basins allowed the formation of these deposits. The best known example is that of the Witwatersrand Basin with its widespread gold–uranium conglomerates (Chapter 18, Fig. 18.10), but other examples are known along the north shore of Lake Huron in Canada (Blind River area), at Serra de Jacobina in Brazil and at localities in Australia and Ghana. The Witwatersrand Supergroup, which contains the gold–uranium conglomerates, has been dated at about 3060–2708 Ma and the Pongola Supergroup of Swaziland, which also carries gold placers, developed on a stabilized craton during the period 3100–2900 Ma. These deposits represent a unique metallogenic event which many feel has not been repeated because a reducing atmosphere was a *sine qua non* for the preservation of the detrital uranium minerals and pyrite, but this view is disputed (Chapter 18).

Uranium is a relatively mobile, lithophile element having an average crustal abundance of 2–4 ppm. As with many other metals, economic concentrations show a distinct time-bound nature in the Precambrian that may well reflect an oxygenation event in the earth's atmosphere at about 2400–1800 Ma ago (Fig. 24.1). Before this time, given the right conditions, large placer deposits with detrital uranium minerals were able to form. After this oxygenation event, however, hexavalent uranium was dissolved during weathering and transported as uranyl complexes. The development of these aqueous solutions permitted the extensive formation of unconformity-associated uranium deposits in the period 1800–1200 Ma. These uranium concentrations are related spatially to palaeoweathering surfaces (pp. 229–232). Unlike the late Archaean palaeoplacers and the Phanerozoic sandstone uranium-type deposits, the Proterozoic unconformity-associated deposits contain extremely high concentrations of uranium, which makes them a much more attractive mining proposition (Marmont 1989). The time-bound nature of uranium placer deposits explains the lack of detrital uranium in the mid Proterozoic, Witwatersrand-like, Tarkwaian gold placers of Ghana. As well as lacking uraninite these have no detrital pyrite, but do carry detrital hematite, which may indicate that an oxidizing atmosphere was by then well developed. (Alternatively it could be pyrite oxidized *in situ*, or hematite derived from Superior type BIF!)

Sedimentary manganese deposits

With the increasing development of cratonization, shallow water, marine environments, within which large manganese deposits could form, were present. The early Proterozoic Kalahari Field, RSA in the Transvaal Supergroup (2500–2100 Ma) rocks is enormous and estimates of about 7500 Mt of plus 30% ore have been published. This, in terms of tonnage, is probably the largest terrestrial manganese field in the world (Roy 1988).

Many of the major Proterozoic manganese deposits were formed between 2100–1700 Ma; these include the West African deposits of Ghana, the Ivory Coast, other neighbouring countries and the large and important Moanda Mine, Gabon.

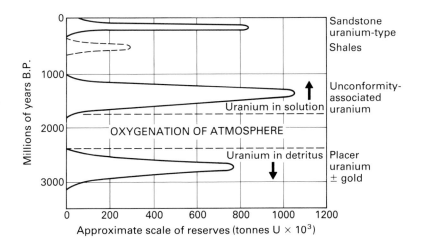

Fig. 24.1 The time-bound character of some major types of uranium deposit. The time period assigned on this figure to the oxygenation of the atmosphere must be considered as approximate. Experts in this field have postulated different time limits. The position of the upper line also varies according to whether we are concerned with the atmosphere and surface waters alone, or include the deep ocean waters, which will have taken longer to become significantly oxygenated. (Adapted from Marmont 1989.)

Sediment-hosted stratiform lead–zinc deposits

By about 1700 Ma ago, the CO_2 content in the hydrosphere had reached a level that permitted the deposition of thick dolomite sequences. In a number of localities these host syngenetic base metal sulphide orebodies, such as those of the Balmat–Edwards and Franklin Furnace districts, USA. In other sedimentary hosts, varying from dolomitic shales to siltstones, there are deposits such as McArthur River (Pb–Zn–Ag) and Mount Isa (Pb–Zn–Ag and separate Cu orebodies) in Australia and Sullivan, Canada. Various exhalative and biogenic origins have been suggested for these ores.

The chromium–nickel–platinum–copper association

The presence of small crustal plates permitted the development of large scale fracture systems and the intrusion at this time of giant dyke-like layered bodies, such as the Great Dyke of Zimbabwe, and enormous layered stratiform igneous complexes like that of the Bushveld in South Africa. Many of these layered intrusions were formed in the period 2900–2000 Ma ago. These are the repositories of enormous quantities of chromium and platinum, with other important by-products (Chapter 10). Though similar intrusions occur in other parts of the world, the great concentration of chromium is in southern Africa, and this has led some workers to postulate the presence of chromium-rich mantle beneath this region.

Titanium–iron association

About the middle of the Proterozoic many anorthosite plutons were emplaced in two linear belts that now lie in the northern and southern hemispheres when plotted on a pre-Permian continental drift reconstruction. A number of these carry ilmenite orebodies, which are exploited in Norway and Canada. This was a unique magmatic event that has not been repeated. It suggests the gathering of reservoirs of magma (in the top of the mantle) that were able to penetrate upwards along deep fractures in the crust. The strength of the crust and the thermal conditions seem to have reached the point where magma could accumulate at the base of the crust in this extensive manner, rather like the accumulation of magma beneath present day rift valleys.

Diamonds

Diamantiferous kimberlites appear for the first time in the Proterozoic. This suggests that the geothermal gradients had decreased considerably permitting the development of thick lithospheric plates, because diamonds, requiring extreme pressure for their formation, cannot crystallize unless the lithosphere is at least 120 km thick.

Banded iron formation (BIF)

The greatest development of BIF occurred during the interval 2500–1900 Ma ago (Goldich 1973). Although this rock type is important in the Archaean, it could not be developed on the large scale seen in the early Proterozoic because stable continental plates were not generally present. Following the development of stable lithospheric plates, BIF could be laid down synchronously over very large areas; this took place possibly in intra-plate basins and certainly on continental shelves. The weathering of basic volcanics in the greenstone belts would have yielded ample iron and silica. If the atmosphere was essentially CO_2-rich, the iron could have travelled largely in ionic solution. It is now suspected that iron-precipitating bacteria may have played an important part in depositing the iron and oxidizing it to the ferric state, as modern iron bacteria are able to oxidize ferrous iron at very low levels of oxygen concentration. Although BIF appears at later times in the Proterozoic, its development is very restricted compared with that in the early Proterozoic and this fall off in importance has been correlated by some workers with the evolution of an oxidizing atmosphere. In the Phanerozoic the place of BIF is taken by the Clinton and Minette ironstones.

Mid–late Proterozoic

High grade linear belts

It has been suggested (Piper 1974, 1975, 1976) that a supercontinent existed through much of Proterozoic time, and Davies & Windley (1976) have plotted the trends of major high grade linear belts on this supercontinent showing that they lie on small circles having a common point of rotation. These linear belts affect middle to late Proterozoic as well as older rocks and include shear belts, mobile belts and linear zones of transcurrent displacements of magnetic and gravity anomaly patterns. They contain some deep dislocations that penetrate right down to the mantle and form channelways for uprising magma. Nickel mineralization occurs in some of these belts, e.g. the Nelson River Gneissic Belt of Manitoba.

Sedimentary copper

Watson (1973) drew attention to the anomalously high concentrations of copper in some late Proterozoic sediments in many parts of the world. These represent the oldest large sedimentary copper accumulations. Examples include the Katanga System of Zambia and Shaba (Chapter 15) and the Belt Series of the north-western USA. These, like the Upper Palaeozoic examples, correlate closely with widespread desert sedimentary environments.

Sedimentary manganese deposits

A second important period of manganese deposition occurred during the late Proterozoic, and manganese-rich sediments were laid down on or along the margins of cratonic blocks. The most important deposits are in Brazil, central India, the USSR and Namibia. Many of these, such as Morro do Urucum, Brazil (c. 900 Ma), are interbedded with BIF. On the other hand, the most characteristic late Proterozoic BIF—the Rapitan Group, Canada— does not contain any manganese deposits and the late Proterozoic manganese oxide deposits in the Penganga Beds, India have no associated iron formation (Roy 1988).

Tin

Watson (1973) observed that tin mineralization does not appear in major quantities in the crust until the late Proterozoic, where it is associated with high level alkaline and peralkaline anorogenic granites and pegmatites. This is particularly the case in Africa where these deposits lie in three north–south belts (Fig. 23.1). Another belt passes through the Rhondônia district of western Brazil.

Evaporite deposits

Sulphate evaporites of significant thickness and extent make their appearance in the early to mid Proterozoic, and are quite common in the late Proterozoic.

Coal

The earliest known coals are found in early to mid Proterozoic rocks where they have formed from compacted algal deposits. The first land-plant derived coals occur in the late Devonian rocks of the northern USSR.

The Phanerozoic

Towards the end of the Proterozoic, a new tectonic pattern developed which gave rise to Phanerozoic fold belts formed by continental drift. There was large scale recycling of oceanic crust, which greatly increased the number and variety of ore-forming environments by producing long chains of island and continental margin arcs, back-arc basins, rift-bordered basins and other features described in Chapter 23. Consequently some Archaean volcanic types and Proterozoic sedimentological types reappear together with a few types dependent perhaps on more evolved plate mechanisms or extreme geochemical evolution of siliceous magmas. The latter process, plus recycling of crustal accumulation, may account for the important development of molybdenum, tin and tungsten ores in the Phanerozoic (Meyer 1981).

Although a few large phosphorite deposits and a large number of small ones are known in the Proterozoic, marked worldwide epochs of phosphorite formation belong largely to the Phanerozoic and commence with a very important epoch around the Proterozoic-Cambrian boundary (Fig. 24.2). This may well be related to more frequent continental fragmentation and drift providing suitable sites for phosphorite development in low latitudes during the Phanerozoic (see pp. 325–326). Obviously the supercontinent which existed through much of Proterozoic time will not have provided many favourable sites for widespread phosphorite deposition, and in the Archaean extensive regions of cratonic crust were not present.

The Proterozoic–Cambrian, phosphorite-forming epoch produced a marked concentration of deposits in a belt extending from Kazakhstan through central Asia into southern China, Vietnam and Australia, but deposits of this epoch are present in all the continents. It is estimated that in 1982 the annual production of phosphorite from deposits of this age was 23.5 Mt and these phosphorite deposits are believed to represent a world resource of around 34 Gt (Cook & Shergold 1986).

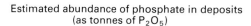

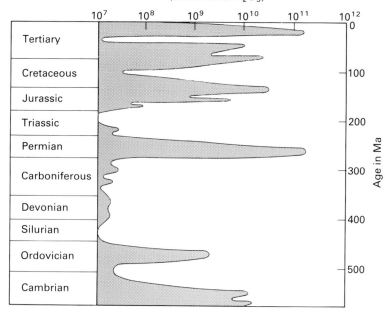

Fig. 24.2 Estimated abundance of phosphate in the world's known deposits of phosphorite (including reserves and resources) plotted against Phanerozoic time. Note that the tonnage scale is logarithmic to keep the peaks on the page! (Modified from Cook & McElhinny 1979.)

Cyprus-type copper–pyrite deposits and sandstone uranium deposits first appear in the Phanerozoic, while podiform chromite, found in the Archaean but not in the Proterozoic (Meyer 1985), becomes much more common. Lead becomes increasingly important in the volcanic-associated massive sulphide deposits. Some of the largest epigenetic metal concentrations of the Phanerozoic are those of the porphyry copper and molybdenum deposits of the continental margin and island arcs. Gold, which shows a peak of epigenetic mineralization in the late Archaean (2.7–2.6 Ga), peaks again in the Mesozoic to Quaternary where virtually all the deposits occur in convergent margin settings (Groves & Foster 1991). The Phanerozoic is also important for residual deposits, e.g. kaolin and bauxite deposits, many of which are Cretaceous–Recent in age.

Much of this Phanerozoic mineralization activity has been covered in the previous chapter. I hope that a study of these two brief chapters will reveal the importance of taking an account of plate tectonic settings, and the effect of continental drift and geological time in designing mineral exploration programmes.

'For as birds are born to fly freely through the air, so are fishes born to swim through the waters, while to other creatures Nature has given the earth that they might live in it, and particularly to man that he might cultivate it and draw out of its caverns metals and other mineral products.' [Georgius Agricola in *De Re Metallica*, 1556]

Appendix / Formulae of some minerals mentioned in the text

Acanthite	Ag_2S
Alabandite	MnS
Almandine	$Fe_3Al_2(SiO_4)_3$
Alunite	$KAl_3(SO_4)_2(OH)_6$
Amblygonite	$(Li,Na)Al(PO_4)(F,OH)$
Anglesite	$PbSO_4$
Anhydrite	$CaSO_4$
Apatite	$Ca_5(F,Cl)(PO_4)_3$
Arsenopyrite	$FeAsS$
Baddeleyite	ZrO_2
Baryte	$BaSO_4$
Bastnäsite	$(Ce,La)(CO_3)F$
Bertrandite	$Be_4Si_2O_7(OH)_2$
Beryl	$Be_3Al_2Si_6O_{18}$
Boehmite	$AlO(OH)$
Bornite	Cu_5FeS_4
Braunite	$(Mn,Si)_2O_3$
Carnotite	$K_2(UO_2)_2(VO_4)_2.3H_2O$
Cassiterite	SnO_2
Celestite	$SrSO_4$
Chalcocite	Cu_2S
Chalcopyrite	$CuFeS_2$
Chromite	$FeCr_2O_4$
Cinnabar	HgS
Coffinite	$(USiO_4)_{1-x}(OH)_{4x}$
Columbite-tantalite	$(Fe,Mn)(Nb,Ta)_2O_6$
Copper	Cu
Covellite	CuS
Diamond	C
Diaspore	$AlO(OH)$
Digenite	Cu_9S_5
Enargite	Cu_3AsS_4
Eucryptite	$LiSiAlO_4$
Famatinite	Cu_3SbS_4
Fluorite	CaF_2
Francolite	$Ca_5(PO_4,CO_3)_3(OH,F)$
Galena	PbS
Garnierite	$(Ni,Mg)_3Si_2O_5(OH)_4$
Gibbsite	$Al(OH)_3$
Gold	Au
Graphite	C
Gypsum	$CaSO_4.2H_2O$
Hematite	Fe_2O_3
Holmquistite	An Li-bearing variant of riebeckite ($NaFeSi_2O_6$)
Illite	$(K,H_3O)(Al,Mg,Fe)_2$ $(Si,Al)_4O_{10}[(OH)_2H_2O]$
Ilmenite	$FeTiO_3$
Jarosite	$KFe_3(SO_4)_2(OH)_6$
Kaolinite	$Al_2Si_2O_5(OH)_4$
Kutnerhorite	$Ca(Fe,Mg,Mn)(CO_3)_2$
Lepidolite	$KLi_2Al(Si_4O_{10})(OH)_2$
Limonite	'Sack' term for brown, hydrous iron oxides such as geothite and lepidocrocite
Löllingite	$FeAs_2$
Magnesite	$MgCO_3$
Magnetite	Fe_3O_4
Manganite	$MnO(OH)$
Marcasite	FeS_2
Microlite	$(Na,Ca)_2Ta_2O_6(O,OH,F)$
Molybdenite	MoS_2
Monazite	$(Ce,La,Y,Th)PO_4$
Montroseite	$VO(OH)$
Muscovite	$KAl_2Si_3O_{10}(OH)_2$
Nepheline	$NaAlSiO_4$
Parisite	$(Ce,La)_2Ca(CO_3)_3F_2$
Pentlandite	$(Fe,Ni)_9S_8$
Petalite	$LiAlSi_4O_{10}$
Pitchblende	UO_2
Pollucite	$CsSi_2AlO_6$
Pyrite	FeS_2
Pyrochlore	$NaCaNb_2O_6F$
Pyrolusite	MnO_2
Pyrophyllite	$Al_2Si_4O_{10}(OH)_2$
Pyrrhotite	$Fe_{1-x}S$
Quartz	SiO_2
Rhodochrosite	$MnCO_3$
Romanechite	$BaMn_9O_{16}(OH)_4$
Rutile	TiO_2
Scheelite	$CaWO_4$
Sericite	Fine-grained muscovite
Serpentine	$Mg_3Si_2O_5(OH)_4$
Siderite	$FeCO_3$
Silver	Ag
Smectite (montmorillonite)	$(Na,Ca)_{0.33}(Al,Mg)_2$ $Si_4O_{10}(OH)_2.nH_2O$
Spessartite	$Mn_3Al_2(SiO_4)_3$
Sphalerite	$(Zn,Fe)S$

Spodumene	$LiAlS_2O_6$	Vallerite	$(Fe,Cu)S_2 \cdot (Mg,Al)(OH)_2$
Stannite	Cu_2FeSnS_4	Wad	'Sack' term for black
Strontianite	$SrCO_3$		manganese oxides and
Talc	$Mg_3Si_4O_{10}(OH)_2$		hydroxides
Tantalite	$(Fe,Mn)(Ta,Nb)_2O_6$	Wolframite	$(Fe,Mn)WO_4$
Tetrahedrite-tennanite	$(Cu,Fe,Ag)_{12}(Sb,As)_4S_{13}$	Wurzite	ZnS
Trona	$Na_2CO_3 \cdot NaHCO_3 \cdot 2H_2O$	Xenotime	YPO_4
Uraninite	UO_2	Zircon	$ZrSiO_4$
Uranothorite	$(Th,U)SiO_4$		

References

Adamek P.M. & Wilson M.R. (1979). The evolution of a uranium province in northern Sweden. *Philos. Trans. R. Soc. London*, **A291**, 355–68.

Adams S.S. (1991) Evolution of genetic concepts for principal types of sandstone uranium deposits in the United States. *Econ. Geol. Monogr.*, **8**, 225–48.

Akande O., Zentilli M. & Reynolds P.H. (1989) Fluid inclusion and stable isotope studies of Pb–Zn–fluorite–barite mineralization in the Lower and Middle Benue Trough, Nigeria. *Miner. Deposita*, **24**, 183–91.

Alapieti T.T. & Lahtinen J.J. (1986) Stratigraphy, petrology, and platinum-group element mineralization of the early Proterozoic Penikat layered intrusion, northern Finland. *Econ. Geol.*, **81**, 1126–36.

Alderton D.H.M. (1978) Fluid inclusion data for lead–zinc ores from south-west England. *Trans. Instn Min. Metall. (Sect. B: Appl. Earth Sci.)*, **87**, B132–5.

Aldous R.T.H. (1986) Copper-rich fluid inclusions in pyroxenes from the Guide Copper Mine, a satellite intrusion of the Palabora Complex, South Africa. *Econ. Geol.*, **81**, 143–55.

Aleva G.J.J. & Dijkstra S. (1986) Host rock lithology as principal criterion for an ITC-adopted classification of mineral deposits. *ITC J.*, **1986-3**, 243–47.

Amade E. (1983) Caractéristiques comparées des quatre principaux 'porphyry copper' de Niugini. *Chron. Rech. Min.*, **472**, 3–22.

Amann, H. (1985) Development of ocean mining in the Red Sea. *Marine Min.*, **5**, 103–16.

Anderson C.A. (1948) Structural control of copper mineralization, Bagdad, Arizona. *Am. Inst. Min. Metall. Eng. Trans.*, **178**, 170–80.

Anderson G.M. (1975) Precipitation of Mississippi Valley-type ores. *Econ. Geol.*, **70**, 937–42.

Anderson G.M. (1977) Thermodynamics and sulfide solubilities. In Greenwood H.J. (ed.), *Application of Thermodynamics to Petrology and Ore Deposits*. Mineralogical Association of Canada, Toronto.

Anderson G.M. (1983) Some geochemical aspects of sulphide precipitation in carbonate rocks. In Kisvarsanyi G., Grant S.K., Pratt W.P. & Koening J.W. (eds), *International Conference on Mississippi Valley Type Lead–Zinc Deposits*, 77–85 Univ. of Missouri-Rolla, Rolla.

Anderson J.A. (1982) Characteristics of leached cappings and techniques of appraisal. In Titley S.R. (ed.), *Advances in Geology of the Porphyry Copper Deposits: Southwestern North America*, 275–95. Univ. Arizona Press, Tucson.

Anderson P. (1980) Regional time–space distribution of porphyry deposits—a decisive test for the origin of metals in magma-related ore deposits. In Ridge J.D. (ed.), *Proceedings of the Fifth Quadrennial IAGOD Symposium*, 35–48. E. Schweizerbart'sche Verlagsbuchhandlung, Stuttgart.

Anderson P. & Guilbert J.M. (1979) The Precambrian massive-sulphide deposits of Arizona—a distinct metallogenic epoch and province. In Ridge J.D. (ed.), *Papers on Mineral Deposits of Western North America*, 39–48 Rep. 33, Nevada Bur. Mines and Geol.

Annels A.E. (1979) Mufulira greywackes and their associated sulphides. *Trans. Instn Min. Metall. (Sect. B: Appl. Earth Sci.)*, **88**, B15-B23.

Annels A.E. (1984) The geotectonic evironment of Zambian copper-cobalt mineralization. *J. Geol. Soc. London*, **141**, 279–89.

Annels A.E. (1989) Ore genesis in the Zambian Copperbelt, with particular reference to the northern sector of the Chambishi Basin. In Boyle R.W., Brown A.C., Jefferson C.W., Jowett E.C. & Kirkham R.V. (eds), *Sediment-hosted Stratiform Copper Deposits*, 427–52. Geol. Assoc. Can. Spec. Pap. 36, Memorial Univ., Newfoundland.

Annels A.E. & Roberts D.E. (1989) Turbidite-hosted gold mineralization at the Dolaucothi Gold Mines, Dyfed, Wales, United Kingdom. *Econ. Geol.*, **85**, 1293–314.

Anon. (1977) *Industrimineral*. Statens Offenliga Utredninger, Stockholm.

Anon. (1979) Nchanga Consolidated Copper Mines. *Eng. Min. J.*, **180** (Nov.), 150.

Anon. (1982) The Rio Tinto-Zinc Corporation PLC Annual Report and Accounts.

Anon. (1983a) Diamonds. *Eng. Min. J.*, **184** (Nov.), 119–23.

Anon. (1983b) Ashton Mining Limited Annual Report and Accounts.

Anon. (1983c) New Zealand: ironsand to steel. *Min. Mag.*, **148**, 346–53.

Anon. (1984) *Min. Mag.* **151**, 292.

Anon. (1985a) *Min. J. (London)*, **7802**, 146.

Anon. (1985b) Synthetic diamond breakthrough. *Ind. Miner.*, **10**, (June).

Anon. (1985b) Hard times for tin mines. *Min. Mag.*, **305**, 125–6.

Anon. (1985c) Where are new developments worth the risk? *World Min. Equip.*, **July**, 44–55.

Anon. (1985d) Politics and mining—when risk becomes reality. *World Min. Equip.*, **August**, 24–6.

Anon. (1987) Strategic minerals and the U.S.A. *Min. Surv.*, **No.1/3**, 3–29.

Anon. (1988a) Diamond exploration venture. *Ind. Miner.*, **252**, 17.

Anon. (1988b) More on Carr Boyd's rare earths. *Ind. Miner.*, **253**, 8–9.

Anon. (1989a) Australian diamonds. *Ind. Miner.*, **257**, 69.

Anon. (1989b) Sri Lankan monazite discovery. *Ind. Miner.*, **248**, 77.

Anon. (1989c) Chromium discovery. *Min. Mag.*, **160**, 12.

Anon. (1989d) CSIRO takes magnesite stake. *Ind. Miner.*, **256**, 11.

Anthony M. (1988) Technical change—the impending revolution. In *Australian Inst. Min. Metall. Sydney Branch, Minerals and Exploration at the Crossroads*, 19–25, Sydney.

Appel P.W.U. (1979) Stratabound copper sulphides in a banded iron-formation and in basaltic tuffs in the early Precambrian Isua Supracrustal Belt, West Greenland. *Econ. Geol.*, **74**, 45–52.

Arndt N.T. & Nisbet E.G. (1982a) What is a komatiite? In Arndt N.T. & Nisbet E.G. (eds), *Komatiites*, 19–27. George Allen & Unwin, London.

Ardnt N.T. & Nisbet E.G. (eds) (1982b) *Komatiites*. George Allen & Unwin, London.

Arndt P. & Lüttig G.W. (eds) (1987) *Mineral Resources' Extraction, Environmental Protection and Land-use Planning in the Industrial and Developing Countries*. E. Schweizerbart'sche Verlagsbuchhandlung, Stuttgart.

Arribas A. & Gumiel P. (1984) First occurrence of a strata-bound Sb–W–Hg deposit in the Spanish Hercynian Massif. In Wauschkuhn A., Kluth C. & Zimmerman R.A. (eds), *Syngenesis and Epigenesis in the Formation of Mineral Deposits*, 468–481 Springer-Verlag, Berlin.

Ashley P.M. & Plimer I.R. (1989) 'Stratiform Skarns'—a re-evaluation of three eastern Australian deposits. *Miner. Deposita*, **24**, 289–98.

Ashley P.M., Dudley R.J., Lesh R.H., Marr J.M. & Ryall A.W. (1988) The Scuddles Cu–Zn prospect, an Archean volcanogenic massive sulphide deposit, Golden Grove District, Western Australia. *Econ. Geol.*, **83**, 918–51.

Atkinson P., Moore J. McM. & Evans A.M. (1982) The Pennine orefields of England with special reference to recent structural and fluid inclusion investigations. *Bull. Bur. Rech. Geol. Minières Sect. 2* **No.2**, 149–56.

Atkinson W.J., Hughes F.E. & Smith C.B. (1984) A review of the kimberlitic rocks of Western Australia. In Kornprobst J. (ed.), *Kimberlites I: Kimberlites and Related Rocks*, 195–224. Elsevier, Amsterdam.

Awramik S.M. & Barghoorn E.S. (1977) The Gunflint microbiota. *Precambrian Res.*, **5**, 121–42.

Aye F. (1982) Contrôles Géologiques des Gîtes Stratiformes de Pb, Zn, Cu, Ag de la Bordure du Bassin de Châteaulin. *Mém. Bur. Rech. Geol. Minières* (Orleans) **120**.

Ayres L.D. & Černý P. (1982) Metallogeny of granitoid rocks in the Canadian Shield. *Can. Mineral.*, **20**, 439–536.

Badham J.P.N. (1978) Slumped sulphide deposits at Avoca, Ireland, and their significance. *Trans. Instn Min. Metall. (Sect. B: Appl. Earth Sci.)*, **87**, B21–B26.

Baker P.E. (1968) Comparative volcanology and petrology of the Atlantic island arcs. *Bull. Volcanol.*, **32**, 189–206.

Baldwin J.A. & Pearce J.A. (1982) Discrimination of productive and nonproductive porphyritic intrusions in Chilean Andes. *Econ. Geol.* **77**, 664–74.

Ball T.K. & Bland D.J. (1985) The Cae Coch volcanogenic massive sulphide deposits, Trefriw, North Wales. *J. geol. Soc. London*, **142**, 889–98.

Ballhaus C.G. & Stumpfl E.F. (1985) Fluid inclusions in Merensky and Bastard Reefs, Western Bushveld Complex (abstr.) *Can. Mineral.*, **23**, 294.

Banks D.A. (1985) A fossil hydrothermal worm assemblage from the Tynagh lead–zinc deposit in Ireland. *Nature*, **313**, 128–31.

Barbier J. (1982) Géochimie en roche sur le Gîte de Scheelite de Salau. *Bull. Bur. Rech. Geol. Minières, Sec. 2*, **No. 1**, 25–44.

Bárdossy G. (1982) *Karst Bauxite: Bauxite Deposits on Carbonate Rocks*. Elsevier, Amsterdam.

Bárdossy G. & Aleva G.J.J. (1990) *Lateritic Bauxites*. Elsevier, Amsterdam.

Barker D.S. (1969) North American felspathoidal rocks in space and time. *Bull. Geol. Soc. Am.*, **80**, 2369–72.

Barker D.S. (1989) Field relations of carbonatites. In Bell, K. (ed.), *Carbonatites*, 38–69. Unwin Hyman, London.

Barnes H.L. (1975) Zoning of ore deposits: types and causes. *Trans. R. Soc. Edinburgh*, **69**, 295–311.

Barnes H.L. (ed.) (1979a) *Geochemistry of Hydrothermal Ore Deposits*, 2nd Ed. Wiley, New York.

Barnes H.L. (1979b) Solubilities of ore minerals. In Barnes H.L. (ed.), *Geochemistry of Hydrothermal Ore Deposits*, 2nd edn, 404–60. Wiley, New York.

Barnes H.L. (1983) Ore-depositing reactions in Mississippi Valley-type deposits. In Kisvarsanyi G., Grant S.K., Pratt W.P. & Koening J.W. (eds), *International Conference on Mississippi Valley Type Lead–Zinc Deposits*, 77–85. Univ. of Missouri-Rolla, Rolla.

Barnes R.G. (1983) Stratiform and stratabound tungsten mineralization in the Broken Hill Block, N.S.W. *J. Geol. Soc. Australia*, **30**, 225–39.

Barnes, S.J. & Barnes, S.-J. (1990) A new interpretation of the Katiniq nickel deposit, Ungava, Northern Quebec. *Econ. Geol.*, **85**, 1269–72.

Barrington, R.C.A. (1986) Outline of the iron ore trade. In Cooke H. & Bailey J. (eds), *Iron Ore Databook 1987*, 15–26. Metal Bulletin Books, London.

Barton, M.D. (1987) Lithophile-element mineralization associated with late Cretaceous two-mica granites in the Great Basin. *Geology*, **15**, 337–40.

Barton P.B. Jr. (1967) Possible role of organic matter in the precipitation of the Mississippi Valley ores. *Econ. Geol. Monogr.*, **3**, 371–78.

Barton P.B. & Skinner B.J. (1979) Sulphide mineral stabilities. In Barnes H.L. (ed.), *Geochemistry of Hydro-*

thermal Ore Deposits, 2nd edn, 278–403. Wiley, New York.

Barton P.B. & Toulmin P. (1963) Sphalerite phase equilibria in the system Fe–Zn–S between 580°C and 850°C. *Econ. Geol.*, **58**, 1191–2.

Batchelor B.C. (1979) Geological characteristics of certain coastal and off-shore placers as essential guides for tin exploration in Sundaland, S.E. Asia. In Yeap C.H. (ed.), *Geology of Tin Deposits*, Bull. Geol. Soc. Malaysia, **II**, 283–314.

Bateman A.M. (1950) *Economic Mineral Deposits*. Wiley, New York.

Baumann L. (1965) Zur Erzführung und regionalen Verbreitung des 'Felsithorizontes' von Halbrücke. *Freiberg. Forschungsn.*, **C186**, 63–81.

Baumann L. (1970). Tin deposits of the Erzgebirge. *Trans. Instn Min. Metall. (Sect. B: Appl. Earth Sci.)*, **79**, B68–B75.

Baumann L. & Krs M. (1967) Paläomagnetische Altersbestimmungen an einigen Mineralparagenesen des Freiberger Lagerstättenbezirkes. *Geologie*, **16**, 765–80.

Bayley R.W. & James H.L. (1973) Precambrian iron-formations of the United States. *Econ. Geol.*, **68**, 934–59.

Beales F.W. & Jackson S.A. (1966) Precipitation of lead-zinc ores in carbonate reservoirs as illustrated by Pine Point Ore Field, Canada. *Trans. Instn Min. Metall. (Sect. B. Appl. Earth Sci.)*, **75**, B278–85.

Beaufort D. & Meunier A. (1983) Petrographic characterization of an argillic hydrothermal alteration containing illite, K-rectorite, K-beidellite, kaolinite and carbonates in a cupromolybdic porphyry at Sibert (Rhône, France). *Bull. mineral.*, **106**, 535–51.

Beckinsale R.D. (1979) Granite magmatism in the tin belt of South-east Asia. In Atherton M.P. & Tarney J. (eds), *Origin of Granite Batholiths*, 34–44 Shiva, Orpington.

Beeson R. (1990) Broken Hill-type lead–zinc deposits—an overview of their occurrence and geological setting. *Trans. Instn Min. Metall. (Sect. B: Appl. Earth Sci.)*, **99**, B163–75.

Benbow J. (1988) Bauxite: aluminating non-metallurgical sectors. *Ind. Miner.*, **252**, 67–83.

Berger, B.R. (1985) Geological and geochemical relationships at the Getchell Mine and vicinity, Humboldt County, Nevada. In Hollister V.F. (ed.), *Discoveries of Epithermal Precious Metal Deposits*, 51–9. Society of Mining Engineers, New York.

Berger B.R. & Bagby W.C. (1991). The geology and origin of Carlin-type gold deposits. In Foster R.P. (ed.), *Gold Metallogeny and Exploration*, 210–48. Blackie, Glasgow.

Berger B.R. & Henley R.W. (1989) Recent advances in the understanding of epithermal gold–silver deposits—with special reference to the western United States. *Econ. Geol. Monogr.*, **6**, 405–23.

Bergman S.C., Dunn D.P. & Krol L.G. (1988) Rock and mineral chemistry of the Linhaisai Minette, Central

Kalimantan, Indonesia and the origin of Borneo diamons. *Can. Miner.*, **26**, 23–44.

Berning, J. (1986) The Rössing uranium deposit, South West Africa/Namibia. In Anhaeusser C.R. & Maske S. (eds), *Mineral Deposits of Southern Africa*, **2**, 1819–32. Geol. Soc. S. Afr., Johannesburg.

Berning J., Cooke R., Hiemstra S.A. & Hoffman U. (1976) The Rössing uranium deposit, South-west Africa. *Econ. Geol.*, **71**, 351–68.

Bernstein L.R. (1986) Geology and mineralogy of the Apex Germanium–gallium Mine, Washington County, Utah. *Bull. 1577 U.S. Geol. Surv.*, Washington.

Besson M., Boyd R., Czamanske G., Foose M., Groves D., van Gruenewaldt G., Naldrett A., Nilsson G., Page N., Papunen H. & Peredery W. (1979) IGCP Project No. 161 and a proposed classification of Ni–Cu–PGE sulphide deposits. *Can. Miner.*, **17**, 143–4.

Best J.L. & Brayshaw A.C. (1985) Flow separation—a physical process for the concentration of heavy mineral within alluvial channels. *J. Geol. Soc. London*, **142**, 747–55.

Best M.C. (1982) *Igneous and Metamorphic Petrology.* Freeman, San Francisco.

Bethke, C.M. (1986) Hyrologic constraints on the genesis of the Upper Mississippi Valley Mineral District from Illinois Basin brines. *Econ. Geol.*, **81**, 233–49.

Bichan R. (1969) Origin of chromite seams in the Hartley Complex of the Great Dyke, Rhodesia. In Wilson H.D.B. (ed.), *Magmatic Ore Deposits*, Econ. Geol. Monogr., **4**, 95–113.

Billingsley P. & Locke A. (1941) Structures of ore districts in the continental framework. *Trans. Am. Inst. Min. (Metall.) Eng.*, **144**, 9–21.

Binda P.L. (1975) Detrital bornite grains in the late Precambrian B Greywacke of Mufulira, Zambia, *Miner. Deposita*, **10**, 101–7.

Bird D. (1988) Boddington gold mine. *Min. Mag.*, **159**, 350–6.

Bird M.I. & Chivas A.R. (1988) Stable-isotope evidence for low-temperature kaolinitic weathering and postformation hydrogen-isotope exchange in Permian kaolinites. *Chem. Geol.*, **72**, 249–65.

Bischoff J.L., Radtke A.S. & Rosenbauer R.J. (1981) Hydrothermal alteration of greywacke by brine and seawater: roles of alteration and chloride complexing on metal solubilization at 200° and 350°C. *Econ. Geol.*, **76**, 659–76.

Bjrlykke A. & Sangster D.F. (1981) An overview of sandstone lead deposits and their relation to red-bed copper and carbonate-hosted lead-zinc deposits. *Econ. Geol.*, **75th Anniv. Vol.**, 179–213.

Blatt H., Middleton G. & Murray R. (1980) *Origin of Sedimentary Rocks.* Prentice-Hall, Englewood Cliffs.

Blissenbach E.B. & Fellerer R. (1973) Continental drift and the origin of certain mineral deposits. *Geol. Rundsch.*, **62**, 812–39.

Blum N. & Puchelt H. (1991) Sedimentary-hosted polymetallic massive sulfide deposits of the Kebrit and Shaban Deeps, Red Sea. *Miner. Deposita*, **26**, 217–27.

Boirat J.M. & Fouquet Y. (1986) Découverte de tubes de vers hydrothermaux fossiles dans un amas sulfuré de l'Eocène Supérieur (Barlo, Ophiolite de Zambalés, Philippines). *Sci. Terre*, **302**, 941–6.

Bonatti E., Guernstein-Honnorez B.-M. & Honnorez J. (1976) Copper–iron sulphide mineralization from the equatorial Mid-Atlantic Ridge. *Econ. Geol.*, **71**, 1515–25.

Bornhorst T.J., Paces J.B., Grant N.K., Obradovich J.D. & Huber N.K. (1988) Age of native copper mineralization, Keweenaw Peninsula, Michigan. *Econ. Geol.*, **83**, 619–25.

Bowden P. (1982) Magmatic evolution and mineralization in the Nigerian Younger Granite Province. In Evans A.M. (ed.), *Metallization Associated with Acid Magmatism*, 51–61, Wiley, Chichester.

Bowles J.F.W. (1985) A consideration of the development of platinum-group minerals in laterite. *Can. Mineral.*, **23**, 296.

Bowman J.R., O'Neil J.R. & Essene E.J. (1985) Contact skarn formation at Elkhorn, Montana. *Am. J. Sci.*, **285**, 621–60.

Bowman J.R., Parry W.T., Kropp W.P. & Kruer S.A. (1987) Chemical and isotopic evolution of hydrothermal solutions at Bingham, Utah. *Econ. Geol.*, **82**, 395–428.

Boyle R.W. (1959) The geochemistry, origin and role of carbon dioxide, sulfur, and boron in the Yellowknife gold deposits Northwest Territories, Canada. *Econ. Geol.*, **54**, 1506–24.

Boyle R.W. (1970) Regularities in wall-rock alteration phenomena associated with epigenetic deposits. In Pouba Z. & Štemprok M. (eds), *Problems of Hydrothermal Ore Deposition*, 233–60. E. Schweizerbart'sche Verlagsbuchandlung, Stuttgart.

Boyle R.W. (1979) The geochemistry of gold and its deposits. *Geol Surv. Canada, Bull.*, **280**.

Boyle R.W. (1991) Auriferous Archean greenstone–sedimentary belts. *Econ. Geol. Monogr.* **8**, 169–91.

Boyle R.W., Brown A.C., Jefferson C.W., Jowett E.C. & Kirkham R.V. (eds) (1989) *Sediment-hosted Stratiform Copper Deposits*, Geol. Assoc. Can. Spec. Pap. 36, Memorial Univ., Newfoundland.

Brady L.L. & Jobson H.E. (1973) An experimental study of heavy-mineral segregation under alluvial-flow conditions. *Prof. Pap. 562-K*, U.S. Geol. Surv., Washington.

Bray C.J., Spooner E.T.C., Golightly J.P. & Saracoglu N. (1982) Carbon and sulphur isotope geochemistry of unconformity-related uranium mineralization, McClean Lake Deposits, Northern Saskatchewan, Canada. *Geol. Soc. Am. Abstr. Vol., New Orleans Meeting*, **451**.

Brigo L. & Omenetto P. (1983) Scheelite-bearing occurrences in the Italian Alps: geotectonic and lithostratigraphic setting. In Schneider H-J. (ed.), *Mineral Deposits of the Alps and of the Alpine Epoch in Europe*, 41–50. Springer-Verlag, Berlin.

Brill, B.A. (1989) Deformation and recrystallization microstructures in deformed ores from the CSA Mine, Cobar, N.S.W., Australia. *J. Struct. Geol.*, **11**, 591–601.

Brimhall G.H. (1979) Lithologic determination of mass transfer mechanisms of multiple-stage porphyry copper mineralization at Butte, Montana: vein formation by hypogene leaching and enrichment of potassium–silicate protore. *Econ. Geol.*, **74**, 556–89.

Brimhall G.H. & Crerar D.A. (1987) Ore fluids: magmatic to supergene. *Rev. Mineral.*, **17**, 235–321.

Brimhall G.H. & Ghiorso M.S. (1983) Origin and ore-forming consequences of the advanced argillic alteration process in hypogene environments by magmatic gas contamination of meteoric fluids. *Econ. Geol.*, **78**, 73–90.

Brimhall G.H., Lewis C.J., Ague J.J., Dietrich W.E., Hampel J., Teague T. & Rix P. (1988) Metal enrichment in bauxites by deposition of chemically mature aeolian dust. *Nature*, **333**, 819–24.

Bristow C.M. (1987a) Society's changing requirements for primary raw materials. *Ind. Miner.* **232**, 59–65.

Bristow C.M. (1987b) World kaolins—genesis, exploration and application. *Ind. Miner.* **238**, 45–59.

Brocoum S.J. & Dalziel I.W.D. (1974) The Sudbury Basin, the Southern Province, the Grenville Front and the Penokean Orogeny. *Bull. Geol. Soc. Am.*, **85**, 1571–80.

Bromley A. (1989) *The Cornubian Orefield.* International Association of Geochemistry and Cosmochemistry, Camborne School of Mines, Camborne, Cornwall.

Brown A.C. (1978) Stratiform copper deposits—evidence for their post-sedimentary origin. *Miner. Sci. Eng.* **10**, 172–81.

Brown K.L. (1986) Gold deposition from geothermal discharges in New Zealand. *Econ. Geol.*, **81**, 979–83.

Browning P., Groves D.I., Blockley J.G. & Rosman K.J.R. (1987) Lead isotope constraints on the age and source of gold mineralization in the Archean Yilgarn Block, Western Australia. *Econ. Geol.*, **82**, 971–86.

Brownlow A.H. (1979) *Geochemistry.* Prentice-Hall, Englewood Cliffs, NJ.

Brunt D.A. (1978) Uranium in Tertiary stream channels, Lake Frome Area, South Australia. *Proc. Australas. Inst. Min. Metall.*, **266**, 79–90.

Bryce J.D., Thompson J.M. & Staff (1958) The Bicroft Operation. *Western Miner and Oil Review*, **31**(4), 79–92.

Bryndzia L.T. (1988) The origin of diaspore and pyrophyllite in the Foxtrap pyrophyllite deposit, Avalon Peninsula, Newfoundland; a reinterpretation. *Econ. Geol.*, **83**, 450–53.

Bryndzia L.T., Scott S.D. & Spry P.G. (1990) Sphalerite and hexagonal pyrrhotite geobarometer: correction in calibration and application. *Econ. Geol.*, **85**, 408–11.

Buchanan D.L. & Rouse J.E. (1984) Role of contamination in the precipitation of sulphides in the Platreef of the Bushveld Complex. In Buchanan D.L. & Jones M.J. (eds), *Sulphide Deposits in Mafic and Ultramafic Rocks*, 141–6. Instn Min. Metall., London.

Bugge J.A.W. (1978) Norway. In Bowie S.H.U., Kvalheim

A. & Haslam H.W. (eds), *Mineral Deposits of Europe Vol. 1 Northwest Europe*, 199–249. Instn Min. Metall. and Mineral. Soc., London.

Burger J.R. (1985) More desert gold: Newmount finds yet another five million oz. at Carlin, Nevada. *Eng. Min. J.* (October) 17.

Burke K. & Dewey J.F. (1973) Plume-generated triple junctions: key indicators in applying plate tectonics to old rocks. *J. Geol.* **81**, 406–33.

Burn R.G. (1971) Localized deformation and recrystallization of sulphides in an epigenetic mineral deposit. *Trans. Instn Min. Metall. (Sect. B: Appl. Earth Sci.)*, **80**, B116–B119.

Burnham C.W. (1959) Metallogenic provinces of the southwestern United States and northern Mexico. *Bull. State Bur. Mines Min. Resources, New Mexico*, **65**, 76 pp.

Burnham C.W. (1979) Magmas and hydrothermal fluids. In Barnes H.L. (ed.), *Geochemistry of Hydrothermal Ore Deposits*, 2nd edn, 71–136. Wiley, New York.

Burnham C.W. & Ohmoto H. (1980) Late stage processes of felsic magmatism. In Ishihari S. & Takenouchi S. (eds.), *Granitic Magmatism and Related Mineralization*, 1–11. Mining Geol. Spec. 8, Soc. Min. Geol. Jpn.

Burnie S.W., Schwarcz H.P. & Crocket J.H. (1972) A sulfur isotopic study of the White Pine Mine, Michigan. *Econ. Geol.*, **67**, 895–914.

Burrows D.R., Wood P.C. & Spooner E.T.C. (1986) Carbon isotope evidence for a magmatic origin for Archaean gold–quartz vein ore deposits. *Nature*, **321**, 851–54.

Bursnall J.T. (ed.) (1989) *Mineralization and Shear Zones*. Short Course Notes, 6, Geol. Assoc. Canada, Montréal.

Burt D.M. (1981) Acidity–salinity diagrams—application to greisen and porphyry deposits. *Econ. Geol.*, **76**, 832–43.

Button A. (1976) Transvaal and Hamersley Basins—review of basin development and mineral deposits. *Miner. Sci. Eng.*, **8**, 262–93.

Callahan W.H. (1967) Some spatial and temporal aspects of the localization of Mississippi Valley–Appalachian type ore deposits. In Brown J.S. (ed.), *Genesis of Stratiform Lead–Zinc–Barite–Fluorite Deposits*, 14–19. Economic Geology Publishing Co., Lancaster, Pennsylvania.

Cameron E.N. & Emerson M.E. (1959) The origin of certain chromite deposits in the eastern part of the Bushveld Complex. *Econ. Geol.*, **54**, 1151–1213.

Cameron E.M. & Hattori K. (1987) Archean gold mineralization and oxidized hydrothermal fluids. *Econ. Geol.*, **82**, 1177–91.

Cameron E. N., Jahns R.H., McNair A.H. & Page L.R. (1949) Internal structure of granitic pegmatites. *Econ. Geol. Monogr.*, **2**, 115.

Campbell J.D. (1953) The Triton gold mine, Reedy, Western Australia. In Edwards A.B. (ed.), *Geology of Australian Ore Deposits*, 195–207. Australas. Inst. Min. Metall, Melbourne.

Candela P.A. & Holland H.D. (1984) A mass transfer model for Cu amd Mo in magmatic hydrothermal systems: the origin of porphyry-type ore deposits *Econ. Geol.*, **81**, 1–19.

Cann J.R. (1970) New model for the structure of the ocean crust. *Nature*, **266**, 928–30.

Cann J.R. & Strens M.R. (1982) Black smokers fuelled by freezing magma. *Nature*, **298**, 147–9.

Card K.D., Gupta V.K., McGrath P.H. & Grant F.S. (1984) The Sudbury structure: its regional geological and geophysical setting. In Pye E.G., Naldrett A.J. & Giblin P.E. (eds), *The Geology and Ore Deposits of the Sudbury Structure*, 25–43, **Spec. Vol. 1**, Ontario Geol. Surv., Toronto.

Carmichael I.S.E., Turner F.J. & Verhoogen J. (1974) *Igneous Petrology*. McGraw-Hill, New York.

Carpenter A.B., Trout M.L. & Pickett E.E. (1974) Preliminary report on the origin and chemical evolution of lead- and zinc-rich oil field brines in central Mississippi. *Econ. Geol.*, **69**, 1191–1206.

Carten R.B. (1986) Sodium–calcium metasomatism: chemical, temporal and spatial relationships at the Yerington, Nevada, porphyry copper deposit. *Econ. Geol.*, **81**, 1495–1519.

Carten R.B., Walker B.M., Geraghty E.P. & Gunow A.J. (1988) Comparison of field-based studies of the Henderson porphyry molybdenum deposit, Colorado, with experimental and theoretical models of porphyry systems. In Taylor R.P. & Strong D.F. (eds), *Recent Advances in the Geology of Granite-related Mineral Deposits*, 351–66, **Spec. Vol. 39**, Can. Inst. Min. Metall., Montréal.

Cassard D., Nicolas A., Rabinovitch M., Moutte J., Leblanc M. & Prinzhofer A. (1981) Structural classification of chromite pods in southern New Caledonia. *Econ. Geol.*, **76**, 805–31.

Cathelineau M. (1982) Signification de la fluorine dans les gisements d'uranium de la chaine hercynienne. *Bull. Bur. Rech. Geol. Minières*, Sect. 2, **4**, 407–13.

Cathles L.M. & Smith A.T. (1983) Thermal constraints on the formation of Mississippi Valley-type lead–zinc deposits and their implications for episodic basin dewatering and deposit genesis. *Econ. Geol.*, **78**, 983–1002.

Cawthorn R.G., Barton J.W. Jr & Viljoen M.J. (1985) Interaction of floor rocks with the Platreef on Overysel, Potgietersrus, Northern Transvaal. *Econ. Geol.*, **80**, 988–1006.

Cayeux L. (1906) Structure et origine probable de minerai de fer magnétique de Dielette (Manche). *C.R. Acad. Sci. Paris*, **142**, 716–8.

Černý P. (1982a) Anatomy and classification of granitic pegmatites. In Černý P. (ed.), *Short Course in Granitic Pegmatites in Science and Industry*, 1–39. Mineral. Assoc. Can. Winnipeg.

Černý P. (1982b) Petrogenesis of granitic pegmatites. In Černý P. (ed.), *Short Course in Granitic Pegmatites in Science and Industry*, 405–461. Mineral. Assoc. Can., Winnipeg.

Černý P. (1982c) The Tanco pegmatite at Bernic Lake, southeastern Manitoba. In Černý P. (ed.), *Short Course in Granitic Pegmatites in Science and Industry*, 527–43 Mineral. Assoc. Can., Winnipeg.

Černý P. (ed.) (1982d) *Short Course in Granitic Pegmatites in Science and Industry*, Mineral. Assoc. Can., Winnipeg.

Černý, P. (1989) Characteristics of pegmatite deposits of tantalum. In Möller P., Černý P. & Saupé F. (eds), *Lanthanides, Tantalum and Niobium*, 195–239. Springer-Verlag, Berlin.

Černý P. & Meintzer R.E. (1988) Fertile granites in the Archean and Proterozoic fields of rare-element pegmatites: crustal environment, geochemistry and petrogenetic relationships. In Taylor R.P. & Strong D.F. (eds), *Recent Advances in the Geology of Granite-related Mineral Deposits*, 170–207, **Spec. Vol. 39**, Can. Inst. Min. Metall., Montréal.

Chadwick B. & Crewe M.A. (1986) Chromite in the Early Archaean Akilia Association (ca. 3000 Ma), Ivisârtoq Region, Inner Godthåbsfjord, West Greenland. *Econ. Geol.*, **81**, 184–91.

Chaloupský J. (1989) Major tectonostratigraphic units of the Bohemian Massif. In Dallmeyer R.D. (ed.), *Terranes in the Circum-Atlantic Paleozoic Orogens*, 101–14, Spec. Pap. 230, Geol. Soc. Am.

Champigny N. & Sinclair A.J. (1982) Cinola gold deposit, Queen Charlotte Island, B.C.—a geochemical case history. In Levinson A.A. (ed.), *Precious Metals in the Northern Cordillera*, 121–37. Association of Exploration Geochemists, Rexdale, Ontario.

Chartrand F.M. & Brown A.C. (1985) The diagenetic origin of stratiform copper mineralization, Coates Lake, Redstone Copper Belt, N.W.T., Canada, *Econ. Geol.*, **80**, 325–43.

Chatterjee A.K. & Strong D.F. (1984) Rare-earth and other element variations in greisens and granites associated with East Kemptville tin deposit, Nova Scotia, Canada. *Trans. Instn Min. Metall. (Sect. B: Appl. Earth Sci.)*, **93**, B59–70.

Chaussidon M., Abarède F. & Sheppard S.M.F. (1987) Sulphur isotope heterogeneity in the mantle from ion microprobe measurements of sulphide inclusions in diamonds. *Nature*, **330**, 242–4.

Chaye d'Albissin M., Guillou J.J. & Letolle, R. (1988) Observations on the paper by F. Velasco *et al*: A Contribution to the Ore Genesis of the Magnesite Deposit of Eugui, Navarra (Spain). *Miner. Deposita*, **23**, 309–12.

Chen W. (1988) Mesozoic and Cenozoic sandstone-hosted copper deposits in South China. *Miner. Deposita*, **23**, 262–7.

Chételat E. de (1947) La Genèse et l'évolution des gisements de nickel de la Nouvelle-Calédonie. *Bull. Soc. Geol. Fr., Ser. 5*, **17**, 105–60.

Chivas R. & Wilkins W.T. (1977) Fluid inclusion studies in relation to hydrothermal alteration and mineralization at the Koloula porphyry copper prospect, Guadalcanal. *Econ. Geol.*, **72**, 153–69.

Christie A.B. & Braithwaite R.L. (1986) Epithermal gold–silver and porphyry copper deposits of the Hauraki Goldfield—a review. In Henley R.W., Hedenquist J.W. & Roberts P.J. (eds), *Guide to the Epithermal (Geothermal) Systems and Precious Metal Deposits of New Zealand*, 129–45. S.G.A. Monogr. Ser. 26, Gebrüder Borntraeger, Berlin.

Clark G.S. (1982) Rubidium–strontium isotope systematics of complex granitic pegmatites. In Černý P. (ed.), *Short Course in Granitic Pegmatites in Science and Industry*, 347–71. Mineral. Assoc. Can., Winnipeg.

Clark J.P. & Reddy B. (1989) Critical and strategic materials. In Carr D.D. & Herz N. (eds), *Concise Encyclopedia of Mineral Resources*, 88–94. Pergamon, Oxford.

Clark R.J. McH., Homeniuk L.A. & Bonnar R. (1982) Uranium geology in the Athabaska and a comparison with other Canadian Proterozoic Basins. *CIM Bull.*, **75** (April), 91–8.

Clark S.H.B., Gallagher M.J. & Poole F.G. (1990) World barite resources: a review of recent production patterns and a genetic classification. *Trans. Instn Min. Metal. (Sect. B: Appl. Earth Sci.)*, **99**, B125–32.

Clarkè M.C.G. (1983) Current Chinese thinking on the South China Tungsten Province. *Trans. Instn Min. Metall. (Sect. B. Appl. Earth Sci.)* **92**, B10–15.

Clement C.R., Skinner E.M.W. & Scott Smith B.H. (1984) Kimberlite redefined. *J. Geol.*, **92**, 223–8.

Cliff R.A., Rickard D. & Blake K. (1990) Isotope systematics of the Kiruna magnetite ores, Sweden. Part 1. Age of the ore. *Econ. Geol.*, **85**, 1770–6.

Coleman R.J. (1975) Savage River magnetite deposits. In Knight C.L. (ed.), *Economic Geology of Australia and Papua New Guinea, 1, Metals*, 598–604. Australas. Inst. Min. Metall., Parkville.

Colley H. & Walsh J.N. (1987) Genesis of Fe–Mn deposits of southwest Viti Levu, Fiji. *Trans. Instn Min. Metall. (Sect. B: Appl. Earth Sci.)*, **96**, B201–12.

Collis L. & Fox R.A. (1985) *Aggregates: Sand, Gravel and Crushed Rock Aggregates for Construction Purposes*. Geol. Soc., London.

Colvine A.C. (ed.) (1983) *The Geology of Gold in Ontario*. Ont. Geol. Surv. Misc. Pap. 110.

Compton J.S., Snyder S.W. & Hodell D.A. (1990) Phosphogenesis and weathering of shelf sediments from the southeastern United States. *Geology*, **18**, 1227–30.

COMRATE (Committee on Mineral Resources and the Environment) (1975) *Mineral Resources and the Environment*. National Academy of Sciences, Washington.

Cook D.R. (1987) A crisis for economic geologists and the future of the society. *Econ. Geol*, **82**, 792–804.

Cook P.J. (1984) Spatial and temporal controls on the formation of phosphate deposits. In Nriagu J.O. & Moore P.B. (eds), *Phosphate Minerals*, 242–74. Springer-Verlag, Berlin.

Cook P.J. & McElhinny M.W. (1979) A reevaluation of the spatial and temporal distribution of sedimentary phosphate deposits in the light of plate tectonics. *Econ. Geol.*, **74**, 315–30.

Cook P.J. & Shergold J.H. (1986) Proterozoic and Cambrian Phosphorites—nature and origin. In Cook P.J. & Shergold J.H. (eds), *Phosphate Deposits of the World, Vol. 1*, 369–86. Cambridge Univ. Press, Cambridge.

Cook S.S. (1988) Supergene Copper Mineralization at the Lakeshore Mine, Pinal County, Arizona. *Econ. Geol.*, **83**, 297–309.

Coveney R.M., Jr. & Nansheng, C. (1991) Ni–Mo–PGE–Au-rich ores in Chinese black shales and speculations on possible analogues in the United States. *Miner. Deposita*, **26**, 83–8.

Cowden A. (1988) Emplacement of komatiite lava flows and associated nickel sulfides at Kambalda, Western Australia. *Econ. Geol.*, **83**, 436–42.

Cowden A. & Woolrich P. (1987) Geochemistry of the Kambalda iron–nickel sulfides: implications for models of sulfide–silicate partitioning. *Can. Mineral.*, **25**, 21–36.

Cox S.F., Etheridge M.A. & Wall V.J. (1987) The role of fluids in syntectonic mass transport, and the location of metamorphic vein-type ore deposits. *Ore Geol. Rev.*, **2**, 65–86.

Craig, J.R. (1989) Ore minerals. In Carr D.D. & Herz N. (eds), *Concise Encyclopedia of Mineral Resources*, 234–41. Pergamon, Oxford.

Craig J.R. & Scott S.D. (1974) Sulphide phase equilibria. In Ribbe P.H. (ed.), *Sulphide Mineralogy*, Mineral. Soc. Am., Short Course Notes, Vol. 1, Ch. 5.

Craig J.R. & Vaughan D.J. (1981) *Ore Microscopy and Ore Petrography*. Wiley, New York.

Criss R.E., Ekren E.B. & Hardyman R.F. (1984) Casto Ring Zone: a 4500 km² fossil hydrothermal system in the Challis Volcanic Field, Central Idaho. *Geology*, **12**, 331–4.

Crocetti C.A. & Holland H.D. (1989) Sulfur–lead isotope systematics and the composition of fluid inclusions in galena from the Viburnum Trend, Missouri. *Econ. Geol.*, **84**, 2196–216.

Crozier R.D. (1986) Lithium: resources and prospects. *Min. Mag.*, **154**, 148–52.

Cuney M. (1978) Geologic environment, mineralogy and fluid inclusions of the Bois Noirs–Limouzat uranium vein, Forez, France. *Econ. Geol.*, **73**, 1567–610.

Czamanske G.K., Haffty J. & Nabbs S.W. (1981) Pt, Pd and Rh analyses and beneficiation of mineralized mafic rocks from the La Perouse Layered Gabbro, Alaska. *Econ. Geol.*, **76**, 2001–11.

Dahlkamp F.J. (1978) Classification of uranium deposits. *Miner. Deposita*, **13**, 83–104.

Dahlkamp F.J. (1984) Characteristics and problematics of the metallogenesis of Proterozoic vein-like type uranium deposits. In Wauschkuhn A., Kluth C. & Zimmermann R.A. (eds), *Syngenesis and Epigenesis in the Formation of Mineral Deposits*, 183–92. Springer-Verlag, Berlin.

Damon P.E., Shafiqullah M. & Clark K.F. (1983) Geochemistry of the porphyry copper deposits and related mineralization of Mexico. *Can. J. Earth Sci.*, **20**, 1052–71.

Danielson M.J. (1975) King Island Scheelite deposits. In Knight C.L. (ed.), *Economic Geology of Australia and Papua New Guinea*, 592–98. Australas. Inst. Min. Metall., Parkville.

Davies F.B. & Windley B.F. (1976) Significance of major Proterozoic high grade linear belts in continental evolution. *Nature*, **263**, 383–5.

Davies, G.H. (1984) *Structural Geology of Rocks and Regions*. Wiley, New York.

Davies, J.F. (1989) Some temporal–spatial aspects of North American porphyry deposits. *Econ. Geol.*, **84**, 2300–6.

Davies, M.H. (1987) Prospects for copper. *Bull. Instn Min. Metall.*, **September**, 3–6.

Davis W.J. & Williams-Jones A.E. (1985) A fluid inclusion study of the porphyry-greisen, tungsten–molybdenum deposit at Mount Pleasant, New Brunswick, Canada. *Miner. Deposita*, **20**, 94–101.

Dawson J.B. (1971) Advances in kimberlite geology. *Earth Sci. Rev.*, 7, 187–214.

Dawson J.B. (1980) *Kimberlites and Their Xenoliths*. Springer-Verlag, Berlin.

Degens E.T. & Ross D.A. (1969) (eds). *Hot Brines and Recent Heavy Metal Deposits in the Red Sea*. Springer, New York.

Degens E.T. & Ross D.A. (1976) Strata-bound metalliferous deposits found in or near active rifts. In Wolf K.H. (ed.), *Handbook of Strata-Bound and Stratiform Ore Deposits, Vol. 4*, 165–202. Elsevier, Amsterdam.

Dejonghe L. & de Walque L. (1981) Pétrologie et géochimie du filon sulfuré de Heure (Belgique) du chapeau de fer associé et de l'encaissant carbonaté. *Bull Bur. Rech. Geol. Minières Sect. 2*, **1980–81**, 165–91.

Dercourt J. & Paquet J. (1985) *Geology: Principles and Methods*. Graham & Trotman, London.

Derkmann K. & Jung R. (1986) Assessing the potential of Burundi's nickel laterites. *Eng. & Min. J.*, **187** (July), 8–9.

Derré C. (1982) Caracteristiques de la distribution des gisements à etain et tungstène dans l'Ouest de L'Europe. *Miner. Deposita*, **17**, 55–77.

De Villiers J.E. (1983) The manganese deposits of Griqualand West, South Africa: some mineralogic aspects. *Econ. Geol.*, **78**, 1108–18.

De Villiers, J.S. (1990) Chromite in the Bushveld Complex –an end-user's perspective. *Miner. Ind. Int.*, **994**, 41–4.

De Vore G.W. (1955) The role of adsorption in the fractionation and distribution of elements. *J. Geol.*, **63**, 159–90.

De Wit, M.J. (1985) *Minerals and Mining in Antarctica*. Clarendon Press, Oxford.

Dicken A.P., Rice C.M. & Harmon R.S. (1986) A strontium and oxygen isotope study of Laramide magmatic and hydrothermal activity near Central City, Colorado. *Econ. Geol.*, **81**, 904–14.

Dietz R.S. (1964) Sudbury structure as an astrobleme. *J. Geol.*, **72**, 412–34.

Dilles J.H. (1987) Petrology of the Yerington Batholith, Nevada: evidence for evolution of porphyry copper ore fluids. *Econ. Geol.*, **82**, 1750–89.

Dimroth E. (1977) Facies models 5—models of physical sedimentation of iron formations. *Geosci. Can.*, **4**, 23–30.

Dines H.G. (1956) The metalliferous mining region of South-west England. *Mem. Geol. Surv. G.B.*, **1**.

Dixon C.J. (1979) *Atlas of Economic Mineral Deposits.* Chapman & Hall, London.

Dmitriev L., Barsukov V. & Udintsev G. (1971) Rift zones of the ocean and the problem of ore-formation. *Proc. IMA-IAGOD Meetings'70*, 65–9. Spec. Issue 3 (IAGOD Vol.), Soc. Min. Geol. Jpn.

Doe B.R. & Delavaux M.H. (1972) Source of lead in southeast Missouri galena ores. *Econ. Geol.*, **67**, 409–25.

Donaldson M.J., Lesher C.M., Groves D.I. & Gresham J.J. (1986) Comparison of Archean dunites and komatiites associated with nickel mineralization in Western Australia: implications for dunite genesis. *Miner. Deposita*, **21**, 296–305.

Dostal J. & Dupuy C. (1987) Gold in late Proterozoic andesites from northwest Africa. *Econ. Geol.*, **82**, 762–6.

Duhovnik J. (1967) Facts for and against a syngenetic origin of the stratiform ore deposits of lead and zinc. In Brown J.S. (ed.), *Genesis of Stratiform Lead–Zinc–Barite–Fluorite Deposits*, 108–25, Econ. Geol. Monogr. 3.

Duke J.M. (1983) Ore deposit models 7, magmatic segregation deposits of chromite. *Geosci. Can.*, **10**, 15–24.

Dulski P., Möller P., Villalpando A. & Schneider H.J. (1982) Correlation of trace element fractionation in cassiterites with the genesis of the Bolivian Metallotect. In Evans A.M. (ed.), *Metallization Associated with Acid Magmatism*, 71–83, Wiley, Chichester.

Dunham A.C. (1986) Mineral resources on Rhum. In Nesbitt R.W. & Nichol I. (eds) *Geology in the Real World—the Kinglsey Dunham Volumes*, 93–9. Instn Min. Metall., London.

Dunham K.C. (1959) Non-ferrous mining potentialities of the northern Pennines. In *Future of Non-ferrous Mining in Great Britain and Ireland*, 115–147. Instn Min. Metall., London.

Dunn J.R. (1989) Portland Cement raw materials. In Carr D.D. & Herz N. (eds), *Concise Encyclopedia of Mineral Resources*, 273–8. Pergamon, Oxford.

Dunnet D. (1971) Some aspects of the Panantarctic cratonic margin in Australia. *Philos. Trans. R. Soc. London*, **A280**, 641–54.

Dunsmore H.E. (1973) Diagenetic processes of lead–zinc emplacement in carbonates. *Trans. Instn Min. Metall. (Sect. B: Appl. Earth Sci.)*, **82**, B168–B173.

Dunsmore H.E. & Shearman D.J. (1977) Mississippi Valley-type lead–zinc orebodies: a sedimentary and diagenetic origin. In *Proceedings of the Forum on Oil and Ore in Sediments*, 189–205. Geology Dept., Imperial College, London.

Dupré B., Blanc G., Boulègue J. & Allègre C. J. (1988) Metal remobilization at a spreading centre studied using lead isotopes. *Nature*, **333**, 165–7.

Durocher M.E. (1983) The nature of hydrothermal alteration associated with the Madsen and Starratt–Olsen gold deposits, Red Lake Area. In Colvine A.C. (ed.), *The Geology of Gold in Ontario*, 123–40, Geol. Surv. Ontario, Misc. Pap., **110**.

Du Toit A.L. (1954) *The Geology of South Africa.* Oliver and Boyd, Edinburgh.

Dybdahl I. (1960) Ilmenite deposits of the Egersund Anorthosite Complex. In Vokes F. (ed.), *Mines in South and Central Norway*, Guide to Excursion No. C10. 21st Int. Geol. Congr., Norden.

Eadington P.J. (1983) A fluid inclusion investigation of ore formation in a tin-mineralized granite, New England, New South Wales. *Econ. Geol.*, **78**, 1204–21.

Eales H.V. & Reynolds I.M. (1985) Cryptic variations within chromites of the Upper Critical Zone, Northwestern Bushveld Complex. *Can. Mineral.*, **23**, 302.

Easthoe C.J. (1978) A fluid inclusion study of the Panguna porphyry copper deposit, Bougainville, Papua New Guinea. *Econ Geol.*, **73**, 721–42.

Eckstrand O.R. (ed.) (1984) *Canadian Mineral Deposit Types: a Geological Synopsis*, 39–42, Geol. Surv. Canada Economic Report **36**, Ottawa.

Edwards A.B. (1952) The ore minerals and their textures. *J. Proc. R. Soc. New South Wales*, **85**, 26–46.

Edwards A.B. (1960) *Textures of the Ore Minerals and their Significance.* Australas. Inst. Min. Metall., Melbourne.

Edwards A.B., Baker G. & Callow K.J. (1956) Metamorphism and metasomatism at King Island scheelite mine. *J. Geol. Soc. Aust.*, **3**, 55–98.

Edwards A.B. & Lyon R.J.P. (1957) Mineralization at Aberfoyle Tin Mine, Rossarden, Tasmania. *Proc. Australas, Inst. Min. Metall.*, **181**, 93–145.

Edwards R. & Atkinson K. (1986) *Ore Deposit Geology.* Chapman & Hall, London.

Eilenberg S. & Carr M.J. (1981) Copper contents of lavas from active volcanoes in El Salvador and adjacent regions in Central America. *Econ. Geol.*, **76**, 2246–8.

Einaudi M.T. (1982) Description of skarns associated with porphyry copper plutons and general features and origins of skarns associated with porphyry copper plutons. In Titley S.R. (ed.), *Advances in Geology of the Porphyry Copper Deposits*, 139–209. Univ. Ariz. Press, Tucson.

Einaudi M.T. & Burt D.M. (1982) Introduction—terminology, classification and composition of skarn deposits. *Econ. Geol.*, **77**, 745–54.

Einaudi M.T., Meinert L.D. & Newberry R.J. (1981) skarn deposits. *Econ Geol.*, **75th Anniv. Vol.**, 317–91.

Eldridge C.S., Barton P.B. Jr & Ohmoto H. (1983) Mineral textures and their bearing on formation of the Kuroko orebodies. In Ohmoto H. & Skinner B.J. (eds), *Kuroko and Related Volcanogenic Massive Sulphide Deposits*, Econ. Geol. Monogr., *5*, 241–81.

Elliott-Meadows S.R. & Appleyard E.C. (1991) The alteration geochemistry and petrology of the Lar Cu–Zn deposit, Lynn Lake Area, Manitoba, Canada. *Econ. Geol.*, **86**, 486–505.

Ellis A.J. (1979) Explored geothermal systems. In Barnes H.L. (ed.), *Geochemistry of Hydrothermal Ore Deposits*, 2nd edn, 632–83. Wiley, New York.

Ellison W. & Makansi J. (1990) World by-product gypsum utilization. *Ind. Miner.* 270, 33–45.

Elmore R.D. (1984) The Copper Harbor Conglomerate: a late Precambrian fining-upward alluvial fan sequence in northern Michigan. *Bull. Geol. Soc. Am.*, *95*, 610–7.

El Shazly, E.M., Webb J.S. & Williams D. (1957) Trace elements in sphalerite, galena and associated minerals from the British Isles. *Trans. Instn Min. Metall.*, *66*, 241–71.

Eriksson S.C. (1989) Phalaborwa: a saga of magmatism, metasomatism and miscibility. In Bell, K. (ed.) *Carbonatites*, 221–54. Unwin Hyman, London.

Eriksson K.A. & Truswell J.F. (1978) Geological processes and atmospheric evolution in the Precambrian. In Tarling D.H. (ed.), *Evolution of the Earth's Crust*, 219–38, Academic Press, London.

Etminan H. & Hoffmann C.F. (1989) Biomarkers in fluid inclusions: a new tool in constraining source regimes and its implications for the genesis of Mississippi Valley-type Deposits. *Geology*, *17*, 19–22.

Evans A.M. (1962) Geology of the Bicroft Uranium Mine, Ontario. PhD thesis, Queen's University, Kingston, Ontario.

Evans A.M. (1966) The development of *Lit-par-lit* Gneiss at the Bicroft Uranium Mine, Ontario. *Can. Mineral.*, *8*, 593–609.

Evans A.M. (1975) Mineralization in geosynclines—the Alpine enigma. *Miner. Deposita*, *10*, 254–60.

Evans A.M. (1976a) Genesis of Irish base-metal deposits. In Wolf K.H. (ed.), *Handbook of Strata-bound and Stratiform Deposits, Vol. 5*, 231–55. Elsevier, Amsterdam.

Evans A.M. (1976b) Mineralization in geosynclines. In Wolf K.H. (ed.), *Handbook of Strata-bound and Stratiform Deposits, Vol. 4*, 1–29. Elsevier, Amsterdam.

Evans A.M. (1980) *An Introduction to Ore Geology*, 1st edn. Blackwell Scientific Publications, Oxford.

Evans A.M. (ed.) (1982) *Metallization Associated with Acid Magmatism, Vol. 6*. Wiley, Chichester.

Evans A.M. & El-Nikhely A. (1982) Some palaeomagnetic dates from the West Cumbrian hematite deposits, England. *Trans. Instn Min. Metall. (Sect. B Appl. Earth Sci.)*, *91*, B41–3.

Evans A.M. & Evans N.D.M. (1977) Some preliminary palaeomagnetic studies of mineralization in the Mendip Orefield. *Trans. Instn Min. Metall. (Sect. B. Appl. Earth Sci.)*, *86*, B149–B151.

Evans A.M. & Maroof S.I. (1976) Basement controls on mineralization in the British Isles. *Min. Mag.*, *134*, 401–11.

Evans A.M., Haslam H.W. & Shaw R.P. (1979) Porphyry style copper–molybdenum mineralization in the Ballachulish Igneous Complex, Argyllshire, with special reference to the fluid inclusions. *Proc. Geol. Assoc.*, *91*, 47–51.

Ewers G.R., Ferguson J. & Donnelly T.H. (1983) The Narbarlek uranium deposit, Northern Territory, Australia: some petrologic and geochemical constraints on genesis. *Econ. Geol.*, *78*, 823–37.

Ewers G.R. & Keays R.R. (1977) Volatile and precious metal zoning in the Broadlands geothermal field, New Zealand. *Econ. Geol.*, *72*, 1337–54.

Farr P. (1984) Beryllium. *Mining Annual Review*, *88*. Min. J. (London).

Feiss P.G. (1978) Magmatic sources of copper in porphyry copper deposits. *Econ. Geol.*, *73*, 397–404.

Feiss P.G. (1989) Metallogenic provinces. In Carr D.D. & Herz N. (eds), *Concise Encyclopedia of Mineral Resources*, 2996–3000. Pergamon Press, Oxford.

Ferguson J. (1980a) Tectonic setting and palaeogeotherms of kimberlites with particular emphasis on southeastern Australia. In Glover J.E. & Groves D.I. (eds), *Kimberlites and Diamonds*, 1–14. Extension Services, University of Western Australia, Nedlands.

Ferguson J. (1980b) Kimberlite and kimberlitic intrusives of southeastern Australia. *Mineral. Mag.*, *43*, 727–31.

Finger F. & Steyrer H.P. (1990) I-type granitoids as indicators of a late Paleozoic convergent ocean–continent margin along the southern flank of the central European Variscan Orogen. *Geology*, *18*, 1207–10.

Fischer, G.J. & Paterson M.S. (1985) Dilatancy and permeability in rock during deformation at high temperature and pressure. *Eos (Trans. Am. Geophys. Union)*, *46*, 1065.

Fleischer V.D., Garlick W.G. & Haldane R. (1976) Geology of the Zambian Copperbelt. In Wolf K.H. (ed.), *Handbook of Strata-Bound and Stratiform Ore Deposits, Vol. 6*, 223–352. Elsevier, Amsterdam.

Fletcher C.J.N. (1984) Strata-bound, vein and breccia-pipe tungsten deposits of South Korea. *Trans. Instn Min. Metall. (Sect. B: Appl. Earth Sci.)*, *93*, B176–84.

Fletcher K. & Couper J. (1975) Greenvale nickel laterite, North Queensland. In Knight C.L. (ed.), *Economic Geology of Australia and Papua New Guinea, I. Metals*, 995–1001. Australas. Inst. Min. Metall., Parkville.

Fonteilles M., Soler P., Demange M., Derré C., Krier-Schellen A.D., Verkaeren J., Guy B. & Zahm A. (1989) The scheelite skarn deposit of Salau, (Ariège, French Pyrenees). *Econ. Geol.*, *84*, 1172–209.

Force E.R. (1991) *Geology of Titanium-mineral Deposits*. Spec. Pap. 259, Geol. Soc. Am., Boulder, Colorado.

Force E.R. & Cannon W.F. (1988) Depositional models for shallow-marine manganese deposits around black shale basins. *Econ. Geol.*, *83*, 93–117.

Förster H. & Jafarzadeh A. (1984) The Chador Malu iron ore deposit (Bafq District, Central Iran)—magnetite filled pipes. *Neues Jahrb. Geol. Palaeontol. Abhandlungen*, **168**, 524–34.

Forsythe D.L. (1971) Vertical zoning of gold–silver tellurides in the Emperor gold mine, Fiji. *Proc. Australas. Inst. Min. Metall.*, **240**, 25–31.

Foster R.P. (ed.) (1984) *GOLD '82: The Geology, Geochemistry and Genesis of Gold Deposits*. Balkema, Rotterdam.

Fournier R.O. (1967) The porphyry copper deposit exposed in the Liberty open-pit mine near Ely, Nevada. Part II. The formation of hydrothermal alteration zones. *Econ. Geol.*, **62**, 207–27.

Fowler A.D. & Doig R. (1983) The age and origin of Grenville Province uraniferous granites and pegmatites *Can. J. Earth Sci.*, **20**, 92–104.

Fox D.J. (1991) Bolivia. *Mining Annual Review*, 51–3. Min. J. (London).

Fox J.S. (1984) Besshi-type volcanogenic sulphide deposits—a review. *CIM. Bull.*, **77** (April), 57–68.

Fralick P.W., Barrett T.J., Jarvis K.E., Jarvis I., Schieders B.R. & Kemp R.V. (1989) Sulphide-facies iron formation at the Archaean Morley occurrence, western Ontario: contrasts with hydrothermal oceanic deposits. *Can. Mineral.*, **27**, 601–16.

Francis P.W., Halls C. & Baker M.C.W. (1983) Relationships between mineralization and silicic volcanism in the central Andes. *Geotherm. J. Volc. Res.*, **18**, 165–90.

Franklin J.M., Lydon J.W. & Sangster D.F. (1981) Volcanic-associated massive sulphide deposits. *Econ. Geol.*, **75th Anniv. Vol.**, 485–627.

Frater K.M. (1985) Mineralization at the Golden Grove Cu–Zn deposit, Western Australia. *Can. J. Earth Sci.*, **22**, 1–26.

Freas R.C. (1989) Lime. In Carr, D.D. & Herz, N. (eds), *Concise Encyclopedia of Mineral Resources*, 189–92. Pergamon, Oxford.

Frietsch R. (1978) On the magmatic origin of iron ores of the Kiruna Type. *Econ. Geol.*, **73**, 478–85.

Fripp R.E.P. (1976) Stratabound gold deposits in Archaean banded iron-formation, Rhodesia. *Econ Geol.*, **71**, 58–75.

Frutos J. (1982) Andean metallogeny related to the tectonic and petrologic evolution of the Cordillera. Some remarkable points. In Amstutz G.C., El Goresy A., Frenzel G., Khuth C., Moh G., Wauschkuhn A. & Zimmerman R.A. (eds.), *Ore Genesis: the State of the Art*, 493–507. Springer-Verlag, Berlin.

Fulp M.S. & Renshaw J.L. (1985) Volcanogenic-exhalative tungsten mineralization of Proterozoic age near Santa Fe, New Mexico, and implications for exploration. *Geology*, **13**, 66–9.

Fyfe W.S. & Henley R.W. (1973) Some thoughts on chemical transport processes, with particular reference to gold. *Mineral. Sci. Eng.*, **5**, 295–303.

Gaál G. & Teixeira J.B.G. (1988) Brazilian gold: metallogeny and structures. *Krystalinikum*, **19**, 43–58.

Gandhi S.S. (1978) Geological setting and genetic aspects of uranium occurrences in the Kaipokok Bay–Big River Area, Labrador. *Econ. Geol.*, **73**, 1492–522.

Garlick W.G. (1989) Genetic interpretation from ore relations to algal reefs in Zambia and Zaïre. In Boyle R.W., Brown A.C., Jefferson C.W., Jowett E.C. & Kirkham R.V. (eds), *Sediment-hosted Stratiform Copper Deposits*, 471–98, Geol. Assoc. Can. Spec. Pap. 36, Memorial Univ., Newfoundland.

Garrels R.M. (1987) A model for the deposition of the microbanded Precambrian iron formations. *Am. J. Sci.*, **287**, 81–106.

Garrels R.M. (1988) Reply. *Am. J. Sci.*, **288**, 669–73.

Garven G. (1985) The role of regional fluid flow in the genesis of the Pine Point Deposit, Western Canada Sedimentary Basin. *Econ. Geol.*, **80**, 307–24.

Gee R.D. (1975) Regional geology of the Archaean nucleii of the Western Australian Shield. In Knight C.L. (ed.), *Economic Geology of Australia and Papua New Guinea, 1. Metals*, Mon. 5, 43–55. Australas. Inst. Min. Metall., Parkville.

Gentilhomme P. (1983) L'évolution des teneurs moyennes des minerais exploités: rétrospective, mécanismes et perspectives. *Chron. Rech. Min.*, **473**, 49–65.

George E., Pagel M. & Dusausoy Y. (1983) α-U_3O_7 à Brousse Broquès (Avreyon, France): une forme quadratique exceptionnelle parmi les oxydes d'uranium naturels. *Terra Cognita*, **3**, 173.

Gervilla F. & Leblanc M. (1990) Magmatic ores in high-temperature Alpine-type lherzolite massifs. *Econ. Geol.*, **85**, 112–32.

Ghazban F., Scharcz H.P. & Ford D.C. (1990) Carbon and sulfur isotope evidence for *in situ* reduction of sulfate, Nanisivik lead–zinc deposits, Northwest Territories, Baffin Island, Canada. *Econ. Geol.*, **85**, 360–75.

Gibson P.C., Noble D.C. & Larson L.T. (1990) Multistage evolution of the Calera epithermal Ag–Au vein system, Orcopampa District, Southern Peru: first results. *Econ. Geol.*, **85**, 1504–19.

Gilluly J. (1932) Geology and ore deposits of the Stockton and Fairfield Quadrangles, Utah. *Prof. Pap.* **173**, *U.S. Geol. Surv.*, Washington.

Gilluly J., Waters A.C. & Woodford A.O. (1959) *Principles of Geology*. Freeman, San Francisco.

Gilmour P. (1982) Grades and tonnages of porphyry copper deposits. In Titley S.R. (ed.), *Advances in Geology of the Porphyry Copper Deposits of Southwestern North America*, 7–36, Univ. of Arizona Press, Tucson.

Gimpel J. (1988) *The Medieval Machine*. Wildwood House, Aldershot.

Ginsburg A.I., Timofeyev I.N. & Feldman L.G. (1979) *Principles of the Geology of Granitic Pegmatites*, Nedra, Moscow, in Russian. (Some aspects of this book are covered in Černý 1982a & 1982c).

Giordano T.H. (1985) A preliminary evaluation of organic

ligands and metal–organic complexing in Mississippi Valley-type ore solutions. *Econ. Geol.*, **80**, 96–106.

Giordano T.H. & Barnes H.L. (1981) Lead transport in Mississippi Valley-type ore solutions. *Econ. Geol.*, **76**, 2200–11.

Gize A.P. & Barnes H.L. (1987) The organic geochemistry of two Mississippi Valley-type lead–zinc deposits. *Econ. Geol.*, **82**, 457–70.

Glasby G.P. (1988) Manganese deposition through geological time: dominance of the post-Eocene deep-sea environment. *Ore Geol. Rev.*, **4**, 135–144.

Glazkovsky A.A., Gorbunov G.I. & Sysoev F.A. (1977) Deposits of nickel. In Smirnov V.I. (ed.), *Ore Deposits of the USSR, Vol. II*, 3–79, Pitman, London.

Gocht W.R., Zantop H. & Eggert R.G. (1988) *International Mineral Economics.* Springer-Verlag, Berlin.

Gold D.P. (1984) A diamond exploration philosophy for the 1980s. *Earth Miner. Sci. Pennsylvania State Univ.*, **53**, 37–42.

Gold D.P., Vallee M. & Charette J.P. (1967) Economic geology and geophysics of the Oka Alkaline Complex, Quebec. *Can. Min. Metall. Bull.*, **60**, 1131–44.

Goldich S.S. (1973) Ages of Precambrian banded iron formations. *Econ. Geol.*, **68**, 1126–34.

Goldsmith J.R. & Newton R.C. (1969) *P–T–X* relations in the system CaCO$_3$–MgCO$_3$ at high temperature and pressures. *Am. J. Sci.*, **267A**, 160–90.

Gole M.J. (1981) Archaean banded iron formations, Yilgarn Block, Western Australia. *Econ. Geol.*, **76**, 1954–74.

Golightly J.P. (1981) Nickeliferous laterite deposits. *Econ Geol.*, **75th Anniv., Vol.**, 710–35.

Goodwin A.M. (1973) Archaean iron-formations and tectonic basins of the Canadian Shield. *Econ. Geol.*, **68**, 915–33.

Goodwin A.M. (1982) Distribution and origin of Precambrian banded iron formations. *Rev. Bras. Geosci.*, **12**, 457–62.

Gouanvic Y. & Gagny C. (1983) Étude d'une aplopegmatite litée à cassiterite et wolframite, magma différencié de l'endogranite de la mine de Santa Comba, (Galice, Espagne). *Bull. Soc. Geol. Fr., Ser. 7*, **25**, 335–48.

Grant J.A. (1986) The isocon diagram—a simple solution to Gresens' equation for metasomatic alteration. *Econ. Geol.*, **81**, 1976–82.

Grant J.N., Halls C., Avila W. & Avila G. (1977) Igneous geology and the evolution of hydrothermal systems in some sub-volcanic tin deposits of Bolivia. In *Volcanic Processes in Ore Genesis*, 117–126, Spec. Pub. No. 7, Geol. Soc. London.

Grant J.N. Halls C., Sheppard S.M.F. & Avila W. (1980) Evolution of the porphyry tin deposits of Bolivia. *Min. Geol.* Special Issue, **8**, 151–73.

Graybeal F.T. & Smith D.M. (1988) Regional distribution of silver deposits on the Pacific Rim. In *Silver– Exploration, Mining and Treatment*, 3–10, Instn Min. Metall., London.

Green C.J.B. (1989) Metal markets and the LME. *Mining. Annual Review*, C5–9. Min. J. (London).

Gregory P.W. & Robinson B.W. (1984) Sulphur isotope studies of the Mt. Molloy, Dianne and O.K. stratiform sulphide deposits, Hodgkinson Province, North Queensland, Australia. *Miner. Deposita*, **19**, 36–43.

Griffiths J. (1989) Olivine. *Ind. Miner.*, **256**, 25–35.

Griffiths J.R. & Godwin C.I. (1983) Metallogeny and tectonics of porphyry copper–molybdenum deposits in British Columbia. *Can. J. Earth Sci.*, **20**, 1000–18.

Gross G.A. (1965) Geology of iron deposits in Canada, I. General geology and evaluation of iron deposits. *Geol. Surv., Can. Econ. Geol. Rep.*, **22**.

Gross G.A. (1970) Nature and occurrence of iron ore deposits. In *Survey of World Iron Ore Resources*, 13–31, United Nations, New York.

Gross G.A. (1980) A classification of iron formations based on depositional environments. *Can. Mineral.*, **18**, 215–22.

Gross G.A. (1986) The metallogenic significance of iron-formation and related stratifer rocks. *J. Geol. Soc. India*, **28**, 92–108.

Gross G.A. (1991) Genetic concepts for iron-formation and associated metalliferous sediments. *Econ. Geol. Monogr.* **8**, 51–81.

Groves D.I. & Foster R.P. (1991) Archaean lode gold deposits. In Foster, R.P. (ed.), *Gold Metallogeny and Exploration*, 63–103. Blackie, Glasgow.

Groves D.I. & Hudson D.R. (1981) The nature and origin of Archaean strata-bound volcanic-associated nickel–iron–copper sulphide deposits. In Wolf K.H. (ed.), *Handbook of Strata-bound and Stratiform Ore Deposits*, 305–410. Elsevier, Amsterdam.

Groves D.I. & Lesher C.M. (eds) (1982) *Regional Geology and Nickel Deposits of the Norseman–Wiluna Belts, Western Australia.* Geology Department and Extension Service, Univ. of Western Australia, East Perth.

Groves D.I. & McCarthy T.S. (1978) Fractional crystallization and the origin of tin deposits in granitoids. *Miner. Deposita*, **13**, 11–26.

Groves D.I. & Solomon M. (1969) Fluid inclusion studies at Mount Bischoff, Tasmania, *Trans. Instn Min. Metall. (Sect. B: Appl. Earth Sci.)*, **78**, B1–B11.

Groves D.I., Solomon M. & Rafter T.A. (1970) Sulphur isotope fractionation and fluid inclusion studies at the Rex Hill Mine, Tasmania. *Econ. Geol.*, **65**, 459–69.

Groves D.I., Lesher C.M. & Gee R.D. (1984) Tectonic setting of sulphide nickel deposits of the Western Australian shield. In Buchanan D.L. & Jones M.J. (eds), *Sulphide Deposits in Mafic and Ultramafic Rocks*, 1–13. Spec. Pub. Instn Min. Metall., London.

Groves D.I., Marchant T., Maske S. & Cawthorn R.G. (1986) Compositions of ilmenites in Fe–Ni–Cu sulfides and host rocks, Insizwa, Southern Africa: proof of coexisting immiscible sulfide and silicate liquids. *Econ. Geol.*, **81**, 725–31.

Groves D.I., Ho S.E., Rock N.M.S., Barley M.E. &

Muggeridge M.T. (1987) Archean cratons, diamond and platinum: evidence for coupled long-lived crust–mantle systems. *Geology*, **15**, 801–5.

Grubb P.L.C. (1973) High-level and low-level bauxitization: a criterion for classification. *Miner. Sci. Eng.*, **5**, 219–31.

Guild P.W. (1978) Metallogenesis in the western United States. *J. Geol. Soc. London*, **135**, 355–76.

Gurney J.J. (1985) A correlation between garnets and diamonds in kimberlites. In Glover J.E. & Harris P.G. (eds), *Kimberlite Occurrence and Origin: a Basis for Conceptual Models in Exploration*, 143–66, Geology Department and University Extension, University of Western Australia, Nedlands.

Gurney J.J. (1990) The diamondiferous roots of our wandering continent. *S. Afr. Geol.*, **93**, 424–37.

Gustafson L.B. & Curtis L.W. (1983) Post-Kombolgie metasomatism at Jabiluka, Northern Territory, Australia, and its significance in the formation of high grade uranium mineralization in Lower Proterozoic rocks. *Econ. Geol.*, **78**, 26–56.

Gustafson L.B. & Hunt J.P. (1975) The porphyry copper deposit at El Salvador, Chile. *Econ. Geol.*, **70**, 857–912.

Gustafson L.B. & Williams N. (1981) Sediment-hosted stratiform deposits of copper, lead and zinc. *Econ. Geol.*, **75th Anniv. Vol.**, 139–78.

Haas, J.L. Jr (1971) The effect of salinity on the maximum thermal gradient of a hydrothermal system at hydrostatic pressure. *Econ. Geol.*, **66**, 940–6.

Haase C.S. (1982) Phase equilibria in metamorphosed iron formations: qualitative $T-X(CO_2)$ petrogenetic grids. *Am. J. Sci.*, **282**, 1623–54.

Hagner A.F. & Collins L.G. (1967) Magnetite ore formed during regional metamorphism. *Econ. Geol.*, **62**, 1034–71.

Halbach P. & 17 co-authors (1989) Probable modern analogue of Kuroko-type massive sulphide deposits in the Okinawa Trough back-arc basin. *Nature*, **438**, 496–9.

Hall A. (1990) Geochemistry of the Cornubian tin province. *Miner. Deposita*, **25**, 1–6.

Hall A.L. (1932) The Bushveld igneous complex of the Central Transvaal. *Geol. Surv. S. Africa Mem.*, **28**, 560.

Hall K.F. (1988) *The Economic Definition of Ore—Cut-off Grades in Theory and Practice.* Mining Journal Books, London.

Hall W.E., Friedman I. & Nash J.T. (1974) Fluid inclusion and light stable isotope study of the Climax molybdenum deposits, Colorado. *Econ. Geol.*, **69**, 884–901.

Hamilton J.M., Delaney G.D., Hauser R.L. & Ransom P.W. (1983) Geology of the Sullivan Deposit, Kimberley, B.C. Canada. In Sangster D.F. (ed.) *Sediment-hosted Stratiform Lead–Zinc Deposits*, Short Course Handbook, Vol. 8, 31–83. Mineralogical Association of Canada. Victoria.

Hannah J.L. & Stein H.J. (1990) Magmatic and hydrothermal processes in ore-bearing systems. In Stein H.J. & Hannah J.L. *Ore-bearing Granite Systems; Petrogenesis and Mineralizing Processes*, 1–10. Geol. Soc. Am. Spec. Paper 246, Boulder, Colorado.

Hannington M.D. & Scott S.D. (1988) Gold and silver potential of polymetallic sulphide deposits on the sea floor. *Mar. Min.*, **7**, 271–82.

Hanor J.S. (1979) The sedimentary genesis of hydrothermal fluids. In Barnes H.L. (ed.), *Geochemistry of Hydrothermal Ore Deposits*, 137–72, Wiley, New York.

Harben P.W. & Bates R.L. (1984) *Geology of the Nonmetallics.* Metal Bulletin, New York.

Harben P.W. & Bates R.L. (1990) *Industrial Minerals Geology and World Deposits*, Metal Bulletin, London.

Harder E.C. & Creig E.W. (1960) Bauxite. In Gillson J.L. *et al.* (eds), *Industrial Minerals and Rocks*, 65–85, Amer. Inst. Min. Eng., New York.

Harris P.G. (1985) Kimberlite volcanism. In Glover J.E. & Harris P.G. (eds), *Kimberlite Occurrence and Origin: a Basis for Conceptual Models in Exploration*, 125–42. Geology Department and University Extension, University of Western Australia, Nedlands.

Harris T.S. (1986) Extender and filler mineral markets in the western United States and Canada. *CIM Bull.*, **May**, 48–51.

Harvey J. (1985) *Mastering Economics.* Macmillan, Basingstoke.

Hatcher M.I. & Bolitho B.C. (1982) The Greenbushes Pegmatite, South-west Australia. In Černý P. (ed.), *Short Course in Granitic Pegmatites in Science and Industry*, 513–525, Mineral. Assoc. Can. Winnipeg.

Hattori K. & Muehlenbachs K. (1980) Marine hydrothermal alteration at a Kuroko ore deposit, Kosaka, Japan. *Contrib. Mineral. Petrol.*, **74**, 285–92.

Hawley J.E. (1962) The Sudbury ores: their mineralogy and origin. *Can. Mineral*, **7**, i-xiv & 1–207.

Hayba D.O., Bethke P.M., Heald P. & Foley N.K. (1986) Geologic, mineralogic and geochemical characteristics of volcanic-hosted epithermal precious metal deposits. In Berger B.R. & Bethke P.M. (eds), *Geology and Geochemistry of Epithermal Systems*, 129–67. Society of Economic Geologists, El Paso.

Haymon R.M., Koski R.A. & Sinclair C. (1984) Fossils of hydrothermal vent worms from Cretaceous sulphide ores of the Samail Ophiolite, Oman. *Science*, **223**, 1407–9.

Heald P., Foley N.K. & Hayba D.O. (1987) Comparative anatomy of volcanic-hosted epithermal deposits: acid-sulfate and adularia–sericite types. *Econ. Geol.*, **82**, 1–26.

Heaton T.H.E. & Sheppard S.M.F. (1977) Hydrogen and oxygen isotope evidence for sea water-hydrothermal alteration and ore deposition, Troodos Complex, Cyprus. In *Volcanic Processes in Ore Genesis*, Spec. Publ. No. 7, Geol. Soc. London.

Hedström P., Simeonov A. & Malmström L. (1989) The Zinkgruvan ore deposit, south-central Sweden: a Proterozoic, proximal Zn–Pb–Ag deposit in distal volcanic facies. *Econ. Geol.*, **84**, 1235–61.

Heinrich C.A., Andrew A.S., Wilkins R.W.T. & Patterson D.J. (1989) A fluid inclusion and stable isotope study of synmetamorphic copper ore formation at Mount Isa, Australia. *Econ. Geol.*, **84**, 529–50.

Hemley J.J. (1959) Some minerological equilibra in the system K_2O–Al_2O_3–SiO_2–H_2O. *Am. J. Sci.*, **257**, 241–70.

Hemley J.J. & Jones W.R. (1964) Chemical aspects of hydrothermal alteration with emphasis on hydrogen metasomatism. *Econ. Geol.*, **59**, 538–69.

Hendry D.A.F., Chivas A.R., Long J.V.P. & Reed S.J.B. (1985) Chemical difference between minerals from mineralizing and barren intrusion from some North American porphyry copper deposits. *Contrib. Mineral. Petrol.*, **89**, 317–29.

Henley R.W. (1986) The geological framework of epithermal deposits. In Berger B.R. & Bethke P.M. (eds), *Geology and Geochemistry of Epithermal Systems*, 1–24. Society of Economic Geologists, El Paso.

Henley R.W. (1991) Epithermal gold deposits in volcanic terranes. In Foster R.P. (ed.), *Gold Metallogeny and Exploration*, 133–64. Blackie, Glasgow.

Henley R.W. & Adams J. (1979) On the evolution of giant gold placers. *Trans. Instn Min. Metall. (Sect. B: Appl. Earth Sci.)*, **88**, B41–50.

Henley R.W. & Ellis A.J. (1983) Geothermal systems ancient and modern: a geochemical review. *Earth Sci. Rev.*, **19**, 1–50.

Henley R.W., Truesdell A.H., Barton P.B. & Whitney J.A. (1984) *Fluid–Mineral Equilibria in Hydrothermal Systems*, Reviews in Econ. Geol. 1, Soc. Econ. Geol.

Hewitt D.F. (1967) *Pegmatite Mineral Resources of Ontario.* Ont. Dept Mines Industrial Mineral report 21, Toronto.

Heyl A.V. (1969) Some aspects of genesis of zinc–lead–barite–fluorite deposits in the Mississippi Valley, U.S.A. *Trans. Instn Min. Metall. (Sect. B. Appl. Earth Sci.)*, **78**, B148–B160.

Heyl A.V. (1972) The 38th Parallel Lineament and its relationship to ore deposits. *Econ. Geol.*, **67**, 879–94.

Heyl A.V. (1983) Some major lineaments reflecting deep-seated fracture zones in the central United States, and mineral districts related to the zones. *Global Tecton. Metallogeny*, **2**, 75–89.

Heyl A.V., Delevaux M.H., Zartman R.E. & Brock M.R. (1966) Isotopic study of galenas from the Upper Mississippi Valley, the Illinois–Kentucky and some Appalachian Valley mineral districts. *Econ. Geol.*, **61**, 933–61.

Heyl A.V., Landis G.P. & Zartman R.E. (1974) Isotopic evidence for the origin of Mississippi Valley-type mineral deposits: a review. *Econ. Geol.*, **69**, 992–1006.

Highley D.E. (1990) Britain's industrial mineral resources and their exploitation. *Erzmetall*, **43** (1), 19–28.

Highley D.E., Slater D. & Chapman G.R. (1988) Geological occurrence of elements consumed in the electronics industry. *Trans. Instn Min. Metall. (Sect. C)*, **97**, C34–42.

Hill R.E.T. & Gole M.J. (1985) Characteristics of centres of Archaean komatiitic volcanism, exemplified by lithologies in the Agnew Area, Yilgarn Block, Western Australia. *Can. Mineral.*, **23**, 327.

Hills E.S. (1953) Tectonic setting of Australian ore deposits. In Edwards A.B. (ed.), *Geology of Australian Ore Deposits*, 41–61, Australas. Inst. Min. Metall., Melbourne.

Hildebrand R.S. (1986) Kiruna-type deposits: their origin and relationship to intermediate subvolcanic plutons in the Great Bear Magmatic Zone, Northwest Canada. *Econ. Geol.*, **81**, 640–59.

Hitzman M.W., Proffett J.M. Jr, Schmidt J.M. & Smith T.E. (1986) Geology and mineralization of the Ambler District, Northwestern Alaska. *Econ. Geol.*, **81**, 1592–1618.

Hodder R.W. & Petruk W. (1982) *Geology of Canadian Gold Deposits.* Spec. Vol. 24, Can. Inst. Min. Metall., Montreal.

Hodgson C.J. (1986) Place of gold ore formation in the geological development of Abitibi Greenstone Belt, Ontario, Canada. *Trans. Instn Min. Metall. (Sect. B: Appl. Earth Sci.)*, **95**, B183–94.

Hoeve J. (1984) Host rock alteration and its application as an ore guide at the Midwest Lake uranium deposit, Northern Saskatchewan. *Can. Min. Metall. Bull.*, **77** (August), 63–72.

Hoffmann P.F. (1989) Precambrian geology and tectonic history of North America. In Bally, A.W. & Palmer, A.R. *The Geology of North America; An Overview*, 447–512, Geol. Soc. Am., Boulder, Colorado.

Holder M.T. & Leveridge B.E. (1986) Correlation of the Rhenohercynian Variscides. *J. Geol. Soc. London*, **143**, 141–7.

Höll R. (1985) Geothermal systems and active ore formation in the Taupo Volcanic Zone, New Zealand. In Germann K. (ed.), *Geochemical Aspects of Ore Formation in Recent and Fossil Sedimentary Environments*, 55–71. Monogr. Ser. on Mineral Deposits 25, Gebrüder Borntraeger, Berlin.

Höll R. & Maucher A. (1976) The strata-bound ore deposits in the eastern Alps. In Wolf K.H. (ed.), *Handbook of Strata-Bound and Stratiform Ore Deposits, Vol. 5*, 1–36. Elsevier, Amsterdam.

Höll R., Ivanova G. & Grinenko V. (1987) Sulfur isotope studies of the Felbertal scheelite deposit, eastern Alps. *Miner. Deposita*, **22**, 301–8.

Hollister V.F. (1975) An appraisal of the nature of some porphyry copper deposits. *Miner. Sci. Eng.*, **7**, 225–33.

Hollister V.F. (1978) *Geology of the Porphyry Copper Deposits of the Western Hemisphere.* Am. Inst. Min. Metall. and Petrol. Eng., New York.

Hollister V.F., Potter R.R. & Baker A.L. (1974) Porphyry type deposits of the Appalachian Orogen. *Econ. Geol.*, **69**, 618–30.

Holzer H.F. (1986) Austria. In Dunning F.W. & Evans A.M. (eds), *Mineral Deposits of Europe Vol. 3: Central* ubble—

Europe, 15–40. Instn Min. Metall. and Miner. Soc., London.

Horikoshi E. (1990) Opening of the Sea of Japan and Kuroko deposit formation. *Miner. Deposita*, **25**, 140–5.

Horikoshi E. & Sato T. (1970) Volcanic activity and ore deposition in the Kosaka Mine. In Tatsumi T. (ed.), *Volcanism and Ore Genesis*, 181–95. University of Tokyo Press, Tokyo.

Hose H.R. (1960) The genesis of bauxites, the ores of aluminium. *21st Int. Geol. Congr.*, **16**, 237–47.

Hosking K.F.G. (1951) Primary ore deposition in Cornwall. *Trans. R. Geol. Soc. Cornwall*, **18**, 309–56.

Howd F.H. & Barnes H.L. (1975) Ore solution chemistry IV. Replacement of marble by sulphides at 450°C. *Econ. Geol.*, **70**, 968–81.

Hsü K.J. (1972) The concept of the geosyncline, yesterday and today. *Trans. Leicester Lit. Philos. Soc.*, **66**, 26–48.

Hsü K.J. (1984) A nonsteady state model for dolomite, evaporite and ore genesis. In Wauschkuhn A., Kluth C. & Zimmermann R.A. (eds), *Syngenesis and Epigenesis in the Formation of Mineral Deposits*, Springer-Verlag, Berlin.

Huppert H.E. Sparks R.S. (1985) Komatiites I: eruption and flow. *J. Petrology*, **26**, 694–725.

Huston D.L. & Large R.R. (1989) A chemical model for the concentration of gold in volcanogenic massive sulphide deposits. *Ore Geol. Rev.*, **4**, 171–200.

Hutchinson R.W. (1980) Massive base metal sulphide deposits as guides to tectonic evolution. In Strangeway D.W. (ed.), *The Continental Crust and Its Mineral Deposits*, 659–684, Geol. Assoc., Canada Spec. Pap. 20.

Hutchinson R.W. (1983) Mineral deposits, time and evolution. *Proc. Denver Region Exploration Geologists Society Symposium—The Genesis of Rock Mountain Ore Deposits: Changes with Time and Tectonics*, 1–9, Denver.

Hutchinson R.W. (1987) Metallogeny of Precambrian gold deposits: space and time relationships. *Econ. Geol.*, **82**, 1993–2007.

Hutchinson R.W. & Viljoen R.P. (1988) Re-evaluation of gold source in Witwatersrand ores. *S. Afr. J. Geol.*, **91**, 157–73.

Hutchison C.S. (1983) *Economic Deposits and Their Tectonic Setting*. Macmillan, London.

Hutchison M.N. & Scott S.D. (1981) Sphalerite geobarometry in the Cu–Fe–Zn–S system. *Econ. Geol.*, **76**, 143–53.

Huyck H.L.O. & Chorey R.W. (1991) Stratigraphic and petrographic comparison of the Creta and Kupferschiefer copper deposits. *Miner. Deposita*, **26**, 132–42.

Ihlen P.M. Trønnes R. & Vokes F.M. (1982) Mineralization, wall rock alteration and zonation of ore deposits associated with the Drammen Granite in the Oslo Region, Norway. In Evans A.M. (ed.), *Metallization Associated with Acid Magmatism*, 111–36, Wiley, Chichester.

Ikingura J.R., Bell K. & Watkinson D.H. (1989) Hydrothermal alteration and oxygen and hydrogen isotope geochemistry of the D-68 Zone Cu–Zn massive sulfide deposit, Noranda District, Quebec, Canada. *Mineral. Petrol.*, **40**, 155–72.

Ilton E.S. & Veblen D.R. (1988) Copper inclusions in sheet silicates from porphyry copper deposits. *Nature*, **234**, 516–8.

Ineson P.R. (1989) *Introduction to Practical Ore Microscopy*. Longman, Harlow.

Irvine T.N. (1977) Origin of chromitite layers in the Muskoka Intrusion and other stratiform intrusions: a new interpretation. *Geology*, **5**, 273–7.

Irvine T.N. & Smith C.H. (1969) Primary oxide minerals in the layered series of the Muskox Intrusion. *Econ. Geol. Monogr.*, **4**, 76–94.

Ivosevic S.W. (1984) *Gold and Silver Handbook*. Ivosevic, Denver.

Ixer R.A. & Townley R. (1979) The sulphide mineralogy and paragenesis of the South Pennine Orefield, England. *Mercian Geol.*, **7**, 51–64.

Jacks J. (1989) Gold. *Mining Annual Review*, C11–21. Min. J. (London).

Jackson E.D. (1961) Primary textures and mineral associations in the ultramafic zone of the Stillwater Complex, Montana. *U.S. Geol. Surv. Prof. Pap.*, **358**.

Jackson E.D. & Thayer T.P. (1972) Some criteria for distinguishing between stratiform, concentric and Alpine peridotite–gabbro complexes. 24th *Int. Geol. Congr.*, **2**, 289–96.

Jackson N.J., Halliday A.N., Sheppard S.M.F. & Mitchell J.G. (1982) Hydrothermal activity in the St Just Mining District Cornwall, England. In Evans A.M. (ed.), *Metallization Associated with Acid Magmatism*, 137–79, Wiley, Chichester.

Jackson R. (1982) *Ok Tedi: the Pot of Gold*. University of Papua New Guinea.

Jackson S.A. & Beales F.W. (1967) An aspect of sedimentary basin evolution; the concentration of Mississippian Valley-type ores during late stages of diagenesis. *Bull. Can. Petrol. Geol.*, **15**, 383–433.

Jacob R.E., Corner B. & Brynard H.J. (1986) The regional geological and structural setting of the uraniferous granitic provinces of southern Africa. In Anhaeusser C.R. & Maske S. (eds), *Mineral Deposits of Southern Africa*, 2, 1807–18. Geol. Soc. S. Afr., Johannesburg.

Jacobsen J.B.E. (1975) Copper deposits in time and space. *Miner. Sci. Eng.*, **7**, 337–71.

Jacobsen J.B.E. & McCarthy T.S. (1976) The copperbearing breccia pipes of the Messina District, South Africa. *Miner. Deposita*, **11**, 33–45.

Jacobsen S.B. & Pimentel-Klose M.R. (1988) A Nd isotope study of the Hamersley and Michipicoten banded iron formations: the source of REE and Fe in Archaean oceans. *Earth Planet. Sci. Let.*, **87**, 29–44.

Jahns R.H. (1955) The study of pegmatites. *Econ. Geol.*, **50th Anniv. vol.**, 1025–130.

Jahns R.H. (1982) Internal evolution of pegmatite bodies. In Černý P. (ed.), *Short Course in Granitic Pegmatites in*

Science and Industry, 293–327. Mineral. Assoc. Can., Winnipeg.

Jahns R.H. & Burnham C.W. (1969) Experimental studies of pegmatite genesis. *Econ. Geol.*, **64**, 843–64.

James H.L. (1954) Sedimentary facies of iron formation. *Econ. Geol.*, **49**, 235–93.

James H.L. (1955) Zones of regional metamorphism in the Precambrian of northern Michigan. *Bull. Geol. Soc. Am.*, **66**, 1455–88.

James H.L. (1983) Distribution of banded iron formations in space and time. In Trendall A.F. & Morris R.C. (eds), *Iron Formation: Facts and Problems*, 471–86. Elsevier, Amsterdam.

James H.L. & Sims P.K. (1973) Precambrian iron-formations of the world. *Econ. Geol.*, **68**, 913–4.

James H.L. & Trendall A.F. (1982) Banded iron formation: distribution in time and palaeoenvironmental significance. In Holland H.D. & Schidlowski M. (eds), *Mineral Deposits and the Evolution of the Biosphere*, 119–218. Springer-Verlag, Heidelberg.

Jaques A.L. & Ferguson J. (1983) Diamondiferous kimberlitic rocks, West Kimberley. *BMR 83*, 44 (Yearbook of the Bureau of Mineral Resources, Canberra).

Jaques A.L., Lewis J.D., Smith C.B., Gregory G.P., Ferguson J., Chappell B.W. & McCulloch M.T. (1984) The diamond-bearing ultrapotassic (lamproitic) rocks of the West Kimberley Region, Western Australia. In Kornprobst J. (ed.), *Kimberlites I: Kimberlites and Related Rocks*, 225–54. Elsevier, Amsterdam.

Jensen M.L. & Bateman A.M. (1979) *Economic Mineral Deposits*. Wiley, New York.

Jones B. (1990) Diamonds a cut above the real thing. *The Times*, **19 July**, 16.

Jones E.M., Rice C.M. & Tweedie J.R. (1987) Lower Proterozoic stratiform sulphide deposits in Loch Maree Group, Gairloch, northwest Scotland. *Trans. Instn Min. Metall. (Sect. B. Appl. Earth Sci.)*, **96**, B128–40.

Jowett E.C. (1987) Formation of sulphide-calcite veinlets in the Kupferschiefer Cu–Ag deposits in Poland by natural hydrofracturing during basin subsidence. *J. Geol.*, **95** 513–26.

Jowett E.C. (1989) Effects of continental rifting on the location and genesis of stratiform copper–silver deposits. In Boyle R.W., Brown A.C., Jefferson C.W., Jowett E.C. & Kirkham R.V. (eds), *Sediment-hosted Stratiform Copper Deposits*, 53–66. Geol. Assoc. Can. Spec. Pap. 36, Memorial Univ., Newfoundland.

Jowett E.C. (1991) The evolution of ideas about the genesis of stratiform copper–silver deposits. In Hutchinson R.W. & Grauch R.I. (eds), *Historical Perspectives of Genetic Concepts and Case Histories of Famous Discoveries*, 117–32, Econ. Geol. Monogr. **8**, Denver.

Kalogeropoulos S.I., Kilias S.P., Bitzios D.C., Nicolaou M. & Both R.A. (1989) Genesis of the Olympias carbonate-hosted Pb–Zn (Au,Ag) sulphide ore deposit, eastern Chalkidiki Peninsula, northern Greece. *Econ. Geol*, **84**, 1210–34.

Kaplan R.S. (1989) Recycling of metals: technology. In Carr D.D. & Herz N. (eds), *Concise Encyclopedia of Mineral Resources*, 296–304. Pergamon Press, Oxford.

Katz M.B. (1987) Graphite deposits of Sri Lanka: a consequence of granulite facies metamorphism. *Miner. Deposita*, **22**, 18–25.

Kay A. & Strong D.F. (1983) Geologic and fluid controls on As–Sb–Au mineralization in the Moretons Harbour Area, Newfoundland. *Econ. Geol.*, **78**, 1590–604.

Keays R.R. (1982) Palladium and iridium in komatiites and associated rocks: application to petrogenetic problems. In Arndt N.T. & Nisbet E.G. (eds), *Komatiites*, 435–57, George Allen & Unwin, London.

Keays R.R. (1984) Archaean gold deposits and their source rocks: the upper mantle connection. In Foster R.P. (ed.), *Gold '82: The Geology, Geochemistry and Genesis of Gold Deposits*, 17–51, Balkema, Rotterdam.

Keays R.R., Nickel E.H., Groves D.I. & McGoldrick P.J. (1982) Iridium and palladium as discriminants of volcanic-exhalative, hydrothermal and magmatic nickel sulfide mineralization. *Econ. Geol.*, **77**, 1535–47.

Kelly A. (1990) The future of metals. *Miner. Ind. Int.*, **996**, 5–15.

Kelly W.C. & Turneaure F.S. (1970) Mineralogy, paragenesis and geothermometry of the tin and tungsten deposits of the eastern Andes, Bolivia. *Econ. Geol.*, **65**, 609–80.

Keqin X. & Jinchu Z. (1988) Time–space distribution of tin/tungsten deposits in South China and controlling factors of mineralization. In Hutchison, C.S. (ed.), *Geology of Tin Deposits in Asia and the Pacific*, 265–77. Springer-Verlag, Berlin.

Kerrich R. (1986) Fluid transport in lineaments. *Philos. Trans. R. Soc. London*, **A317**, 219–51.

Kerrich R. (1989) Geodynamic setting and hydraulic regimes: shear zone hosted mesothermal gold deposits. In Bursnall J.T. (ed.), *Mineralization and Shear Zones*, 89–128. Short Course Notes, 6, Geol. Assoc. Can., Montreal.

Kerrich R. & Fryer B.J. (1979) Archaean precious metal hydrothermal systems, Dome Mine, Abitibi Greenstone H Belt. II. REE and oxygen isotope relations. *Can. J. Earth Sci.*, **16**, 440–58.

Kesler S.E. (1968) Contact-localized ore formation at the Memé Mine, Haiti. *Econ. Geol.*, **63**, 541–52.

Kesler S.E. (1973) Copper, molybdenum and gold abundances in porphyry copper deposits, *Econ. Geol.*, **68**, 106–112.

Kesler S.E., Gesink J.A. & Haynes F.M. (1989) Evolution of mineralizing brines in the East Tennessee Mississippi Valley-type ore field. *Geology*, **17**, 466–9.

Khitarov N.I., Malinin S.P., Lebedev Ye.B. & Shibayeva N.P. (1982) The distribution of Zn, Cu, Pb and Mo between a fluid phase and a silicate melt of granitic composition at high temperatures and pressures. *Geochem. Int.*, **19** (4), 123–36.

Kimberley M.M. (1989) Exhalative origins of iron formations. *Ore Geol. Rev.*, **5**, 13–145.

King H.F., McMahon D.W. & Bujtor G.J. (1982) A guide to the understanding of ore reserve estimation. *Proc. Australas, Inst. Min. Metall. (Suppl.)*, **281**, 1–21.

Kinnaird J.A. & Bowden P. (1991) Magmatism and mineralization associated with Phanerozoic anorogenic plutonic complexes of the African Plate. In Kampunzu A.B. & Lubala R.T. (eds), *Magmatism in Extensional Structural Settings*, 410–87. Springer-Verlag, Heidelberg.

Kirkham R.V. (1989) Distribution, settings and genesis of sediment-hosted stratiform copper deposits. In Boyle R.W., Brown A.C., Jefferson C.W., Jowett E.C. & Kirkham R.V. (eds), *Sediment-hosted stratiform copper deposits*, 3–38, Geol. Assoc. Can. Spec. Pap. 36, Memorial Univ., Newfoundland.

Klein T.L. & Criss R.E. (1988) An oxygen isotope and geochemical study of meteoric–hydrothermal systems at Pilot Mountain and selected other localities, Carolina Slate Belt. *Econ. Geol.*, **83**, 801–21.

Klemd R. & Hallbauer D.K. (1987) Hydrothermally altered peraluminous Archaean granites as a provenance model for Witwatersrand sediments. *Miner. Deposita*, **22**, 227–35.

Knittel U. & Burton C.K. (1985) Polillo Island (Philippines): molybdenum mineralization in an island arc. *Econ. Geol.*, **80**, 2013–8.

Kontak D.J. & Clark A.K. (1985) Exploration criteria for Sn and W mineralization in the Cordillera Oriental of SE Peru. In Taylor R.P. & Strong D.F. (eds), *Granite-related Mineral Deposits*, 173–8. CIM Geology Division, Halifax, Canada.

Kooiman G.J.A., McLeod M.J. & Sinclair W.D. (1986) Porphyry tungsten–molybdenum orebodies, polymetallic veins and replacement bodies, and tin-bearing greisen zones in the Fire Tower Zone, Mount Pleasant, New Brunswick. *Econ. Geol.*, **81**, 1356–73.

Köppel V. & Schroll E. (1988) Pb-isotope evidence for the origin of lead in strata-bound Pb–Zn deposits in Triassic carbonates of the eastern and southern Alps. *Miner. Deposita*, **23**, 96–103.

Kretschmar U. & Scott S.D. (1976) Phase relations involving arsenopyrite in the system Fe–As–S and their application. *Can. Mineral.*, **14**, 364–86.

Krs M. & Stovičková N. (1966) Palaeomagnetic investigation of hydrothermal deposits in the Jáchymov (Joachimsthal) Region, Western Bohemia. *Trans. Inst. Min. Metall. (Sect. B, Appl. Earth Sci.)*, **75**, B51–B57.

Krupp R.E. & Seward T.M. (1987) The Rotokawa geothermal system, New Zealand: an active epithermal ore-depositing environment. *Econ. Geol.*, **82**, 1109–29.

Krupp R.E. & Seward T.M. (1990) Transport and deposition of metals in the Rotokawa geothermal system, New Zealand. *Miner. Deposita*, **25**, 73–81.

Kucha H. (1982) Platinum-group metals in the Zechstein copper deposits, Poland. *Econ. Geol.*, **77**, 1578–91.

Kuhns R.J. & Baitis H.W. (1987) Preliminary study of the

Turner Albright Zn–Cu–Ag–Au–Co massive sulfide deposit, Josephine County, Oregon. *Econ. Geol.*, **82**, 1362–76.

Kullerud G. (1953) The FeS–ZnS system: a geological thermometer. *Nor. Geol. Tiddskr.*, **32**, 61–147.

Kužvart M. & Böhmer M. (1986) *Prospecting and Exploration of Mineral Deposits*. Academia, Prague.

Kwak T.A.P. (1978a) Mass balance relationships and skarn-forming processess at the King Island scheelite deposits, King Island, Tasmania, Australia. *Am. J. Sci.*, **278**, 943–68.

Kwak T.A.P. (1978b) The conditions of formation of the King Island scheelite contact skarn, King Island, Tasmania, Australia. *Am. J. Sci.*, **278**, 969–99.

Kwak T.A.P. (1986) Fluid inclusions in skarns (carbonate replacement deposits). *J. Metamorphic Geol.*, **4**, 363–84.

Kwak T.A.P. (1987) *W–Sn Skarn Deposits*. Elsevier, Amsterdam.

Lambert I.B., Knutson J., Donnelly T.H. & Etminan H. (1987) Stuart Shelf–Adelaide Geosyncline Copper Province, South Australia. *Econ. Geol.*, **82**, 108–23.

Lamey C.A. (1966) *Metallic and Industrial Mineral Deposit*. McGraw-Hill, New York.

Landis G.P. & Rye R.O. (1974) Geologic, fluid inclusion, and stable isotope studies of the Pasto Buena tungsten-base metal ore deposit, northern Peru. *Econ. Geol.*, **69**, 1025–59.

Lane K.F. (1988) *The Economic Definition of Ore—Cut-off Grades in Theory and Practice*. Mining Journal Books, London.

Lang A.H. (1970) Prospecting in Canada. *Geol. Surv. Canada, Econ. Geol. Rep.*, **7**.

Lapham D.M. (1968) Triassic magnetite and diabase at Cornwall, Pennsylvania. In Ridge J.D. (ed.), *Ore Deposits of the United States 1933–1967, Vol. 1*, 72–94. Am. Inst. Min. Metall. Petrol. Eng., New York.

Large D.E. (1983) Sediment-hosted massive sulphide lead–zinc deposits: an empirical model. In Sangster D.F. (ed.), *Short Course in Sediment-hosted Stratiform Lead-Zinc Deposits*, 1–29, Mineral Assoc. Can., Victoria.

Large R.R., Herrmann W. & Corbett K.D. (1987) Base metal exploration of the Mount Read Volcanics, western Tasmania: Pt. I. Geology and exploration, Elliott Bay. *Econ. Geol.*, **82**, 267–90.

Larter R.C.L., Boyce A.J. & Russell M.J. (1981) Hydrothermal pyrite chimneys from the Ballynoe baryte deposit, Silvermines, County Tipperary, Ireland. *Mineral Deposita*, **16**, 309–18.

Lawrence L.J. (1972) The thermal metamorphism of a pyritic sulphide ore. *Econ. Geol.*, **67**, 487–96.

Laznicka P. (1976) Porphyry copper and molybdenum deposits of the USSR and their plate tectonic settings. *Trans. Instn Min. Metall. (Sect. B: Appl. Earth Sci.)*, **85**, B14-B32.

Laznicka P. (1985) *Empirical Metallogeny*. Elsevier, Amsterdam.

Leach D.L. & Rowan E.L. (1986) Genetic link between

Ouachita Fold Belt tectonism and the Mississippi Valley-type lead-zinc deposits of the Ozarks. *Geology*, **14**, 931–5.

Leake R.C., Fletcher C.J.N., Haslam H.W., Khan B. & Shakirullah (1989) Origin and tectonic setting of stratabound tungsten mineralization within the Hindu Kush, Pakistan. *J. Geol. Soc. London*, **146**, 1003–16.

Leat P.T., Jackson S.E., Thorpe R.S. & Stillman C.J. (1986) Geochemistry of bimodal basalt-subalkaline/peralkaline rhyolite provinces within the southern British Caledonides. *J. Geol. Soc. London*, **143**, 259–73.

Le Bas M.J. (1977) *Carbonatite–Nephelinite Volcanism.* Wiley, London.

Lefond S.J. (ed.) (1983) *Industrial Minerals and Rocks.* Am. Inst. Min. Metall. and Petrol. Eng., New York.

Lehmann B. (1985) Formation of the strata-bound Kellhuani tin deposits, Bolivia. *Miner. Deposita*, **20**, 169–76.

Lehmann B. (1990) *Metallogeny of Tin. Lecture Notes in Earth Sciences, 32.* Springer-Verlag, Berlin.

Lericolais G., Berne S., Hamzah Y., Lallier S., Mulyadi W., Robach F. & Sujitno S. (1987) High-resolution seismic and magnetic exploration for tin deposits in Bangka, Indonesia. *Mar. Min.*, **6**, 9–21.

Leroy J. (1978) The Margnac and Fanay uranium deposits of the La Crouzille District (Western Massif Central, France): geological and fluid inclusion studies. *Econ. Geol.*, **73**, 1611–34.

Leroy J. (1984) Episyénitisation dans le Gisement d'Uranium du Bernardan (Marche): Comparaison avec des gisements similaires du Nord-Ouest du Massif Central Français. *Miner. Deposita*, **19**, 26–35.

Leventhal J.S. (1990) Organic matter and thermochemical sulfate reduction in the Viburnum Trend, southeast Missouri. *Econ. Geol.*, **85**, 622–32.

Levin E.M., Robbins C.R. & McMurdie H.F. (1969) *Phase Diagrams for Ceramicists.* Am. Ceram. Soc., Columbus, Ohio.

Lin I.J. (1989) Vermiculite and perlite. *Ind. Miner.*, **262**, 43–9.

Lincoln T.N. (1981) The redistribution of copper during low-grade metamorphism of the Karmutsen Volcanics, Vancouver Island, British Columbia. *Econ. Geol.*, **76**, 2147–61.

Lindberg P.A. (1985) A volcanogenic interpretation for massive sulphide origin, West Shasta District, California. *Econ. Geol.*, **80**, 2240–54.

Lindgren H. (1985) Diagenesis and primary migration in Upper Jurassic Claystone source rocks in the North Sea. *Am. Asso. Petrol. Geol.*, **69**, 525–36.

Lindgren W. (1913) (second edition, 1933). *Mineral Deposits.* McGraw-Hill, New York.

Lindgren W. (1922) A suggestion for the terminology of certain mineral deposits. *Econ. Geol.*, **17**, 292–4.

Lindgren W. (1924) Contact metamorphism at Bingham, Utah. *Geol. Soc. Am. Bull.*, **35**, 507–34.

Lofty G.J., Hillier J.A., Burton E.M., Mitchell D.C., Cooke S.A. & Linley K.A. (1989) *World Mineral Statistics.* British Geological Survey, Keyworth.

London D. (1984) Experimental phase equilibria in the system $LiAlSiO_4$–SiO_2–H_2O: a petrogenetic grid for lithium-rich pegmatites. *Am. Mineral.*, **69**, 995–1004.

London D. (1986) Holmquistite as a guide to pegmatitic rare metal deposits. *Econ. Geol.*, **81**, 704–12.

London D. (1987) Internal differentiation of rare-element pegmatites: effects of boron, phosphorus and fluorine. *Geochim. Cosmochim. Acta*, **51**, 403–20.

London D. (1990) Internal differentiation of rare-element pegmatites; a synthesis of recent research. In Stein H.J. & Hannah J.L. (eds), *Ore-bearing Granite Systems; Petrogenesis and Mineralizing Processess,* 35–50. Spec. Pap. 246, Geol. Soc. Am., Boulder, Colorado.

London D., Morgan G.B.VI & Hervig R.L. (1989) Vapor-undersaturated experiments with Macusani glass + H_2O at 200 MPa and the internal differentiation of granitic pegmatites. *Contrib. Mineral Petrol.*, **102**, 1–17.

Lonsdale P. & Becker K. (1985) Hydrothermal plumes, hot springs and conductive heat flow in the Southern Trough of Guaymas Basin. *Earth Planet. Sci. Lett*, **73**, 211–25.

Lowell J.D. (1974) Regional characteristics of porphyry copper deposits of the Southwest. *Econ. Geol.*, **69**, 601–17.

Lowell J.D. & Guilbert J.M. (1970) Lateral and vertical alteration mineralization zoning in porphyry ore deposits. *Econ. Geol.*, **65**, 373–408.

Lumbers S.B. (1979) The Grenville Province of Ontario, 5th Ann. Meeting Int. Union Geol. Sci., Subcomm. Precamb. Stratigraphy. *Geol. Surv. Minnesota Univ. Minnesota Guidebook Ser.*, **13**, 1–35.

Lund K., Snee L.W. & Evans K.V. (1986) Age and genesis of precious metal deposits, Buffalo Hump District, Central Idaho: implications for depth of emplacement of quartz veins. *Econ. Geol.*, **81**, 990–6.

Lusk J., Campbell F.A. & Krouse H.R. (1975) Application of sphalerite geobarometry and sulphur isotope geothermometry to ores of the Quemont Mine, Noranda, Quebec. *Econ. Geol.*, **70**, 1070–83.

Lydon J.W. (1989) Volcanogenic massive sulphide deposits parts 1 & 2. In Roberts R.G. & Sheahan P.A. (eds) *Ore Deposit Models,* 145–81. Geol. Assoc. Can., Memorial Univ., Newfoundland.

Lyons J.L. (1988) Volcanogenic iron oxide deposits, Cerro de Mercado and Vicinity, Durango, Mexico. *Econ. Geol.*, **83**, 1886.

Macdonald E.H. (1983) *Alluvial Mining.* Chapman & Hall, London.

Macfarlane A.W. & Petersen U. (1990) Pb isotopes of the Hualgayoc area, northern Peru: implications for metal provenance and genesis of a cordilleran polymetallic mining district. *Econ. Geol.*, **85**, 1303–27.

Machamer J.F. (1987) A working classification of manganese deposits. *Min. Mag.*, **157**, 348–51.

Madu, B.E., Nesbitt B.E. & Muehlenbachs K. (1990) A

mesothermal gold–stibnite–quartz vein occurrence in the Canadian Cordillera. *Econ. Geol.*, **85**, 1260–8.

Mainwaring P.R. & Naldrett A.J. (1977) Country rock assimilation and the genesis of Cu–Ni sulphides in the Water Hen Intrusion, Duluth Complex, Minnesota. *Econ. Geol.*, **72**, 1269–84.

Malyutin R.S. & Sitkovskiy I.N. (1968) Structural features of the Gyumushlug lead–zinc deposit. *Geol. Rudn. Mestorozhd*, **10**, 96–9. (In Russian.)

Manning D.A.C. (1984) Volatile control of tungsten partitioning in granitic melt–vapour systems. *Trans. Instn Min. Metall. (Sect. B Appl. Earth Sci.)*, **93**, B185–94.

Manning D.A.C. (1986) Contrasting types of Sn–W mineralization in Peninsular Thailand and SW England. *Miner. Deposita*, **21**, 44–52.

Marcoux E. (1982) Étude géologique et métallogénique du district plombozincifère de Pontivy (Massif armoricain, France). *Bull. Bur. Rech. Geol. Minères* (2), Sect. II, (**1**), 1–24.

Mariano A.N. (1989) Economic geology of rare earth minerals. In Lipin B.R. & McKay G.A. (eds), *Geochemistry and Mineralogy of Rare Earth Elements, Reviews in Mineralogy, 21*, 309–38. Mineral. Soc. Am., Washington.

Marmont S. (1983) The role of felsic intrusions in gold mineralization. In Colvine A.C. (ed.), *The Geology of Gold in Ontario*, 38–47, Geol. Surv. Ont. Misc. Pap., **110**.

Marmont S. (1989) Unconformity-type uranium deposits. In Roberts R.G. & Sheahan P.A. (eds), *Ore Deposit Models*, 103–15. Geol. Assoc. Can., Memorial Univ., Newfoundland.

Martin J.E. & Allchurch P.D. (1975) Perservance nickel deposit, Agnew. In Knight C.L. (ed.), *Economic Geology of Australia and Papua New Guinea–1, Metals*, Monogr. 5, 149–155. Australas. Inst. Min. Metall., Parkville.

Martin P.L. (1966) Structural analysis of the Chisel Lake Orebody. *CIM*, Bull., **59**, 630–36.

Martini J.E.J. (1986) Stratiform gold mineralization in palaeosol and ironstone of early Proterozoic age, Transvaal Sequence, South Africa. *Miner. Deposita*, **21**, 306–12.

Mason A.A.C. (1953) The Vulcan tin mine. In Edwards A.B. (ed.), *Geology of Australian Ore Deposits*, 718–721, Australas. Inst. Min. Metall., Melbourne.

Mason D.R. & Feiss P.G. (1979) On the relationship between whole rock chemistry and porphyry copper mineralization. *Econ. Geol.*, **74**, 1506–10.

Mather K. & Mather B. (1991) *Aggregates. Centenial Spec. Vol. 3*, 323–32. Geol. Soc. Am., Tucson.

Mattinen P.R. & Bennett G.H. (1986) The Green Mountain massive sulphide deposit. Besshi-style mineralization within the California Foothills Copper–Zinc Belt. *J. Geochem. Explor.*, **25**, 185–200.

Maynard J.B. (1983) *Geochemistry of Sedimentary Ore Deposits*, Springer-Verlag, New York.

McCarl H.N. (1989) Prices of industrial minerals: history.

In Carr D.D. & Herz N. (eds), *Concise Encyclopedia of Mineral Resources*, 285–8. Pergamon Press, Oxford.

McClay K.R. (1983) Deformation of stratiform lead–zinc deposits. In Sangster D.F. (ed.), *Short Course in Sediment-Hosted Stratiform Lead–Zinc Deposits*, 283–309. Mineral. Assoc. Can., Toronto.

McConchie D. (1984) A depositional environment for the Hamersley Group: palaeogeography and geochemistry. In Muhling J.R., Groves D.I. & Blake T.S. (eds), *Archaean and Proterozoic Basins of the Pilbara, Western Australia: Evolution and Mineralization Potential*, 144–77, **Pub. No. 9**, Geol. Dept & Univ. Extension, Univ. of Western Australia, Nedlands.

McGee E.S. (1988) Potential for diamond in kimberlites from Michigan and Montana as indicated by garnet xenocryst compositions. *Econ. Geol.*, **83**, 428–32.

McGoldrick P.J. & Keays R.R. (1990) Mount Isa copper and lead–zinc–silver ores: coincidence or cogenesis? *Econ. Geol.*, **85**, 641–50.

McKelvey V.E. (1973) Mineral resource estimates and public policy. *Prof. Pap. U.S. Geol. Surv.*, **820**, 9–19.

McKibben M.A., Andes J.P. & Williams A.E. (1988) Active ore formation at a brine interface in metamorphosed deltaic lacustrine sediments: the Salton Sea Geothermal System, California. *Econ. Geol.*, **83**, 511–23.

McKinstry H.E. (1948) *Mining Geology*. Prentice-Hall, New York.

McMichael B. (1989) Chromite. *Ind. Miner.* **257**, 25–45.

McMillan W.J. & Panteleyev A. (1989) Porphyry copper deposits. In Roberts R.G. & Sheahan P.A. (eds), *Ore Deposit Models*, 45–58. Geol. Assoc. Can., Memorial Univ., Newfoundland.

McVey H. (1989) Industrial minerals—can we live without them? *Ind. Miner.*, **259**, 74–5.

Megaw P.K.M., Ruiz J. & Titley S.R. (1988) High-temperature, carbonate-hosted Ag–Pb–Zn(Cu) deposits of northern Mexico. *Econ. Geol.*, **83**, 1856–85.

Meinart L.D. (1983) Variability of skarn deposits: guides to exploration. In Boardman S.J. (ed.) *Revolution in the Earth Sciences*, 301–15. Kendall/Hunt, Dubuque, Iowa.

Melcher G.C. (1966) The carbonatites of Jacupiranga, Sao Paulo, Brazil. In Tuttle O.F. & Gittins J. (eds), *Carbonatites*, 169–81. Wiley, New York.

Melvin J.L. (ed.) (1991) *Evaporites, Petroleum and Mineral Resources*. Elsevier, Amsterdam.

Mertie J.B. (1969) *Economic Geology of the Platinum Metals*. Prof. Pap. 630, U.S. Geol. Surv., Washington.

Meyer C. (1981) Ore-forming processes in geologic history. *Econ. Geol.*, **75th Anniv. Vol.**, 6–41.

Meyer C. (1985) Ore metals through geologic history. *Science*, **227**, 1421–8.

Meyer C. & Hemley J.J. (1967) Wall rock alteration. In Barnes H.L. (ed.), *Geochemistry of Hydrothermal Ore Deposits*, 166–235. Holt, Rinehart and Winston, New York.

Meyer C., Shea E.P., Goodard Jr. C.C. & Staff (1968) Ore deposits at Rutte, Montana. In Ridge J.R. (ed.), *Ore De-*

posits of the United States, 1933–1967, Vol. II, 1373–416. Am. Inst. Min. Metall and Petrol Eng., New York.

Meyer H.O.A. (1985) Genesis of diamond: a mantle saga. *Am. Mineral.*, **70**, 344–55.

Michel D. (1987) Concentration of gold in *in situ* laterites from Mato Grosso. *Miner. Deposita*, **22**, 185–9.

Milledge H.J., Mendelssohn M.J., Seal M., Rouse J.E., Swart P.K. & Pillinger C.T. (1983) Carbon isotopic variation in spectral type II diamonds. *Nature*, **303**, 79102.

Miller C.F. & Bradfish L.J. (1980) An inner cordilleran belt of muscovite-bearing plutons. *Geology*, **8**, 412–6.

Miller R.G. & O'Nions R.K. (1985) Sources of Precambrian chemical and clastic sediments. *Nature*, **314**, 325–30.

Ming-guo Z. & Windley B.F. (1989) Banded iron formation in high-grade gneisses in northern China and implications for early crustal growth. *(Trans. Instn Min. Metall. (Sect. B: Appl. Earth Sci.)*, **98**, B32–4.

Minter W.E.L. (1991) Ancient placer gold deposits. In Foster R.P. (ed.), *Gold Metallogeny and Exploration, 283–308*. Blackie, Glasgow.

Minter W.E.L., Feather C.E. & Glatthaar C.W. (1988) Sedimentological and mineralogical aspects of the newly discovered Witwatersrand placer deposit that reflect Proterozoic weathering. Welkem Gold Field, South Africa. *Econ. Geol.*, **83**, 481–91.

Mitcham T.W. (1974) Origin of breccia pipes, *Econ. Geol.*, **69**, 412–13.

Mitchell A.H.G. (1973) Metallogenic belts and angle of dip of Benioff Zones. *Nature (London) Phys. Sci.*, **245**, 49–52.

Mitchell A.H.G. & Garson M.S. (1972) Relationship of porphyry copper and circum-Pacific tin deposits to palaeo-Benioff zones. *Trans. Instn Min. Metall. (Sec. B: Appl. Earth Sci.)*, **81**, B10–B25.

Mitchell A.H.G. & Garson M.S. (1976) Mineralization at plate boundaries. *Miner. Sci. Eng.*, **8**, 129–69.

Mitchell A.H.G. & Garson M.S. (1981) *Mineral Deposits and Global Tectonic Settings*. Academic Press, London.

Mitchell A.H.G. & Reading H.G. (1986) Sedimentation and Tectonics. In Reading H.G. (ed.), *Sedimentary Environments and Facies*, 471–51. Blackwell Scientific Publications, Oxford.

Moine B., Fortune J.P., Moreau P. & Viguier F. (1989) Comparative mineralogy, geochemistry, and conditions of formation of two metasomatic talc and chlorite deposits: Trimouns (Pyrenees, France) and Rabenwald (Eastern Alps, Austria). *Econ. Geol.*, **84**, 1398–1416.

Möller P. (1985) Development and application of the Ga/Ge-Geo thermometer for sphalerite from sediment-hosted deposits. In Germann K. (ed.), *Geochemical Aspects of Ore Formation in Recent and Fossil Sedimentary Environments*, 15–30. Gebrüder Borntraeger, Berlin.

Möller P. (1989) *Magnesite*. Monogr. 28, Soc. for Geol. Appl. to Miner. Deposits. Gebrüder Borntraeger, Berlin.

Möller P., Černý P. & Saupé F. (1989) *Lanthanides, Tantalum and Niobium*, Springer-Verlag, Heidelberg.

Moorbath S., O'Nions R.L. & Pankhurst R.J. (1973) Early Archaean age for the Isua iron formation, West Greenland. *Nature*, **245**, 138–9.

Moore A.C. (1973) Carbonatites and Kimberlites in Australia: a review of the evidence. *Miner. Sci. Eng.*, **5**, 81–91.

Moore J.M. (1982) Mineral zonation near the granite batholiths of south-west and northern England and some geothermal analogues. In Evans A.M. (ed.), *Metallization Associated with Acid Magmatism*, 229–41. Wiley, Chichester.

Morgan J.D. (1989) Stockpiling. In Carr D.D. & Herz N. (eds), *Concise Encyclopedia of Mineral Resources*, 343–56. Pergamon Press, Oxford.

Moritz R.P., Crocket J.H. & Dickin A.P. (1990) Source of lead in the gold-bearing quartz–fuchsite vein at the Dome Mine, Timmins Area, Ontario, Canada. *Miner. Deposits*, **25**, 272–80.

Morteani G. (1989) Mg-metasomatic type sparry magnesites of Entachen Alm, Hockfilzen/Bürglkopf and Spiessnägel (Austria). In Möller P. (ed.), *Magnesite*, 105–13. Gebrüder Borntraeger, Berlin.

Morton R.L. & Franklin J.M. (1987) Two-fold classification of Archean volcanic-associated massive sulfide deposits. *Econ. Geol.*, **82**, 1057–63.

Morton R.L. & Nebel M.L. (1984) Hydrothermal alternation of felsic volcanic rocks at the Helen Siderite Deposit, Wawa, Ontario. *Econ. Geol.*, **79**, 1319–33.

Mosch E. & Becker M. (1985) Largest cassiterite flotation plant in the world under test operation at Altenberg. *Min. Mag.*, **153**, 531–7.

Mosig R.W. (1980) Morphology of indicator minerals as a guide to proximity of source. In Glover J.E. and Groves D.I. (eds), *Kimberlites and Diamonds*, 81–88. Extension Service, Univ. of Western Australia, Nedlands.

Munha J., Barriga F.J.A.S. & Kerrich R. (1986) High $^{18}O/^{16}O$ ore-forming fluids in volcanic-hosted base metal massive sulphide deposits: geologic, $^{18}O/^{16}O$, and D/H evidence from the Iberian Pyrite belt; Crandon, Wisconsin; and Blue Hill, Maine. *Econ. Geol.*, **81**, 530–52.

Muntean J.L., Kesler S.E., Russell N. & Polanco J. (1990) Evolution of the Monte Negro Acid Sulfate Au–Ag deposit, Pueblo Viejo, Dominican Republic: important factors in grade development. *Econ. Geol.*, **85**, 1738–58.

Naldrett A.J. (1973) Nickel sulphide deposits—their classification and genesis, with special emphasis on deposits of volcanic association. *Can. Inst. Min. Metall. Trans.*, **76**, 183–201.

Naldrett A.J. (1981) Nickel sulphide deposits: Classification, composition and genesis. *Econ. Geol.*, **75th Anniv. Vol.**, 628–85.

Naldrett A.J. (1989) *Magmatic Sulphide Deposits*. Oxford University Press, New York.

Naldrett A.J. & Cabri L.J. (1976) Ultramafic and related

mafic rocks: their classification and genesis with special reference to the concentration of nickel sulphides and platinium group elements. *Econ. Geol.*, **71**, 1131–58.

Naldrett A.J. & Campbell I.A.H. (1982) Physical and chemical constraints on genetic models for komatiite-related Ni-sulphide deposits. In Arndt N.T. & Nisbet E.G. (eds), *Komatiites*, 423–34. George Allen & Unwin, London.

Naldrett A.J. & Hewins R.H. (1984) The main mass of the Sudbury Igneous Complex. In Pye E.G., Naldrett A.J. & Giblin P.E. (eds), *The Geology and Ore Deposits of the Sudbury Structure*, 235–51, Spec. Vol. 1. Geol. Surv. Ontario, Toronto.

Naldrett A.J., Duke J.M., Lightfoot P.C. & Thompson J.F.H. (1984) Quantitative modelling of the segregation of magmatic sulphides: an exploration guide. *Can. Min. Metall. Bull.*, **77** (April), 46–56.

Nash J.T. (1976) Fluid inclusion petrology—Data from porphyry copper deposits and applications to exploration. U.S. Geol. Surv. Prof. Pap., 907-D.

Nash J.T., Granger H.C. & Adams S.S. (1981) Geology and concepts of genesis of important types of uranium deposits. *Econ. Geol.*, **75th Anniv. Vol.**, 63–116.

Natarajan W.K. & Mukerjee M.M. (1986) A note on the auriferous banded iron-formation of Kolar Schist Belt. *J. Geol. Soc. India*, **88**, 218–22.

Nesbitt B.E. (1991) Phanerozoic gold deposits in tectonically active continental margins. In Foster R.P. (ed.), *Gold Metallogeny and Exploration*, 104–32. Blackie, Glasglow.

Nesbitt B.E., Muehlenbachs K. & Murowchick J.B. (1989) Genetic implications of stable isotope characteristics of mesothermal Au deposits and related Sb and Hg deposits in the Canadian Cordillera. *Econ. Geol.*, **84**, 1489–506.

Newberry R.J., Burns L.E. Swanson S.E. & Smith T.E. (1990) Comparative petrologic evolution of the Sn and W granites of the Fairbanks–Circle Area, interior Alaska. In Stein H.J. & Hannah J.L. *Ore-bearing Granite Systems; Petrogenesis and Mineralizing Processes, 121–42* Geol. Soc. Am. Spec. Paper 246, Boulder, Colorado.

Nickel K.B. & Green D.H. (1985) Empirical geothermobarometry for garnet peridotites and implications for the nature of the lithosphere, kimberlites and diamonds, *Earth Planet, Sci. Lett.*, **73**, 158–170.

Nisbet B.W., Devlin S.P. & Joyce P.J. (1983) Geology and suggested genesis of cobalt–tungsten mineralization at Mt. Cobalt, North Western Queensland. *Proc. Australas Inst. Min. Metall.*, **287**, 9–17.

Nixon P.H. (ed.) (1973) *Lesotho Kimberlites*. Lesotho National Development Corporation, Maseru.

Nixon P.H. (1980a). The morphology and mineralogy of diamond pipes. In Glover J.E. & Groves D.I. (eds.), *Kimberlites and Diamonds*, 32–47. Extension Services, Univ. of Western Australia, Nedlands.

Nixon P.H. (1980b) Regional diamond exploration—theory and practice. In Glover J.E. & Groves D.I. (eds),

Kimberlites and Diamonds, 64–80. Extension Service, Univ. of Western Australia, Nedlands.

Nixon P.H., Boyd F.R. & Boctor N.Z. (1983) East Griqualand kimberlites. *Trans. Geol Soc. S. Afr.* **86**, 221–36.

Noble J.A. (1980) Two metallogenic maps for North America *Geol. Rundsch.*, **69**, 594–609.

Noble S.R., Spooner E.T.C. & Harris F.R. (1984) The Logtung large tonnage low-grade W (scheelite)—Mo porphyry deposit, south-central Yukon Territory. *Econ. Geol.*, **79**, 848–68.

Noetstaller R. (1988) *Industrial Minerals: a Technical Review*. The World Bank, Washington.

Norman D.I. & Trangcotchasan Y. (1982) Mineralization and fluid inclusion study of the Yod Nam Tin Mine, Southern Thailand. In Evans A.M. (ed.), *Metallization Associated with Acid Magmatism*, 261–72. Wiley, Chichester.

Northrop H.R. & Goldhaber M.B. (eds) (1990) Genesis of the tabular-type vanadium–uranium deposits of the Henry Basin, Utah. *Econ. Geol.*, **85**, 215–69.

Norton J.J. (1983) Sequence of mineral assemblages in differentiated granitic pegmatites. *Econ. Geol.*, **78**, 854–74.

Notholt A.J.G. (1979) The economic geology and development of igneous phosphate deposits in Europe and the USSR. *Econ. Geol.*, **74**, 339–50.

Notholt A.J.G., Highley D.E. & Deans T. (1990) Economic minerals in carbonatites and associated alkaline igneous rocks. *Trans. Instn Min. Metall. (Sect. B: Appl. Earth Sci.)* **99**, B59–80.

Nyström J.O. (1985) Apatite iron ores of the Kiruna Field, northern Sweden: magmatic textures and carbonatitic affinity. *Geol. Fören Förh.*, **107**, 133–41.

Nyström J.O. & Henriquez F. (1989) Dendritic magnetite and miniature diapir-like concentrations of apatite: two magmatic features of the Kirunavaara iron ore. *Geol. Fören. Förh.*, **111**, 53–64.

O'Driscoll E.S.T. (1986) Observations of the lineament-ore relations. *Philos. Trans. R. Soc. London*, **A317**, 195–218.

O'Driscoll M. (1988a) Rare earths. *Ind. Miner.*, **254**, 21–55.

O'Driscoll M. (1988b) Burnt lime/dolime. *Ind. Miner.*, **248**, 23–51.

O'Driscoll M. (1990) Fine carbonate fillers. *Ind. Miner.*, **276**, 21–43.

Ohmoto H. (1986) Stable isotope geochemistry of ore deposits. In Valley J.W., Taylor H.P. Jr & O'Neil J.R. (eds), *Stable Isotopes in High Temperature Geological Processes. Reviews in Mineralogy*, **16**, 491–559. Miner. Soc. Am., Washington.

Ohmoto H. & Rye R.O. (1974) Hydrogen and oxygen isotopic compositions of fluid inclusions in the Kuroko deposits, Japan. *Econ. Geol.*, **69**, 947–53.

Ohmoto H. & Skinner B.J. (1983) *The Kuroko and Related Volcanogenic Massive Sulphide Deposits*. Econ. Geol, Monogr, 5.

Olade M.A. (1980) Plate tectonics and metallogeny of intracontinental rifts and aulacogens in Africa—A review. In Ridge J.D. (ed.), *Proc. 5th Quadriennial IAGOD Symp.*, **Vol.1**, 81–9. E. Schweizer-bart'sche Verlagbuchhandlung, Stuttgart.

Olson J.C. & Pray L.C. (1954) The Mountain Pass rare-earth deposits. In *Geology of Southern California*, Chap. VIII. *Mineral Deposits and Mineral Industry*, Bull. 170, 23–29. Division of Mines, State of California.

Olson J.C., Shawe D.R., Pray L.C. & Sharp W.N. (1954) Rare-earth mineral deposits of the Mountain Pass District, San Bernardino County, California. *U.S. Geol. Surv., Prof., Pap.*, **261**.

Open University S333 Course Team (1976) *Porphyry Copper Case Study.* Open University, Milton Keynes.

Oreskes N. & Einaudi M.T. (1990) Origin of rare earth element-enriched hematite breccias at the Olympic Dam Cu–U–Au–Ag Deposit, Roxby Downs, South Australia. *Econ. Geol.*, **85**, 1–28.

Osmani I.A., Stott G.M., Sanborn-Barrie M. & Williams H.R. (1989) Recognition of regional shear zones in south-central and northwestern Superior Province of Ontario and their economic significance. In Bursnall J.T. (ed.) *Mineralization and Shear Zones*, 199–218. Short Course Notes, 6, Geol. Assoc. Can., Montréal.

Ostwald J. (1981) Evidence for a biogeochemical origin of the Groote Eylandt manganese ores. *Econ. Geol.*, **76**, 556–67.

Oszczepalski S. (1989) Kupferschiefer in southwestern Poland: sedimentary environments, metal zoning, and ore controls. In Boyle R.W., Brown A.C., Jefferson C.W., Jowett E.C. & Kirkham R.V. (eds), *Sediment-hosted Stratiform Copper Deposits*, 571–600. Geol. Assoc. Can. Spec. Pap. 36, Memorial Univ., Newfoundland.

Oudin E. & Constantinou G. (1984) Black smoker chimney fragments in Cyprus sulphide deposits. *Nature*, **308**, 349–53.

Owen H.B. & Whitehead S. (1965) Iron ore deposits of Iron Knob and the Middleback Ranges. In McAndrew J. (ed.), *Geology of Australian Ore Deposits*, 301–8. Austral as. Inst. Min. Metall., Melbourne.

Page R.W. & McDougall I. (1972) Ages of mineralization in gold and prophyry copper deposits in the New Guinea Highlands. *Econ. Geol.*, **67**, 1034–48.

Pagel M. (1975) Détermination des conditions physico-chimiques de la silicification diagénétique des grès Athabaska (Canada) au moyen des inclusions fluides. *C.R. Acad. Sci. Paris, Ser. D.*, **280**, 2301–4.

Pagel M. & Jafferezic H. (1977) Analyses chimiques des saumures des inclusions du quartz et de la dolomite du gisement d'uranium de Rabbit Lake (Canada). *C.R. Acad. Sci. Paris, Ser. D.*, **284**, 113–6.

Pagel M., Poty B. & Sheppard S.M.F. (1980) Contributions to some Saskatchewan uranium deposits mainly from fluid inclusions and isotopic data. In Ferguson J. & Goleby A.B. (eds), *Uranium in the Pine Creek Geosyncline*, 639–54. Int. Atom. Energy Agency, Vienna.

Pagel M., Wheatley K. & Ey F. (1986) The origin of the Carswell circular structure. *Geol. Assoc. Can. Spec. Pap.*, **29**, 213–23.

Palabora Mining Company Limited Mine Geological and Mineralogical Staff (1976) The geology and the economic deposits of copper, iron and vermiculite in the Palabora Igneous Complex: a brief review. *Econ. Geol.*, **71**, 177–92.

Parák T. (1975) Kiruna iron ores are not 'intrusive magmatic ores of the Kiruna type.' *Econ. Geol.*, **70**, 1242–58.

Parák T. (1985) Phosphorus in different types of ore, sulfides in the iron deposits, and the type and origin of ores at Kiruna. *Econ. Geol.*, **80**, 646–65.

Parák T. (1991) Volcanic sedimentary rock-related metallogenesis in the Kiruna–Skellefte belt of northern Sweden. *Econ. Geol. Monogr.*, **8**, 20–50.

Park C.F. (1961) A magnetite 'flow' in northern Chile. *Econ. Geol.*, **56**, 431–41.

Park C.F. Jr. & MacDiarmid R.A. (1975) *Ore Deposits.* Freeman, San Francisco.

Park R.G. (1988) *Geological Structures and Moving Plates.* Blackie, Glasgow.

Partington G.A. (1990) Environment and structural controls on the intrusion of the giant rare metal Greenbushes Pegmatite, Western Australia. *Econ. Geol.*, **85**, 437–56.

Pašava J. (1991) Comparison between the distribution of PGE in black shales from the Bohemian Massif (CSFR) and other black shale occurrences. *Miner. Deposita*, **26**, 99–103.

Pavlov N.V. & Grigor'eva I.I. (1977) Deposits of chromium. In Smirnov V.I. (ed.), *Ore Deposits of the USSR*, **1**, 179–236. Pitman London.

Pearce J.A., Lippard S.J. & Roberts S. (1984) Characteristics and tectonic significance of suprasubduction zone ophiolites. In Kokelaar B.P. & Howells M.F. (eds), *Marginal Basin Geology*, 77–94. Blackwell Scientific Publications, Oxford.

Pearson D.G., Davies G.R., Nixon P.H. & Milledge H.J. (1989) Graphitized diamonds from a peridotite massif in Morocco and implications for anomalous diamond occurrences. *Nature*, **338**, 60–2.

Pesquera A. & Velasco F. (1989) The Arditurri Pb–Zn–F–Ba Deposit (Cinco Villas Massif, Basque, Pyrenees): a deformed and metamorphosed stratiform deposit. *Miner. Deposita*, **24**, 199–209.

Peterson C.D., Komar P.D. & Scheidegger K.F. (1986) Distribution, geometry and origin of heavy mineral placer deposits on Oregon beaches. *J. Sediment. Petrol.*, **56**, 67–77.

Peterson U. (1970) Metallogenic provinces in South America. *Geol. Rundsch.*, **59**, 834–97.

Petrascheck W.E. (1965) Typical features of metallogenic provinces. *Econ. Geol.*, **60**, 1620–34.

Phillips D., Onstott T.C. & Harris J.W. (1989) $^{40}Ar/^{39}Ar$ laser-probe dating of diamond inclusions from the Premier Kimberlite. *Nature*, **340**, 460–2.

Phillips G.N. & Groves D.I. (1983) The nature of Archaean gold-bearing fluids as deduced from gold deposits of Western Australia. *J. Geol. Soc. Aust.*, **30**, 25–39.

Phillips G.N., Groves D.I. & Clark M.E. (1983) The importance of host-rock mineralogy in the location of Archaean epigenetic gold deposits. In De Villiers J.P.R. & Cawthorn P.A. (eds), *ICAM 81*, Spec. Publ. Geol. Soc. S. Afr., **7**, 79–86.

Phillips G.N., Groves D.I. & Martyn J.E. (1984) An epigenetic origin for Archaean banded iron formation hosted gold deposits. *Econ. Geol.*, **79**, 162–71.

Phillips W.J. (1972) Hydraulic fracturing and mineralization. *J. Geol. Soc. London*, **128**, 337–59.

Phillips W.J. (1973) Mechanical effects of retrograde boiling and its probable importance in the formation of some porphyry ore deposits. *Trans. Instn Min. Metall. (See B: Appl. Earth Sci.)*, **82**, B90-B98.

Phillips W.J. (1986) Hydraulic fracturing effects in the formation of mineral deposits. *Trans. Instn Min. Metall. (Sect. B: Appl. Earth Sci.)*, **95**, B17–24.

Piper J.D.A. (1974) Proterozoic crustal distribution, mobile belts and apparent polar movement, *Nature*, **251**, 381–4.

Piper J.D.A. (1974) Proterozoic supercontinent: time duration and the Grenville Problem. *Nature*, **256**, 519–20.

Piper J.D.A. (1976) Palaeomagnetic evidence for a Proterozoic supercontinent. *Philos. Trans. R. Soc. London.*, **A280**, 469–90.

Pirajno F. & Jacob R.E. (1987) Sn–W metallogeny in the Damara Orogen, South West Africa/Namibia. *S. Afr. J. Geol.*, **90**, 239–55.

Piroshco D.W. & Hodgson C.J. (1988) Relationship of hydrothermal alteration to structures and stratigraphy at the Coniaurum Gold Mine, northern Ontario. *Can. J. Earth Sci.*, **25**, 2028–40.

Pisutha-Arnond V. & Ohmoto H. (1983) Thermal history, chemical and isotopic compositions of the ore-forming fluids responsible for the Kuroko massive sulfide deposits in the Hokuroku district of Japan. In Ohmoto H. & Skinner B.J. (eds), *Kuroko and Related Volcanogenic Massive Sulfide Deposits,* Econ. Geol. Monogr. 5, 523–58.

Pitcher W.S. (1983) Granite: typology, geological environments and melting relationships. In Atherson M.P. & Gribbe C.D. (eds), *Migmatites Melting and Metamorphism,* 277–85. Shiva, Nantwich.

Plimer I.R. (1985) Broken Hill Pb–Zn–Ag deposit—a product of mantle metasomatism. *Miner. Deposita*, **20**, 147–53.

Plimer I.R. (1986) Sediment-hosted exhalative Pb–Zn deposits—products of contrasting ensialic rifting. *Trans. Geol. Soc. S. Afr.*, **89**, 57–73.

Plimer I.R. (1987a) Fundamental parameters for the formation of granite-related tin deposits. *Geol. Rundsch.*, **76**, 23–40.

Plimer I.R. (1987b) The association of tourmalinite with stratiform scheelite deposits. *Miner. Deposita*, **22**, 282–91.

Plimer I. (1988) Broken Hill, Australia and Bergsdalen, Sweden—why God and Mammon bless the Antipodes. *Geol. Mijnbouw*, **67**, 265–78.

Pohl W. (1989) Comparative geology of magnesite deposits and occurrences. In Möller P. (ed.), *Magnesite,* Monograph 28, 1–13, Soc. for Geol. Appl. to Miner. Deposits. Gebrüder Borntraeger, Berlin.

Pohl W., Amouri M., Kolli O., Scheffer R. & Zachmann D. (1986) A new genetic model for the North African metasomatic siderite deposits. *Miner. Deposita*, **21**, 228–33.

Pollard P.J., Taylor R.G. & Cuff C. (1988) Genetic modelling of greisen-style tin systems. In Hutchison C.S. (ed.), *The Geology of Tin Deposits in Asia and the Pacific,* 59–72. Springer-Verlag, Berlin.

Pollard P.J., Taylor R.G. & Tate N.M. (1989) Textural evidence for quartz and feldspar dissolution as a mechanism of formation for Maggs Pipe, Zaaiplaats Tin Mine, South Africa. *Miner. Deposita*, **24**, 210–8.

Poty B. & Pagel M. (1988) Fluid inclusions related to uranium deposits; a review. *J. Geol. Soc. London*, **145**, 157–62.

Pouba Z. & Ilavský J. (1986) Czechoslovakia. In Dunning F.W. & Evans A.M. (eds), *Mineral Deposits of Europe, Vol. 3,* 117–73. Instn Min. Metall. and Mineral. Soc., London.

Pracejus B., Bolton B.R. & Frakes L.A. (1988) Nature and development of supergene manganese deposits, Groote Eylandt, Northern Territory, Australia. *Ore Geol. Rev.*, **4**, 71–98.

Premoli C. (1985) The future of large, low grade, hard-rock tin deposits. *Natural Resour. Forum*, **9**, 107–119.

Prendergast M.D. (1987) The chromite ore field of the Great Dyke, Zimbabwe. In Stowe C.W. (ed.), *Evolution of Chromium Ore Fields,* 89–108. Van Nostrand Reinhold, New York.

Prendergast M.D. (1990) Platinum-group minerals and hydrosilicate 'Alteration' in Wedza–Mimosa Platinum Deposit, Great Dyke. Zimbabwe—genetic and metallurgical implications. *Trans. Instn Min. Metall. (Sect. B, Appl. Earth Sci.)*, **99**, B91–105.

Prentice J.E. (1990) *Geology of Construction Materials.* Chapman & Hall, London.

Preto V.A. (1978) Setting and genesis of uranium mineralization at Rexspar. *Can. Inst. Min. Metall., Bull.*, **71**, 82–8.

Pretorius D.A. (1975) The depositional environment of the Witwatersrand Goldfields: a chronological review of speculations and observations. *Miner. Sci. Eng.*, **7**, 18–47.

Pretorius D.A. (1981) Gold and uranium in quartz-pebble conglomerates. *Econ. Geol.*, **75th Anniv Vol.**, 117–38.

Pretorius D.A. (1991) The sources of Witwatersrand gold and uranium: a continued difference of opinion. *Econ. Geol. Monogr.* **8**, 139–63.

Prouhet J.-P. Les minéralisations de skarns des Pyrénées Françaises. (Field guide to tungsten mineralization in the Pyrenees.) *Bur. Rech. Geol. Minières* (in French).

Pye E.G., Naldrett A.J. & Giblin P.E. (eds) (1984) *The Geology and Ore Deposits of the Sudbury Structure*. Spec. Vol. 1, Geol. Surv. Ontario, Toronto.

Pye K. (1988) Bauxites gathering dust. *Nature*, **333**, 800–1.

Qiusheng Z. (1988) Early Proterozoic tectonic styles and associated mineral deposits of the North China Platform. *Precam. Res.* **39**, 1–29.

Rackley R.I. (1976) Origin of Western States-type uranium mineralization. In Wolf K.H. (ed.), *Handbook of Strata-Bound and Stratiform Deposits, Vol. 7*, 89–156. Elsevier, Amsterdam.

Radkevich E.A. (1972) The metallogenic zoning in the Pacific ore belt. 24th *Int. Geol. Congr.*, **4**, 52–59.

Radtke A.S. (1985) *Geology of the Carlin Gold Deposit, Nevada*. Prof. Pap. 1267, U.S. Geol. Survey, Washington.

Raedeke L.D. & Vian R.W. (1985) A three-dimensional view of mineralization in the Stillwater J-M Reef. *Can. Mineral.*, **23**, 312.

Raman P.K. (1986) Decisive controls in the formation of the east coast bauxite deposits of India. *J. Geol. Soc. India*, **28**, 1–8.

Ramboz C. & Charef A. (1988) Temperature, pressure, burial history and paleohydrology of the Les Malines, Pb–Zn deposit: reconstruction from aqueous inclusions in barite. *Econ. Geol.*, **83**, 784–800.

Ramdohr P. (1969) *The Ore Minerals and Their Intergrowths*. Pergamon Press, Oxford.

Ramsay J.G. & Huber M.I. (1987) *The Techniques of Modern Structural Geology Vol. 2*. Academic Press, London.

Ransome F.L. (1919) The copper deposits of Ray and Miami, Arizona. *U.S. Geol. Surv. Prof. Pap.*, **115**.

Rankin A.H. & Graham M.J. (1988) Na, K and Li contents of mineralizing fluids in the Northern Pennine Orefield, England and their genetic significance. *Trans. Instn Min. Metall. (Sect. B: Appl. Earth Sci.)*, **97**, B99–107.

Ranta D.E., Ward A.D. & Ganster M.W. (1984) Ore zoning applied to geologic reserve estimation of molybdenum deposits. In Erickson A.J. (ed.), *Applied Mining Geology*, 83–114. Am. Inst. Min. Metall. and Petrol. Eng., New York.

Ravenhurst C.E., Reynolds P.H., Zentilli M., Krueger H.W. & Blenkinsop J. (1989) Formation of Carboniferous Pb–Zn and barite mineralization from basin-derived fluids, Nova Scotia, Canada. *Econ. Geol.*, **84**, 1471–88.

Raybould J.G. (1978) Tectonic controls on Proterozoic stratiform copper mineralization. *Trans. Instn Min. Metall. (Sect. B: Appl. Earth Sci.)*, **87**, B79-B86.

Reading H.G. (1986) *Sedimentary Environments and Facies*. Blackwell Scientific Publications, Oxford.

Reedman A.J., Colman T.B., Campbell S.D.G. & Howells M.F. (1985) Volcanogenic mineralization related to the Snowdon Volcanic Group (Ordovician), Gwynedd, North Wales. *J. Geol. Soc. London*, **142**, 875–88.

Reedman J.H. (1984) Resources of phosphate, niobium, iron and other elements in residual soils over the Sukulu carbonatite complex, southeastern Uganda. *Econ. Geol.*, **79**, 716–24.

Reid I. & Frostick L.E. (1985) Role of settling, entrainment and dispersive equivalence and of interstice trapping in placer formation. *J. Geol. Soc. London*, **142**, 739–46.

Ren S.K., Eggleton R.A. & Walshe J.L. (1988) The formation of hydrothermal cookeite in the breccia pipes of the Ardlethan Tin Field, New South Wales, Australia. *Can. Mineral.*, **26**, 407–12.

Reynolds I.M. (1985) The nature and origin of titaniferous-rich layers in the upper zone of the Bushveld Complex: a review and synthesis. *Econ. Geol.*, **80**, 1089–108.

Reynolds R.L. & Goldhaber M.B. (1978) Origin of a South Texas roll-type uranium deposit: I. alteration of iron-titanium oxide minerals. *Econ. Geol.*, **73**, 1677–89.

Richards J.R. & Pidgeon R.T. (1983) Some age measurements on micas from Broken Hill, Australia. *J. Geol. Soc. Aust.*, **10**, 664–78.

Richardson C.K., Rye, R.O. & Wasserman M.D. (1988) The chemical and thermal evolution of the fluids in the Cave-in-Rock Fluorspar District, Illinois: stable isotope systematics at the Deardorff Mine. *Econ. Geol.*, **83**, 765–83.

Richardson J.M.G., Spooner E.T.C. & McAuslan D.A. (1982) The East Kemptville Tin Deposit, Nova Scotia: an example of a large tonnage, low grade, greisen-hosted deposit in the endocontact of a granite batholith. In *Current Research, Part B*, Geo. Surv. Can. Pap. 82-1Bm, 27–32.

Richardson J.M., Bell K., Watkinson D.H. & Blenkinsop J. (1990) Genesis and fluid evolution of the East Kemptville greisen-hosted tin mine, southwestern Nova Scotia, Canada. In Stein H.J. & Hannah J.L. (eds), *Ore-bearing Granite Systems; Petrogenesis and Mineralizing Processes*, 181–203. Spec. Pap. 246, Geol. Soc. Am., Boulder, Colorado.

Richardson S.H. (1986) Latter-day origin of diamonds of eclogitic paragenesis, *Nature*, **322**, 623–7.

Richardson S.H., Gurney J.J., Erlank A.J. & Harris J.W. (1984) Origin of diamonds in old enriched mantle. *Nature*, **310**, 198–202.

Richardson S.H., Erlank A.J., Harris J.W. & Hart S.R. (1990) Eclogitic diamonds of Proterozoic age from Cretaceous kimberlites. *Nature*, **346**, 54–6.

Rickard D. (1987) Proterozoic volcanogenic mineralization styles. In Pharaoh T.C., Beckinsale R.D. & Rickard D. (eds), *Geochemistry and Mineralization of Proterozoic Volcanic Suites*, Geol. Soc. London, Spec. Pub. 33, 23–35.

Rickard D.T., Willden M.Y., Marde Y. & Ryhage R. (1975) Hydrocarbons associated with lead–zinc ores at Laisvall, Sweden. *Nature*, **255**, 131–3.

Rimskaya-Korsakova O.M. (1964) Genesis of the Kovdor

iron ore deposit (Kola Peninsula). *Int. Geol. Rev.*, **6**, 1735–46.

Ringwood A.E. (1974) The petrological evolution of island arc systems. *J. Geol. Soc. London*, **130**, 183–204.

Ripley E.M. (1990) Se/S ratios of the Virginia Formation and Cu–Ni sulfide mineralization in the Babbitt Area, Duluth Complex, Minnesota. *Econ. Geol.*, **85**, 1935–40.

Rittenhouse G. (1943) Transportation and deposition of heavy minerals. *Geol. Soc. Am. Bull.*, **54**, 1725–80.

Robb L.J. & Meyer F.M. (1990) The nature of the Witwatersrand hinterland: conjectures on the source area problem. *Econ. Geol.*, **85**, 511–36.

Robert F. & Brown A.C. (1986a) Archean gold-bearing quartz veins at the Sigma Mine, Abitibi Greenstone Belt, Quebec: Part I. Geologic relations and formation of the vein system. *Econ. Geol.*, **81**, 578–92.

Robert, F. & Brown A.C. (1986b) Archean gold-bearing quartz veins at the Sigma Mine, Abitibi Greenstone Belt, Quebec: Part II. *Econ. Geol.*, **81**, 593–616.

Robertson D.S. (1962) Thorium and uranium variations in the Blind River ores. *Econ. Geol.*, **57**, 1175–84.

Robinson A. & Spooner E.T.C. (1984) Can the Elliot Lake uraninite-bearing quartz pebble conglomerates be used to place limits on the oxygen content of the Early Proterozoic atmosphere? *J. Geol. Soc. London*, **141**, 221–8.

Robinson B.W. & Ohmoto H. (1973) Mineralogy, fluid inclusions and stable isotopes of the Echo Bay U–Ni–Ag–Cu deposits, Northwest Territories, Canada. *Econ. Geol.*, **68**, 635–56.

Robinson D.N. (1978) The characteristics of natural diamonds and their interpretation. *Miner. Sci. Eng.*, **10**, 55–72.

Rock N.S. (1991) *Lamprophyres*. Blackie, Glasgow.

Roedder E. (1972) Composition of fluid inclusions. *U.S. Geol. Surv. Prof. Pap.*, 440-JJ.

Roedder E. (1984) *Fluid Inclusions. Reviews in Mineralogy Vol. 12*, Miner. Soc. Am., Resten, Virginia.

Roedder E. & Bodnar R.J. (1980) Geologic pressure determinations from fluid inclusion studies. *Ann. Rev. Earth Planet. Sci.*, **8**, 263–301.

Roedder E. & Howard K.W. (1988) Taolin Zn-Pb-fluorite deposits, People's Republic of China: an example of some problems in fluid inclusion research on mineral deposits. *J. Geol. Soc. London*, **145**, 163–74.

Roehl P.O. (1981) Dilation brecciation—proposed mechanism of fracturing, petroleum expulsion, and dolomitization in Monterey Formation, California. *Am. Assoc. Petrol. Geol. Bull.*, **65**, 980–1.

Rogers N. & Hawkesworth C. (1984) New date for diamonds. *Nature*, **310**, 187–8.

Rona P.A. (1988) Hydrothermal mineralization at oceanic ridges. *Can. Mineral.*, **26**, 431–465.

Rona P.A., Klinkhammer G., Nelson T.A., Trefry J.H. & Elderfield H. (1986) Black smokers, massive sulphides and vent biota at the Mid-Atlantic Ridge. *Nature*, **321**, 33–7.

Ronov A.B. (1964) Common tendencies in the chemical evolution of the earth's crust, ocean and atmosphere. *Geochem. Int.*, **1**, 713–37.

Roscoe S.M. (1968) Huronian rocks and uraniferous conglomerates of the Canadian Shield. *Geol. Surv. Can. Pap.*, **68–40**.

Rose A.W. & Burt D.M. (1979) Hydrothermal alteration. In Barnes H.L. (ed.), *Geochemistry of Hydrothermal Ore Deposits*, 2nd edn, 173–235. Wiley, New York.

Ross J.R. & Hopkins G.M.F. (1975) Kambalda nickel sulphide deposits. In Knight C.L. (ed.), *Economic Geology of Australia and Papua New Guinea-1, Metals*. Monog, 5, 100–121. Australas. Inst. Min. Metall., Parkville.

Routhier P. (1963) *Les Gisements Métallifères, Vol. 1*. Masson et Cie, Paris.

Rowland S.M. (1986) *Structural Analysis and Synthesis*. Blackwell Scientific Publications, Palo Alto.

Roy S. (1976) Ancient manganese deposits. In Wolf K.H. (ed.), *Handbook of Strata-Bound and Stratiform Deposits, Vol. 7*, 395–474. Elsevier, Amsterdam.

Roy S. (1981) *Manganese Deposits*. Academic Press, London.

Roy, S. (1988) Manganese metallogenesis: a review. *Ore Geol. Rev.*, **4**, 155–70.

Rubey W.W. (1933) The size distribution of heavy minerals within a waterlain sandstone. *J. Sediment. Petrol.*, **3**, 3–29.

Ruckmick J.C. (1963) 'The iron ores of Cerro Bolivar, Venezuela *Econ. Geol.*, **58**, 218–36.

Ruiz J., Kelly W.C. & Kaiser C.J. (1985) Strontium isotopic evidence for the origin of barites and sulfides from the Mississippi Valley-type ore deposits in southeast Missouri—a discussion. *Econ. Geol.*, **80**, 773–5.

Rumble D., Duke E.F. & Hoering T.L. (1986) Hydrothermal graphite in New Hampshire: evidence of carbon mobility during regional metamorphism. *Geology*, **14**, 452–5.

Russell A. (1987) Phosphate rock: trends in processing and production. *Ind. Miner.*, **240**, 25–59.

Russell A. (1988a) Ball and plastic clays. *Ind. Miner.*, **253**, 27–47.

Russell A. (1988b) Graphite: current shortfalls in flake supply. *Ind. Miner.*, **255**, 23–43.

Russell A. (1988c) Bikita minerals. *Ind. Miner.*, **249**, 63–71.

Russell M.J. (1983) Major sediment-hosted exhalative zinc and lead deposits: formation from hydrothermal convection cells that deepen during crustal extension. In Sangster D.F., (ed.), *Short Course in Sediment-hosted Stratiform Lead–Zinc Deposits*, 251–82. Mineral. Assoc. Can. Victoria.

Russell M.J. & Skauli H. (1991) A history of theoretical developments in carbonate-hosted base metal deposits and a new tri-level enthalpy classificaton. *Econ. Geol. Monogr.*, **8**, 96–116.

Russell M.J., Solomon M. & Walshe J.L. (1981) The genesis of sediment-hosted, exhalative zinc and lead deposits. *Mineral Deposita*, **16**, 113–27.

Russell N., Seaward M., Rivera J.A., McCurdy K., Kesler S.E. & Cloke P.L. (1981) Geology and geochemistry of the Pueblo Viejo gold–silver oxide ore deposit, Dominican Republic. *Trans. Instn Min. Metall. (Sect. B: Appl. Earth Sci.)*, **90**, B153–62.

Rye R.O. & Ohmoto H. (1974) Sulfur and carbon isotopes and ore genesis: a review. *Econ. Geol.*, **69**, 826–42.

Rye R.O. & Sawkins F.J. (1974) Fluid inclusion and stable isotope studies on the Casapalca Ag–Pb–Zn–Cu deposit, central Andes, Peru. *Econ. Geol.*, **69**, 181–205.

Saager R., Meyer M. & Muff R. (1982) Gold distribution in supracrustal rocks from Archaean greenstone belts of southern Africa and from Paleozoic ultramafic complexes of the European Alps: metallogenic and geochemical implications. *Econ. Geol.*, **77**, 1–24.

Saager R., Oberthür T. & Tomschi H-P. (1987) Geochemistry and mineralogy of banded iron-formation-hosted gold mineralization in the Gwanda Greenstone Belt, Zimbabwe. *Econ. Geol.*, **82**, 2017–32.

Samoilov V.S. & Plyusnin G.S. (1982) The source of material for rare-earth carbonatites. *Geochem. Int.*, **19**, 13–25.

Sangster D.F. (1976) Carbonate-hosted lead–zinc deposits. In Wolf K.H. (ed.), *Handbook of Strata-Bound and Stratiform Deposits, Vol. 6*, 447–56. Elsevier, Amsterdam.

Sangster D.F. (1983a) *Short Course in Sediment-hosted Stratiform Lead–Zinc Deposits.* Mineral. Assoc. Can., Victoria.

Sangster D.F. (1983b) Mississippi Valley-type deposits: a geological mélange. In Kisvarsanyi G., Grant S.K., Pratt W.P. & Koenig J.W. (eds), *International Conference on Mississippi Valley Type Lead–Zinc Deposits*, 7–19, Univ. of Missouri-Rolla, Rolla.

Sangster D.F. (1990) Mississippi Valley-type and sedex lead–zinc deposits: a comparative examination. *Trans. Instn Min. Metall. (Sect. B: Appl. Earth Sci.)*, **99**, B21–42.

Sangster D.F. & Scott S.D. (1976) Precambrian, stratabound, massive Cu–Zn–Pb sulphide ores in North America. In Wolf K.H. (ed.), *Handbook of Strata-Bound and Stratiform Ore Deposits, Vol. 6,* 129–222. Elsevier, Amsterdam.

Sato T. (1977) Kuroko deposits: their geology, geochemistry and origin. In *Volcanic Processes in Ore Genesis,* Spec. Publ. 7, Geol. Soc. London.

Saupé F. (1990) Geology of the Almadén mercury deposit, Province of Ciudad Real, Spain. *Econ. Geol.*, **85**, 482–510.

Sawkins F.J. (1972) Sulphide ore deposits in relation to plate tectonics. *J. Geol.*, **80**, 377–97.

Sawkins F.J. (1984 & 1990) *Metal Deposits in Relation to Plate Tectonics.* Springer-Verlag, Berlin.

Sawkins, F.J. (1986) Some thoughts on the genesis of Kuroko-type deposits. In Nesbitt R.W. & Nichol I. (eds), *Geology in the Real World—the Kingsley Dunham Volume,* 387–94. Instn Min. and Metall., London.

Sawkins F.J. & Rye D.M. (1974) Relationship of Homestake-type gold deposits to iron-rich Precambrian sedimentary rocks. *Trans. Instn Min. Metall. (Sect. B: Appl. Earth Sci.)*, **83**, B56–9.

Sawkins F.J. & Scherkenbach D.A. (1981) High copper content of fluid inclusions in quartz from northern Sonora: implications for ore-genesis theory. *Geology*, **9**, 37–40.

Sawlowicz Z. (1990) Primary copper sulphides from the Kupferschiefer, Poland. *Miner. Deposita*, **25**, 262–71.

Schiffries C.M. & Skinner B.J. (1987) The Bushveld hydrothermal system: field and petrologic evidence. *Am. J. Sci.*, **287**, 566–95.

Schmid H. (1984) China—the magnesite giant. *Ind. Miner.*, **203**, 27–45.

Schmid H. (1987) Turkey's Salda Lake. *Ind. Miner.*, **239**, 19–31.

Schneiderhöhn H. (1955) *Erzlagerstätten.* Gustav Fischer-Verlag, Stuttgart.

Schreiber B.C. (1986) Arid shorelines and evaporites. In Reading H.G. (ed.), *Sedimentary Environments and Facies,* 189–228. Blackwell Scientific Publications, Oxford.

Schuiling R.D. (1967) Tin belts on continents around the Atlantic Ocean. *Econ. Geol.*, **62**, 540–50.

Schwarz M.O. & Surjono (1990) Greisenization and albitization at the Tikus Tin–tungsten Deposit, Belitung, Indonesia. *Econ. Geol.*, **85**, 691–713.

Scott P.W. (1987) The exploration and evaluation of industrial rocks and minerals. *Annual Review 1987,* 19–28, Irish Assoc. for Econ. Geol., Dublin.

Scott S.D. (1973) Experimental calibration of the sphalerite geobarometer. *Econ. Geol.*, **68**, 466–74.

Scott S.D. (1974) Experimental methods in sulphide synthesis. In Ribbe P.H. (ed.), *Sulphide Mineralogy,* Mineral. Soc. Am. Short Course Notes, Vol. 1, Ch. 4.

Scott S.D. & Barnes H.L. (1971) Sphalerite geothermometry and geobarometry. *Econ. Geol.*, **66**, 653–69.

Scott Smith B.H. & Skinner E.M.W. (1984) Diamondiferous lamproites. *J. Geol.*, **92**, 433–38.

Scratch R.B., Watson G.P., Kerrich R. & Hutchinson R.W. (1984) Fracture-controlled antimony–quartz mineralization, Lake George deposit, New Brunswick: mineralogy, geochemistry, alteration and hydrothermal regimes. *Econ. Geol.*, **79**, 1159–86.

Seasor R.W. & Brown A.C. (1989) Syngenetic and diagenetic concepts at the White Pine Copper Deposit, Michigan. In Boyle R.W., Brown A.C., Jefferson C.W., Jowett E.C. & Kirkham R.V. (eds), *Sediment-hosted Stratiform Copper Deposits,* 257–67. Geol. Assoc. Can. Spec. Pap. 36, Memorial Univ., Newfoundland.

Seccombe P.K. (1990) Fluid inclusion and sulphur isotope evidence for syntectonic mineralisation at the Elura Mine, southeastern Australia. *Miner. Deposita*, **25**, 304–12.

Seedorff E. (1988) Cyclic development of hydrothermal mineral assemblages related to multiple intrusions at the Henderson porphyry molybdenum deposit, Colorado.

In Taylor R.P. & Strong D.F. (eds) *Recent Advances in the Geology of Granite-related Mineral Deposits,* 367–93, Spec. Vol. 39. Can. Inst. Min. Metall., Montréal.

Selby J. (1991) Olympic Dam: a giant Australian ore deposit. *Geol. Today,* Jan.–Feb., 24–7.

Select Committee (1982) Memorandum by the Inst. Geol. Sci., Strategic Minerals. In *Strategic Minerals Select Committee,* HL Rep; 1981–2 (217) xii.

Selkman S.O. (1984) A deformation study at the Saxerget sulphide deposit, Sweden. *Geol. Förh, Fören.* **106,** 235–44.

Selley R.C. (1976) *An Introduction to Sedimentology.* Academic Press, London.

Seward T.M. (1991) The hydrothermal geochemistry of gold. In Foster R.P. (ed.), *Gold Metallogeny and Exploration,* 37–62. Blackie, Glasgow.

Seward T.M. & Sheppard D.S. (1986) Waimangu Geothermal Field. *Monograph Series on Mineral Deposits, 26,* 81–91. Gebrüder Borntraeger, Berlin.

Shackley, D. (1990) The Hungarian perlite industry. *Ind. Miner.,* **271,** 61–5.

Sharp Z.D., Essene E.J. & Kelly W.C. (1985) A re-examination of the arsenopyrite geothermometer; pressure considerations and applications to natural assemblages. *Can. Mineral.,* **23,** 517–34.

Shaw A.L. & Guilbert J.M. (1990) Geochemistry and metallogeny of Arizona peraluminous granitoids with reference to Appalachian and European occurrences. In Hannah J.L. & Stein H.J. (eds), *Ore-bearing Granite Systems; Petrogenesis and Mineralizing Processes,* 317–54. Geol. Soc. Am. Spec Paper 246, Boulder, Colorado.

Shaw D.M. (1954) Trace elements in pelitic rocks. *Geol. Soc. Am. Bull.,* **65,** 1151–82.

Shaw S.E. (1968) Rb–Sr isotopic studies of the mine sequence rocks at Broken Hill. In Radmanovich M. & Woodcock J.T. (eds), *Broken Hill Mines—1968,* 185–98, Australas. Inst. Min. Metall. Mongr. Ser. 3.

Shearman D.J. (1966) Origin of marine evaporites by diagenesis. *Trans. Instn Min. Metall. (Sect. B: Appl. Earth Sci.),* **79,** B155–62.

Shenberger D.M. & Barnes H.L. (1989) Solubility of gold in aqueous sulfide solutions from 150 to 350°C. *Geochim. Cosmochim. Acta,* **53,** 269–78.

Shepherd T.J. (1977) Fluid inclusion study of the Witwatersrand gold–uranium ores. *Philos. Trans. R. Soc. London,* **286A,** 549–65.

Shepherd T.J. & Allen P.M. (1985) Metallogenesis in the Harlech Dome, North Wales: a fluid inclusion interpretation. *Miner. Deposita,* **20,** 159–68.

Shepherd T.J., Rankin A.H. & Alderton D.H.M. (1985) *A Practical Guide to Fluid Inclusion Studies.* Blackie, Glasgow.

Sheppard S.M.F. (1977) Identification of the origin of ore-forming solutions by the use of stable isotopes. In *Volcanic Processes in Ore Genesis,* Spec. Publ. 7, Geol. Soc. London.

Sheppard, S.M.F. (1986) Characterization and isotopic

variations in natural waters. In Valley J.W., Taylor H.P. Jr & O'Neil J.R. (eds), *Stable Isotopes in High Temperature Geological Processes, Reviews in Mineralogy, 16,* 165–83. Min. Soc. Am., Washington.

Shilo N.A., Milov A.P. & Sobolev A.P. (1983) Mesozoic granitoids of north-east Asia. *Mem. Geol. Soc. Am.,* **159,** 149–57.

Shimazaki H. & MacLean W.H. (1976) An experimental study on the partition of zinc and lead between the silicate and sulphide liquids. *Miner. Deposita,* **11,** 125–32.

Shmakin B.M. (1983) Geochemistry and origin of granitic pegmatites. *Geochem. Int.,* **20**(6), 1–8.

Sibson R.H., Moore J. McM. & Rankin A.H. (1975) Seismic pumping—a hydrothermal fluid transport mechanism. *J. Geol. Soc. London,* **131,** 653–9.

Siddaiah N.S. & Rajamani V. (1989) The geologic setting, mineralogy, geochemistry, and genesis of gold deposits of the Archean Kolar Schist Belt, India. *Econ. Geol.,* **84,** 2155–72.

Silberman M.L. (1985) Geochemistry of hydrothermal mineralization and alteration: Tertiary epithermal precious-metal deposits in the Great Basin. In Tooker E.W. (ed.), *Geologic Characteristics of Sediment- and Volcanic-hosted Disseminated Gold Deposits—Search for an Occurrence Model,* 55–70, U.S. Geol. Survey Bull., 1646.

Sillitoe R.H. (1972a) Formation of certain massive sulphide deposits at sites of sea-floor spreading. *Trans. Instn Min. Metall. (Sect. B: Appl. Earth Sci.),* **81,** B141–B148.

Sillitoe R.H. (1972b) A plate tectonic model for the origin of porphyry copper deposits. *Econ. Geol.,* **67,** 184–97.

Sillitoe R.H. (1973) The tops and bottoms of porphyry copper deposits. *Econ. Geol.,* **68,** 799–815.

Sillitoe R.H. (1976) Andean mineralization: a model for the metallogeny of convergent plate margins. In Strong D.F. (ed.), *Metallogeny and Plate Tectonics,* 59–100, Geol. Assoc. Can. Spec. Pap. 14.

Sillitoe R.H. (1980a) Types of porphyry molybdenum deposits. *Min. Mag.,* **142,** 550–553.

Sillitoe R.H. (1980b) Strata-bound ore deposits related to Infracambrian rifting along northern Gondwanaland. In Ridge J.D. (ed.), *Proc. 5th IAGOD Symp. Vol. 1,* 163–72. E. Schweizerbart'sche Verlagsbuchhandlung, Stuttgart.

Sillitoe R.H. (1981a) Regional aspects of the Andean porphyry copper belt in Chile and Argentina. *Trans. Inst. Min. Metall. (Sect. B Appl. Earth Sci.),* **90,** B15–36.

Sillitoe R.H. (1981b) Ore deposits in cordilleran and island arc settings. *Ariz. Geol. Soc. Dig.* **14,** 49–70, cited by Sawkins (1984).

Sillitoe R.H. (1991a) Gold metallogeny of Chile—an introduction. *Econ. Geol.,* **86,** 1187–1205.

Sillitoe R.H. (1991b) Intrusion-related gold deposits. In Foster R.P. (ed.), *Gold Metallogeny and Exploration,* 165–209. Blackie, Glasgow.

Sillitoe R.H., Halls C. & Grant J.N. (1975) Porphyry tin deposits in Bolivia. *Econ. Geol.*, **70**, 913–27.

Sillitoe R.H. & Hart S.R. (1984) Lead-isotope signatures of porphyry copper deposits in oceanic and continental settings, Colombian Andes. *Geochim. Cosmochim. Acta*, **48**, 2135–42.

Silva K.K.M.W. (1987) Mineralization and wall-rock alteration at the Bogala Graphite Deposit, Bulathkohupitiya, Sri Lanka. *Econ. Geol.*, **82**, 1710–22.

Simmons S.F., Gemmell J.B. & Sawkins F.J. (1988) The Santo Niño silver–lead–zinc vein, Fresnillo District, Zacatecas, Mexico: part II. physical and chemical nature of ore-forming solutions. *Econ. Geol.*, **83**, 1619–41.

Simmons S.F., Sawkins F.J. & Schlutter D.J. (1987) Mantle-derived helium in two Peruvian hydrothermal ore deposits. *Nature*, **329**, 429–32.

Simpson P.R. & Bowles J.F. (1977) Uranium mineralization of the Witwatersrand and Dominion Reef Systems. *Philos. Trans. R. Soc. London*, **A286**, 527–48.

Simpson P.R., Youxun G. & Bingzang G. (1987) Metallogeny, magmatism and structure in Jiangxi Province China. *Trans. Instn Min. Metall. (Sect. B: Appl. Earth Sci.)*, **96**, B77–83.

Sinclair A.J., Drummond A.D., Carter N.C. & Dawson K.M. (1982) A preliminary analysis of gold and silver grades of porphyry-type deposits in western Canada. In Levinson A.A. (ed.), *Precious Metals in the Northern Cordillera*, 157–72. Assoc. Explor. Geochem., Rexdale.

Siva Siddaiah N. & Rajamani V. (1988) Geochemistry and origin of gold mineralization in the Kolar Schist Belt. *J. Geol. Soc. India*, **31**, 142–3.

Skinner B.J. (1979) The many origins of hydrothermal mineral deposits. In Barnes H.L. (ed.), *Geochemistry of Hydrothermal Ore Deposits*, 2nd edn, 1–21. Wiley, New York.

Skinner E.M.W., Smith C.B., Bristow J.W., Scott Smith B.H. & Dawson J.B. (1985) Proterozoic kimberlites and lamproites and a preliminary age for the Argyle lamproite pipe, Western Australia. *Trans. Geol. Soc. S. Afr.*, **88**, 335–40.

Skirrow R. & Coleman M.L. (1982) Origin of sulphur and geothermometry of hydrothermal sulphides from the Galapagos Rift 86°W. *Nature*, **299**, 142–4.

Slade M.E. (1989) Prices of metals: history. In Carr D.D. & Herz N. (eds), *Concise Encyclopedia of Mineral Resources*, 288–92. Pergamon Press, Oxford.

Slingerland R.L. (1984) Role of hydraulic sorting in the origin of fluvial placers. *J. Sediment. Petrol.*, **54**, 137–50.

Smirnov V.I. (1976) *Geology of Mineral Deposits*. MIR, Moscow.

Smith A. (1989) Canada in perspective: the international scene. *Br. Geol.*, **15**, 136–9.

Smith A.D., Fox J.S., Farqhar R. & Smith P. (1990) Geochemistry of Ordovician Køli Group basalts associated with Besshi-type Cu–Zn deposits from the Southern Trondheim and Sulitjelma Mining Districts of Norway. *Miner. Deposita*, **25**, 15–24.

Smith C.S. (1964) Some elementary principles of polycrystalline microstructure. *Metall. Rev.*, **9**, 1–48.

Smith M. (1987) Recycled roads cut the cost of materials. *New Sci.*, **30 July**, 34.

Smith T.E., Miller P.M. & Huang C.H. (1982) Solidification and crystallization of a stanniferous granitoid pluton, Nova Scotia, Canada. In Evans A.M. (ed.), *Metallization Associated with Acid Magmatism, Vol. 6*, 301–20. Wiley, Chichester.

Snee L.W., Sutter J.F. & Kelly W.C. (1988) Thermochronology of economic mineral deposits: dating the stages of mineralization at Panasqueira, Portugal, by high-precision $^{40}Ar/^{39}Ar$ age spectrum techniques on muscovite. *Econ. Geol.*, **83**, 335–53.

Sokolov G.A. & Grigor'ev V.M. (1977) Deposits of iron. In Smirnov V.I. (ed.), *Ore Deposits of the USSR*, 7–113. Pitman, London.

Solomon M. (1976) 'Volcanic' massive sulphide deposits and their host rocks—a review and an explanation. In Wolf K.H. (ed.), *Handbook of Strata-Bound and Stratiform Ore Deposits, Vol. 6*, 21–54. Elsevier, Amsterdam.

Solomon M. & Walshe J.L. (1979) The formation of massive sulphide deposits on the sea floor. *Econ. Geol.*, **74**, 797–813.

Solomon M., Walshe J.L. & Eastoe C.J. (1987) Experiments on convection and their relevance to the genesis of massive sulphide deposits. *Aust. J. Earth Sci.*, **34**, 311–23.

Souch B.E., Podolsky T. & Geological Staff (1969) The sulphide ores of Sudbury: their particular relationship to a distinctive inclusion-bearing facies of the nickel irruptive. In Wilson H.D.B. (ed.), *Magmatic Ore Deposits—A symposium*, 252–61, Econ. Geol. Monogr. 4.

South B.C. & Taylor B.E. (1985) Stable isotope geochemistry and metal zonation at the Iron Mountain Mine, West Shasta District, California. *Econ. Geol.*, **80**, 2177–95.

Spooner E.T.C. (1977) Hydrodynamic model for the origin of ophiolitic cupriferous pyrite ore deposits of Cyprus. In *Volcanic Process in Ore Genesis*, Spec. Publ. 7, Geol. Soc. London.

Spooner E.T.C., Bray C.J. & Chapman H.J. (1977) A sea water source for the hydrothermal fluid which formed the ophiolitic cupriferous pyrite ore deposits of the Troodos Massif, Cyprus. *J. Geol. Soc. London*, **134**, 395.

Springer J. (1983) Invisible gold. In Colvine A.C. (ed.), *The Geology of Gold in Ontario*, 240–50, Geol. Surv. Ont., Misc. Pap. 110.

Stacey J.S., Zartman R.E. & Nkomo I.T. (1968) A lead isotope study of galenas and selected feldspars from mining districts in Utah. *Econ. Geol.* **63**, 796–814.

Stanton R.L. (1972) *Ore Petrology*. McGraw-Hill, New York.

Stanton R.L. (1978) Mineralization in island arcs with particular reference to the south-west Pacific region. *Proc. Australas. Inst. Min. Metall.*, **268**, 9–19.

Stanton R.L. (1986) Stratiform ores and geological pro-cessess. *Trans. Instn Min. Metall. (Sect. B, Appl. Earth Sci.)*, **95**, B165–78.

Stanworth C.W. & Badham J.P.N. (1984) Lower Protero-zoic red beds, evaporites and secondary sedimentary uranium deposits from the East Arm, Great Slave Lake, Canada. *J. Geol. Soc. London*, **141**, 235–42.

Steele I.M., Bishop F.C., Smith J.V. & Windley B.F. (1977) The Fiskenæsset Complex, West Greenland, Part III, *Grønlands Geol. Unders. Bull*, **124.**

Stein H.J. (1985) Genetic traits of Climax-type granites and molybdenum mineralization, Colorado Mineral Belt. In Taylor R.P. & Strong D.F. (eds)., *Granite-related Mineral Deposits*, 242–7. Can. Inst. Min. Geology Division, Halifax, Canada.

Stein H.J. (1988) Genetic traits of Climax-type granites and molybdenum mineralization, Colorado Mineral Belt. In Taylor R.P. & Strong D.F. (eds) *Recent Advances in the Geology of Granite-related Mineral Deposits*, 394–401. Spec. Vol. 39, Can. Inst. Min. Metall., Montreal.

Štemprok M. (1985) Vertical extent of greisen mineraliz-ation in the Krusnehory/Erzgebirge Granite Pluton of central Europe. In Halls C. (ed.), *High Heat Production (HHP) Granites, Hydrothermal Circulation and Ore Genesis.* 383–91. Instn Min. Metall., London.

Štemprock M. (1987) Greisenization (a review). *Geol. Rundsch.*, **76**, 169–75.

Stone M. & Exley C.S. (1986) High heat production granites of southwest England and their associated mineralization: a review. *Trans. Instn Min. Metall. (Sect. B: Appl Earth Sci.)*, **95**, B25–36.

Stowe C.W. (ed.) (1987) *Evolution of Chromiun Ore Fields.* Van Nostrand Reinhold, New York.

Strauss G.H. & Beck J.S. (1990) Gold Mineralisations in the SW Iberian Pyrite Belt. *Miner. Deposita*, **25**, 237–45.

Streckeisen A.L. (1976) To each plutonic rock its proper name. *Earth Sci. Rev.*, **12**, 1–33.

Strens M.R., Cann D.L. & Cann J.R. (1987) A thermal balance model of the formation of sedimentary-exhalative lead–zinc deposits. *Econ. Geol.*, **82**, 1192–1203.

Strong D.F. (1981) Ore deposit models: 5. A model for granophile mineral deposits. *Geosci. Can.*, **8**, 155–60.

Stuckey J.L. (1967) *Pyrophyllite Deposits in North Caro-lina.* North Carolina Dept Conservation and Develop-ment, Div. Min. Res., Bull. 80.

Susak N.J. & Crerar D.A. (1982) Factors controlling mineral zoning in hydrothermal ore deposits. *Econ. Geol.*, **77**, 476–82.

Sutherland D.G. (1985) Geomorphological controls on the distribution of placer deposits. *J. Geol. Soc. London*, **142**, 727–31.

Suttill K.R. (1988) Cerro Rico de Potosi. *Eng. Min. J.*, **189** (March), 50–3.

Sverjensky D.A. (1984) Oilfield brines as ore-forming solution. *Econ. Geol.*, **79**, 23–37.

Sweeney M.A. & Binda P.L. (1989) The role of diagenesis in the formation of the Konkola Cu–Co Orebody of the Zambian Copperbelt. In Boyle R.W., Brown A.C., Jefferson C.W., Jowett E.C. & Kirkham R.V. (eds), *Sediment-hosted Stratiform Copper Deposits*, 499–518. Geol. Assoc. Can. Spec. Pap. 36, Memorial Univ., Newfoundland.

Symons D.T.A. & Sangster D.F. (1991) Palaeomagnetic age of the central Missouri barite deposits and its genetic implications. *Econ. Geol.*, **86**, 1–12.

Symons R. (1961) Operation at Bikita minerals (Private) Ltd., Southern Rhodesia. *Trans. Instn Min. Metall.*, **71**, 129–72.

Tanelli G. & Lattanzi P. (1985) The cassiterite-polymetallic sulfide deposits of Dachang (Guangxi, People's Republic of China). *Miner. Deposita*, **20**, 102–6.

Tarney J., Dalziel I. & De Wit M. (1976) Marginal basin 'Rocas Verdes' complex from S. Chile: a model for Archaean greenstone belt formation. In Windley B.F. (ed.), *The Early History of the Earth*, 131–46. Wiley, London.

Taylor H.K. (1989) Ore reserves—a general overview. *Miner. Ind. Int.*, **990**, 5–12.

Taylor R.G. (1979) *Geology of Tin Deposits.* Elsevier, Amsterdam.

Taylor R.G. & Pollard P.J. (1988) Pervasive hydrothermal alteration in tin-bearing granites and implications for the evolution of ore-bearing magmatic fluids. In Taylor R.P. & Strong D.F. (eds), *Recent Advances in the Geology of Granite-related Mineral Deposits*, 86–95. Can. Inst. Min. Metall., Montréal.

Taylor R.P. (1981) Isotope geology of the Bakircay Por-phyry Copper Prospect, northern Turkey. *Miner. Depos-ita*, **16**, 375–90.

Taylor R.P. & Fryer B.J. (1982) Rare earth element geochemistry as an aid to interpretating hydrothermal ore deposits. In Evans A.M. (ed.), *Metallization Associ-ated with Acid Magmatism*, 357–65. Wiley, Chichester.

Taylor R.P. & Fryer B.J. (1983) Strontium isotope geochemistry of the Santa Rita Porphyry Copper De-posits, New Mexico. *Econ. Geol.*, **78**, 170–4.

Taylor R.P. & Strong D.F. (1985) *Granite-related Mineral Deposits.* Can. Inst. Min. Metall. Geology Division, Halifax, Canada.

Taylor S. (1984) Structural and paleotopographic controls of lead–zinc mineralization in the Silvermines orebod-ies, Republic of Ireland. *Econ. Geol.*, **79**, 529–48.

Taylor S. & Andrew C.J. (1978) Silvermines Orebodies, County Tipperary, Ireland. *Trans. Instn Min. Metall.* (Sect. B: Appl. Earth Sci.), **87**, B111–24.

Taylor S.R. (1955) The origin of some New Zealand metamorphic rocks as shown by their major and trace element compositions. *Geochim. Cosmochim. Acta*, **8**, 182–97.

Thalhammer O.A.R., Stumpfl E.F. & Jahoda R. (1989) The Mittersill Scheelite Deposit, Austria. *Econ. Geol.*, **84**, 1153–71.

Thayer T.P. (1964) Principal features and origin of podiform chromite deposits, and some observations on the Guleman–Soridag District, Turkey. *Econ. Geol.*, **59**, 1497–524.

Thayer T.P. (1967) Chemical and structural relations of ultramafic and feldspathic rocks in Alpine intrusive complexes. In Wyllie P.J. (ed.), *Ultramafic and Related Rocks*, 222–39. Wiley, New York.

Thayer T.P. (1969a) Gravity differentiation and magmatic re-emplacement of podiform chromite deposits. *Econ. Geol. Monogr.*, **4**, 132–46.

Thayer T.P. (1969b) Peridotite-gabbro complexes as keys to petrology of mid-ocean ridges. *Geol. Soc. Am. Bull.*, **80**, 1515–22.

Thayer T.P. (1971) Authigenic, polygenic and allogenic ultramafic and gabbroic rocks as hosts for magmatic ore deposits. *Geol. Soc. Aust., Spec. Publ.*, **3**, 239–51.

Thayer T.P. (1973) *Chromium*. In Probst D.A. & Pratt W.P. (eds), *United States Mineral Resources*, 111–121, Prof. Pap. 820, U.S. Geol. Surv., Washington.

Thein J. (1990) Paleogeography and geochemistry of the 'Cenomo-Turonian' Formations in the Manganese District of Imini (Morocco) and their relation to ore deposition. *Ore Geol. Rev.*, **5**, 257–91.

Theis N.J. (1979) Uranium-bearing and associated minerals in their geochemical and sedimentological context. Elliot Lake, Ontario. *Geol. Surv. Can. Bull.*, **304**.

Theodore T.G. (1977) Selected copper-bearing skarns and epizonal granitic-intrusions in the south-western United States. *Geol. Soc. Malaysia Bull.*, **9**, 31–50.

Theodore T.G., Howe S.S., Blake D.W. & Wotruba P.R. (1986) Geochemical and fluid zonation in the skarn environment at the Tomboy–Minnie gold deposits, Lander County, Nevada. *J. Geochem. Explor.*, **25**, 99–128.

Thomas A.V., Bray C.J. & Spooner E.T.C. (1988) A discussion of the Jahns–Burnham proposal for the formation of zoned granitic pegmatites using solid–liquid–vapour inclusions from the Tanco Pegmatite, S.E. Manitoba, Canada. *Trans. Roy. Soc. Edinburgh, Earth Sci.*, **79**, 299–315.

Thurlow J.G. (1977) Occurrences, origin and significance of mechanically transported sulphide ores at Buchans, Newfoundland. In *Volcanic Processes in Ore Genesis*, Spec. publ. 7, Geol. Soc. London.

Tilsley J.E. (1981) Ore deposit models: 3. Genetic considerations relating to some uranium deposits, *Geosci. Can.*, **8**, 3–7.

Titley S.R. (1978) Copper, molybdenum and gold content of some porphyry copper systems of the southwestern and western Pacific. *Econ. Geol.*, **73**, 977–81.

Titley S.R. (ed.) (1982a), *Advances in Geology of the Porphyry Copper Deposits*. Univ. Arizona Press, Tucson.

Titley S.R. (1982b) The style and progress of mineralization and alteration in porphyry copper systems. In Titley S.R. (ed.) *Advances in Geology of the Porphyry Copper Deposits: Southwestern North America*, 93–116. Univ. Arizona Press, Tucson.

Titley S.R. & Beane R.E. (1981) Porphyry copper deposits. *Econ. Geol.*, **75th Anniv. Vol.**, 214–69.

Tooker E.W. (ed.) (1985), *Geologic Characteristics of Sediment- and Volcanic-hosted Disseminated Gold Deposits—Search for an Occurrence Model*. U.S. Geol. Survey, Bull. 1646.

Touray J-C. & Guilhaumou (1984) Characterization of H_2S bearing fluid inclusions. *Bull. Mineral.*, **107**, 181–8.

Touray J-C., Marcoux E., Hubert P. & Proust D (1989) Hydrothermal processes and ore-forming fluids in the Le Bourneix Gold Deposit, central France. *Econ. Geol.*, **84**, 1328–39.

Trendall A.F. (1968) Three great basins of Precambrian iron formation: a systematic comparison. *Bull. Geol. Soc. Am.*, **79**, 1527–44.

Trendall A.F. (1973) Iron formations of the Hamersley Group of Western Australia: type examples of varved Precambrian evaporites. In *Genesis of Precambrian Iron and Manganese Deposits*, 257–70, Proc. Kiev Symp. 1970, Unesco, Paris.

Trendall A.F. & Blockley J.G. (1970) The iron formation of the Precambrian Hamersley Group, Western Australia. *Bull. Geol. Surv. W. Aust.*, **119**.

Trocki L.K., Curtis D.B., Gancarz A.J. & Banar J.C. (1984) Ages of major uranium mineralization and lead loss in the Key Lake Uranium Deposit, Northern Saskatchewan, Canada. *Econ. Geol.*, **79**, 1378–86.

Trueman D.L. & Černý P. (1982) Exploration for rare-element granitic pegmatites. In Černý P. (ed), *Short Course in Granitic Pegmatites in Science and Industry*, 463–93. Mineral. Assoc. Can., Winnipeg.

Trueman D.L., Pedersen J.C., de St Jorre L. & Smith D.G.W. (1988) The Thorr Lake rare-metal deposits, Northwest Territories. In Taylor R.P. & Strong D.F. (eds) *Recent Advances in the Geology of Granite-related Mineral Deposits*, 280–290. Spec. Vol. 39, Can. Inst. Min. Metall., Montreal.

Truesdell A.H. (1984) Chemical geothermometers for geothermal exploration. In Henley R.W., Truesdell A.H. & Barton P.B. Jr. (eds), *Reviews in Economic Geology*, **1**, 31–44. Soc. Econ. Geol.

Tucker M.E. (1988) *Sedimentary Petrology*. Blackwell Scientific Publications, Oxford.

Turneaure F.S. (1955) Metallogenic provinces and epochs. In Bateman A.M (ed.), *Econ. Geol.*, **50th Anniv. Vol.**, 38–98

Turneaure F.S. (1960) A comparative study of major ore deposits of central Bolivia. Parts I and II, *Econ. Geol.*, **55**, 217–254 and 574–806.

Ullmer E. (1985) The results of exploration for unconformity-type uranium deposits in east central Minnesota. *Econ. Geol.*, **80**, 1425–35.

Urabe T. (1987) The effect of pressure on the partitioning ratios of lead and zinc between vapor and rhyolitic melts. *Econ. Geol.*, **82**, 1049–52.

Usui A., Yuasa M., Yokota S., Nohara M., Nishimura A. & Murakami F. (1986) Submarine hydrothermal manganese deposits from the Ogasawara (Bonin) Arc, off the Japan Islands. *Mar. Geol.*, **73**, 311–22.

Valley J.W., Taylor H.P. Jr. & O'Neil J.R. (eds) (1986) Stable isotopes in high temperature geological processes. *Reviews in Mineralogy*, **16**, Mineral Soc. Am., Washington.

Vance R.K. & Condie K.C. (1987) Geochemistry of footwall alteration associated with the early Proterozoic United Verde Massive Sulfide Deposit, Jerome, Arizona. *Econ. Geol.*, **82**, 571–86.

Van Gruenewaldt G. (1977) The mineral resources of the Bushveld Complex. *Miner. Sci. Eng.*, **9**, 83–95.

Van Gruenewaldt G., Sharpe M.R. & Hatton C.J. (1985) The Bushveld Complex: introduction and review. *Econ. Geol.*, **80**, 803–12.

Van Kauwenbergh S.J. & McClellan G.H. (1990) Comparative geology and mineralogy of the southeastern United States and Togo phosphorites. In Notholt A.J.G. & Jarvis I. (eds), *Phosphorite Research and Development*, 139–55. Geological Society, London.

Van Rensberg W.C.J. (1986) *Strategic Minerals Vol. 1.* Prentice-Hall, Englewood Cliffs.

Van Staal C.R. & Williams P.F. (1984) Structure, origin and concentration of the Brunswick 12 and 6 Orebodies. *Econ. Geol.*, **79**, 1669–92.

Varentsov I.M. (1964) *Sedimentary Manganese Ores.* Elsevier, Amsterdam.

Varentsov I.M. & Rakhmanov V.P. (1977) Deposits of manganese. In Smirnov V.I. (ed.), *Ore Deposits of the USSR, Vol. 1*, 114–78. Pitman, London.

Vartiainen H. & Parma H. (1979) Geological characteristics of the Sokli Carbonatite Complex, Finland. *Econ. Geol.*, **74**, 1296–306.

Vaughan D.J., Sweeney M., Friedrich G., Diedel R. & Haranczyk C. (1989) The Kupferschiefer: an overview with an appraisal of the different types of mineralization. *Econ. Geol.*, **84**, 1003–27.

Vearncombe J.R., Cheshire P.E., de Beer J.H., Killick A.M., Mallinson W.S., McCourt S. & Stettler E.H. (1988) Structures related to the Antimony Line, Murchison Schist Belt, Kaapraal Craton, South Africa, *Tectonophysics*, **154**, 285–308.

Velasco F., Pesquera A., Arce R. & Olmedo F. (1987) A contribution to the ore genesis of the magnesite deposit of Eugui, Navarra (Spain). *Miner. Deposita*, **22**, 33–41.

Vermaak C.F. (1976) The Merensky Reef—thoughts on its environment and genesis. *Econ. Geol.*, **71**, 1270–98.

Verwoerd W.J. (1964) South African carbonatites and their probable mode of origin. *Ann. Univ. Stellenbosch., Ser. A*, **41**, 115–233.

Verwoerd W.J. (1986) Mineral deposits associated with carbonatites and alkaline rocks. In Anhaeusser C.R. & Maske S. (eds), *Mineral Deposits of Southern Africa*, 2, 2173–91. Geol. Soc. S. Afr., Johannesburg.

Veselý T. (1985) Jáchymov Uranium Deposit. *Krystalinikum*, **18**, 133–65.

Vokes F.M. (1968) Regional metamorphism of the Palae-

ozoic geosynclinal sulphide ore deposits of Norway. *Trans. Instn Min. Metall. (Sect. B: Appl. Earth Sci.)*, **77**, B53–9.

Vokes F.M. (1987) Caledonian stratabound sulphide ores and the factors affecting them. *Geol. Surv. Finland, Special Pap.*, **1**, 15–26.

Wallace S.R. (1991) Model development: porphyry molybdenum deposits. *Econ. Geol. Monogr.*, **8**, 207–24.

Wallace S.R., MacKenzie W.B., Blair R.G. & Muncaster N.K. (1978) Geology of the Urad and Henderson Molybdenite Deposits, Clear Creek County, Colorado, with a section on a comparison of these deposits with those at Climax, Colorado. *Econ. Geol.*, **73**, 325–68.

Wallis R.H., Saracoglu N., Brummer J.J. & Golightly J.P. (1984) The geology of the McClean Uranium Deposits, Northern Saskatchewan. *Can. Min. Metall. Bull.*, **77**, (April), 69–96.

Watson J.V. (1973) Influence of crustal evolution on ore deposition. *Trans. Instn Min. Metall. (Sect. B: Appl. Earth Sci.)*, **82**, B107–14.

Watson J.V. (1976) Mineralization in Archaean provinces. In Windley B.F. (ed.), *The Early History of the Earth*, 443–53. Wiley, London.

Weiss S.A. (1977) *Manganese: the other uses.* Metal Bulletin Books, London.

Weissberg B.G., Browne P.R.L. & Seward T.M. (1979) Ore metals in active geothermal systems. In Barnes H.L. (ed.), *Geochemistry of Hydrothermal Ore Deposits*, 2nd edn, 738–80. Wiley, New York.

Wenkui G. (1988) The problems of tin metallogeny. In Hutchison C.S. (ed.), *Geology of Tin Deposits in Asia and the Pacific*, 50–8. Springer-Verlag, Berlin.

Wesolowski D., Cramer J.J. & Ohmoto H. (1988) Scheelite mineralization in skarns adjacent to Devonian granitoids at King Island, Tasmania. In Taylor R.P. & Strong D.F. (eds), *Recent Advances in the Geology of Granite-related Mineral Deposits*, 234–50. Spec. Vol. 39, Can. Inst. Min. Metall, Montréal.

Westra G. & Keith S.B. (1981) Classification and genesis of stockwork molybdenum deposits. *Econ. Geol.*, **76**, 844–73.

Wheat T.A. (1987) Advanced ceramics in Canada. *Bull. Can. Inst. Min. Metall.*, **80** (April), 43–8.

White D. (1989) A marriage of convenience? *RTZ Review*, **10** (July), 8–10.

White D.E. (1974) Diverse origins of hydrothermal ore fluids. *Econ. Geol.*, **69**, 954–73.

White D.E. (1981) Active geothermal systems and hydrothermal ore deposits. In Skinner B.J. (ed.), *Econ. Geol.*, **75th Anniv. Vol.**, 392–423.

White W.H., Bookstrom A.A., Kamilli R.J., Ganster M.W., Smith R.P., Ranta D.E. & Steininger R.C. (1981) Character and origin of Climax-type molybdenum deposits. *Econ. Geol.*, **75th Anniv. Vol.**, 270–316.

White W.S. (1971) A paleohydrologic model for mineralization of the White Pine Copper Deposits, northern Michigan. *Econ. Geol.*, **66**, 1–13.

Whitney J.A. (1975) Vapor generation in a quartz mon-zonite magma: a synthetic model with application to porphyry copper deposits. *Econ. Geol.*, **70**, 346–58.

Whitney J.A., Hemley J.J. & Simon F.O. (1985) The concentration of iron in chloride solutions equilibrated with synthetic granitic compositions: the sulphur-free system. *Econ. Geol.*, **80**, 444–60.

Wickremeratne W.S. (1986) Preliminary studies on the offshore occurrences of monazite-bearing heavy mineral placers, south-western Sri Lanka. *Mar. Geol.*, **72**, 1–9.

Wilde A.R., Mernagh T.P., Bloom M.S. & Hoffman C.F. (1989) Fluid inclusion evidence on the origin of some Australian unconformity-related uranium deposits. *Econ. Geol.*, **84**, 1627–42.

Williams D. (1969) Ore deposits of volcanic affiliation. In James C.H. (ed.), *Sedimentary Ores Ancient and Modern (Revised)*, 197–206. Spec. Publ. 1, Geol. Dept., Leicester.

Williams D., Stanton R.L. & Rambaud F. (1975) The Planes–San Antonio Pyritic Deposit of Rio Tinto, Spain: its nature, environment and genesis. *Trans. Instn Min. Metall. (Sect. B, Appl. Earth Sci.)*, **84**, B73–82.

Williams H.R. & Williams R.A. (1977) Kimberlites and plate tectonics in West Africa. *Nature*, **270**, 507–8.

Williams P.J. (1990) Evidence for a late metamorphic origin of disseminated gold mineralization in Grenville gneisses at Calumet, Quebec. *Econ. Geol.*, **85**, 164–71.

Williams-Jones A.E. (1986) Low-temperature metamorphism of the rocks surrounding Les Mines Gaspé, Quebec: implications for mineral exploration. *Econ. Geol.*, **81**, 466–70.

Wilson A.F. (1983) The economic significance of non-hydrothermal transport of gold, and of the accretion of large gold nuggets in laterite and other weathering profiles in Australia. In De Villiers J.P.R. & Cawthorn P.A. (eds.), *ICAM 81*, 229–34, Spec. Publ. 7, Geol. Soc. S. Afr.

Wilson A.H. & Treadoux M. (1990) Lateral and vertical distribution of platinum-group elements and petrogenic controls on the sulfide mineralization in the P1 Pyroxenite Layer of the Darwendale Subchamber of the Great Dyke, Zimbabwe. *Econ. Geol.*, **85**, 556–84.

Wilson D.G. (1989) Recycling of demolition wastes. In Carr D.D. & Herz N. (eds), *Concise Encyclopedia of Mineral Resources*, 294–6. Pergamon Press, Oxford.

Wilson J.T. (1949) Some major structures of the Canadian Shield. *Can. Inst. Min. Metall. Bull.*, **52**, 231–42.

Wilson M. (1989) *Igneous Petrogenesis.* Unwin Hyman, London.

Wilton D.H.C. (1985) REE and background Au/Ag evidence concerning the origin of hydrothermal fluids in the Cape Ray Electrum Deposits, southwestern Newfoundland. *Can. Min. Metall. Bull.*, **78**, (February), 48–59.

Windley B.F. (1984) *The Evolving Continents.* Wiley, London.

Winkler H.G.F. (1979) *Petrogenesis of Metamorphic Rocks.* Springer-Verlag, New York.

Winward K. (1975) Quaternary coastal sediments. In Marham N.L. & Basden H. (eds.), *The Mineral Deposits of New South Wales*, 595–621. Geol. Surv., New South Wales, Dept. of Mines, Sydney.

Wolf K.H. (1981) Terminologies, structuring and classifications in ore and host-rock petrology. In Wolf K.H. (ed.), *Handbook of Strata-bound and Stratiform Ore Deposits*, 1–337. Elsevier, Amsterdam.

Wolfe J.A. (1984) *Mineral Resources—A World Review.* Chapman & Hall, New York.

Wood S.A., Crerar D.A. & Borcsik M.P. (1987) Solubility of the assemblage pyrite–pyrrhotite–magnetite–s p h a l e r i t e – g a l e n a – g o l d – s t i b n i t e – b i s m u t h i n i t e –argentite–molybdenite in H_2O–NaCl–CO_2 solutions from 200°C to 350°C. *Econ. Geol.*, **82**, 1864–87.

Woodall R. (1979) Gold—Australia and the world. In Glover J.E. & Groves D.I. (eds), *Gold Mineralization*, 1–34. Geology Department and Extension Service, Univ. Western Australia, Nedlands.

Woodall R. (1984) Success in mineral exploration: confidence in science and ore deposit models. *Geosci. Can.*, **11**, 127–32.

Woolley A.R. (1989) The spatial and temporal distribution of carbonatites. In Bell K. (ed.), *Carbonatites*, 15–37. Unwin Hyman, London.

Wright J.B. & McCurry P. (1973) Magmas, mineralization and seafloor spreading. *Geol. Rundsch.*, **62**, 116–25.

Wu Y. & Beales F. (1981) A reconnaissance study by palaeomagnetic methods of the age of mineralization along the Viburnum Trend, southeast Missouri. *Econ. Geol.*, **76**, 1879–94.

Yardley B.W.D. (1983) Quartz veins and devolatilization during metamorphism. *J. Geol. Soc. London*, **140**, 657–63.

Younquist W. (1990) *Mineral Resources and the Destinies of Nations.* National Book Company, Portland, Oregon.

Zachrisson E. (1984) Lateral metal zonation and stringer zone development, reflecting fissure-controlled exhalations at the Stekenjokk-Levi Strata-bound Sulphide Deposit, central Scandinavian Caledonides. *Econ. Geol.*, **79**, 1643–59.

Zantop H. (1981) Trace elements in volcanogenic manganese oxides and iron oxides: the San Francisco manganese deposit, Jalisco, Mexico. *Econ. Geol.*, **76**, 545–55.

Zierenberg R.A. & Shanks W.C. III, (1988) Isotopic studies of epigenetic features in metalliferous sediment, Atlantis II Deep, Red Sea. *Can. Mineral.*, **26**, 737–53.

Index

References to figures appear in *italic* type and references to tables appear in **bold** type.